Advanced Theories and Concepts of Physics

Advanced Theories and Concepts of Physics

Edited by
Adaline Cerny

www.willfordpress.com

Published by Willford Press,
118-35 Queens Blvd., Suite 400,
Forest Hills, NY 11375, USA

ISBN: 978-1-68285-646-8

Cataloging-in-Publication Data

Advanced theories and concepts of physics / edited by Adaline Cerny.
p. cm.
Includes bibliographical references and index.
ISBN 978-1-68285-646-8
1. Physics. 2. Physical sciences. I. Cerny, Adaline.
QC21.3 .A38 2019
530--dc23

For information on all Willford Press publications
visit our website at www.willfordpress.com

Contents

Preface

Physics is a fundamental science that studies matter and its motion through space-time. The concepts of force and energy are vital to this field. Classical mechanics, thermodynamics, statistical mechanics, electromagnetism, quantum mechanics and special relativity are the fundamental theories of physics. Classical and modern physics represent the two approaches to understanding the diverse phenomena in the universe. Classical physics describes the systems whose length scales are larger than atomic scales. It also studies motions at speeds much smaller than the speed of light. In the study of atomic and subatomic scales and speeds approaching the speed of light, the domain of modern physics is employed. This book outlines the advanced theories and concepts of physics in detail. It includes some of the vital pieces of work being conducted across the world, on various topics related to this field. Coherent flow of topics, student-friendly language and extensive use of examples make this book an invaluable source of knowledge.

The researches compiled throughout the book are authentic and of high quality, combining several disciplines and from very diverse regions from around the world. Drawing on the contributions of many researchers from diverse countries, the book's objective is to provide the readers with the latest achievements in the area of research. This book will surely be a source of knowledge to all interested and researching the field.

In the end, I would like to express my deep sense of gratitude to all the authors for meeting the set deadlines in completing and submitting their research chapters. I would also like to thank the publisher for the support offered to us throughout the course of the book. Finally, I extend my sincere thanks to my family for being a constant source of inspiration and encouragement.

Editor

1

Ambitwistor strings and reggeon amplitudes in $\mathcal{N}=4$ SYM

L.V. Bork [a,b,*], A.I. Onishchenko [c,d,e]

[a] Institute for Theoretical and Experimental Physics, Moscow, Russia
[b] The Center for Fundamental and Applied Research, All-Russia Research Institute of Automatics, Moscow, Russia
[c] Bogoliubov Laboratory of Theoretical Physics, Joint Institute for Nuclear Research, Dubna, Russia
[d] Moscow Institute of Physics and Technology (State University), Dolgoprudny, Russia
[e] Skobeltsyn Institute of Nuclear Physics, Moscow State University, Moscow, Russia

ARTICLE INFO

Editor: N. Lambert

Keywords:
Ambitwistor strings
Super Yang-Mills theory
Reggeon amplitudes

ABSTRACT

We consider the description of reggeon amplitudes (Wilson lines form factors) in $\mathcal{N}=4$ SYM within the framework of four dimensional ambitwistor string theory. The latter is used to derive scattering equations representation for reggeon amplitudes with multiple reggeized gluons present. It is shown, that corresponding tree-level string correlation function correctly reproduces previously obtained Grassmannian integral representation of reggeon amplitudes in $\mathcal{N}=4$ SYM.

1. Introduction

The behavior of scattering amplitudes in high energy or Regge limit is determined by the positions of singularities of their partial wave amplitudes in the complex angular momentum plane. Already an account for leading pole singularities, so called Regge poles, allows to construct phenomenologically successful models. In particular, to explain the experimentally observed asymptotic rise of total cross-section at high energies the Regge pole with the quantum numbers of the vacuum and even parity – the Pomeron was introduced. Later it was realized that in relativistic theory Regge poles should be supplemented with Regge cuts, which could be understood as coming from the exchanges of two or more Regge poles. Next, following the pioneering work of Gribov [1] the reggeon field theory describing interactions between various reggeons and physical particles was developed [1–3]. Subsequent development of these ideas within the context of quantum chromodynamics, in particular resummation of leading high energy logarithms $(\alpha_s \ln s)^n$ to all orders in strong coupling constant (LLA resummation) with the help of Balitsky–Fadin–Kuraev–Lipatov (BFKL) equation [4–8], showed the LLA reggeization of QCD scattering amplitudes. The corresponding Regge pole was identified with reggeized gluon. The latter has quantum numbers of the ordinary gluon and Regge trajectory $j(t)$ passing through unity at $t=0$. Today BFKL equation is known at next-to-leading-logarithmic-approximation (NLLA) [9,10] and the reggeization of QCD amplitudes is also proven at NLLA [11]. In general, the amplitudes with reggeized gluons (also known as gauge invariant off-shell amplitudes) [12–17] arise either in the study of multi-Regge kinematics [18–21] or within the context of k_T or high-energy factorization [22–25].

On the other hand, recently Roiban, Spradlin and Volovich (RSV) based on Witten's twistor string theory [26] got the description of $\mathrm{N}^{k-2}\mathrm{MHV}_n$ $\mathcal{N}=4$ SYM tree level amplitudes in terms of integrals over the moduli space of degree $k-1$ curves in super twistor space [27,28]. Subsequent generalization of RSV result by Cachazo, He and Yuan (CHY) led to the discovery of so called *scattering equations* [29–33]. Within the latter tree level $\mathcal{N}=4$ SYM amplitudes are written in terms of integrals (localized on the solutions of mentioned scattering equations) over the marked points on the Riemann sphere. Subsequently the CHY formulae together with their loop level generalization (see [34] and references therein) were derived from *ambitwistor string theory* [35,36].

Another very close research direction is related to the representation of $\mathrm{N}^{k-2}\mathrm{MHV}_n$ scattering amplitudes in terms of integrals over Grassmannians [37–42]. The latter naturally unifies different BCFW [43,44] representations both for tree level amplitudes and loop level integrands [37,38] and is ultimately connected to the integrable structure behind $\mathcal{N}=4$ SYM S-matrix [45–49].

* Corresponding author.
E-mail addresses: borkleonid@gmail.com (L.V. Bork), onish@bk.ru (A.I. Onishchenko).

The use of mentioned representations (Grassmannian, RSV, scattering equations and so on) of $N^{k-2}MHV_n$ amplitudes (i.e. the full tree level S-matrix of $\mathcal{N}=4$ SYM) provides us with relatively compact analytical expressions for n-point tree level amplitudes, which in their turn could be used to compute corresponding loop level amplitudes via modern unitarity based methods both at high orders of perturbation theory and/or with large number of external particles in $\mathcal{N}=4$ SYM and other field theories including QCD (see for a review [50]). It is important to note that these results was almost impossible to obtain by standard Feynman diagram methods.

The aim of the present work is to extend the recently obtained results for the usual $\mathcal{N}=4$ SYM scattering amplitudes and to derive scattering equations representation for reggeon amplitudes in $\mathcal{N}=4$ SYM from four dimensional ambitwistor string theory. At a moment, there are already scattering equations representations for the form factors of operators from stress-tensor operator supermultiplet and scalar operators of the form $\text{Tr}(\phi^m)$ [51,52]. Also some formulae were extended to Standard Model amplitudes [53]. Besides, there are several results for the Grassmannian integral representation of form factors of operators from stress-tensor operator supermultiplet [54–57] and reggeon amplitudes (form factors[1] of Wilson line insertions) [58,59], see also [60] for a recent interesting duality for Wilson loop form factors. A very close research direction is the twistor and Lorentz harmonic chiral superspace formulation of form factors and correlation functions developed in [61–67], see the discussion in conclusion.

This paper is organized as follows. First, in section 2 we introduce necessary definitions for reggeon amplitudes (Wilson lines form factors). Next, in section 3 after recalling some of the basic facts of four dimensional ambitwistor string theory we proceed with the construction of string vertex operator for reggeized gluon and derive scattering equations representation for reggeon amplitudes from corresponding string correlation functions. Finally, in section 4 we come with our conclusion.

2. Reggeon amplitudes and Wilson lines

To describe amplitudes with reggeized gluons it is convenient to use the representation of the latter in terms of Wilson line operators as in [15]:

$$\mathcal{W}_p^c(k) = \int d^4x e^{ix\cdot k} \times \text{Tr}\left\{\frac{1}{\pi g} t^c \,\mathcal{P}\exp\left[\frac{ig}{\sqrt{2}}\int_{-\infty}^{\infty} ds\; p\cdot A_b(x+sp)t^b\right]\right\}. \quad (1)$$

Here t^c is $SU(N_c)$ generator,[2] k ($k^2 \neq 0$) is the reggeized gluon momentum and p is its direction or polarization vector, such that $p^2=0$, $p\cdot k=0$. The momentum and polarization vector of the reggeized gluon could be related to each other through so called k_T - decomposition of momentum k:

$$k^\mu = xp^\mu + k_T^\mu, \quad x \in [0,1]. \quad (2)$$

Note, that such decomposition could be also parametrized by an auxiliary light-cone four-vector q^μ, so that

$$k_T^\mu(q) = k^\mu - x(q)p^\mu \quad \text{with} \quad x(q) = \frac{q\cdot k}{q\cdot p} \text{ and } q^2=0. \quad (3)$$

Using the fact, that the transverse momentum k_T^μ is orthogonal to both p^μ and q^μ vectors one can decompose it into the basis of two "polarization" vectors[3] [12]:

$$k_T^\mu(q) = -\frac{\kappa}{2}\frac{\langle p|\gamma^\mu|q]}{[pq]} - \frac{\kappa^*}{2}\frac{\langle q|\gamma^\mu|p]}{\langle qp\rangle}$$
$$\text{with} \quad \kappa = \frac{\langle q|\not{k}|p]}{\langle qp\rangle}, \; \kappa^* = \frac{\langle p|\not{k}|q]}{[pq]}. \quad (4)$$

It is easy to check, that $k^2 = -\kappa\kappa^*$ and both κ and κ^* variables are independent of auxiliary four-vector q^μ [12].

Both usual and color ordered[4] reggeon amplitudes with n reggeized and m usual on-shell gluons could be then written in terms of form factors with multiple Wilson line insertions as [15]:

$$\mathcal{A}_{m+n}\left(1^\pm,\ldots,m^\pm,g^*_{m+1},\ldots,g^*_{n+m}\right)$$
$$= \langle\{k_i,\epsilon_i,c_i\}_{i=1}^m|\prod_{j=1}^n \mathcal{W}_{p_{m+j}}^{c_{m+j}}(k_{m+j})|0\rangle, \quad (5)$$

here asterisk denotes an off-shell gluon and p, k, c are its direction, momentum and color index. Next $\langle\{k_i,\epsilon_i,c_i\}_{i=1}^m| = \bigotimes_{i=1}^m\langle k_i,\varepsilon_i,c_i|$ and $\langle k_i,\varepsilon_i,c_i|$ denotes on-shell gluon state with momentum k_i, polarization vector ε_i^- or ε_i^+ and color index c_i, p_i is the direction of the i'th ($i=1,\ldots,n$) off-shell gluon and k_i is its off-shell momentum. For the case when only reggeized gluons are present (correlation function of Wilson line operators) we have:

$$\mathcal{A}_{0+n}\left(g_1^*\ldots g_n^*\right) = \langle 0|\mathcal{W}_{p_1}^{c_1}(k_1)\ldots\mathcal{W}_{p_n}^{c_n}(k_n)|0\rangle. \quad (6)$$

In the case of $\mathcal{N}=4$ SYM we may also consider other on-shell states from $\mathcal{N}=4$ supermultiplet. The most convenient way to do so is to consider color ordered superamplitudes defined on $\mathcal{N}=4$ on-shell momentum superspace:

$$A^*_{m+n}\left(\Omega_1,\ldots,\Omega_m,g^*_{m+1},\ldots,g^*_{n+m}\right)$$
$$= \langle\Omega_1\ldots\Omega_m|\prod_{j=1}^n \mathcal{W}_{p_{m+j}}(k_{m+j})|0\rangle, \quad (7)$$

where $\langle\Omega_1\Omega_2\ldots\Omega_m| \equiv \bigotimes_{i=1}^m\langle 0|\Omega_i$ and Ω_i ($i=1,\ldots,m$) denotes $\mathcal{N}=4$ on-shell chiral superfield [69]:

$$\Omega = g^+ + \tilde{\eta}_A\psi^A + \frac{1}{2!}\tilde{\eta}_A\tilde{\eta}_B\phi^{AB} + \frac{1}{3!}\tilde{\eta}_A\tilde{\eta}_B\tilde{\eta}_C\epsilon^{ABCD}\bar{\psi}_D$$
$$+ \frac{1}{4!}\tilde{\eta}_A\tilde{\eta}_B\tilde{\eta}_C\tilde{\eta}_D\epsilon^{ABCD}g^-. \quad (8)$$

Here, g^+, g^- denote creation/annihilation operators of gluons with $+1$ and -1 helicities, ψ^A, $\bar{\psi}_A$ stand for creation/annihilation operators of four Weyl spinors with negative helicity $-1/2$ and four Weyl spinors with positive helicity correspondingly, while ϕ^{AB} denote creation/annihilation operators for six scalars (anti-symmetric in the $SU(4)_R$ R-symmetry indices AB). All $\mathcal{N}=4$ SYM fields transform in the adjoint representation of $SU(N_c)$ gauge group. The $A^*_{m+n}\left(\Omega_1,\ldots,g^*_{n+m}\right)$ superamplitude is then the function of the following kinematic[5] and Grassmann variables

$$A^*_{m+n}\left(\Omega_1,\ldots,g^*_{m+n}\right)$$
$$= A^*_{m+n}\left(\{\lambda_i,\tilde{\lambda}_i,\tilde{\eta}_i\}_{i=1}^m;\{k_i,\lambda_{p,i},\tilde{\lambda}_{p,i}\}_{i=m+1}^{m+n}\right), \quad (9)$$

[1] See the precise definition in section 2.

[2] The color generators are normalized as $\text{Tr}(t^at^b) = \delta^{ab}$.

[3] Here we used the helicity spinor decomposition of light-like four-vectors p and q.

[4] Here we are dealing with color ordered amplitudes for simplicity. The usual amplitudes are then obtained using color decomposition, see for example [58,68].

[5] We used helicity spinor decomposition of on-shell particles momenta.

and encodes in addition to amplitudes with gluons also amplitudes with other on-shell states similar to the case of usual on-shell superamplitudes [50].

3. Ambitwistor string correlation functions

As we already mentioned in Introduction our aim here is to derive scattering equations representation for reggeon amplitudes in $\mathcal{N}=4$ SYM using four dimensional ambitwistor string theory [36], see also [70] for further details. The target space of the latter is given by projective ambitwistor space $\mathbb{PA}$:

$$\mathbb{PA} = \{(Z,W)\in\mathbb{T}\times\mathbb{T}^*|\, Z\cdot W=0\}/\{Z\cdot\partial_Z - W\cdot\partial_W\}, \quad (10)$$

with $\mathbb{T}$ and $\mathbb{T}^*$ denoting twistor and dual twistor spaces. Next, $Z=(\lambda_\alpha,\mu^{\dot\alpha},\chi^r)\in\mathbb{T}=\mathbb{C}^{4|4}$, $W=(\tilde\mu,\tilde\lambda,\tilde\chi)\in\mathbb{T}^*$ and $Z\cdot W=\lambda_\alpha\tilde\mu^\alpha+\mu^{\dot\alpha}\tilde\lambda_{\dot\alpha}+\chi^r\tilde\chi_r$ $(r=1,\ldots,4)$. The corresponding worldsheet theory consists from the worldsheet spinors (Z,W) taking values in $\mathbb{T}\times\mathbb{T}^*$ together with $\mathrm{GL}(1,\mathbb{C})$ gauge field a serving as a Lagrange multiplier for the constraint $Z\cdot W=0$. In the conformal gauge its action is given by

$$S=\frac{1}{2\pi}\int_\Sigma W\cdot\bar\partial Z - Z\cdot\bar\partial W + aZ\cdot W + S_J, \quad (11)$$

where $\bar\partial = d\bar\sigma\partial_{\bar\sigma}$ ($\sigma,\bar\sigma$ are some local holomorphic and anti-holomorphic coordinates on Riemann surface Σ) and S_J denotes the action for the $\mathfrak{su}(N_c)$ worldsheet Kac–Moody current algebra $J\in\Omega^0(\Sigma,K_\Sigma\otimes\mathfrak{su}(N_c))$. K_Σ, as usual, is the canonical bundle on the surface Σ and the other worldsheet fields take values in[6]

$$Z\in\Omega^0(\Sigma,K_\Sigma^{1/2}\otimes\mathbb{T}), \quad (12)$$
$$W\in\Omega^0(\Sigma,K_\Sigma^{1/2}\otimes\mathbb{T}^*), \quad (13)$$
$$a\in\Omega^{0,1}(\Sigma), \quad (14)$$

To calculate string scattering amplitudes we need vertex operators. There are two equivalent representations for integrated vertex operator used to describe on-shell states [36]:

$$\mathcal{V}_a=\int\frac{ds_a}{s_a}\bar\delta^2(\lambda_a-s_a\lambda)e^{is_a\left([\mu\,\tilde\lambda_a]+\chi^r\tilde\eta_{ar}\right)}J\cdot T_a, \quad (15)$$

$$\widetilde{\mathcal{V}}_a=\int\frac{ds_a}{s_a}\bar\delta^{2|4}(\tilde\lambda_a-s_a\tilde\lambda|\tilde\eta_a-s_a\tilde\chi)e^{is_a\langle\tilde\mu\,\lambda_a\rangle}J\cdot T_a, \quad (16)$$

where $\bar\delta(z)=\bar\partial(1/2\pi iz)$. We would like to stress here, that both vertex operators contain all sixteen on-shell states of $\mathcal{N}=4$ SYM. To obtain N^{k-2}MHV on-shell scattering amplitudes one may use for example the correlation function of k $\widetilde{\mathcal{V}}$ operators and $n-k$ $\mathcal{V}$ operators [36]:

$$A_{k,n}=\langle\widetilde{\mathcal{V}}_1\ldots\widetilde{\mathcal{V}}_k\mathcal{V}_{k+1}\ldots\mathcal{V}_n\rangle. \quad (17)$$

This correlation function is not hard to calculate[7] and we get [36]:

$$A_{n,k}=\int\frac{1}{\mathrm{Vol\,GL}(2,\mathbb{C})}\prod_{a=1}^n\frac{ds_a d\sigma_a}{s_a(\sigma_a-\sigma_{a+1})}\times\prod_{p=k+1}^n\bar\delta^2(\lambda_p-s_p\lambda(\sigma_p))\times\prod_{i=1}^k\bar\delta^{2|4}(\tilde\lambda_i-s_i\tilde\lambda(\sigma_i),\tilde\eta_i-s_i\tilde\chi(\sigma_i)), \quad (18)$$

where

$$\lambda(\sigma)=\sum_{i=1}^k\frac{s_i\lambda_i}{\sigma-\sigma_i}, \quad \tilde\lambda(\sigma)=\sum_{p=k+1}^n\frac{s_p\tilde\lambda_p}{\sigma-\sigma_p},$$
$$\tilde\chi(\sigma)=\sum_{p=k+1}^n\frac{s_p\tilde\eta_p}{\sigma-\sigma_p}. \quad (19)$$

In terms of homogeneous coordinates on Riemann sphere $\sigma_\alpha=\frac{1}{s}(1,\sigma)$ the same result is written as [36]:

$$A_{n,k}=\int\frac{1}{\mathrm{Vol\,GL}(2,\mathbb{C})}\prod_{a=1}^n\frac{d^2\sigma_a}{(a\,a+1)}\prod_{p=k+1}^n\delta^2(\lambda_p-\lambda(\sigma_p))\times\prod_{i=1}^k\bar\delta^{2|4}(\tilde\lambda_i-\tilde\lambda(\sigma_i),\tilde\eta_i-\tilde\chi(\sigma_i)) \quad (20)$$

with $(i\,j)=\sigma_{i\alpha}\sigma_j^\alpha$ and

$$\lambda(\sigma)=\sum_{i=1}^k\frac{\lambda_i}{(\sigma\,\sigma_i)}, \quad \tilde\lambda(\sigma)=\sum_{p=k+1}^n\frac{\tilde\lambda_p}{(\sigma\,\sigma_p)},$$
$$\tilde\chi(\sigma)=\sum_{p=k+1}^n\frac{\tilde\eta_p}{(\sigma\,\sigma_p)}. \quad (21)$$

The scattering equations are then straightforwardly follow from the arguments of delta functions:

$$k_a\cdot P(\sigma_a)=\lambda_a^\alpha\tilde\lambda_a^{\dot\alpha}P_{\alpha\dot\alpha}(\sigma_a)=\lambda_a^\alpha\tilde\lambda_a^{\dot\alpha}\lambda_\alpha(\sigma_a)\tilde\lambda_{\dot\alpha}(\sigma_a)=0. \quad (22)$$

To describe reggeon amplitudes we also need the ambitwistor string vertex operator for reggeized gluon. The latter could be obtained from the pullback of corresponding ambitwistor space wave function to string theory worldsheet.[8] The required ambitwistor space wave function could be easily found using a representation of corresponding reggeon amplitudes with $n+1$ legs in terms of convolutions of particle–particle-reggeon PPR vertexes (minimal off-shell amplitudes in the language of [58,59]) with on-shell amplitudes with $n+2$ legs.[9] This construction also comes naturally [74] by noting that gluing (introducing loop integration) PPR vertex to the on-shell amplitude we get one-loop reggeon amplitude whose leading singularity (extracted by maximally cutting the loop) gives us the corresponding tree level reggeon amplitude.

It should be noted, that this gluing procedure is applicable to any representation of on-shell scattering amplitudes, not only given by ambitwistor string correlation function. In the latter case however the mentioned gluing procedure could be used to reconstruct the ambitwistor string vertex operator for reggeized gluon (Wilson line operator insertion). This way we get

$$\mathcal{V}^{\mathrm{WL}}_{n,n+1}=\int\prod_{i=n}^{n+1}\frac{d^2\lambda_i d^2\tilde\lambda_i}{\mathrm{Vol}[\mathrm{GL}(1)]}d^4\tilde\eta_i A^*_{2,2+1}(\Omega_n,\Omega_{n+1},g^*)\Big|_{\lambda\to-\lambda}\times\mathcal{V}_n\mathcal{V}_{n+1}\Big|_{T^aT^b\to if^{abc}T^c\to T^c}, \quad (23)$$

Here, c denotes the color index of the reggeized gluon and we have used projection of tensor product of two adjoint on-shell

[6] Powers of the canonical bundle denote corresponding field conformal weights.

[7] See [36] for details.

[8] For the previous work on reggeon string vertexes within superstring theory see [71,72] and references therein.

[9] The form of the minimal off-shell vertex could be also obtained from the symmetry arguments along the lines of [73].

gluon color representations onto reggeized gluon adjoint color representation. The minimal PPR vertex $A^*_{2,2+1}(\Omega_n, \Omega_{n+1}, g^*)$ is given by [58]:

$$A^*_{2,2+1}(\Omega_n, \Omega_{n+1}, g^*) = \frac{\delta^4(k + \lambda_n\tilde{\lambda}_n + \lambda_{n+1}\tilde{\lambda}_{n+1})}{\kappa^*} \frac{\delta^4\left(\tilde{\eta}_n\langle p\, n+1\rangle + \tilde{\eta}_{n+1}\langle p\, n\rangle\right)}{\langle p\, n\rangle\langle n\, n+1\rangle\langle n+1\, p\rangle}, \quad (24)$$

where $p = \lambda_p\tilde{\lambda}_p$ is the reggeized gluon direction and κ^* was defined in Section 2 when introducing k_T decomposition of the reggeized gluon momentum k. It should be noted, that each of $\mathcal{V}$ vertex operators above could be exchanged for $\widetilde{\mathcal{V}}$ operator and thus the above representation for reggeized gluon vertex operator is not unique. The ambitwistor string vertex operator we got is non-local by construction. The latter property is expected as Wilson line is non-local object by itself. Also, the solution of scattering equations with off-shell leg represented by two on-shell ones (kinematical part) as in (23) does not lead to coordinates of points z_n, z_{n+1} on Riemann sphere being close and hence the OPE expansion of the product of two on-shell vertex operators $\mathcal{V}_n$, $\mathcal{V}_{n+1}$ as in [71, 72] could be used only in multi-Regge kinematics limit. From the viewpoint of worldsheet theory the Wilson line string vertex operator we got is composite operator build from on-shell vertexes $\mathcal{V}_n$, $\mathcal{V}_{n+1}$ and is on the same footing as composite operators in any other QFT. Note, that it has the same good worldsheet properties as the simple product of two $\mathcal{V}_n$, $\mathcal{V}_{n+1}$ operators and the gluing with minimal PPR vertex $A^*_{2,2+1}(\Omega_n, \Omega_{n+1}, g^*)$ may be considered as the corresponding projector or wave function in target projective ambitwistor space $\mathbb{PA}$. Now, performing integrations[10] over helicity spinors $\lambda_i, \tilde{\lambda}_i$ we get (the projection operator $\partial^4_{\eta_p}$ acting on $\mathcal{V}_n\mathcal{V}_{n+1}$ is assumed)

$$\mathcal{V}^{\rm WL}_{n,n+1} = \frac{\langle \xi p\rangle}{\kappa^*}\int\frac{d\beta_2}{\beta_2}\int\frac{d\beta_1}{\beta_1}\frac{1}{\beta_1^2\beta_2}\mathcal{V}_n\mathcal{V}_{n+1}\Big|_{T^aT^b\to if^{abc}T^c\to T^c}, \quad (25)$$

where

$$\lambda_n = \underline{\lambda}_n + \beta_2\underline{\lambda}_{n+1},$$

$$\tilde{\lambda}_n = \beta_1\underline{\tilde{\lambda}}_n + \frac{(1+\beta_1)}{\beta_2}\underline{\tilde{\lambda}}_{n+1}, \qquad \tilde{\eta}_n = -\beta_1\underline{\tilde{\eta}}_n, \quad (26)$$

$$\lambda_{n+1} = \underline{\lambda}_{n+1} + \frac{(1+\beta_1)}{\beta_1\beta_2}\underline{\lambda}_n,$$

$$\tilde{\lambda}_{n+1} = -\beta_1\underline{\tilde{\lambda}}_{n+1} - \beta_1\beta_2\underline{\lambda}_n, \qquad \tilde{\eta}_{n+1} = \beta_1\beta_2\underline{\tilde{\eta}}_n, \quad (27)$$

with

$$\underline{\lambda}_n = \lambda_p,\ \underline{\tilde{\lambda}}_n = \frac{\langle\xi|k}{\langle\xi p\rangle},\ \underline{\tilde{\eta}}_n = \tilde{\eta}_p;$$

$$\underline{\lambda}_{n+1} = \lambda_\xi,\ \underline{\tilde{\lambda}}_{n+1} = \frac{\langle p|k}{\langle\xi p\rangle},\ \underline{\tilde{\eta}}_{n+1} = 0, \quad (28)$$

where $\lambda_\xi \equiv \langle\xi|$ is some arbitrary spinor. In practical calculations it is useful to identify it with the spinor λ_q coming from helicity spinor decomposition of auxiliary momentum q arising in k_T decomposition of reggeized gluon momentum k.

The reggeon amplitude with one reggeized gluon and n on-shell final states is then given by the following ambitwistor string correlation function:

$$A^*_{k,n+1} = \left\langle\widetilde{\mathcal{V}}_1\ldots\widetilde{\mathcal{V}}_k\mathcal{V}_{k+1}\ldots\mathcal{V}_n\mathcal{V}^{\rm WL}_{n+1,n+2}\right\rangle. \quad (29)$$

[10] See [55,58] for details.

Evaluating first string correlator of on-shell vertexes with the help of (18) we get

$$A^*_{k,n+1} = \frac{\langle\xi p\rangle}{\kappa^*}\int\frac{d\beta_2}{\beta_2}\int\frac{d\beta_1}{\beta_1}\frac{1}{\beta_1^2\beta_2}\frac{1}{{\rm Vol\,GL}(2,\mathbb{C})} \times\int\prod_{a=1}^{n+2}\frac{ds_a d\sigma_a}{s_a(\sigma_a-\sigma_{a+1})}\prod_{p=k+1}^{n+2}\bar{\delta}^2(\lambda_p - s_p\lambda(\sigma_p)) \times\prod_{i=1}^{k}\bar{\delta}^{2|4}(\tilde{\lambda}_i - s_i\tilde{\lambda}(\sigma_i), \tilde{\eta}_i - s_i\tilde{\chi}(\sigma_i)). \quad (30)$$

Now using unity decomposition as in [41]:

$$1 = \frac{1}{{\rm Vol\,GL}(k)}\int d^{k\times(n+2)}C \times d^{k\times k}L\,(\det L)^{n+2}\delta^{k\times(n+2)}\left(C - L\cdot C^V[s,\sigma]\right), \quad (31)$$

where the integral over L matrix is the integral over GL(k) transformations and $C^V[\sigma]$ denotes the Veronese map from $(\mathbb{C}^2)^{n+2}/{\rm GL}(2)$ to $G(k, n+2)$ Grassmannian [41] (see also [51]):

$$C^V[s,\sigma] = \begin{pmatrix} \vdots & \vdots & \cdots & \vdots \\ \sigma^V[s_1,\sigma_1] & \sigma^V[s_2,\sigma_2] & \cdots & \sigma^V[s_{n+2},\sigma_{n+2}] \\ \vdots & \vdots & \cdots & \vdots \end{pmatrix},$$

$$\sigma^V[s,\sigma] \equiv \begin{pmatrix} \xi \\ \xi\sigma \\ \vdots \\ \xi\sigma^{k-1} \end{pmatrix}, \quad (32)$$

where [28,51]:

$$\xi_i = s_i^{-1}\prod_{j=1, j\neq i}^{k}(\sigma_j - \sigma_i)^{-1}, \qquad i\in(1,k) \quad (33)$$

$$\xi_i = s_i\prod_{j=1}^{k}(\sigma_j - \sigma_i)^{-1}, \qquad i\in(k+1, n+2). \quad (34)$$

Next, integrating (30) over s_a and σ_a we get

$$A^*_{k,n+1} = \frac{\langle\xi p\rangle}{\kappa^*}\int\frac{d\beta_2}{\beta_2}\int\frac{d\beta_1}{\beta_1}\frac{1}{\beta_1^2\beta_2}\frac{1}{{\rm Vol\,GL}(k)} \times\int d^{k\times(n+2)}C\,F(C)\,\delta^{k\times 2}(C\cdot\tilde{\lambda})\delta^{k\times 4}(C\cdot\tilde{\eta}) \times\delta^{(n+2-k)\times 2}(C^\perp\cdot\lambda), \quad (35)$$

where

$$F(C) = \int\frac{1}{{\rm Vol\,GL}(2,\mathbb{C})} \times\prod_{a=1}^{n+2}\frac{ds_a d\sigma_a}{s_a(\sigma_a-\sigma_{a+1})}d^{k\times k}L\,\delta^{k\times(n+2)}\left(C - L\cdot C^V[s,\sigma]\right), \quad (36)$$

and

$$\delta^{k\times 2}(C\cdot\tilde{\lambda}) \equiv \prod_{a=1}^{k}\delta^2\left(\sum_{i=1}^{n}C_{ai}\tilde{\lambda}_i\right),$$

$$\delta^{(n+2-k)\times 2}(C^\perp\cdot\lambda) \equiv \prod_{b=k+1}^{n+2}\delta^2\left(\sum_{j=1}^{n+2}C^\perp_{bj}\lambda\right),$$

$$\delta^{k\times 4}(C\cdot\tilde{\eta}) \equiv \prod_{a=1}^{k} \hat{\delta}^4\left(\sum_{i=1}^{n} C_{ai}\tilde{\eta}_i\right). \quad (37)$$

Here, $C^\perp$ is the matrix defined by the identity $C\cdot(C^\perp)^T = 0$ and it is assumed that all matrix manipulations are performed after $GL(k)$ gauge fixing. Also note that the above delta functions should be thought as $\delta(x) = 1/x$ and the corresponding contour integral will then compute the residue at $x=0$ [75].

By construction $F(C)$ contains $(k-2)\times(n-k)$ delta function factors forcing integral over C matrix to have Veronese form [41]. In general $F(C)$ is a rather complicated rational function of the minors of C matrix. For example, for $k=3$ and $n+2=6$ it is given by [41,76]:

$$F(C) = \frac{(135)}{(123)(345)(561)}\frac{1}{S},$$
$$S = (123)(345)(561)(246) - (234)(456)(612)(351). \quad (38)$$

Here[11] $(i_1\ldots i_k)$ is minor of C matrix constructed from $i_1,\ldots,i_k$ columns of C. The integral over $d^{k\times(n+2)}C/\mathrm{Vol\,GL}(k,\mathbb{C})$ can then be reduced to multidimensional contour integral over $(k-2)(n-k)$ complex variables τ and evaluated by taking residues. However, at the end after all technicalities (which are highly non-trivial and interesting in their own turn [41,76]) it could be shown that $F(C)$ may be chosen in the form

$$F(C) = \frac{1}{(1\cdots k)(2\cdots k+1)\cdots(n+2\cdots k-1)}. \quad (39)$$

Next, let us rewrite (35) in the form (the proper choice of integration contour $\Gamma^{tree}_{k,n+2}$ is implemented [41,77])

$$A^*_{k,n+1} = \frac{\langle\xi p\rangle}{\kappa^*}\int\frac{d^{k\times(n+2)}C}{\mathrm{Vol}[GL(k)]}\frac{d\beta_1 d\beta_2}{\beta_1\beta_2^2}$$
$$\times\frac{\delta^{k\times 2}\left(C'\cdot\underline{\underline{\tilde{\lambda}}}\right)\delta^{k\times 4}\left(C'\cdot\underline{\underline{\tilde{\eta}}}\right)\delta^{(n+2-k)\times 2}\left(C'^{\perp}\cdot\underline{\underline{\lambda}}\right)}{(1\cdots k)(2\cdots k+1)\cdots(n+2\cdots k-1)}, \quad (40)$$

where

$$C'_{n+1} = -\beta_1 C_{n+1} + \beta_1\beta_2 C_{n+2},$$
$$C'^{\perp}_{n+1} = C^\perp_{n+1} + \frac{1+\beta_1}{\beta_1\beta_2}C^\perp_{n+2},$$
$$C'_{n+2} = -\beta_1 C_{n+2} + \frac{1+\beta_1}{\beta_2}C_{n+1},$$
$$C'^{\perp}_{n+2} = C^\perp_{n+2} + \beta_2 C^\perp_{n+1}, \quad (41)$$

and

$$\underline{\underline{\lambda}}_i = \lambda_i,\quad i=1,\ldots n,\quad \underline{\underline{\lambda}}_{n+1} = \lambda_p,\quad \underline{\underline{\lambda}}_{n+2} = \xi$$
$$\underline{\underline{\tilde{\lambda}}}_i = \tilde{\lambda}_i,\quad i=1,\ldots n,\quad \underline{\underline{\tilde{\lambda}}}_{n+1} = \frac{\langle\xi|k}{\langle\xi p\rangle},\quad \underline{\underline{\tilde{\lambda}}}_{n+2} = -\frac{\langle p|k}{\langle\xi p\rangle},$$
$$\underline{\underline{\tilde{\eta}}}_i = \tilde{\eta}_i,\quad i=1,\ldots n,\quad \underline{\underline{\tilde{\eta}}}_{n+1} = \tilde{\eta}_p,\quad \underline{\underline{\tilde{\eta}}}_{n+2} = 0. \quad (42)$$

Now, introducing inverse C-matrix transformation

$$C_{n+1} = C'_{n+1} + \beta_2 C'_{n+2},$$
$$C_{n+2} = \frac{1+\beta_1}{\beta_1\beta_2}C'_{n+1} + C'_{n+2}, \quad (43)$$

minors of C-matrix containing both $n+1$ and $n+2$ columns when rewritten in terms of minors of C'-matrix acquire extra $-\frac{1}{\beta_1}$ factor. For example, for $(n+1\cdots k-2)$ minor we have

$$(n+1\cdots k-2) = -\frac{1}{\beta_1}(n+1\cdots k-2)'. \quad (44)$$

On the other hand, minors containing either $n+1$ or $n+2$ column transform as

$$(n+2\,1\cdots k-1)$$
$$= \frac{1+\beta_1}{\beta_1\beta_2}(n+1\,1\cdots k-1)' + (n+2\,1\cdots k-1)', \quad (45)$$
$$(n-k+2\cdots n+1)$$
$$= (n-k+2\cdots n+1)' + \beta_2(n-k+2\cdots n\,n+2)', \quad (46)$$

while all other minors remain unchanged $(\cdots) = (\cdots)'$. Now, going to the integral over C' matrix and accounting for the transition Jacobian $\left(-\frac{1}{\beta_1}\right)^k$ we get

$$A^*_{k,n+1} = -\frac{\langle\xi p\rangle}{\kappa^*}\int\frac{d^{k\times(n+2)}C'}{\mathrm{Vol}[GL(k)]}\frac{d\beta_1 d\beta_2}{\beta_1\beta_2}\delta^{k\times 2}\left(C'\cdot\underline{\underline{\tilde{\lambda}}}\right)\delta^{k\times 4}\left(C'\cdot\underline{\underline{\tilde{\eta}}}\right)$$
$$\times\,\delta^{(n+2-k)\times 2}\left(C'^{\perp}\cdot\underline{\underline{\lambda}}\right)$$
$$\times\left[(1\cdots k)'\cdots(n+2\cdots k-1)'\right.$$
$$\times\left(1+\beta_2\frac{(n-k+2\cdots n\,n+2)'}{(n-k+2\cdots n\,n+1)'}\right)$$
$$\times\left.\left(\beta_1\beta_2 + (1+\beta_1)\frac{(n+1\,1\cdots k-1)'}{(n+2\,1\cdots k-1)'}\right)\right]^{-1}.$$

Next, taking first residue at $\beta_2 = 0$ and then at $\beta_1 = -1$ (i.e. considering corresponding residual form) we recover our previous result for reggeon amplitude with one reggeized gluon [58] (here we again assume $\partial^4_{\eta_p}$ projection operator acting on Grassmannian integral):

$$A^*_{k,n+1} = \int_{\Gamma^{tree}_{k,n+2}}\frac{d^{k\times(n+2)}C'}{\mathrm{Vol}[GL(k)]}Reg.$$
$$\times\frac{\delta^{k\times 2}\left(C'\cdot\underline{\underline{\tilde{\lambda}}}\right)\delta^{k\times 4}\left(C'\cdot\underline{\underline{\tilde{\eta}}}\right)\delta^{(n+2-k)\times 2}\left(C'^{\perp}\cdot\underline{\underline{\lambda}}\right)}{(1\cdots k)'\cdots(n+1\cdots k-2)'(n+2\,1\cdots k-1)'}, \quad (47)$$

with[12]

$$Reg. = \frac{\langle\xi p\rangle}{\kappa^*}\frac{(n+2\,1\cdots k-1)'}{(n+1\,1\cdots k-1)'}. \quad (48)$$

We have also verified that direct evaluation of (30) reproduces all particular off-shell amplitudes considered as examples in [58].

Finally we need to perform inverse operation, that is to reduce integral in (47) to the integral over $G(2,n+2)$ Grassmannian. This is again done with the help of Veronese map [41,51]. Using resolution of unity (31), fixing $GL(k)$ gauge to enforce rational form of scattering equations [36] and performing integration over C' matrix our Grassmannian integral representation (47) takes the form of scattering equations representation we are looking for:

[11] We hope there will be no confusion with previous definition $(i\,j) = \sigma_{i\alpha}\sigma_j^\alpha$ used in $d^2\sigma_a$ integrals over homogeneous coordinates on Riemann sphere before.

[12] The *Reg.* notation is chosen because this ratio of minors regulates soft holomorphic limit with respect to external kinematical variables associated with reggeized gluon [58].

$$\int \prod_{a=1}^{n+2} \frac{d^2\sigma_a}{(a\,a+1)} \frac{Reg.^V}{\text{Vol}\,\text{GL}(2,\mathbb{C})} \prod_{p=k+1}^{n+2} \bar{\delta}^2(\underline{\underline{\lambda}}_p - \underline{\underline{\lambda}}(\sigma_p))$$

$$\times \prod_{i=1}^{k} \delta^{2|4}(\underline{\underline{\tilde{\lambda}}}_i - \underline{\underline{\tilde{\lambda}}}(\sigma_i), \underline{\underline{\eta}}_i - \underline{\underline{\tilde{\chi}}}(\sigma_i)), \tag{49}$$

where

$$Reg.^V = \frac{\langle \xi p \rangle}{\kappa^*} \frac{(k n+1)}{(k n+2)} \tag{50}$$

and we have also performed the transition to homogeneous coordinates on Riemann sphere. The doubly underlined functions are defined as

$$\underline{\underline{\lambda}} = \sum_{i=1}^{k} \frac{\underline{\underline{\lambda}}_i}{(\sigma\,\sigma_i)}, \quad \underline{\underline{\tilde{\lambda}}} = \sum_{p=k+1}^{n+2} \frac{\underline{\underline{\tilde{\lambda}}}_p}{(\sigma\,\sigma_p)}, \quad \underline{\underline{\tilde{\chi}}} = \sum_{p=k+1}^{n+2} \frac{\underline{\underline{\tilde{\eta}}}_p}{(\sigma\,\sigma_p)}. \tag{51}$$

The result for the case of reggeon amplitudes with multiple reggeized gluons A^*_{m+n} could be obtained along the same lines. For example, in the case with first m particles on-shell and last n being reggeized gluons we would get:

$$\int \prod_{a=1}^{n+2} \frac{d^2\sigma_a}{(a\,a+1)} \frac{Reg.^V(m+1,\ldots,m+n)}{\text{Vol}\,\text{GL}(2,\mathbb{C})} \prod_{p=k+1}^{m+2n} \bar{\delta}^2(\underline{\underline{\lambda}}_p - \underline{\underline{\lambda}}(\sigma_p))$$

$$\times \prod_{i=1}^{k} \delta^{2|4}(\underline{\underline{\tilde{\lambda}}}_i - \underline{\underline{\tilde{\lambda}}}(\sigma_i), \underline{\underline{\eta}}_i - \underline{\underline{\tilde{\chi}}}(\sigma_i)), \tag{52}$$

where

$$Reg.^V(m+1,\ldots,m+n) = \prod_{j=1}^{n} Reg^V.(j+m),$$

$$Reg.^V(j+m) = \frac{\langle \xi_j p_j \rangle}{\kappa_j^*} \frac{(k\,2j-1+m)}{(k\,2j+m)}, \tag{53}$$

and external kinematical variables are defined as

$$\underline{\underline{\lambda}}_i = \lambda_i, \quad i = 1,\ldots m,$$

$$\underline{\underline{\lambda}}_{m+2j-1} = \lambda_{p_j}, \quad \underline{\underline{\lambda}}_{m+2j} = \xi_j, \quad j = 1,\ldots n,$$

$$\underline{\underline{\tilde{\lambda}}}_i = \tilde{\lambda}_i, \quad i = 1,\ldots m,$$

$$\underline{\underline{\tilde{\lambda}}}_{m+2j-1} = \frac{\langle \xi_j | k_{m+j}}{\langle \xi_j p_j \rangle}, \quad \underline{\underline{\tilde{\lambda}}}_{m+2j} = -\frac{\langle p_j | k_{m+j}}{\langle \xi_j p_j \rangle}, \quad j = 1,\ldots n,$$

$$\underline{\underline{\tilde{\eta}}}_i = \tilde{\eta}_i, \quad i = 1,\ldots m,$$

$$\underline{\underline{\tilde{\eta}}}_{m+2j-1} = \tilde{\eta}_{p_j}, \quad \underline{\underline{\tilde{\eta}}}_{m+2j} = 0, \quad j = 1,\ldots n. \tag{54}$$

Note, that it is possible to rewrite (52) as an integral over $Gr(k, m+2n)$ Grassmannian coinciding with our previous result [59].

At the end of this section we want to make a short comment on how these results could be incorporated into modern approaches to the computation of scattering amplitudes in gauge theories. The insertion of our new $\mathcal{V}^{\text{WL}}_{i,i+1}$ vertex operator can be reformulated in the language of BCFW recursion as an application of some simple linear integral operator to individual BCFW terms for on-shell scattering amplitudes This should be much more simple approach compared to the use of analog of BCFW recursion for the off-shell gauge invariant amplitudes themselves [12–15]. Next, considered tree level off-shell gauge invariant amplitudes can be further used in some variations of generalized unitarity approach at loop level (see for a review [50]).

4. Conclusion

In this paper we presented results for scattering equations representations for reggeon amplitudes in $\mathcal{N}=4$ SYM derived from four dimensional ambitwistor string theory. The presented derivation could be also easily generalized to the case of tree level form factors of local operators and loop integrands of reggeon amplitudes, which will be the subject of our forthcoming publication [74].

As by product we found an easy and convenient gluing procedure (linear integral operator) allowing us to obtain required reggeon amplitude expressions from already known on-shell amplitudes. The construction of string vertex operator for reggeized gluon was inspired by the mentioned gluing procedure. This gluing procedure can also be formulated in momentum twistor space which rises an interesting question about its relation to recent developments in the computation of form factors [61–63] within twistor approach to $\mathcal{N}=4$ SYM [78,79].

It would be extremely interesting to consider pullbacks of composite operators defined on twistor or Lorentz harmonic chiral superspace [61–67] to construct corresponding string vertex operators. We hope that along these lines we will be able to get scattering equations representation for arbitrary local composite operators.

Having obtained scattering equations representations one may wonder what is the most efficient way to get final expressions for amplitudes with given numbers of reggeized gluons and other on-shell states. In the case of usual on-shell amplitudes we know that they could be obtained through the computations of global residues by the methods of computational algebraic geometry [80–82], see also [83]. It would be interesting to see how this procedure works in the case of reggeon amplitudes considered here.

Finally, we should also develop methods for computing loop corrections to reggeon amplitudes together with their loop level generalization of scattering equations representation. Besides, it is extremely interesting to see how the presented approach works in gravity and supergravity theories, where we have a well developed framework for reggeon amplitudes based on high-energy effective lagrangian, see [84,85] and references therein.

Acknowledgements

The authors would like to thank D.I. Kazakov, L.N. Lipatov and Yu-tin Huang for interesting and stimulating discussions. This work was supported by RSF grant #16-12-10306.

References

[1] V.N. Gribov, A Reggeon diagram technique, Sov. Phys. JETP 26 (1968) 414–422, Zh. Eksp. Teor. Fiz. 53 (1967) 654.

[2] H.D.I. Abarbanel, J.B. Bronzan, R.L. Sugar, A.R. White, Reggeon field theory: formulation and use, Phys. Rep. 21 (1975) 119–182.

[3] M. Baker, K.A. Ter-Martirosian, Gribov's Reggeon calculus: its physical basis and implications, Phys. Rep. 28 (1976) 1–143.

[4] L.N. Lipatov, Reggeization of the vector meson and the vacuum singularity in nonabelian gauge theories, Sov. J. Nucl. Phys. 23 (1976) 338–345, Yad. Fiz. 23 (1976) 642.

[5] E.A. Kuraev, L.N. Lipatov, V.S. Fadin, Multi-Reggeon processes in the Yang–Mills theory, Sov. Phys. JETP 44 (1976) 443–450, Zh. Eksp. Teor. Fiz. 71 (1976) 840.

[6] V.S. Fadin, E.A. Kuraev, L.N. Lipatov, On the Pomeranchuk singularity in asymptotically free theories, Phys. Lett. B 60 (1975) 50–52.

[7] E.A. Kuraev, L.N. Lipatov, V.S. Fadin, The Pomeranchuk singularity in nonabelian gauge theories, Sov. Phys. JETP 45 (1977) 199–204, Zh. Eksp. Teor. Fiz. 72 (1977) 377.

[8] I.I. Balitsky, L.N. Lipatov, The Pomeranchuk singularity in quantum chromodynamics, Sov. J. Nucl. Phys. 28 (1978) 822–829, Yad. Fiz. 28 (1978) 1597.

[9] V.S. Fadin, L.N. Lipatov, BFKL pomeron in the next-to-leading approximation, Phys. Lett. B 429 (1998) 127–134, arXiv:hep-ph/9802290.

[10] M. Ciafaloni, G. Camici, Energy scale(s) and next-to-leading BFKL equation, Phys. Lett. B 430 (1998) 349–354, arXiv:hep-ph/9803389.
[11] V.S. Fadin, R. Fiore, M.G. Kozlov, A.V. Reznichenko, Proof of the multi-Regge form of QCD amplitudes with gluon exchanges in the NLA, Phys. Lett. B 639 (2006) 74–81, arXiv:hep-ph/0602006.
[12] A. van Hameren, BCFW recursion for off-shell gluons, J. High Energy Phys. 07 (2014) 138, arXiv:1404.7818 [hep-ph].
[13] A. van Hameren, M. Serino, BCFW recursion for TMD parton scattering, J. High Energy Phys. 07 (2015) 010, arXiv:1504.00315 [hep-ph].
[14] K. Kutak, A. Hameren, M. Serino, QCD amplitudes with 2 initial spacelike legs via generalised BCFW recursion, J. High Energy Phys. 02 (2017) 009, arXiv:1611.04380 [hep-ph].
[15] P. Kotko, Wilson lines and gauge invariant off-shell amplitudes, J. High Energy Phys. 07 (2014) 128, arXiv:1403.4824 [hep-ph].
[16] A. van Hameren, P. Kotko, K. Kutak, Multi-gluon helicity amplitudes with one off-shell leg within high energy factorization, J. High Energy Phys. 12 (2012) 029, arXiv:1207.3332 [hep-ph].
[17] A. van Hameren, P. Kotko, K. Kutak, Helicity amplitudes for high-energy scattering, J. High Energy Phys. 01 (2013) 078, arXiv:1211.0961 [hep-ph].
[18] L.N. Lipatov, Gauge invariant effective action for high-energy processes in QCD, Nucl. Phys. B 452 (1995) 369–400, arXiv:hep-ph/9502308.
[19] L.N. Lipatov, Small x physics in perturbative QCD, Phys. Rep. 286 (1997) 131–198, arXiv:hep-ph/9610276.
[20] R. Kirschner, L.N. Lipatov, L. Szymanowski, Effective action for multi-Regge processes in QCD, Nucl. Phys. B 425 (1994) 579–594, arXiv:hep-th/9402010.
[21] R. Kirschner, L.N. Lipatov, L. Szymanowski, Symmetry properties of the effective action for high-energy scattering in QCD, Phys. Rev. D 51 (1995) 838–855, arXiv:hep-th/9403082.
[22] L.V. Gribov, E.M. Levin, M.G. Ryskin, Semihard processes in QCD, Phys. Rep. 100 (1983) 1–150.
[23] S. Catani, M. Ciafaloni, F. Hautmann, High-energy factorization and small x heavy flavor production, Nucl. Phys. B 366 (1991) 135–188.
[24] J.C. Collins, R.K. Ellis, Heavy quark production in very high-energy hadron collisions, Nucl. Phys. B 360 (1991) 3–30.
[25] S. Catani, F. Hautmann, High-energy factorization and small x deep inelastic scattering beyond leading order, Nucl. Phys. B 427 (1994) 475–524, arXiv:hep-ph/9405388.
[26] E. Witten, Perturbative gauge theory as a string theory in twistor space, Commun. Math. Phys. 252 (2004) 189–258, arXiv:hep-th/0312171.
[27] R. Roiban, M. Spradlin, A. Volovich, On the tree level S matrix of Yang–Mills theory, Phys. Rev. D 70 (2004) 026009, arXiv:hep-th/0403190.
[28] M. Spradlin, A. Volovich, From twistor string theory to recursion relations, Phys. Rev. D 80 (2009) 085022, arXiv:0909.0229 [hep-th].
[29] F. Cachazo, S. He, E.Y. Yuan, Scattering equations and Kawai–Lewellen–Tye orthogonality, Phys. Rev. D 90 (6) (2014) 065001, arXiv:1306.6575 [hep-th].
[30] F. Cachazo, S. He, E.Y. Yuan, Scattering of massless particles in arbitrary dimensions, Phys. Rev. Lett. 113 (17) (2014) 171601, arXiv:1307.2199 [hep-th].
[31] F. Cachazo, S. He, E.Y. Yuan, Scattering of massless particles: scalars, gluons and gravitons, J. High Energy Phys. 07 (2014) 033, arXiv:1309.0885 [hep-th].
[32] F. Cachazo, S. He, E.Y. Yuan, Einstein–Yang–Mills scattering amplitudes from scattering equations, J. High Energy Phys. 01 (2015) 121, arXiv:1409.8256 [hep-th].
[33] F. Cachazo, S. He, E.Y. Yuan, Scattering equations and matrices: from Einstein to Yang–Mills, DBI and NLSM, J. High Energy Phys. 07 (2015) 149, arXiv:1412.3479 [hep-th].
[34] Y. Geyer, L. Mason, R. Monteiro, P. Tourkine, Two-loop scattering amplitudes from the Riemann sphere, Phys. Rev. D 94 (12) (2016) 125029, arXiv:1607.08887 [hep-th].
[35] L. Mason, D. Skinner, Ambitwistor strings and the scattering equations, J. High Energy Phys. 07 (2014) 048, arXiv:1311.2564 [hep-th].
[36] Y. Geyer, A.E. Lipstein, L.J. Mason, Ambitwistor strings in four dimensions, Phys. Rev. Lett. 113 (8) (2014) 081602, arXiv:1404.6219 [hep-th].
[37] N. Arkani-Hamed, F. Cachazo, C. Cheung, J. Kaplan, A duality for the S matrix, J. High Energy Phys. 03 (2010) 020, arXiv:0907.5418 [hep-th].
[38] N. Arkani-Hamed, J.L. Bourjaily, F. Cachazo, A.B. Goncharov, A. Postnikov, J. Trnka, Scattering Amplitudes and the Positive Grassmannian, Cambridge University Press, 2012, arXiv:1212.5605 [hep-th].
[39] N. Arkani-Hamed, J.L. Bourjaily, F. Cachazo, S. Caron-Huot, J. Trnka, The all-loop integrand for scattering amplitudes in planar N = 4 SYM, J. High Energy Phys. 01 (2011) 041, arXiv:1008.2958 [hep-th].
[40] N. Arkani-Hamed, F. Cachazo, C. Cheung, The Grassmannian origin of dual superconformal invariance, J. High Energy Phys. 03 (2010) 036, arXiv:0909.0483 [hep-th].
[41] N. Arkani-Hamed, J. Bourjaily, F. Cachazo, J. Trnka, Unification of residues and Grassmannian dualities, J. High Energy Phys. 01 (2011) 049, arXiv:0912.4912 [hep-th].
[42] L.J. Mason, D. Skinner, Dual superconformal invariance, momentum twistors and Grassmannians, J. High Energy Phys. 11 (2009) 045, arXiv:0909.0250 [hep-th].
[43] R. Britto, F. Cachazo, B. Feng, New recursion relations for tree amplitudes of gluons, Nucl. Phys. B 715 (2005) 499–522, arXiv:hep-th/0412308.
[44] R. Britto, F. Cachazo, B. Feng, E. Witten, Direct proof of tree-level recursion relation in Yang–Mills theory, Phys. Rev. Lett. 94 (2005) 181602, arXiv:hep-th/0501052.
[45] J.M. Drummond, J.M. Henn, J. Plefka, Yangian symmetry of scattering amplitudes in N = 4 super Yang–Mills theory, J. High Energy Phys. 05 (2009) 046, arXiv:0902.2987 [hep-th].
[46] J.M. Drummond, J. Henn, G.P. Korchemsky, E. Sokatchev, Dual superconformal symmetry of scattering amplitudes in N = 4 super-Yang–Mills theory, Nucl. Phys. B 828 (2010) 317–374, arXiv:0807.1095 [hep-th].
[47] D. Chicherin, S. Derkachov, R. Kirschner, Yang–Baxter operators and scattering amplitudes in N = 4 super-Yang–Mills theory, Nucl. Phys. B 881 (2014) 467–501, arXiv:1309.5748 [hep-th].
[48] R. Frassek, N. Kanning, Y. Ko, M. Staudacher, Bethe ansatz for Yangian invariants: towards super Yang–Mills scattering amplitudes, Nucl. Phys. B 883 (2014) 373–424, arXiv:1312.1693 [math-ph].
[49] N. Beisert, A. Garus, M. Rosso, Yangian symmetry and integrability of planar N = 4 super-Yang–Mills theory, arXiv:1701.09162 [hep-th].
[50] H. Elvang, Y.-t. Huang, Scattering amplitudes, arXiv:1308.1697 [hep-th].
[51] A. Brandhuber, E. Hughes, R. Panerai, B. Spence, G. Travaglini, The connected prescription for form factors in twistor space, J. High Energy Phys. 11 (2016) 143, arXiv:1608.03277 [hep-th].
[52] S. He, Z. Liu, A note on connected formula for form factors, J. High Energy Phys. 12 (2016) 006, arXiv:1608.04306 [hep-th].
[53] S. He, Y. Zhang, Connected formulas for amplitudes in standard model, arXiv:1607.02843 [hep-th].
[54] L.V. Bork, A.I. Onishchenko, On soft theorems and form factors in $\mathcal{N} = 4$ SYM theory, J. High Energy Phys. 12 (2015) 030, arXiv:1506.07551 [hep-th].
[55] R. Frassek, D. Meidinger, D. Nandan, M. Wilhelm, On-shell diagrams, Graßmannians and integrability for form factors, J. High Energy Phys. 01 (2016) 182, arXiv:1506.08192 [hep-th].
[56] M. Wilhelm, Form Factors and the Dilatation Operator in $\mathcal{N} = 4$ Super Yang–Mills Theory and Its Deformations, PhD thesis, Humboldt U., Berlin, 2016, arXiv:1603.01145 [hep-th].
[57] L.V. Bork, A.I. Onishchenko, Grassmannians and form factors with $q^2 = 0$ in $\mathcal{N} = 4$ SYM theory, J. High Energy Phys. 12 (2016) 076, arXiv:1607.00503 [hep-th].
[58] L.V. Bork, A.I. Onishchenko, Wilson lines, Grassmannians and gauge invariant off-shell amplitudes in N = 4 SYM, arXiv:1607.02320 [hep-th].
[59] L.V. Bork, A.I. Onishchenko, Grassmannian integral for general gauge invariant off-shell amplitudes in N = 4 SYM, arXiv:1610.09693 [hep-th].
[60] D. Chicherin, P. Heslop, G.P. Korchemsky, E. Sokatchev, Wilson loop form factors: a new duality, arXiv:1612.05197 [hep-th].
[61] L. Koster, V. Mitev, M. Staudacher, M. Wilhelm, Composite operators in the twistor formulation of N = 4 supersymmetric Yang–Mills theory, Phys. Rev. Lett. 117 (1) (2016) 011601, arXiv:1603.04471 [hep-th].
[62] L. Koster, V. Mitev, M. Staudacher, M. Wilhelm, All tree-level MHV form factors in $\mathcal{N} = 4$ SYM from twistor space, J. High Energy Phys. 06 (2016) 162, arXiv:1604.00012 [hep-th].
[63] L. Koster, V. Mitev, M. Staudacher, M. Wilhelm, On form factors and correlation functions in twistor space, arXiv:1611.08599 [hep-th].
[64] D. Chicherin, E. Sokatchev, $\mathcal{N} = 4$ super-Yang–Mills in LHC superspace part I: classical and quantum theory, J. High Energy Phys. 02 (2017) 062, arXiv:1601.06803 [hep-th].
[65] D. Chicherin, E. Sokatchev, N = 4 super-Yang–Mills in LHC superspace. Part II: Non-chiral correlation functions of the stress-tensor multiplet, J. High Energy Phys. 03 (2017) 048, arXiv:1601.06804 [hep-th].
[66] D. Chicherin, E. Sokatchev, Demystifying the twistor construction of composite operators in N = 4 super-Yang–Mills theory, arXiv:1603.08478 [hep-th].
[67] D. Chicherin, E. Sokatchev, Composite operators and form factors in N = 4 SYM, arXiv:1605.01386 [hep-th].
[68] L.J. Dixon, Calculating scattering amplitudes efficiently, in: QCD and Beyond. Proceedings, Theoretical Advanced Study Institute in Elementary Particle Physics, TASI-95, Boulder, USA, June 4–30, 1995, 1996, pp. 539–584, arXiv:hep-ph/9601359.
[69] V.P. Nair, A current algebra for some gauge theory amplitudes, Phys. Lett. B 214 (1988) 215–218.
[70] Y. Geyer, Ambitwistor Strings: Worldsheet Approaches to Perturbative Quantum Field Theories, PhD thesis, Oxford U., Inst. Math., 2016, arXiv:1610.04525 [hep-th].
[71] M. Ademollo, A. Bellini, M. Ciafaloni, Superstring Regge amplitudes and emission vertices, Phys. Lett. B 223 (1989) 318–324.
[72] M. Ademollo, A. Bellini, M. Ciafaloni, Superstring Regge amplitudes and graviton radiation at Planckian energies, Nucl. Phys. B 338 (1990) 114–142.
[73] N. Arkani-Hamed, F. Cachazo, J. Kaplan, What is the simplest quantum field theory?, J. High Energy Phys. 09 (2010) 016, arXiv:0808.1446 [hep-th].
[74] L.V. Bork, A.I. Onishchenko, Four dimensional ambitwistor strings and form factors of local and Wilson line operators, arXiv:1704.04758 [hep-th].

[75] N. Arkani-Hamed, J. Bourjaily, F. Cachazo, J. Trnka, Local spacetime physics from the Grassmannian, J. High Energy Phys. 01 (2011) 108, arXiv:0912.3249 [hep-th].
[76] D. Nandan, A. Volovich, C. Wen, A Grassmannian etude in NMHV minors, J. High Energy Phys. 07 (2010) 061, arXiv:0912.3705 [hep-th].
[77] L. Dolan, P. Goddard, Gluon tree amplitudes in open twistor string theory, J. High Energy Phys. 12 (2009) 032, arXiv:0909.0499 [hep-th].
[78] L.J. Mason, D. Skinner, The complete planar S-matrix of $N = 4$ SYM as a Wilson loop in twistor space, J. High Energy Phys. 12 (2010) 018, arXiv:1009.2225 [hep-th].
[79] T. Adamo, M. Bullimore, L. Mason, D. Skinner, Scattering amplitudes and Wilson loops in twistor space, J. Phys. A 44 (2011) 454008, arXiv:1104.2890 [hep-th].
[80] M. Søgaard, Y. Zhang, Scattering equations and global duality of residues, Phys. Rev. D 93 (10) (2016) 105009, arXiv:1509.08897 [hep-th].
[81] J. Bosma, M. Søgaard, Y. Zhang, The polynomial form of the scattering equations is an H-basis, Phys. Rev. D 94 (4) (2016) 041701, arXiv:1605.08431 [hep-th].
[82] Y. Zhang, Lecture notes on multi-loop integral reduction and applied algebraic geometry, arXiv:1612.02249 [hep-th], 2016.
[83] S. Weinzierl, Tales of 1001 gluons, arXiv:1610.05318 [hep-th], 2016.
[84] L.N. Lipatov, Effective action for the Regge processes in gravity, Phys. Part. Nucl. 44 (2013) 391–413, arXiv:1105.3127 [hep-th].
[85] L.N. Lipatov, Euler–Lagrange equations for the Gribov reggeon calculus in QCD and in gravity, Int. J. Mod. Phys. A 31 (28–29) (2016) 1645011.

2

Extended solutions for the biadjoint scalar field

Pieter-Jan De Smet, Chris D. White *

Centre for Research in String Theory, School of Physics and Astronomy, Queen Mary University of London, 327 Mile End Road, London E1 4NS, UK

ARTICLE INFO

Editor: N. Lambert

ABSTRACT

Biadjoint scalar field theories are increasingly important in the study of scattering amplitudes in various string and field theories. Recently, some first exact nonperturbative solutions of biadjoint scalar theory were presented, with a pure power-like form corresponding to isolated monopole-like objects located at the origin of space. In this paper, we find a novel family of extended solutions, involving non-trivial form factors that partially screen the divergent field at the origin. All previous solutions emerge as special cases.

1. Introduction

The study of (quantum) field theories remains a highly active research area, given that such theories describe the four fundamental forces in nature. It is widely believed that these forces may emerge as the low energy limit of an underlying framework, which may not necessarily be a field theory (e.g. it may be a string theory). It is thus interesting to elucidate common structures between theories, and also to investigate how different types of theory are related to each other. In this context, so-called *biadjoint scalar theory* has recently been increasingly studied, whose Lagrangian is given by

$$\mathcal{L} = \frac{1}{2}\partial^\mu \Phi^{aa'} \partial_\mu \Phi^{aa'} + \frac{y}{3} f^{abc} \tilde{f}^{a'b'c'} \Phi^{aa'} \Phi^{bb'} \Phi^{cc'}. \quad (1)$$

Here the (un)primed indices are adjoint indices associated with two different Lie groups, with structure constants f^{abc} and $\tilde{f}^{a'b'c'}$ respectively, and y is a coupling constant. Although this theory does not seem to be directly physically applicable by itself, there is increasing evidence that it underlies the dynamics in more relevant theories. For example, perturbative scattering amplitudes in this theory can be used as building blocks for amplitudes in non-abelian gauge theories [1,2]. The latter are related to amplitudes in gravity theories by the *double copy* of refs. [3–5], thus providing a ladder of theories related by taking the field $\Phi^{aa'}$ of eq. (1), and replacing either of its adjoint indices with Lorentz (spacetime) indices. This same procedure can be applied to exact classical solutions as well as perturbative amplitudes [6–13], where again the biadjoint theory plays a crucial role. Perturbative radiative solutions have also been considered recently [14–16], as well as amplitudes in curved space [17]. A natural framework for unifying the description of amplitudes in biadjoint, gauge and gravity theories is the CHY equations of refs. [18–21], which have been shown to emerge from a string theory in ambitwistor space [22–27], itself a limit of conventional string theory [28]. Loop level aspects of biadjoint theories in this formalism have been recently explored in ref. [29]. For related studies, see also refs. [30–37].

Given the wide range of instances in which the biadjoint scalar theory appears, it is clearly worth studying this theory in its own right. All of the above examples of its use involve perturbative solutions of the equation of motion, which from eq. (1) is found to be

$$\partial_\mu \partial^\mu \Phi^{aa'} - y f^{abc} \tilde{f}^{a'b'c'} \Phi^{bb'} \Phi^{cc'} = 0. \quad (2)$$

Instead, one may consider exact nonlinear solutions, which may involve inverse powers of the coupling y. A number of solutions of this type were recently presented in ref. [38], consisting of singular point-like disturbances located at the origin. The first of these has spherical symmetry, and is applicable when the two Lie groups are identified:

$$\Phi^{ab} = -\frac{2\delta^{ab}}{y T_A r^2}. \quad (3)$$

Here r is the radial space coordinate,[1] and we define

$$f^{abc} f^{dbc} = T_A \delta^{ad}. \quad (4)$$

* Corresponding author.
E-mail addresses: pejedees@yahoo.com (P.-J. De Smet), christopher.white@qmul.ac.uk (C.D. White).

[1] We use the metric $(+, -, -, -)$ throughout.

Further solutions are possible if one takes the common group to be $SU(2)$, which allows mixing between spacetime and adjoint indices, analogous to known non-perturbative solutions in non-abelian gauge theories [39–44]. In particular, ref. [38] presented a one-parameter family of solutions with an axial form in the internal space:

$$\Phi^{ab} = \frac{1}{yr^2}\left[-k\left(\delta^{ab} - \frac{x^a x^b}{r^2}\right) \pm \sqrt{2k-k^2}\,\frac{\epsilon^{abd}x^d}{r}\right], \tag{5}$$

where $0 \leq k \leq 2$ if $\Phi^{ab} \in \mathbb{R}$. Like eq. (3), this has a pure power-like behaviour in r. Indeed, this is the only possibility if an axial form is imposed [38], such that it tempting to think that the spectrum of nonperturbative solutions in biadjoint scalar theory is much simpler than that of non-abelian gauge theories. The aim of this paper, however, is to show that extended solutions do in fact exist. That is, it is possible to dress the power-like divergence in r with a non-trivial form factor, such that all previously found solutions emerge as special cases. The new solutions are still singular at the origin (a consequence of Derrick's theorem for scalar field theories [45], which prohibits finite energy). However, one solution in particular will exhibit an interesting screening behaviour, such that the strength of the divergence in energy is ameliorated. Our results will be useful in future studies of biadjoint scalar theory, including the issue of whether or not double copy-like relationships can be generalised beyond the perturbative sector.

The structure of the paper is as follows. In section 2, we present extended solutions for the case in which both Lie groups in the biadjoint scalar Lagrangian are the same. In section 3, we apply similar techniques to construct an extended solution for the case in which both Lie groups are $SU(2)$. We discuss our results and conclude in section 4.

2. Extended solutions for common Lie groups

In this section, we consider the Lagrangian of eq. (1), but where both sets of structure constants are associated with the same Lie group. We may then make a similar ansatz to that made in ref. [38], namely

$$\Phi^{ab} = \delta^{ab} S(r), \quad S(r) = \frac{\bar{S}(r)}{y T_A}. \tag{6}$$

Substituting this in eq. (2) yields a second-order non-linear differential equation for $\bar{S}(r)$:

$$\frac{1}{r^2}\frac{d}{dr}\left(r^2\frac{d\bar{S}(r)}{dr}\right) + \bar{S}^2(r) = 0. \tag{7}$$

We now look for extended solutions in which the r^{-2} divergence of the field at the origin, of eq. (3), is dressed by a finite form factor. To this end, we may write

$$K(r) = 1 + r^2\bar{S}(r), \tag{8}$$

where $K(r)$ is finite for all r. Equation (7) then becomes

$$r^2K''(r) - 2rK'(r) + K^2(r) - 1 = 0. \tag{9}$$

We can study this further by defining

$$r = e^{-\xi} \tag{10}$$

so that eq. (9) becomes

$$\frac{\partial^2 K}{\partial \xi^2} + 3\frac{\partial K}{\partial \xi} - 1 + K^2 = 0. \tag{11}$$

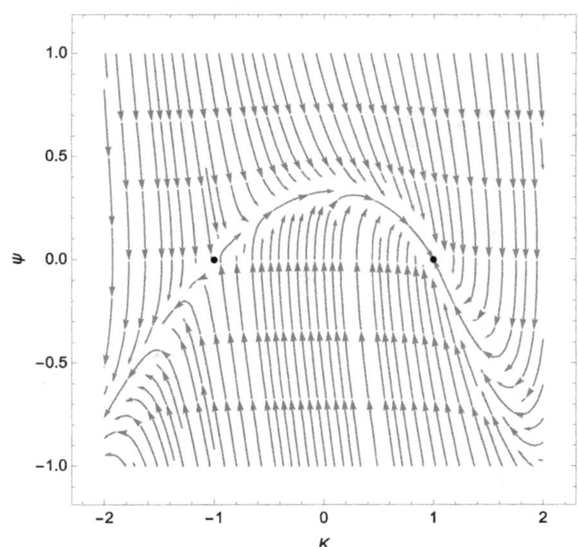

Fig. 1. The integral curves of the vector field defined by (14). The stationary points $(\pm 1, 0)$ are marked in red. (For interpretation of the references to colour in this figure legend, the reader is referred to the web version of this article.)

We may transform this into a more recognisable form by setting

$$\frac{\partial K}{\partial \xi} = -3w(K), \tag{12}$$

so that eq. (11) implies

$$w(K)w'(K) - w(K) + \frac{-1+K^2}{9} = 0. \tag{13}$$

This is an Abel equation of the second kind. Unlike some equations of this type, there appears to be no analytic solution in terms of known functions (e.g. it is not listed in the well-known compendium of ref. [46]). Instead, we may revert back to eq. (11) and analyse it using a method similar to that used by Wu and Yang [44] to construct nonperturbative solutions in pure Yang–Mills theory. First, one may turn eq. (11) into two coupled first-order differential equations as follows:

$$\left(\frac{\partial K}{\partial \xi}, \frac{\partial \psi}{\partial \xi}\right) = \left(\psi, -3\psi + 1 - K^2\right). \tag{14}$$

This defines a vector field in the (K, ψ) plane, where the curves that are tangent to this vector field are the solutions we desire. We show a plot of these curves in Fig. 1, and solutions for $K(\xi)$ that are finite for all values of r correspond to curves which are bounded in the plane. By inspection, and noting that the vector field is zero at $(K, \psi) = (\pm 1, 0)$, there are three possibilities:

1. $K(r) = 1$: this corresponds to $\bar{S}(r) = 0$, and hence the trivial solution $\Phi^{ab} = 0$.
2. $K(r) = -1$: this yields $\bar{S}(r) = -2/r^2$, which is the solution already found in ref. [38], and reported here in eq. (3).
3. $K(r) \to \pm 1$ as $\xi \to \pm\infty$ respectively. This is a non-trivial solution corresponding to the single curve that flows from $(-1, 0)$ to $(+1, 0)$ in Fig. 1. It has not been previously obtained.

Let us analyse the new solution yet further. In fact, this is a one-parameter family of solutions: if $(K(\xi), \psi(\xi))$ solve eq. (14), then $(K(\xi - \xi_0), \psi(\xi - \xi_0))$, for arbitrary constant ξ_0, are also solutions. Translating back to the spatial coordinate r, this implies that if $K(r)$ solves eq. (11), so does the rescaled function $K(r/\lambda)$, for general constant λ.

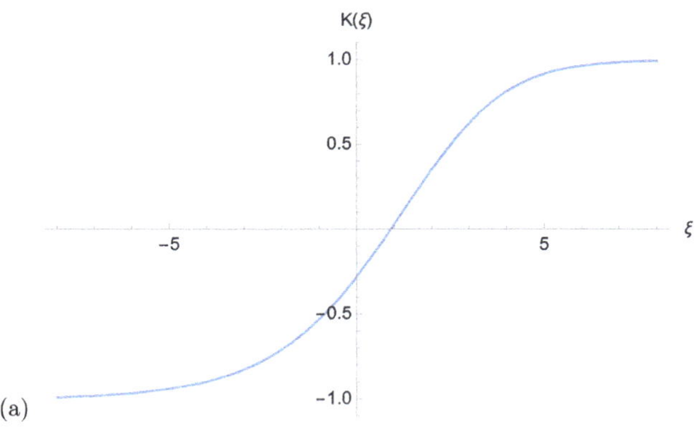

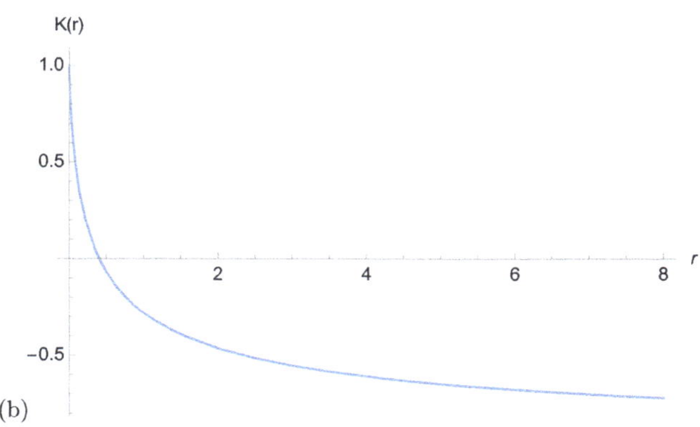

Fig. 2. (a) Numerical solution of (11) with asymptotic behaviour given by (17). We have chosen $A=1$; other choices of A will lead to the same graph but translated along the ξ-axis. (b) The behaviour of K as a function of r.

Although a complete analytic solution for $K(\xi)$ is not possible, we can examine its behaviour in asymptotic limits. For $\xi \to -\infty$ one may write

$$K(\xi) = -1 + f(\xi).$$

Upon substituting this into eq. (11) and neglecting terms quadratic and higher in $f(\xi)$, one finds the general solution

$$f(\xi) = Ae^{\left(\frac{\sqrt{17}}{2}-\frac{3}{2}\right)\xi} + Be^{\left(-\frac{3}{2}-\frac{\sqrt{17}}{2}\right)\xi}. \tag{15}$$

The boundary condition $f(\xi) \to 0$ as $\xi \to -\infty$ then requires $B=0$. Similarly, for $\xi \to +\infty$ we may substitute

$$K(\xi) = 1 + g(\xi)$$

in eq. (11) and ignore terms quadratic and higher in $g(\xi)$, leading to the general solution

$$g(\xi) = A'e^{-\xi} + B'e^{-2\xi}, \tag{16}$$

where the first term gives the dominant behaviour. We conclude that

$$K(\xi) \simeq \begin{cases} -1 + Ae^{\left(\frac{\sqrt{17}}{2}-\frac{3}{2}\right)\xi}, & \xi \to -\infty \\ +1 + A'e^{-\xi}, & \xi \to +\infty \end{cases}, \tag{17}$$

or, translating back to the function $\bar{S}(r)$,

$$\bar{S}(r) \simeq \begin{cases} \frac{1}{r^2}\left[-2 + Ar^{-\frac{\sqrt{17}}{2}+\frac{3}{2}}\right], & r \to \infty \\ \frac{A'}{r}, & r \to 0. \end{cases} \tag{18}$$

The two parameters A and A' are not independent. As discussed above, all solutions for $K(r)$ constitute a rescaling of a canonical solution. The scaling is fixed by choosing either A or A', such that the other parameter also becomes fixed. For the particular choice $A=1$, we show a numerical solution for $K(\xi)$ in Fig. 2.

The simple power-like solution of eq. (3) is reminiscent of a Coulomb solution in gauge theory, with a correspondingly divergent field at the origin.[2] The new solution found here consists of dressing the previous solution with a form factor, such that the divergence at $r \to 0$ is partially screened. That is, the r^{-2} behaviour is softened to r^{-1}, and by choosing the rescaling parameter λ in $K(r/\lambda)$, this screening can be made as smooth or abrupt as desired. The softening of the divergence in the field is also reflected in the energy of the solution. From eq. (1), the Hamiltonian density of the biadjoint scalar theory is found to be [38]

$$\mathcal{H} = \frac{1}{2}\left[(\dot{\Phi}^{aa'})^2 + \nabla\Phi^{aa'}\cdot\nabla\Phi^{aa'}\right] - \frac{y}{3}f^{abc}\tilde{f}^{a'b'c'}\Phi^{aa'}\Phi^{bb'}\Phi^{cc'}, \tag{19}$$

which for the ansatz of eq. (6) evaluates to

$$\mathcal{H} = \frac{\mathcal{N}}{y^2T_A^2}\left(\frac{1}{2}\bar{S}'(r)^2 - \frac{1}{3}\bar{S}(r)^3\right), \tag{20}$$

with $\mathcal{N}$ the dimension of the (common) Lie group. The energy is divergent due to the singularity at the origin, but can be evaluated by implementing a short-distance cutoff r_0, corresponding to a *charge radius*:

$$E = 4\pi\int_{r_0}^{\infty} dr\, r^2\, \mathcal{H}. \tag{21}$$

For the solution with the asymptotic behaviour of eq. (18), this gives (as $r_0 \to 0$)

$$E \simeq \frac{\mathcal{N}}{y^2T_A^2}\frac{2\pi(A')^2}{r_0}, \tag{22}$$

which is a much softer divergence than the energy associated with eq. (3), that behaves as $E \sim r_0^{-3}$ [38].

3. Extended solutions for a common group of *SU*(2)

Having succeeded in constructing an extended solution for the case in which both gauge groups are identical, let us see if further such solutions are possible upon restricting each gauge group to be $SU(2)$, for which one may make a similar ansatz to that used in ref. [38]

$$\Phi^{ab} = \frac{1}{r^2}\left(A(r)\delta^{ab} + B(r)\frac{x^a x^b}{r^2} + C(r)\epsilon^{abd}\frac{x^d}{r}\right). \tag{23}$$

Introducing $\bar{A} = A/y$ etc., eq. (2) then implies the coupled non-linear differential equations

$$r^2\bar{A}'' - 2r\bar{A}' + 2\bar{A}\bar{B} + 2\bar{A}^2 + 2\bar{A} + 2\bar{B} = 0; \tag{24}$$

$$r^2\bar{B}'' - 2r\bar{B}' - 2\bar{A}\bar{B} - 4\bar{B} + 2\bar{C}^2 = 0; \tag{25}$$

$$r^2\bar{C}'' - 2r\bar{C}' + 2\bar{A}\bar{C} + 2\bar{B}\bar{C} = 0. \tag{26}$$

[2] The fact that the power associated with the divergence is different to the Coulomb solution can be traced to the fact that the coupling constant y is dimensionful [38].

To simplify this we make a further ansatz, namely we write

$$\bar{A}(r) = c_1 f(r) + c_2 \tag{27}$$

$$\bar{B}(r) = c_3 f(r) + c_4 \tag{28}$$

$$\bar{C}(r) = c_5 f(r) + c_6, \tag{29}$$

and then we look for special values of the constants $c_1, \ldots, c_6$ so that the three equations (24)-(26) reduce to a single differential equation for $f(r)$. We thus find two new types of solution that were not presented previously in ref. [38].

3.1. Multiple power-like solution

The first new solution is obtained upon choosing

$$\bar{A}(r) = -1, \quad \bar{C}(r) = 0, \tag{30}$$

in which case $\bar{B}(r)$ satisfies the linear homogeneous differential equation

$$r^2 \bar{B}''(r) - 2r\bar{B}'(r) - 2\bar{B}(r) = 0, \tag{31}$$

whose general solution is

$$\bar{B}(r) = r^{3/2}\left(b_1 r^{\frac{\sqrt{17}}{2}} + b_2 r^{-\frac{\sqrt{17}}{2}}\right), \tag{32}$$

where b_1 and b_2 are arbitrary constants. The full solution for Φ^{ab} is then

$$\begin{aligned}\Phi^{ab} &= \frac{1}{yr^2}\left[-\delta^{ab} + \left(b_1 r^{\frac{3}{2}+\frac{\sqrt{17}}{2}} + b_2 r^{\frac{3}{2}-\frac{\sqrt{17}}{2}}\right)\frac{x^a x^b}{r^2}\right] \\ &\simeq \frac{1}{yr^2}\left[-\delta^{ab} + \left(b_1 r^{3.562} + b_2 r^{-0.562}\right)\frac{x^a x^b}{r^2}\right].\end{aligned} \tag{33}$$

Here the second term in the square brackets has a part which diverges more rapidly at the origin than the r^{-2} behaviour of the solutions of ref. [38], and a term which is softer. If one requires a finite energy upon integrating to infinity, however, one must set $b_1 = 0$. The energy, from eq. (21), is then found to be

$$E = 4\pi\left(\frac{1}{4}\left(1+\sqrt{17}\right)b_2^2 r_0^{-\sqrt{17}} - 2b_2 r_0^{-\frac{3}{2}-\frac{\sqrt{17}}{2}} + \frac{8}{3r_0^3}\right), \tag{34}$$

where again r_0 is a small-distance cutoff. For small r_0, the energy diverges as $E \sim r_0^{-\sqrt{17}} \simeq r_0^{-4.123}$ (assuming $b_2 \neq 0$). Interestingly, the solution of eq. (33) involves the same power of $r^{\frac{3}{2}-\frac{\sqrt{17}}{2}}$ as appears in the asymptotic behaviour for the extended solution of the previous section, eq. (18).

3.2. Extended solutions

We have also found extended solutions of eqs. (24)-(26), that do not have a pure power-like form. For these solutions, one has

$$\bar{A}(r) = -1 + \frac{1}{2}(c^2-1)\bar{B}(r), \quad \bar{C}(r) = c\bar{B}(r), \tag{35}$$

where c is a constant, and $B(r)$ satisfies

$$r^2\bar{B}''(r) - 2r\bar{B}'(r) + (c^2+1)\bar{B}^2(r) - 2\bar{B}(r) = 0. \tag{36}$$

Upon writing

$$\bar{B}(r) = \frac{1}{c^2+1}\left(K(r)+1\right), \tag{37}$$

one finds that $K(r)$ satisfies eq. (9). Thus, an extended solution for Φ^{ab} is given by

$$\Phi^{ab} = \frac{1}{yr^2}\left[-\delta^{ab} + \frac{K(r)+1}{c^2+1}\left(\frac{c^2-1}{2}\delta^{ab} + \frac{x^a x^b}{r^2} + c\,\epsilon^{abd}\frac{x^d}{r}\right)\right]. \tag{38}$$

In section 2 we found three solutions for $K(r)$, which lead to the following cases:

1. $K(r) = 1$: in this case, eq. (38) reduces to

$$\Phi^{ab} = \frac{1}{yr^2}\left[-\frac{2}{c^2+1}\left(\delta^{ab} - \frac{x^a x^b}{r^2}\right) + \frac{2c}{c^2+1}\epsilon^{abd}\frac{x^d}{r}\right]. \tag{39}$$

   By replacing

$$c \to \pm\sqrt{\frac{2-k}{k}}, \tag{40}$$

   we see that eq. (39) is the same as the solution of eq. (5), that was already presented in ref. [38].
2. $K(r) = -1$: in this case, eq. (38) reduces to

$$\Phi^{ab} = -\frac{1}{yr^2}\delta^{ab},$$

   which is the same as eq. (3) for the special case in which both gauge groups are $SU(2)$, so that $T_A = 2$.
3. The most general solution has $K(r)$ given by the function of Fig. 2, with asymptotic limits given by eq. (18). This solution is new.

The extended solution no longer has the pure-axial property of eq. (5), consistent with the fact noted in ref. [38] that extended axial solutions cannot exist. The energy of the solution in case 3 above, in terms of the usual short-distance cutoff r_0, is found to have the leading behaviour (as $r_0 \to 0$)

$$E \simeq \frac{32\pi}{y^2}\frac{1}{(1+c^2)r_0^3}. \tag{41}$$

Note that the extended solution of section 2 (where again $T_A = 2$) can also be obtained from eq. (38), by choosing $c \to \infty$:

$$\Phi^{ab} \xrightarrow{c\to\infty} \frac{K(r)-1}{2yr^2}\delta^{ab}. \tag{42}$$

Finally, we note that a complex solution is also possible. By choosing $c^2 = -1$ in eq. (36), this reduces to eq. (31), whose general solution is given by eq. (32). One then obtains

$$\Phi^{ab} = \frac{1}{yr^2}\left[-\delta^{ab} + b_2 r^{\frac{3}{2}-\frac{\sqrt{17}}{2}}\left(-\delta^{ab} + \frac{x^a x^b}{r^2} \pm i\epsilon^{abd}\frac{x^d}{r}\right)\right], \tag{43}$$

where we have again set $b_1 \to 0$ to ensure that the energy is bounded at spatial infinity.

4. Conclusion

Biadjoint scalar field theory occurs in a number of contexts, and plays an intriguing role in determining the dynamics of perturbative scattering amplitudes in gauge and gravity theories. The nonperturbative properties of this theory remain relatively unexplored, and in this paper we have presented a number of new solutions involving inverse powers of the coupling constant. Unlike the first solutions presented in ref. [38], the results of the present paper have an extended structure, indicating that the spectrum of nonperturbative solutions of biadjoint scalar theory is much richer than has previously been suggested. Interestingly, the solutions presented here include those in which the strong divergent behaviour of the field at the origin is partially screened.

All known examples of the double copy between gauge and gravity theories (or the zeroth copy between gauge and biadjoint scalar theories) involve positive powers of the coupling constant, rather than negative powers as in this paper. It is thus not at all clear how the double copy can be extended to the nonperturbative sector. Nevertheless, our hope is that by collecting a catalogue of such solutions, one may be able to formulate a suitable copy procedure, which would then dramatically extend the remit of the double copy and related ideas. The simplest nonperturbative copy procedure would be expected between biadjoint and gauge theories, which motivates the present study. Furthermore, the fact that very little is so far known about biadjoint theories, despite their cropping up in a number of different contexts, makes these solutions of interest in themselves.

There are a number of avenues for further work. Firstly, it may be possible to find more solutions of the biadjoint scalar theory, including non-static solutions. Secondly, one may imagine coupling the biadjoint scalar to a gauge field, as occurs in some applications of the double copy. Finally, the question of whether the nonperturbative solutions found here and in ref. [38] can themselves be copied to gauge theory or gravity deserves further attention.

Acknowledgements

CDW is supported by the UK Science and Technology Facilities Council (STFC), through grant number ST/P000754/1.

References

[1] N. Bjerrum-Bohr, P.H. Damgaard, R. Monteiro, D. O'Connell, Algebras for amplitudes, J. High Energy Phys. 1206 (2012) 061, arXiv:1203.0944.

[2] C.R. Mafra, Berends–Giele recursion for double-color-ordered amplitudes, J. High Energy Phys. 07 (2016) 080, arXiv:1603.09731.

[3] Z. Bern, J. Carrasco, H. Johansson, New relations for gauge-theory amplitudes, Phys. Rev. D 78 (2008) 085011, arXiv:0805.3993.

[4] Z. Bern, J.J.M. Carrasco, H. Johansson, Perturbative quantum gravity as a double copy of gauge theory, Phys. Rev. Lett. 105 (2010) 061602, arXiv:1004.0476.

[5] Z. Bern, T. Dennen, Y.-t. Huang, M. Kiermaier, Gravity as the square of gauge theory, Phys. Rev. D 82 (2010) 065003, arXiv:1004.0693.

[6] R. Monteiro, D. O'Connell, C.D. White, Black holes and the double copy, J. High Energy Phys. 1412 (2014) 056, arXiv:1410.0239.

[7] A. Luna, R. Monteiro, D. O'Connell, C.D. White, The classical double copy for Taub–NUT spacetime, Phys. Lett. B 750 (2015) 272–277, arXiv:1507.01869.

[8] A. Luna, R. Monteiro, I. Nicholson, D. O'Connell, C.D. White, The double copy: Bremsstrahlung and accelerating black holes, arXiv:1603.05737.

[9] A.K. Ridgway, M.B. Wise, Static spherically symmetric Kerr–Schild metrics and implications for the classical double copy, arXiv:1512.02243.

[10] A. Anastasiou, L. Borsten, M.J. Duff, L.J. Hughes, S. Nagy, Yang–Mills origin of gravitational symmetries, Phys. Rev. Lett. 113 (23) (2014) 231606, arXiv:1408.4434.

[11] L. Borsten, M.J. Duff, Gravity as the square of Yang–Mills?, Phys. Scr. 90 (2015) 108012, arXiv:1602.08267.

[12] A. Anastasiou, L. Borsten, M.J. Duff, M.J. Hughes, A. Marrani, S. Nagy, M. Zoccali, Twin supergravities from Yang–Mills theory squared, Phys. Rev. D 96 (2) (2017) 026013, arXiv:1610.07192.

[13] A. Anastasiou, L. Borsten, M.J. Duff, A. Marrani, S. Nagy, M. Zoccali, Are all supergravity theories Yang–Mills squared?, arXiv:1707.03234.

[14] W.D. Goldberger, A.K. Ridgway, Radiation and the classical double copy for color charges, Phys. Rev. D 95 (12) (2017) 125010, arXiv:1611.03493.

[15] W.D. Goldberger, S.G. Prabhu, J.O. Thompson, Classical gluon and graviton radiation from the bi-adjoint scalar double copy, arXiv:1705.09263.

[16] A. Luna, R. Monteiro, I. Nicholson, A. Ochirov, D. O'Connell, N. Westerberg, C.D. White, Perturbative spacetimes from Yang–Mills theory, J. High Energy Phys. 04 (2017) 069, arXiv:1611.07508.

[17] T. Adamo, E. Casali, L. Mason, S. Nekovar, Scattering on plane waves and the double copy, arXiv:1706.08925.

[18] F. Cachazo, S. He, E.Y. Yuan, Scattering of massless particles: scalars, gluons and gravitons, arXiv:1309.0885.

[19] F. Cachazo, S. He, E.Y. Yuan, Scattering of massless particles in arbitrary dimension, arXiv:1307.2199.

[20] F. Cachazo, S. He, E.Y. Yuan, Scattering equations and KLT orthogonality, arXiv: 1306.6575.

[21] F. Cachazo, S. He, E.Y. Yuan, Scattering in three dimensions from rational maps, J. High Energy Phys. 1310 (2013) 141, arXiv:1306.2962.

[22] L. Mason, D. Skinner, Ambitwistor strings and the scattering equations, J. High Energy Phys. 1407 (2014) 048, arXiv:1311.2564.

[23] Y. Geyer, A.E. Lipstein, L.J. Mason, Ambitwistor strings in 4-dimensions, Phys. Rev. Lett. 113 (2014) 081602, arXiv:1404.6219.

[24] E. Casali, Y. Geyer, L. Mason, R. Monteiro, K.A. Roehrig, New ambitwistor string theories, J. High Energy Phys. 11 (2015) 038, arXiv:1506.08771.

[25] Y. Geyer, L. Mason, R. Monteiro, P. Tourkine, Loop integrands for scattering amplitudes from the Riemann sphere, Phys. Rev. Lett. 115 (12) (2015) 121603, arXiv:1507.00321.

[26] Y. Geyer, L. Mason, R. Monteiro, P. Tourkine, One-loop amplitudes on the Riemann sphere, J. High Energy Phys. 03 (2016) 114, arXiv:1511.06315.

[27] Y. Geyer, L. Mason, R. Monteiro, P. Tourkine, Two-loop scattering amplitudes from the Riemann sphere, Phys. Rev. D 94 (12) (2016) 125029, arXiv: 1607.08887.

[28] E. Casali, P. Tourkine, On the null origin of the ambitwistor string, J. High Energy Phys. 11 (2016) 036, arXiv:1606.05636.

[29] H. Gomez, C. Lopez-Arcos, P. Talavera, One-loop Parke–Taylor factors for quadratic propagators from massless scattering equations, arXiv:1707.08584.

[30] M. Chiodaroli, Simplifying amplitudes in Maxwell–Einstein and Yang–Mills–Einstein supergravities, arXiv:1607.04129, 2016.

[31] L. de la Cruz, A. Kniss, S. Weinzierl, Relations for Einstein–Yang–Mills amplitudes from the CHY representation, Phys. Lett. B 767 (2017) 86–90, arXiv: 1607.06036.

[32] G.L. Cardoso, S. Nagy, S. Nampuri, A double copy for $\mathcal{N}=2$ supergravity: a linearised tale told on-shell, J. High Energy Phys. 10 (2016) 127, arXiv:1609.05022.

[33] C.R. Mafra, O. Schlotterer, Non-abelian Z-theory: Berends–Giele recursion for the α'-expansion of disk integrals, J. High Energy Phys. 01 (2017) 031, arXiv: 1609.07078.

[34] J.J.M. Carrasco, C.R. Mafra, O. Schlotterer, Abelian Z-theory: NLSM amplitudes and α'-corrections from the open string, J. High Energy Phys. 06 (2017) 093, arXiv:1608.02569.

[35] S. Mizera, Inverse of the string theory KLT kernel, J. High Energy Phys. 06 (2017) 084, arXiv:1610.04230.

[36] M. Campiglia, L. Coito, S. Mizera, Can scalars have asymptotic symmetries?, arXiv:1703.07885.

[37] H. Johansson, J. Nohle, Conformal gravity from gauge theory, arXiv:1707.02965.

[38] C.D. White, Exact solutions for the biadjoint scalar field, Phys. Lett. B 763 (2016) 365–369, arXiv:1606.04724.

[39] M.K. Prasad, C.M. Sommerfield, An exact classical solution for the 't Hooft monopole and the Julia–Zee dyon, Phys. Rev. Lett. 35 (1975) 760–762.

[40] E.B. Bogomolny, Stability of classical solutions, Sov. J. Nucl. Phys. 24 (1976) 449, Yad. Fiz. 24 (1976) 861.

[41] B. Julia, A. Zee, Poles with both magnetic and electric charges in nonabelian gauge theory, Phys. Rev. D 11 (1975) 2227–2232.

[42] G. 't Hooft, Magnetic monopoles in unified gauge theories, Nucl. Phys. B 79 (1974) 276–284.

[43] A.M. Polyakov, Isomeric states of quantum fields, Zh. Eksp. Teor. Fiz. 68 (1975) 1975.

[44] T.T. Wu, C.-N. Yang, Some solutions of the classical isotopic gauge field equations, in: Properties of Matter Under Unusual Conditions, 1967.

[45] G.H. Derrick, Comments on nonlinear wave equations as models for elementary particles, J. Math. Phys. 5 (1964) 1252–1254.

[46] V. Zaitsev, A. Polyanin, Handbook of Exact Solutions for Ordinary Differential Equations, CRC Press, 2002.

Fock-space projection operators for semi-inclusive final states

Robert Dickinson [a], Jeff Forshaw [a], Peter Millington [b,*]

[a] Consortium for Fundamental Physics, School of Physics and Astronomy, University of Manchester, Manchester M13 9PL, United Kingdom
[b] School of Physics and Astronomy, University of Nottingham, Nottingham NG7 2RD, United Kingdom

ARTICLE INFO

ABSTRACT

We present explicit expressions for Fock-space projection operators that correspond to realistic final states in scattering experiments. Our operators automatically sum over unobserved quanta and account for non-emission into sub-regions of momentum space.

Editor: B. Grinstein

Keywords:
Quantum field theory
Projection operators

1. Introduction

When calculating matrix elements for scattering processes, it is necessary to sum over all final states that contribute to an observable, which often necessitates summing over unmeasured quanta. The classic example is the computation of the cross-section for $e^+e^- \to$ hadrons, in which infra-red singularities cancel between the virtual gluon corrections and corresponding zero-energy real gluon emissions (at the level of the squared matrix element) by the Kinoshita–Lee–Nauenberg theorem [1,2] (see also Refs. [3,4]). Infrared divergences can in fact be avoided at the amplitude level (see e.g. Refs. [5–8]), by absorbing unobserved emissions into a redefinition of the asymptotic states. In this paper, we instead pursue the direct calculation of probabilities and focus on effect operators that correspond to the measurement of general semi-inclusive final states. These effect operators have the virtue that unobserved emissions simply do not enter the calculation.

The probability $\mathbb{P}$ that a system, described by some density operator ρ, will register an outcome, described by some effect operator E, is

$$\mathbb{P} = \mathrm{Tr}(E\rho)\,. \tag{1}$$

Furthermore, if the measurement is performed at time t_f and the system is known to be described at time t_i by the density operator ρ_i then, in the Interaction Picture,

$$\rho_f = U_{fi}\,\rho_i\,U^\dagger{}_{fi}\,, \tag{2}$$

where

$$U_{fi} = \mathrm{T}\exp\left(\frac{1}{i}\int_{t_i}^{t_f} \mathrm{d}t\; H_{\mathrm{int}}(t)\right) \tag{3}$$

is the unitary time-evolution operator and H_{int} is the interaction Hamiltonian. If the initial state is a pure state, i.e. $\rho_i = |i\rangle\langle i|$, the probability takes the form

$$\mathbb{P} = \langle i|\,(U^\dagger{}_{fi}\,E\,U_{fi})\,|i\rangle\,. \tag{4}$$

Of course, if the measurement also corresponds to a pure state, i.e. $E = |f\rangle\langle f|$, we obtain the usual squared matrix element

$$\mathbb{P} = |\,\langle f|\,U_{fi}\,|i\rangle\,|^2\,. \tag{5}$$

However, we may compute Eq. (4) directly by treating E as an operator. We then view Eq. (4) as an "in-in" expectation value, which can be written in the form [9]

$$\mathbb{P} = \sum_{j=0}\int_{t_i}^{t_f} \mathrm{d}t_1\,\mathrm{d}t_2\,\ldots\,\mathrm{d}t_j\;\Theta_{12\ldots j}\,\langle i|\mathcal{F}_j|i\rangle\,, \tag{6}$$

where

* Corresponding author.
E-mail addresses: robert.dickinson-2@manchester.ac.uk (R. Dickinson), jeff.forshaw@manchester.ac.uk (J. Forshaw), p.millington@nottingham.ac.uk (P. Millington).

$$\mathcal{F}_0 = E\,, \tag{7a}$$

$$\mathcal{F}_j = \tfrac{1}{i}\Big[\mathcal{F}_{j-1}, H_{\rm int}(t_j)\Big]\,, \tag{7b}$$

and $\Theta_{ijk\ldots} \equiv 1$ if $t_i > t_j > t_k\ldots$ and zero otherwise.

Whilst the explicit consideration of effect operators is ubiquitous in the description of measurement processes in quantum mechanics, they have, to our knowledge, been ignored in the context of particle physics. In what follows, we will present expressions for effect operators corresponding to general semi-inclusive measurements.

Our operators will be projection operators in Fock space and they all have the feature that unobserved quanta do not appear explicitly. For example, the effect operator corresponding to the inclusive cross-section for $e^+e^- \to$ one $q\bar{q}$ pair + anything is simply

$$E = |q,\bar{q}\rangle\langle q,\bar{q}| \otimes \mathbb{I}_{e^+} \otimes \mathbb{I}_{e^-} \otimes \mathbb{I}_\gamma \otimes \mathbb{I}_g\,, \tag{8}$$

where the sums over unobserved final-state electrons, positrons, photons and gluons appear as unit operators in their respective Fock spaces, which trivially commute through the structure in Eq. (6). These implicit summations over unobserved quanta are not present at the amplitude level, and this motivates further development of techniques along the lines of Ref. [9] aimed at directly computing probabilities in quantum field theory.

2. Projection operators in Fock space: bosonic case

It is a well-known result in quantum optics that the vacuum projection operator can be written as the exponential of the photon number operator (see e.g. Refs. [10–12]):

$$\begin{aligned} E^{(0)}_{\mathbb{R}^3} &\equiv \mathbb{I} + \sum_{j=1}^{\infty}\frac{(-1)^j}{j!} : (N_{\mathbb{R}^3})^j : \\ &= : \mathrm{e}^{-N_{\mathbb{R}^3}} : \\ &= |0\rangle\langle 0|\,, \end{aligned} \tag{9}$$

where the number operator

$$N_{\mathcal{R}} \equiv \sum_\lambda \int_{\mathcal{R}} \frac{\mathrm{d}^3\mathbf{k}}{(2\pi)^3 2E}\, a^\dagger_\lambda(\mathbf{k}) a_\lambda(\mathbf{k}) \tag{10}$$

counts the number of quanta in a region $\mathcal{R}$ of momentum space, i.e.

$$N_{\mathcal{R}}\,|\mathbf{k}_1\ldots\mathbf{k}_N\rangle = n\,|\mathbf{k}_1\ldots\mathbf{k}_N\rangle\,, \tag{11}$$

where

$$n = \sum_{a=1}^{N} \mathbf{1}_{\mathcal{R}}(\mathbf{k}_a) \tag{12}$$

and $\mathbf{1}_A(x)$ denotes the indicator function of set A, which is 1 if $x \in A$ and 0 otherwise. The colons indicate normal ordering. The sum is over all physical polarizations λ, if the projection is to be independent of polarization, or it could be over some subset of all allowed polarizations. Moreover, the region of momentum space need not be common to all polarizations, i.e. $\mathcal{R} \to \mathcal{R}_\lambda$. For ease of notation, we suppress the polarization indices that are needed to fully specify Fock states.

Whilst $E^{(0)}_{\mathbb{R}^3}$ is the projection operator corresponding to zero quanta (anywhere in configuration space), we can show that

$$E^{(0)}_{\mathcal{R}} \equiv : \mathrm{e}^{-N_{\mathcal{R}}} : \tag{13}$$

is the projection operator corresponding to zero quanta in the region $\mathcal{R}$, i.e.

$$E^{(0)}_{\mathcal{R}}\,|\mathbf{k}_1\ldots\mathbf{k}_N\rangle = \begin{cases} |\mathbf{k}_1\ldots\mathbf{k}_N\rangle & \text{if zero quanta in } \mathcal{R}\,, \\ 0 & \text{otherwise}\,. \end{cases} \tag{14}$$

The proof of this result, and of those that follow, is contained in the appendix. For $\mathcal{R} = \mathbb{R}^3$, we project out the vacuum state, as in Eq. (9). For the opposite extreme $\mathcal{R} = \emptyset$ (the empty set), the effect operator is just the unit operator, i.e. $E^{(0)}_\emptyset = \mathbb{I}$, and the measurement is inclusive over all final states [cf. Eq. (4)]. Everywhere in between, we automatically sum over states with zero quanta inside $\mathcal{R}$ and *any number* of quanta outside $\mathcal{R}$.

These non-emission operators are specific cases of a more general projection operator:

$$E^{\{j_a\}}_{\{\mathcal{R}_a \subseteq \mathcal{R}\}} \equiv : \left[\prod_a \frac{1}{j_a!}(N_{\mathcal{R}_a})^{j_a}\right] \mathrm{e}^{-N_{\mathcal{R}}} : \,. \tag{15}$$

This operator projects onto the subspace of states in which exactly $\sum_a j_a$ quanta have momenta in $\mathcal{R}$, distributed so that exactly j_a quanta have momenta in each disjoint subset $\mathcal{R}_a \subseteq \mathcal{R}$. Again, there is no restriction on quanta lying outside of $\mathcal{R}$. The special case of

$$E^{(j)}_{\mathcal{R}} \equiv : \frac{1}{j!}(N_{\mathcal{R}})^j\,\mathrm{e}^{-N_{\mathcal{R}}} : \tag{16}$$

projects onto exactly j particles in $\mathcal{R}$ and resembles the operator form of the photon counting distribution in quantum optics (see e.g. Ref. [13]).

To illustrate Eq. (15), we might consider the simple case where one quantum has momentum in the range $\mathbf{k} \to \mathbf{k} + \mathrm{d}^3\mathbf{k}$ and there are no other quanta anywhere, i.e. $\mathcal{R} = \mathbb{R}^3$. In this case, the projection operator is

$$\begin{aligned} E^{(1)}_{\mathcal{R}_1 \subset \mathbb{R}^3} &= : N_{\mathcal{R}_1}\mathrm{e}^{-N_{\mathbb{R}^3}} : \\ &= \frac{\mathrm{d}^3\mathbf{k}}{(2\pi)^3 2E} : a^\dagger(\mathbf{k}) a(\mathbf{k})\,|0\rangle\langle 0| : \\ &= \frac{\mathrm{d}^3\mathbf{k}}{(2\pi)^3 2E}\,|\mathbf{k}\rangle\langle\mathbf{k}|\,. \end{aligned} \tag{17}$$

With $j_a = 1\ \forall\ a$, Eq. (15) could be employed in situations where the observable final state has the form of n particles with given momenta $\mathbf{k}_a \to \mathbf{k}_a + \mathrm{d}^3\mathbf{k}_a$, accompanied by any number of undetectable particles below a given energy and/or transverse momentum threshold.

Since these projection operators share a common eigenbasis — the Fock basis — they mutually commute and can be combined straightforwardly. For example, $E^{(j)}_{\mathcal{R}_1} E^{(k)}_{\mathcal{R}_2}$ projects onto states with exactly j quanta in $\mathcal{R}_1$ *and* exactly k quanta in $\mathcal{R}_2$, regardless of whether $\mathcal{R}_1$ and $\mathcal{R}_2$ are disjoint.

We may now construct a projection operator $\mathrm{d}E_{\mathcal{R},v_n}(V)$ for an n-particle final state satisfying a constraint of the form $V \le v_n(\mathbf{k}_1,\ldots,\mathbf{k}_n) \le V + \mathrm{d}V$, which is symmetric under interchange of any two momenta, and inclusive of particles outside region $\mathcal{R}$. In the low-density regime in which Fock state occupation numbers rarely exceed unity, this is

$$\begin{aligned} \frac{\mathrm{d}E_{\mathcal{R},v_n}}{\mathrm{d}V} &= \left[\prod_{i=1}^{n}\int_{\mathcal{R}} \frac{\mathrm{d}^3\mathbf{k}_i}{(2\pi)^3 2E_i}\right]\frac{1}{n!}\,\delta(v_n(\{\mathbf{k}_i\}) - V) \\ &\quad\times : \left[\prod_{i=1}^{n} a^\dagger(\mathbf{k}_i) a(\mathbf{k}_i)\right]\mathrm{e}^{-N_{\mathcal{R}}} : \,. \end{aligned} \tag{18}$$

Where a single choice of particle number n is not appropriate, we may bring the particle number into the constraint function $v(\{\mathbf{k}_i\}; n) = v_n(\mathbf{k}_1, \ldots, \mathbf{k}_n)\ \forall\, n$, and define

$$\frac{\mathrm{d}E_{\mathcal{R},v}}{\mathrm{d}V} = \sum_n \frac{\mathrm{d}E_{\mathcal{R},v_n}}{\mathrm{d}V} . \tag{19}$$

3. Projection operators in Fock space: fermionic case

The projection operators for fermions are analogous to the bosonic case. We may regard the sum over λ in Eq. (10) to be inclusive of particle ($b_s^\dagger(\mathbf{k})b_s(\mathbf{k})$), anti-particle ($d_s^\dagger(\mathbf{k})d_s(\mathbf{k})$) and spin states (indexed by s), i.e. $N_{\mathcal{R}} \to N_{\mathcal{R}} + \bar{N}_{\bar{\mathcal{R}}}$, where

$$N_{\mathcal{R}} = \sum_s \int_{\mathcal{R}} \frac{\mathrm{d}^3\mathbf{k}}{(2\pi)^3 2E}\, b_s^\dagger(\mathbf{k})b_s(\mathbf{k}) , \tag{20a}$$

$$\bar{N}_{\bar{\mathcal{R}}} = \sum_s \int_{\bar{\mathcal{R}}} \frac{\mathrm{d}^3\mathbf{k}}{(2\pi)^3 2E}\, d_s^\dagger(\mathbf{k})d_s(\mathbf{k}) . \tag{20b}$$

As was true of the polarization sum, the regions $\mathcal{R}$ and $\bar{\mathcal{R}}$ need not be common to all spin projections, i.e. $\mathcal{R} \to \mathcal{R}_s$ and $\bar{\mathcal{R}} \to \bar{\mathcal{R}}_s$. The anti-commutativity of the fermion creation and annihilation operators is accounted for in the definition of normal ordering:

$$: b_s^\dagger(\mathbf{k})b_s(\mathbf{k}) : = +b_s^\dagger(\mathbf{k})b_s(\mathbf{k}) , \tag{21a}$$

$$: b_s(\mathbf{k})b_s^\dagger(\mathbf{k}) : = -b_s^\dagger(\mathbf{k})b_s(\mathbf{k}) , \tag{21b}$$

with analogous expressions holding for the anti-fermion operators $d_s^\dagger(\mathbf{k})$ and $d_s(\mathbf{k})$.

For a general product of j operators, we find

$$: \prod_{a=1}^{j} b_{s_a}^\dagger(\mathbf{k}_a)b_{s_a}(\mathbf{k}_a) : = (-1)^{j(j-1)/2} \prod_{a=1}^{j} b_{s_a}^\dagger(\mathbf{k}_a) \prod_{b=1}^{j} b_{s_b}(\mathbf{k}_b) . \tag{22}$$

The normal ordering has given rise to an overall factor of $(-1)^{j(j-1)/2}$. However, after acting on a given state with the annihilation operators, the order of the creation operators is reversed relative to the original state. Using anti-commutation to recover the original order, we pick up an additional factor of $(-1)^{j(j-1)/2}$, with the result that there is no overall sign relative to the bosonic case. We can account for this directly at the level of Eq. (22) by re-ordering the creation operators, picking up the same additional factor of $(-1)^{j(j-1)/2}$:

$$: \prod_{a=1}^{j} b_{s_a}^\dagger(\mathbf{k}_a)b_{s_a}(\mathbf{k}_a) : = \prod_{a=j}^{1} b_{s_a}^\dagger(\mathbf{k}_a) \prod_{b=1}^{j} b_{s_b}(\mathbf{k}_b) . \tag{23}$$

The behaviour of the normal-ordered products of fermion number operators is therefore identical to that of the normal-ordered boson number operators described previously. This can also be understood by virtue of the fact that fermionic number operators are *commutative* not anti-commutative.

As an example, the operator projecting onto the subspace of states in which there are exactly j fermions (of any spin) and zero anti-fermions in $\mathcal{R}$ is

$$E_{\mathcal{R}}^{(j,0)} \equiv : \frac{1}{j!}(N_{\mathcal{R}})^j e^{-N_{\mathcal{R}}-\bar{N}_{\mathcal{R}}} : = : \frac{1}{j!}(N_{\mathcal{R}})^j e^{-N_{\mathcal{R}}} : \otimes : e^{-\bar{N}_{\mathcal{R}}} : . \tag{24}$$

In all cases, the projection operators of a given degree of freedom are built from the corresponding number operator. The results presented here may therefore be generalized readily to include additional gauge structure, multiple flavours or higher-spin representations, simply by accounting for summations over the additional quantum numbers. Since the number operators of different degrees of freedom mutually commute — for fermions as well as bosons — their projection operators may be combined straightforwardly by tensor multiplication. One can then imagine constructing semi-inclusive projection operators able to deal with final states of any content and complexity by combining those of different species across various disjoint and/or overlapping regions of momentum space.

4. Conclusions

These projection operators have the interesting property that unobserved quanta never appear in the calculation. This may have a significant impact upon the way in which we deal with infrared divergences in gauge theories. In order to take advantage of this property, we must compute probabilities *directly*, bypassing amplitude-level calculations altogether. Were we to revert to the latter, we would need to break the projection operators apart again, reintroducing the explicit sums over unobserved emissions that we intend to avoid. It remains to develop technology that makes tractable the explicit calculation of these probabilities, perhaps building on the results of Ref. [9] and the earlier ideas of Ref. [14] by exploiting the connection to the path-integral approach of the in-in (or closed-time-path) formalism.

Acknowledgements

The work of PM is supported by STFC Grant No. ST/L000393/1 and a Leverhulme Trust Research Leadership Award.

Appendix A. Proofs of results quoted in the main text

It is useful to be able to compute the eigenvalues of normal-ordered products of number operators. The eigenvalue equations themselves have identical forms for bosonic and fermionic number operators, and we will suppress all but the momentum dependence of states for conciseness. The first non-trivial example is

$$: N_{\mathcal{R}_1}N_{\mathcal{R}_2} : |\mathbf{k}_1 \ldots \mathbf{k}_N\rangle = (n_1 n_2 - n_{12})\, |\mathbf{k}_1 \ldots \mathbf{k}_N\rangle , \tag{A.1}$$

where n_i counts the number of quanta lying in $\mathcal{R}_i$ and n_{12} counts the number of quanta lying in the overlapping region $\mathcal{R}_1 \cap \mathcal{R}_2$. Similarly,

$$: N_{\mathcal{R}_1}N_{\mathcal{R}_2}N_{\mathcal{R}_3} : |\mathbf{k}_1 \ldots \mathbf{k}_N\rangle = (n_1 n_2 n_3 - n_{12}n_3 - n_{13}n_2 - n_{23}n_1 + 2n_{123})\, |\mathbf{k}_1 \ldots \mathbf{k}_N\rangle . \tag{A.2}$$

These are the simplest examples of the more general formula:

$$: N_{\mathcal{R}_1}N_{\mathcal{R}_2} \ldots N_{\mathcal{R}_p} : |\mathbf{k}_1 \ldots \mathbf{k}_N\rangle = \left[\prod_{r=1}^{p} n_r\right]\left[1 + \sum_{i<j} \frac{(-1)n_{ij}}{n_i n_j} + \sum_{i<j<k} \frac{(-1)(-2)n_{ijk}}{n_i n_j n_k} + \sum_{i<j<k<l} \frac{(-1)(-2)(-3)n_{ijkl}}{n_i n_j n_k n_l} + \ldots\right.$$

$$+ \sum_{i<j,i<k<l} \frac{(-1)n_{ij}(-1)n_{kl}}{n_i n_j n_k n_l}$$
$$+ \sum_{i<j,k<l<m} \frac{(-1)(-2)n_{ijk}(-1)n_{lm}}{n_i n_j n_k n_l n_m} + \ldots$$
$$+ \left. \frac{(-1)^{p-1}(p-1)! n_{12\ldots p}}{n_1 n_2 \ldots n_p} \right] |\mathbf{k}_1 \ldots \mathbf{k}_N\rangle , \qquad \text{(A.3)}$$

in which a sum is listed for every integer partition of p.

The eigenvalue of this normal-ordered product of number operators counts the total number of ways to select p quanta from the set specified by the state $|\mathbf{k}_1 \ldots \mathbf{k}_N\rangle$ such that one quantum is in each of the regions $\mathcal{R}_i$. If the regions are nested, such that $\mathcal{R}_1 \subseteq \mathcal{R}_2 \subseteq \ldots$, Eq. (A.3) reduces to

$$: \prod_{i=1}^{p} (N_{\mathcal{R}_i}) : |\mathbf{k}_1 \ldots \mathbf{k}_N\rangle = \left[\prod_{i=1}^{p} \left(n_i - (i-1)\right) \right] |\mathbf{k}_1 \ldots \mathbf{k}_N\rangle , \qquad \text{(A.4)}$$

and, if all the $\mathcal{R}_i$ are identical, this becomes

$$: (N_{\mathcal{R}})^p : |\mathbf{k}_1 \ldots \mathbf{k}_N\rangle = \begin{cases} \frac{n!}{(n-p)!} |\mathbf{k}_1 \ldots \mathbf{k}_N\rangle & \text{if } n \geq p , \\ 0 & \text{otherwise} . \end{cases} \qquad \text{(A.5)}$$

We now consider more than one sequence of nested regions in the case that the regions in different sequences are disjoint. If we have j_1 copies of region $\mathcal{R}_1$, j_2 copies of region $\mathcal{R}_2$ and so on, with $\mathcal{R}_i \cap \mathcal{R}_j = \emptyset \ \forall \, i \neq j$, then

$$: \prod_a (N_{\mathcal{R}_a})^{j_a} : |\mathbf{k}_1 \ldots \mathbf{k}_N\rangle$$
$$= \begin{cases} \left[\prod_a \frac{n_a!}{(n_a - j_a)!} \right] |\mathbf{k}_1 \ldots \mathbf{k}_N\rangle & \text{if } n_a \geq j_a \ \forall \, a , \\ 0 & \text{otherwise} . \end{cases} \qquad \text{(A.6)}$$

The product form of the eigenvalues is a consequence of the fact that the operator factorizes into mutually commuting operators of the form given in Eq. (A.5).

Now we consider a set of disjoint regions within a superset. Let us augment the case of the previous paragraph with k copies of a region $\mathcal{R} \supset \mathcal{R}_i$. After selecting j_a particles from each disjoint region $\mathcal{R}_a$, the number of particles remaining in $\mathcal{R}$ is $n_x \equiv n - \sum_a j_a$. The number of ways of selecting these k particles is then $n_x!/(n_x - k)!$, and

$$: \left[\prod_a (N_{\mathcal{R}_a})^{j_a} \right] (N_{\mathcal{R}})^k : |\mathbf{k}_1 \ldots \mathbf{k}_N\rangle$$
$$= \begin{cases} \frac{n_x!}{(n_x - k)!} \left[\prod_a \frac{n_a!}{(n_a - j_a)!} \right] |\mathbf{k}_1 \ldots \mathbf{k}_N\rangle & \text{if } n_a \geq j_a \ \forall \, a \text{ and } n_x \geq k , \\ 0 & \text{otherwise} . \end{cases} \qquad \text{(A.7)}$$

These results for the action of normal-ordered products of the number operator are the key to proving the results quoted in the main text.

Specifically, using Eq. (A.5), we can go ahead and prove Eq. (14):

$$E_{\mathcal{R}}^{(0)} \, |\mathbf{k}_1 \ldots \mathbf{k}_N\rangle = \sum_{p=0}^{n} \frac{(-1)^p \, n!}{p!(n-p)!} |\mathbf{k}_1 \ldots \mathbf{k}_N\rangle$$
$$= \lim_{x \to -1} (1+x)^n \, |\mathbf{k}_1 \ldots \mathbf{k}_N\rangle$$
$$= \begin{cases} |\mathbf{k}_1 \ldots \mathbf{k}_N\rangle & \text{if } n = 0 , \\ 0 & \text{otherwise} . \end{cases} \qquad \text{(A.8)}$$

A proof of Eq. (15), using Eq. (A.7) with $n_x \equiv n - \sum_a j_a$, runs as follows:

$$E_{\{\mathcal{R}_a \subseteq \mathcal{R}\}}^{\{j_a\}} \, |\mathbf{k}_1 \ldots \mathbf{k}_N\rangle$$
$$= \sum_{k=0}^{\infty} \frac{(-1)^k}{k!} : \left[\prod_a \frac{1}{j_a!} (N_{\mathcal{R}_a})^{j_a} \right] (N_{\mathcal{R}})^k : |\mathbf{k}_1 \ldots \mathbf{k}_N\rangle$$
$$= \sum_{k=0}^{n_x} \frac{(-1)^k}{k!} \frac{n_x!}{(n_x - k)!} \left[\prod_a \frac{n_a!}{j_a!(n_a - j_a)!} \right] |\mathbf{k}_1 \ldots \mathbf{k}_N\rangle \qquad \text{(A.9)}$$

provided $n_a \geq j_a \ \forall a$, and zero otherwise. The eigenvalue may be written

$$\left[\prod_a \binom{n_a}{j_a} \right] \sum_{k=0}^{n_x} \binom{n_x}{k} (-1)^k = \left[\prod_a \binom{n_a}{j_a} \right] \lim_{x \to -1} (1+x)^{n_x} , \qquad \text{(A.10)}$$

which vanishes unless $n_x = 0$. Since $n \geq \sum_a n_a \geq \sum_a j_a = n - n_x$, this implies $n_a = j_a \ \forall \, a$. Hence,

$$E_{\{\mathcal{R}_a \subseteq \mathcal{R}\}}^{\{j_a\}} \, |\mathbf{k}_1 \ldots \mathbf{k}_N\rangle$$
$$= \begin{cases} |\mathbf{k}_1 \ldots \mathbf{k}_N\rangle & \text{if } n_a = j_a \ \forall \, a \text{ and } n = \sum_a j_a , \\ 0 & \text{otherwise} . \end{cases} \qquad \text{(A.11)}$$

References

[1] T. Kinoshita, J. Math. Phys. 3 (1962) 650.
[2] T.D. Lee, M. Nauenberg, Phys. Rev. 133 (1964) B1549.
[3] F. Bloch, A. Nordsieck, Phys. Rev. 52 (1937) 54.
[4] S. Weinberg, Phys. Rev. 140 (1965) B516.
[5] V. Chung, Phys. Rev. 140 (1965) B1110.
[6] P.P. Kulish, L.D. Faddeev, Theor. Math. Phys. 4 (1970) 745, Teor. Mat. Fiz. 4 (1970) 153.
[7] S. Catani, M. Ciafaloni, Nucl. Phys. B 249 (1985) 301.
[8] D.A. Forde, A. Signer, Nucl. Phys. B 684 (2004) 125.
[9] R. Dickinson, J. Forshaw, P. Millington, Phys. Rev. D 93 (2016) 065054.
[10] W.H. Louisell, Quantum Statistical Properties of Radiation, John Wiley & Sons, New York, 1973, p. 158.
[11] H.-y. Fan, J. Opt. B, Quantum Semiclass. Opt. 5 (2003) R147.
[12] P. Blasiak, A. Horzela, K.A. Penson, A.I. Solomon, G.H.E. Duchamp, Am. J. Phys. 75 (2007) 639–646.
[13] J.R. Klauder, E.C.G. Sudarshan, Fundamentals of Quantum Optics, Benjamin, New York, 1968, pp. 168–178.
[14] R. Dickinson, J. Forshaw, P. Millington, B. Cox, J. High Energy Phys. 1406 (2014) 049.

4

Generalized pure Lovelock gravity

Patrick Concha [a,*], Evelyn Rodríguez [b]

[a] *Instituto de Física, Pontificia Universidad Católica de Valparaíso, Casilla 4059, Valparaiso, Chile*
[b] *Departamento de Ciencias, Facultad de Artes Liberales, Universidad Adolfo Ibáñez, Av. Padre Hurtado 750, Viña del Mar, Chile*

ARTICLE INFO

Editor: M. Cvetič

ABSTRACT

We present a generalization of the n-dimensional (pure) Lovelock Gravity theory based on an enlarged Lorentz symmetry. In particular, we propose an alternative way to introduce a cosmological term. Interestingly, we show that the usual pure Lovelock gravity is recovered in a matter-free configuration. The five and six-dimensional cases are explicitly studied.

1. Introduction

The most natural generalization of General Relativity (GR) in d dimensions satisfying the criteria of general covariance and leading to second order field equations for the metric is given by the Lanczos–Lovelock (LL) gravity theory [1,2]. In the differential forms language the LL action can be written as the most general d-form invariant under local Lorentz transformations, constructed out of the spin connection ω^{ab}, the vielbein e^a and their exterior derivatives [3,4],

$$S_{LL} = \int \sum_{p=0}^{[d/2]} \alpha_p \epsilon_{a_1 a_2 \cdots a_d} R^{a_1 a_2} \cdots R^{a_{2p-1} a_{2p}} e^{a_{2p+1}} \cdots e^{a_d} , \tag{1}$$

where $R^{ab} = d\omega^{ab} + \omega^a{}_c \omega^{cb}$ is the Lorentz curvature two-form and the α_p coefficients are not fixed from first principles.

Different numbers of degrees of freedom emerge depending on the value of the arbitrary coefficients. In particular, the higher curvature terms can produce degenerate sectors with no degrees of freedom. Such degeneracy can be avoided with particular choices of the α_p constants. In particular, as explained in Ref. [5], there are mainly two ways to avoid degenerate sectors. One of them consist in restrict the theory to have a unique degenerate vacuum which leads to a family of gravity theories [6] labeled by the integer k which represents the highest power of curvature. Interestingly, the α_p constants can be fixed requiring that the theory has the maximum possible number of degrees of freedom [7]. Then the LL Lagrangian is a Chern–Simons (CS) form [8–10] in odd dimensions which is gauge invariant under the $(A)dS$ symmetry. In even dimensions, the LL Lagrangian can be written as a Born–Infeld (BI) gravity Lagrangian [11,12] which is locally invariant under a Lorentz subalgebra.

Another way of fixing the α_p constants avoiding degeneracy is to demand that there is non-degenerate vacuum. Such requirement leads to the pure Lovelock (PL) theory [5,13] which consists only in two terms of the full Lovelock Lagrangian,

$$S_{PL} = \int \left(\alpha_0 \mathcal{L}_0 + \alpha_p \mathcal{L}_p \right) , \tag{2}$$

with

$$\mathcal{L}_0 = \epsilon_{a_1 \ldots a_d} e^{a_1} \cdots e^{a_d} , \tag{3}$$

$$\mathcal{L}_p = \epsilon_{a_1 \ldots a_d} R^{a_1 a_2} \cdots R^{a_{2p-1} a_{2p}} e^{a_{2p+1}} \cdots e^{a_d} . \tag{4}$$

The coefficients are fixed in terms of the gravitational constant κ and the cosmological constant Λ,

$$\alpha_0 = -\frac{2\Lambda\kappa}{d!} = -\frac{(\mp 1)^p \kappa}{d\,(d-2p-1)!\,\ell^{2p}} , \tag{5}$$

$$\alpha_p = \frac{\kappa}{(d-2p)!} . \tag{6}$$

With this particular choice, the PL theory has a unique non-degenerate $(A)dS$ vacuum in odd dimensions and admits non-degenerate vacua in even dimensions. Additionally, the black holes (BH) solutions of the PL theory behave asymptotically like the AdS-Schwarzschild ones [5,14]. Of particular interest are the BH solutions of the maximal pure Lovelock since the thermodynamical parameters are universal in terms of horizon radius. Recently, a Hamiltonian analysis has shown that the maximum possible number of degrees of freedom of the PL case is the same as in the Einstein–Hilbert (EH) gravity [15].

* Corresponding author.
E-mail addresses: patrick.concha@pucv.cl (P. Concha), evelyn.rodriguez@edu.uai.cl (E. Rodríguez).

The supersymmetric version of the general LL theory is unknown, except in the EH case and in odd dimensions when the LL action can be seen as a CS supergravity action for the *AdS* superalgebra. The construction of a super LL gravity action, and in particular for a super PL, remains a difficult task. Indeed, there is no clarity in which terms should be considered in the action in order to guarantee the supersymmetric invariance of the theory. A discussion about a five-dimensional supergravity action for the EH term coupled to a Lovelock term can be found in Ref. [16]. More recently, the authors of Refs. [17,18] suggested that the supersymmetric version of a PL theory could emerge as a particular limit of a supergravity theory. The procedure that could be used is not new and has already been used to relate GR with different gravity theories [19–23].

As shown in Refs. [17,18], the PL Lagrangian can be recovered as a particular limit of CS and BI like Lagrangians constructed out the $\mathfrak{C}_k$ family [27]. Although the procedure presented in [17,18] can be reproduced in any spacetime dimension d, the obtention of the PL action requires a large amount of extra fields in higher dimensions. Additionally in order to recover the PL dynamics, it is necessary to impose additional restrictions on the fields.

Here we present a generalized pure Lovelock (GPL) gravity theory which leads to the PL action and its dynamics in a matter-free configuration limit without further considerations. In particular, the field content of the theory is spacetime dimension independent. Interestingly, the GPL action allows us to introduce alternatively a generalized cosmological term which generalizes the result obtained in Ref. [37] to higher dimensions. Moreover, we show that the GPL gravity action corresponds to a particular case of a more generalized Lovelock (GL) gravity theory. We also show that there is a particular choice of the coefficients appearing in the GL action leading in odd and even dimensions to a CS and BI like gravity, respectively.

2. Generalized Lovelock gravity action

A generalization of the Lanczos–Lovelock gravity action can be performed enlarging the Lorentz symmetry. A Generalized Lovelock (GL) gravity action can be written as the most general d-form invariant under local Lorentz-like transformations, constructed with the spin connection ω^{ab}, the vielbein e^a, a Lorentz-like field k^{ab} and their exterior derivatives, without the Hodge dual,

$$S_{GL} = \int \sum_{p=0}^{[d/2]} \sum_{m=0}^{p} \alpha_p \binom{p}{p-m} L_{GL}^{(p)}, \tag{7}$$

where

$$L_{GL}^{(p)} = \epsilon_{a_1 a_2 \cdots a_d} R^{a_1 a_2} \cdots R^{a_{2m-1} a_{2m}} F^{a_{2m+1} a_{2m+2}} \cdots F^{a_{2p-1} a_{2p}} e^{a_{2p+1}} \cdots e^{a_d}, \tag{8}$$

with

$$R^{ab} = d\omega^{ab} + \omega^a{}_c \omega^{cb}, \tag{9}$$

$$F^{ab} = dk^{ab} + \omega^a{}_c k^{cb} - \omega^b{}_c k^{ca} + k^a{}_c k^{cb}. \tag{10}$$

Here the α_p coefficients are arbitrary constants which are not fixed from first principles. The possible permutations of the curvature 2-forms R^{ab} and F^{ab} appearing in the action are reflected in the coefficients $\binom{p}{p-m}$.

Let us note that for $p = 0$ the Lagrangian reproduces the cosmological constant term, while for $p = 1$ the GL Lagrangian is the Einstein–Hilbert term plus the coupling of F^{ab} with the vielbeins. This is an important difference with the usual Lovelock–Cartan gravity theory which contains General Relativity as a particular case. However, a matter-free configuration ($k^{ab} = 0$) allows to recover the usual Lovelock gravity action. In particular, the Einstein–Hilbert term is recovered for $p = 1$ and $k^{ab} = 0$.

The dynamical content is obtained considering the variation of the GL action with respect to $(e^a, \omega^{ab}, k^{ab})$,

$$\delta S_{GL} = \int \delta e^a \mathcal{E}_a + \delta\omega^{ab} \mathcal{E}_{ab} + \delta k^{ab} \mathcal{K}_{ab} = 0, \tag{11}$$

modulo boundary terms. The field equations are given by

$$\mathcal{E}_a = \sum_{p=0}^{\left[\frac{d-1}{2}\right]} \sum_{m=0}^{p} \alpha_p (d-2p) \mathcal{E}_a^p = 0, \tag{12}$$

$$\mathcal{E}_{ab} = \sum_{p=1}^{\left[\frac{d-1}{2}\right]} \sum_{m=0}^{p} \alpha_p m (d-2p) \mathcal{E}_{ab}^p = 0, \tag{13}$$

$$\mathcal{K}_{ab} = \sum_{p=1}^{\left[\frac{d-1}{2}\right]} \sum_{m=0}^{p} \alpha_p (p-m)(d-2p) \mathcal{E}_{ab}^p = 0, \tag{14}$$

where

$$\mathcal{E}_a^p \equiv \binom{p}{p-m} \epsilon_{ab_1 \cdots b_{d-1}} R^{b_1 b_2} \cdots R^{b_{2m-1} b_{2m}} F^{b_{2m+1} b_{2m+2}} \cdots F^{b_{2p-1} b_{2p}} e^{b_{2p+1}} \cdots e^{b_{d-1}}, \tag{15}$$

$$\mathcal{E}_{ab}^p \equiv \binom{p}{p-m} \epsilon_{aba_3 \cdots a_d} R^{a_3 a_4} \cdots R^{a_{2m-1} a_{2m}} F^{a_{2m+1} a_{2m+2}} \cdots F^{a_{2p-1} a_{2p}} R^{a_{2p+1}} e^{a_{2p+2}} \cdots e^{a_d}. \tag{16}$$

Here $R^a = D_\omega e^a + k^a{}_b e^b$ with $D_\omega = d + \omega$ the Lorentz covariant exterior derivative. In particular, the Bianchi identities $D_\omega R^{ab} = 0$ and $D_{\omega+k} F^{ab} + k^a{}_c R^{cb} - k^b{}_c R^{ca} = 0$ along with $d^2 = 0$, assure that the field equations involve only first derivatives of e^a, ω^{ab} and k^{ab}.

Let us note that the variation of the action under the spin connection ω^{ab} and the Lorentz-like field k^{ab} imply the same field equation and then the $(d-1)$-forms $\mathcal{E}_{ab}$ and $\mathcal{K}_{ab}$ coincide. On the other hand, analogously to the usual Lovelock Cartan gravity, the $(d-1)$-form $\mathcal{E}_a$ is independent of the $(d-1)$-forms $\mathcal{E}_{ab}$.

Furthermore, using the Bianchi identities for the curvature 2-forms one can show that the following relation holds

$$D\mathcal{E}_a^p = (d-1-2p) e^b \mathcal{E}_{ba}^{p+1}, \tag{17}$$

for $0 \leq p \leq \left[\frac{d-1}{2}\right]$. Then, as in Refs. [7,23] we have that the previous identity leads to

$$D\mathcal{E}_a = \sum_{p=1}^{\left[\frac{d+1}{2}\right]} \sum_{m=0}^{p} \alpha_{p-1} (d+2-2p)(d+1-2p) e^b \mathcal{E}_{ba}^p, \tag{18}$$

which by consistency with $\mathcal{E}_a = 0$ must also be zero. Besides, we can see that the following product

$$e^b \mathcal{E}_{ba} = \sum_{p=1}^{\left[\frac{d-1}{2}\right]} \sum_{m=0}^{p} \alpha_p m (d-2p) e^b \mathcal{E}_{ba}^p \tag{19}$$

must also vanish by consistency with $\mathcal{E}_{ab} = 0$. One can easily check that the same result apply for the product $e^b \mathcal{K}_{ba}$.

Thus, fixing the α_p coefficients could lead to different numbers of degrees of freedom depending on additional constraints of the form $e^b \mathcal{E}_{ba}^p = 0$. Following the same arguments of Ref. [7], there is a particular choice in odd dimensions allowing to avoid additional restrictions and such that $\mathcal{E}_a$ and $\mathcal{E}_{ab}$ (or $\mathcal{K}_{ab}$) are independent.

2.1. Chern–Simons gravity and $\mathfrak{C}_4$ algebra

The odd-dimensional GL action (7) reproduces a Chern–Simons action for a particular Lie algebra, known as the $\mathfrak{C}_4$ algebra[1] [24–27], when the α_p's are fixed to the following values:

$$\alpha_0 = \frac{\kappa}{d\ell^d}, \tag{20}$$

$$\alpha_p = \alpha_0 \frac{(2n-1)(2\gamma)^p}{(2n-2p-1)}\binom{n-1}{p}, \tag{21}$$

where γ is related to the cosmological constant,

$$\gamma = -sgn(\Lambda)\frac{\ell^2}{2}, \tag{22}$$

and ℓ is a length parameter related to the cosmological constant.

The $d = 2n-1$ CS Lagrangian is given by

$$L_{CS}^{\mathfrak{C}_4} = \kappa \epsilon_{a_1 a_2 \cdots a_{2n-1}} \sum_{p=0}^{n-1}\sum_{m=0}^{p} \ell^{2(p-n)+1} c_p \binom{p}{p-m} \times R^{a_1 a_2} \ldots R^{a_{2m-1}a_{2m}} F^{a_{2m+1}a_{2m+2}} \ldots F^{a_{2p-1}a_{2p}} e^{a_{2p+1}} \ldots e^{a_{2n-1}}, \tag{23}$$

with

$$c_p = \frac{1}{2(n-p)-1}\binom{n-1}{p}. \tag{24}$$

Let us note that the $d = 3$ CS form reproduces the generalized CS gravity theory presented in [25] applying an appropriate change of basis where the $AdS \oplus Lorentz$ structure is manifested. The black hole solution of the three-dimensional CS gravity theory based on this symmetry have been recently studied in [28].

It is important to clarify that, unlike the GL Lagrangian, the CS $(2n-1)$-form is invariant not only under Lorentz-like transformation,

$$\delta e^a = e^b \rho_b{}^a + e^b \lambda_b{}^a, \tag{25}$$

$$\delta\omega^{ab} = D_\omega \rho^{ab}, \tag{26}$$

$$\delta k^{ab} = D_\omega \lambda^{ab} + k^a{}_c \rho^{cb} - k^b{}_c \rho^{ca} + k^a{}_c \lambda^{cb} - k^b{}_c \lambda^{ca}, \tag{27}$$

but also under a local $\mathfrak{C}_4$ boost

$$\delta e^a = D_\omega \rho^a + k^a{}_b \rho^a, \tag{28}$$

$$\delta\omega^{ab} = 0, \tag{29}$$

$$\delta k^{ab} = \rho^a e^b - \rho^b e^a, \tag{30}$$

where the $\mathfrak{C}_4$ gauge parameter is given by

$$\rho = \frac{1}{2}\rho^{ab} J_{ab} + \frac{1}{2}\lambda^{ab} Z_{ab} + \frac{1}{\ell}\rho^a P_a. \tag{31}$$

In particular, the generators of the $\mathfrak{C}_4$ algebra satisfy the following commutation relations:

$$[J_{ab}, J_{cd}] = \eta_{bc} J_{ad} - \eta_{ac} J_{bd} - \eta_{bd} J_{ac} + \eta_{ad} J_{bc}, \tag{32}$$

$$[J_{ab}, Z_{cd}] = \eta_{bc} Z_{ad} - \eta_{ac} Z_{bd} - \eta_{bd} Z_{ac} + \eta_{ad} Z_{bc}, \tag{33}$$

$$[Z_{ab}, Z_{cd}] = \eta_{bc} Z_{ad} - \eta_{ac} Z_{bd} - \eta_{bd} Z_{ac} + \eta_{ad} Z_{bc}, \tag{34}$$

$$[J_{ab}, P_c] = \eta_{bc} P_a - \eta_{ac} P_b, \quad [P_a, P_b] = Z_{ab}, \tag{35}$$

$$[Z_{ab}, P_c] = \eta_{bc} P_a - \eta_{ac} P_b, \tag{36}$$

Such symmetry can be obtained as a deformation of the Maxwell symmetry and belongs to a generalized family of Lie algebras denoted by $\mathfrak{C}_k$ [27]. At the fermionic level, a recent application has been developed in the three-dimensional CS context where a (p,q) *AdS*-Lorentz supergravity model was presented [29].

2.2. Born–Infeld gravity and Lorentz-like symmetry

The even-dimensional case requires an alternative approach since eqs. (18) and (19) have not the same number of terms as in the odd-dimensional case.

Following the same procedure introduced in Refs. [7,23], one can note that there is a particular choice of the α_p's

$$\alpha_0 = \frac{\kappa}{d\ell^d}, \tag{37}$$

$$\alpha_p = \alpha_0 (2\gamma)^p \binom{n}{p}, \tag{38}$$

which reproduces a Born–Infeld (BI) like gravity action with $0 \leq p \leq n$. As in the CS case, γ is related to the cosmological constant $\gamma = -sgn(\Lambda)\frac{\ell^2}{2}$. With these coefficients the GL Lagrangian takes the BI-like form

$$L_{BI}^{\mathcal{L}_{\mathfrak{C}_4}} = \kappa \epsilon_{a_1 a_2 \cdots a_{2n}} \sum_{p=0}^{n}\sum_{m=0}^{p} \frac{\ell^{2p-2n}}{2n}\binom{n}{p}\binom{p}{p-m} \times R^{a_1 a_2} \ldots R^{a_{2m-1}a_{2m}} F^{a_{2m+1}a_{2m+2}} \ldots F^{a_{2p-1}a_{2p}} e^{a_{2p+1}} \ldots e^{a_{2n}}. \tag{39}$$

In particular, the Lagrangian can be rewritten in a reduced form,

$$L_{BI}^{\mathcal{L}_{\mathfrak{C}_4}} = \frac{\kappa}{2n}\epsilon_{a_1 a_2 \cdots a_{2n}} \bar{F}^{a_1 a_2} \ldots \bar{F}^{a_{2n-1}a_{2n}}, \tag{40}$$

where

$$\bar{F}^{ab} = R^{ab} + F^{ab} + \frac{1}{\ell^2} e^a e^b, \tag{41}$$

is the $\mathfrak{C}_4$ curvature. With this form, the Lagrangian corresponds to the Pfaffian of the 2-form $\bar{F}^{ab}$ and can be rewritten similarly to the Born–Infeld electrodynamics Lagrangian,

$$L_{BI}^{\mathcal{L}_{\mathfrak{C}_4}} = 2^{n-1}(n-1)!\sqrt{\det\left(R^{ab} + F^{ab} + \frac{1}{\ell^2} e^a e^b\right)}. \tag{42}$$

It is important to emphasize that the BI-like gravity Lagrangian is only off-shell invariant under a Lorentz-like subalgebra $\mathcal{L}_{\mathfrak{C}_4}$ of the $\mathfrak{C}_4$ algebra. This can be clarified through the Levi-Civita symbol $\epsilon_{a_1 a_2 \cdots a_{2n}}$ in (40) which consists in the only non-vanishing component of the Lorentz-like invariant tensor of rank n, namely

$$\langle \bar{Z}_{a_1 a_2} \cdots \bar{Z}_{a_{2n-1}a_{2n}} \rangle = \frac{2^{n-1}}{n}\epsilon_{a_1 a_2 \cdots a_{2n}}. \tag{43}$$

Here $\bar{Z}_{ab} = J_{ab} + Z_{ab}$ satisfy the $\mathfrak{C}_4$ commutation relations (32)–(34). Such choice of the invariant tensor breaks the full $\mathfrak{C}_4$ symmetry to its Lorentz-like subgroup $\mathcal{L}_{\mathfrak{C}_4}$.

Recently, diverse BI-like gravity theories have been constructed with different purposes [18,21,22]. At the supersymmetric level, similar constructions have been done based on the MacDowell–Mansouri formalism [30–36].

---

[1] Also known as Poincaré semi-simple extended algebra.

3. Generalized pure Lovelock gravity action

An alternative way of fixing the α_p's can be implemented such that a generalized pure Lovelock (GPL) gravity action can be constructed. The new coefficients are fixed in terms of the gravitational constant κ and the cosmological constant Λ in the same way as in Refs. [5,13],

$$\alpha_0 = -\frac{2\Lambda\kappa}{d!} = -\frac{(\mp 1)^p\,\kappa}{d\,(d-2p-1)!\ell^{2p}}\,, \tag{44}$$

$$\alpha_p = \frac{\kappa}{(d-2p)!}\,. \tag{45}$$

Such generalization contains only two term of the full GL series (given by eq. (7)),

$$S_{GPL} = \int \left(\alpha_0 \mathcal{L}_0 + \alpha_p \mathcal{L}_p\right)\,, \tag{46}$$

where

$$\mathcal{L}_0 = \epsilon_{a_1 a_2 \ldots a_d} e^{a_1} e^{a_2} \cdots e^{a_d}\,, \tag{47}$$

$$\mathcal{L}_p = \sum_{m=0}^{p} \binom{p}{p-m} \epsilon_{a_1 a_2 \ldots a_d} R^{a_1 a_2} \cdots R^{a_{2m-1} a_{2m}} F^{a_{2m+1} a_{2m+2}} \cdots F^{a_{2p-1} a_{2p}} e^{a_{2p+1}} \cdots e^{a_d}\,. \tag{48}$$

Interestingly, the p-order term in the curvature 2-forms reproduces a generalized cosmological term in any spacetime dimension d. This particular case of the generalized Lovelock series has been first introduced in four dimensions in Refs. [33,37]. Subsequently in Ref. [18], a generalized cosmological constant term has been obtained in even dimensions as a particular configuration limit of a BI-like gravity theory. Thus, the generalized pure Lovelock action presented here, through the Lorentz-like gauge field k^{ab}, allows us to introduce alternatively a cosmological term in arbitrary dimensions.

Let us note that the Einstein–Hilbert term appears only in the case $p=1$ along with the following term:

$$\epsilon_{a_1 a_2 a_3 \ldots a_d} F^{a_1 a_2} e^{a_3} \cdots e^{a_d}\,. \tag{49}$$

On the other hand, topological densities are obtained in even dimensions for $p = d/2$,

$$\mathcal{L}_{d/2} = \sum_{m=0}^{p} \binom{d/2}{d/2-m} \epsilon_{a_1 a_2 \ldots a_d} R^{a_1 a_2} \cdots R^{a_{2m-1} a_{2m}} F^{a_{2m+1} a_{2m+2}} \cdots F^{a_{d-1} a_d}\,. \tag{50}$$

These terms are related to Euler type characteristic classes. Then, we have that for $1 \le p \le \frac{d-2}{2}$ the even-dimensional GPL gravity action reproduces truly dynamical actions.

Interestingly, let us note that in a matter-free configuration $k^{ab}=0$ the GPL action (46) reduces to the pure Lovelock action,

$$S_{PL} = \int \left(\alpha_0 \mathcal{L}_0 + \alpha_p \mathcal{L}_p\right)\,, \tag{51}$$

with

$$\mathcal{L}_0 = \epsilon_{a_1 a_2 \ldots a_d} e^{a_1} e^{a_2} \cdots e^{a_d}\,, \tag{52}$$

$$\mathcal{L}_p = \epsilon_{a_1 a_2 \ldots a_d} R^{a_1 a_2} \cdots R^{a_{2p-1} a_{2p}} e^{a_{2p+1}} \cdots e^{a_d}\,. \tag{53}$$

Obtaining the PL gravity action in a particular matter-free configuration limit is not new and has already been presented in Refs. [17, 18]. Nevertheless, the techniques considered in [17,18] require an excessive amount of new extra fields as the spacetime dimension grows. In our case, the field content of the theory does not depend of the spacetime dimension d. Furthermore, one can show that the PL dynamics [5,13,14] can also be reproduced in a matter-free configuration leading appropriately to the pure Lovelock gravity theory. In fact considering $k^{ab}=0$, we have

$$\alpha_0 \epsilon_{a_1 a_2 \cdots a_d} e^{a_1} e^{a_2} \cdots e^{a_{d-1}} + \alpha_p \epsilon_{a_1 a_2 \cdots a_d} R^{a_1 a_2} \cdots R^{a_{2p-1} a_{2p}} e^{a_{2p+1}} \cdots e^{a_{d-1}} = 0\,, \tag{54}$$

$$\alpha_p \epsilon_{a_1 a_2 \cdots a_d} R^{a_3 a_4} \cdots R^{a_{2p-1} a_{2p}} T^{a_{2p+1}} e^{a_{2p+2}} \cdots e^{a_d} = 0\,, \tag{55}$$

where $T^a = D_\omega e^a$. Unlike the procedure presented in previous works, the truly PL dynamics is recovered here without imposing any identification on the fields.

One could obtain the same result considering k^{ab} as the true spin-connection and ω^{ab} as the new extra-field, but it is straightforward to see that the curvatures (9)–(10) reproduces the usual Lorentz curvature only when ω^{ab} is identified as the true spin-connection one-form.

Thus, we have obtained the PL theory considering not only an appropriate limit in the GPL action (46) but also a right dynamical limit without any identification of the fields.

3.1. The five-dimensional case

The five-dimensional generalized pure Lovelock gravity reproduces two diverse actions depending on the value of p. Indeed, for $p=1$ the GPL action reduces to

$$S_{GPL}^{p=1} = \int_{M_5} \epsilon_{abcde}\left[\alpha_0 e^a e^b e^c e^d e^e + \alpha_1 \left(R^{ab} e^c e^d e^e + F^{ab} e^c e^d e^e\right)\right]. \tag{56}$$

Here α_0 and α_1 are given by eqs. (44)–(45). The $p=1$ GPL action can be seen as the coupling of a generalized cosmological term $\mathcal{L}_{\tilde{\Lambda}}$ to the Einstein–Hilbert term,

$$S_{GPL}^{p=1} = \int_{M_5} \mathcal{L}_{\tilde{\Lambda}} + \mathcal{L}_{EH}\,, \tag{57}$$

where

$$\mathcal{L}_{\tilde{\Lambda}} = \alpha_0 \epsilon_{abcde} e^a e^b e^c e^d e^e + \alpha_1 \epsilon_{abcde}\left(D_\omega k^{ab} e^c e^d e^e + k^a{}_f k^{fb} e^c e^d e^e\right). \tag{58}$$

One can see that the matter-free configuration limit ($k^{ab}=0$) leads to the $p=1$ pure Lovelock action

$$S_{PL}^{p=1} = \int_{M_5} \epsilon_{abcde}\left(\alpha_0 e^a e^b e^c e^d e^e + \alpha_1 R^{ab} e^c e^d e^e\right), \tag{59}$$

where the α_p coefficients are identical to the PL ones. Let us note that the $p=1$ PL action corresponds to the standard General Relativity action in presence of a cosmological constant term. The obtention of GR in a matter-free configuration limit of the GPL theory is a desirable feature in order to generalize gravity since it should satisfy the correspondence principle. Furthermore, in a matter-free configuration, the field equations read

$$\epsilon_{abcde}\left(\alpha_0 e^a e^b e^c e^d + \alpha_1 R^{ab} e^c e^d\right)\delta e^e = 0\,, \tag{60}$$

$$\epsilon_{abcde}\left(\alpha_1 T^c e^d e^e\right)\delta\omega^{ab} = 0\,, \tag{61}$$

$$\epsilon_{abcde}\left(\alpha_1 T^c e^d e^e\right)\delta k^{ab} = 0\,, \tag{62}$$

which correspond to the appropriate $p=1$ PL dynamics described in Refs. [5,13,14].

On the other hand, the $p=2$ case does not contain the EH term and the GPL action is given by

$$S_{GPL}^{p=2}=\int_{M_5}\epsilon_{abcde}\left[\alpha_0 e^a e^b e^c e^d e^e+\alpha_2\left(R^{ab}R^{cd}e^e+R^{ab}F^{cd}e^e\right.\right.$$
$$\left.\left.+F^{ab}F^{cd}e^e\right)\right], \tag{63}$$

which can be seen as the coupling of a generalized cosmological term $\mathcal{L}_{\hat{\Lambda}}$ to a Lanczos–Lovelock term $\mathcal{L}_{LL}$,

$$S_{GPL}^{p=2}=\int_{M_5}\mathcal{L}_{\hat{\Lambda}}+\mathcal{L}_{LL}. \tag{64}$$

The $\mathcal{L}_{\hat{\Lambda}}$ term includes the usual cosmological term plus additional terms depending on the Lorentz-like field k^{ab},

$$\mathcal{L}_{\hat{\Lambda}}=\alpha_0\epsilon_{abcde}e^a e^b e^c e^d e^e+\alpha_1\epsilon_{abcde}\left(R^{ab}Dk^{cd}e^e+Dk^{ab}Dk^{cd}e^e\right), \tag{65}$$

with $D=d+\omega+k$.

As in the $p=1$ case, the $p=2$ PL theory is recovered in a matter-free configuration limit,

$$S_{PL}^{p=2}=\int_{M_5}\epsilon_{abcde}\left[\alpha_0 e^a e^b e^c e^d e^e+\alpha_2 R^{ab}R^{cd}e^e\right], \tag{66}$$

while the field equations considering $k^{ab}=0$ reproduce the $p=2$ PL dynamics,

$$\epsilon_{abcde}\left(\alpha_0 e^a e^b e^c e^d+\alpha_1 R^{ab}R^{cd}\right)\delta e^e=0, \tag{67}$$

$$\epsilon_{abcde}\left(\alpha_1 R^{cd}T^e\right)\delta\omega^{ab}=0, \tag{68}$$

$$\epsilon_{abcde}\left(\alpha_1 R^{cd}T^e\right)\delta k^{ab}=0. \tag{69}$$

Let us note that the value and the sign of α_0 is different for any value of p and thus α_0 is distinct for $p=1$ and $p=2$. In particular, for even value of p the α_0 coefficient has a negative sign which makes the pure Lovelock theory to have a unique nondegenerate *ds* and *AdS* vacuum [15]. Interestingly, no further considerations on the α_p constants or in the fields must be imposed in order to obtain appropriately the pure Lovelock theory. A similar procedure has been considered in Ref. [17] in order to recover the five-dimensional PL theory. However, the obtention of the PL action and dynamics in [17] required the introduction not only of four new extra-fields but also appropriate identifications on the extra-fields.

3.2. The six-dimensional case

The six-dimensional generalized pure Lovelock action also reproduces two diverse gravity actions depending on the value of p. Each case describes an alternative way to introduce a cosmological term. Let us note that only $p=1$ and $p=2$ reproduce non-trivial actions meanwhile the $p=3$ case does not correspond to a GPL action since it leads to topological terms. For $d\geq 7$, a $p=3$ GPL action can be constructed.

The $p=1$ GPL action consists in the Einstein–Hilbert term plus a six-dimensional generalized cosmological term,

$$S_{GPL}^{p=1}=\int_{M_6}\epsilon_{abcdef}\left[\alpha_0 e^a e^b e^c e^d e^e e^f+\alpha_1\left(R^{ab}e^c e^d e^e e^f\right.\right.$$
$$\left.\left.+F^{ab}e^c e^d e^e e^f\right)\right]. \tag{70}$$

The GPL action can be rewritten explicitly as

$$S_{GPL}^{p=1}=\int_{M_6}\mathcal{L}_{\hat{\Lambda}}+\mathcal{L}_{EH}, \tag{71}$$

where

$$\mathcal{L}_{\hat{\Lambda}}=\alpha_0\epsilon_{abcdef}e^a e^b e^c e^d e^e e^f+\alpha_1\epsilon_{abcdef}\left(D_\omega k^{ab}e^c e^d e^e e^f\right.$$
$$\left.+k^a{}_f k^{fb}e^c e^d e^e e^f\right). \tag{72}$$

The $p=1$ GPL corresponds to one of the simplest generalizations of the GR theory. Interestingly GR in presence of the cosmological constant, which corresponds to the $p=1$ pure Lovelock action, emerges considering a matter-free configuration limit ($k^{ab}=0$),

$$S_{PL}^{p=1}=\int_{M_6}\epsilon_{abcdef}\left(\alpha_0 e^a e^b e^c e^d e^e e^f+\alpha_1 R^{ab}e^c e^d e^e e^f\right). \tag{73}$$

The $p=1$ PL action always corresponds to the standard General Relativity action in presence of a cosmological constant. Moreover, in a matter-free configuration, the field equations read

$$\epsilon_{abcdef}\left(\alpha_0 e^a e^b e^c e^d e^e+\alpha_1 R^{ab}e^c e^d e^e\right)\delta e^f=0, \tag{74}$$

$$\epsilon_{abcdef}\left(\alpha_1 T^c e^d e^e e^f\right)\delta\omega^{ab}=0, \tag{75}$$

$$\epsilon_{abcdef}\left(\alpha_1 T^c e^d e^e e^f\right)\delta k^{ab}=0, \tag{76}$$

which describe appropriately the $p=1$ PL dynamics.

On the other hand, the $p=2$ GPL action describes an alternative way to introduce a cosmological term,

$$S_{GPL}^{p=2}=\int_{M_6}\epsilon_{abcdef}\left[\alpha_0 e^a e^b e^c e^d e^e e^f+\alpha_2\left(R^{ab}R^{cd}e^e e^f\right.\right.$$
$$\left.\left.+R^{ab}F^{cd}e^e e^f+F^{ab}F^{cd}e^e e^f\right)\right], \tag{77}$$

which can be seen as the coupling of a generalized cosmological term $\mathcal{L}_{\hat{\Lambda}}$ to a Gauss–Bonnet term $\mathcal{L}_{GB}$,

$$S_{GPL}^{p=2}=\int_{M_6}\mathcal{L}_{\hat{\Lambda}}+\mathcal{L}_{GB}. \tag{78}$$

Here

$$\mathcal{L}_{\hat{\Lambda}}=\alpha_0\epsilon_{abcde}e^a e^b e^c e^d e^e e^f$$
$$+\alpha_1\epsilon_{abcdef}\left(R^{ab}Dk^{cd}e^e e^f+Dk^{ab}Dk^{cd}e^e e^f\right), \tag{79}$$

with $D=d+\omega+k$.

Considering $k^{ab}=0$ we recover the six-dimensional $p=2$ PL theory,

$$S_{PL}^{p=2}=\int_{M_6}\epsilon_{abcdef}\left[\alpha_0 e^a e^b e^c e^d e^e e^f+\alpha_2 R^{ab}R^{cd}e^e e^f\right], \tag{80}$$

while the field equations read in a matter-free configuration limit

$$\epsilon_{abcdef}\left(\alpha_0 e^a e^b e^c e^d e^f + \alpha_1 R^{ab} R^{cd} e^e\right)\delta e^f = 0\,, \tag{81}$$

$$\epsilon_{abcdef}\left(\alpha_1 R^{cd} T^e e^f\right)\delta\omega^{ab} = 0\,, \tag{82}$$

$$\epsilon_{abcdef}\left(\alpha_1 R^{cd} T^e e^f\right)\delta k^{ab} = 0\,, \tag{83}$$

which correspond to the appropriate $p = 1$ PL dynamics [5,13,14].

As in odd-dimensions, no further considerations have to be imposed in order to obtain appropriately the pure Lovelock theory. A similar procedure has been considered in Ref. [18] in order to recover the even-dimensional PL theory from a Born–Infeld like gravity theory. Nevertheless, the obtention of the PL theory in Ref. [18] requires much more conditions and an excessive amount of extra fields.

4. Discussion

In the present work, we have presented a generalized Lovelock gravity theory introducing an additional field which enlarge the symmetry to a Lorentz-like algebra. Interestingly, a generalized pure Lovelock theory is obtained fixing the α_p coefficients which consists only in two terms of full generalized Lovelock action. The generalized PL action considered here allows us to present an alternative way of introducing a generalized cosmological term. Our result generalizes the four-dimensional case presented in Ref. [37] to arbitrary dimensions.

In addition, the usual pure Lovelock theory is recovered in a matter-free configuration of the GPL theory. Unlike Refs. [17,18], not only the PL action is recovered but also the right PL dynamics is directly obtained in the matter-free configuration limit. Such limit is considered without imposing any identifications on the fields. Besides the field content of the GL gravity theory is independent of the spacetime dimension avoiding excessive number of terms in higher dimensions.

The results obtained here, along with the ones presented in [17, 18], could be useful in order to construct a supersymmetric extension of the pure Lovelock theory. Furthermore, the same procedure could be applied in other (super)gravities in order to establish explicit relations between non-trivial (super)gravity actions. In particular it would be interesting to explore the existence of a configuration limit in order to derive the CJS supergravity. In Refs. [38, 39], it has been suggested that the Maxwell like superalgebras could be useful for such task.

Additionally, there are interesting features of the Lovelock formalism which deserve to be explored in our generalized Lovelock theory. Of particular relevance in the AdS/CFT context are the black hole solutions of the Lovelock gravity [40–42]. On the other hand, various problems in the Lovelock gravity can be solved exactly [43] leading to a particular interest in the effect of higher-curvature terms in the holography context [44,45]. Moreover, matter conformally-coupled to gravity can be seen as an extension of the Lovelock gravity theory [46]. In this model, interesting problems related to the black hole geometry can be solved exactly [47] allowing to study Hawking–Page phase transitions [48,49]. Further interesting studies about the Lovelock gravity theory can be found in Refs. [50–54].

Finally, it would be worth exploring our generalization to the quasi-topological gravity which consist in higher curvature gravity [55–63]. The field equations of such theory reduce intriguingly to second order differential equations for spherically symmetric spacetimes and have exact solutions similar to the Lovelock ones. Such interesting behavior is not unique but appears in a bigger family of theories that contains the Lovelock and the quasi-topological theories, as well as the recent Einsteinian cubic gravity theory [64–66] as particular examples [67–69].

Acknowledgements

This work was supported by the Chilean FONDECYT Projects No. 3170437 (P.C.) and No. 3170438 (E.R.). The authors wish to thank N. Merino and R. Durka for enlightening discussions and comments.

References

[1] D. Lovelock, The Einstein tensor and its generalizations, J. Math. Phys. 12 (1971) 498.

[2] C. Lanczos, The four-dimensionality of space and the Einstein tensor, J. Math. Phys. 13 (1972) 874.

[3] B. Zumino, Gravity theories in more than four dimensions, Phys. Rep. 137 (1986) 109.

[4] C. Teitelboim, J. Zanelli, Dimensionally continued topological gravitation theory in Hamiltonian form, Class. Quantum Gravity 4 (1987) L125.

[5] N. Dadhich, J.M. Pons, K. Prabhu, On the Static Lovelock black holes, Gen. Relativ. Gravit. 45 (2013) 1131, arXiv:1201.4994 [gr-qc].

[6] J. Crisostomo, R. Troncoso, J. Zanelli, Black hole scan, Phys. Rev. D 62 (2000) 084013, arXiv:hep-th/0003271.

[7] R. Troncoso, J. Zanelli, Higher dimensional gravity, propagating torsion and AdS gauge invariance, Class. Quantum Gravity 17 (2000) 4451, arXiv:hep-th/9907109.

[8] A.H. Chamseddine, Topological gauge theory of gravity in five-dimensions and all odd dimensions, Phys. Lett. B 223 (1989) 291.

[9] A.H. Chamseddine, Topological gravity and supergravity in various dimensions, Nucl. Phys. B 346 (1990) 213.

[10] J. Zanelli, Chern–Simons forms in gravitation theories, Class. Quantum Gravity 29 (2012) 133001, arXiv:1208.3353 [hep-th].

[11] M. Bañados, C. Teitelboim, J. Zanelli, Lovelock–Born–Infeld theory of gravity, in: H. Falomir, E. Gamboa-Saraví, P. Leal, F. Schaposnik (Eds.), J.J. Giambiagi Festschrift, World Scientific, Singapore, 1991.

[12] S. Deser, G.W. Gibbons, Born–Infeld–Einstein actions?, Class. Quantum Gravity 15 (1998) L35, arXiv:hep-th/9803049.

[13] R.G. Cai, N. Ohta, Black holes in pure Lovelock gravities, Phys. Rev. D 74 (2006) 064001, arXiv:hep-th/0604088.

[14] N. Dadhich, J.M. Pons, K. Prabhu, Thermodynamical universality of the Lovelock black holes, Gen. Relativ. Gravit. 44 (2012) 2595, arXiv:1110.0673 [gr-qc].

[15] N. Dadhich, R. Durka, N. Merino, O. Miskovic, Dynamical structure of pure Lovelock gravity, Phys. Rev. D 93 (2016) 064009, arXiv:1511.02541 [hep-th].

[16] S. Deser, J. Franklin, Canonical analysis and stability of Lanczos–Lovelock gravity, Class. Quantum Gravity 29 (2012) 072001, arXiv:1110.6085 [gr.qc].

[17] P.K. Concha, R. Durka, C. Inostroza, N. Merino, E.K. Rodríguez, Pure Lovelock gravity and Chern–Simons theory, Phys. Rev. D 94 (2016) 024055, arXiv:1603.09424 [hep-th].

[18] P.K. Concha, N. Merino, E.K. Rodríguez, Lovelock gravity from Born–Infeld gravity theory, Phys. Lett. B 765 (2017) 395, arXiv:1606.07083.

[19] J.D. Edelstein, M. Hassaine, R. Troncoso, J. Zanelli, Lie-algebra expansions, Chern–Simons theories and the Einstein–Hilbert Lagrangian, Phys. Lett. B 640 (2006) 278, arXiv:hep-th/0605174.

[20] F. Izaurieta, P. Minning, A. Perez, E. Rodríguez, P. Salgado, Standard general relativity from Chern–Simons gravity, Phys. Lett. B 678 (2009) 213, arXiv:0905.2187 [hep-th].

[21] P.K. Concha, D.M. Peñafiel, E.K. Rodríguez, P. Salgado, Even-dimensional general relativity from Born–Infeld gravity, Phys. Lett. B 725 (2013) 419, arXiv:1309.0062 [hep-th].

[22] P.K. Concha, D.M. Peñafiel, E.K. Rodríguez, P. Salgado, Chern–Simons and Born–Infeld gravity theories and Maxwell algebras type, Eur. Phys. J. C 74 (2014) 2741, arXiv:1402.0023 [hep-th].

[23] P.K. Concha, D.M. Peñafiel, E.K. Rodríguez, P. Salgado, Generalized Poincare algebras and Lovelock–Cartan gravity theory, Phys. Lett. B 742 (2015) 310, arXiv:1405.7078 [hep-th].

[24] D.V. Soroka, V.A. Soroka, Semi-simple extension of the (super)Poincaré algebra, Adv. High Energy Phys. 2009 (2009) 234147, arXiv:hep-th/0605251.

[25] J. Diaz, O. Fierro, F. Izaurieta, N. Merino, E. Rodríguez, P. Salgado, O. Valdivia, A generalized action for (2 + 1)-dimensional Chern–Simons gravity, J. Phys. A 45 (2012) 255207, arXiv:1311.2215 [gr-qc].

[26] P. Salgado, S. Salgado, $\mathfrak{so}(D-1,1)\otimes\mathfrak{so}(D-1,2)$ algebras and gravity, Phys. Lett. B 728 (2013) 5.

[27] P.K. Concha, R. Durka, N. Merino, E.K. Rodríguez, New family of Maxwell like algebras, Phys. Lett. B 759 (2016) 507, arXiv:1601.06443 [hep-th].

[28] S. Hoseinzadeh, A. Rezaei-Aghdam, (2+1)-dimensional gravity from Maxwell and semi-simple extension of the Poincaré gauge symmetric models, Phys. Rev. D 90 (2014) 084008, arXiv:1402.0320 [hep-th].

[29] P.K. Concha, O. Fierro, E.K. Rodríguez, Inönü–Wigner contraction and D=2+1 supergravity, Eur. Phys. J. C 77 (2017) 48, arXiv:1611.05018 [hep-th].

[30] S.W. MacDowell, F. Mansouri, Unified geometric theory of gravity and supergravity, Phys. Rev. Lett. 38 (1977) 739.
[31] P.K. Townsend, P. van Nieuwenhuizen, Geometrical interpretation of extended supergravity, Phys. Lett. B 67 (1977) 439.
[32] P.K. Concha, E.K. Rodríguez, N=1 supergravity and Maxwell superalgebras, J. High Energy Phys. 1409 (2014) 090, arXiv:1407.4635 [hep-th].
[33] P.K. Concha, E.K. Rodríguez, P. Salgado, Generalized supersymmetric cosmological term in N=1 supergravity, J. High Energy Phys. 08 (2015) 009, arXiv:1504.01898 [hep-th].
[34] L. Andrianopoli, R. D'Auria, N=1 and N=2 pure supergravities on a manifold with boundary, J. High Energy Phys. 08 (2014) 012, arXiv:1405.2010 [hep-th].
[35] P.K. Concha, M.C. Ipinza, L. Ravera, E.K. Rodríguez, On the supersymmetric extension of Gauss–Bonnet like gravity, J. High Energy Phys. 09 (2016) 007, arXiv:1607.00373 [hep-th].
[36] D.M. Peñafiel, L. Ravera, On the hidden Maxwell superalgebra underlying D=4 supergravity, arXiv:1701.04234 [hep-th].
[37] J.A. Azcarraga, K. Kamimura, J. Lukierski, Generalized cosmological term from Maxwell symmetries, Phys. Rev. D 83 (2011) 124036, arXiv:1012.4402 [hep-th].
[38] P.K. Concha, E.K. Rodríguez, Maxwell superalgebras and Abelian emigroup expansion, Nucl. Phys. B 886 (2014) 1128, arXiv:1405.1334 [hep-th].
[39] P.K. Concha, O. Fierro, E.K. Rodríguez, P. Salgado, Chern–Simons supergravity in D=3 and Maxwell superalgebra, Phys. Lett. B 750 (2015) 117, arXiv:1507.02335 [hep-th].
[40] R.G. Cai, K.S. Soh, Topological black holes in the dimensionally continued gravity, Phys. Rev. D 59 (1999) 044013, arXiv:gr-qc/9808067.
[41] R.G. Cai, Gauss–Bonnet black holes in AdS spaces, Phys. Rev. D 65 (2002) 084014, arXiv:hep-th/0109133.
[42] R.G. Cai, A note on thermodynamics of black holes in Lovelock gravity, Phys. Lett. B 582 (2004) 237, arXiv:hep-th/0311240.
[43] D.G. Boulware, S. Deser, String generated gravity models, Phys. Rev. Lett. 55 (1985) 2566.
[44] D.G. Boulware, H. Liu, R.C. Myers, S. Shenker, S. Yaida, The viscosity bound and causality violation, Phys. Rev. Lett. 100 (2008) 191601, arXiv:0802.3318 [hep-th].
[45] X.O. Camanho, J.D. Edelstein, J.M. Sánchez De Santos, Lovelock theory and the AdS/CFT correspondence, Gen. Relativ. Gravit. 46 (2014) 1637, arXiv:1309.6483 [hep-th].
[46] J. Oliva, S. Ray, Conformal couplings of a scalar field to higher curvature terms, Class. Quantum Gravity 29 (2012) 205008, arXiv:1112.4112 [gr-qc].
[47] G. Giribet, M. Leoni, J. Oliva, S. Ray, Hairy black holes sources by a conformally coupled scalar field in D dimensions, Phys. Rev. D 89 (2014) 085040, arXiv:1401.4987 [hep-th].
[48] G. Giribet, A. Goya, J. Oliva, Different phases of hairy black holes in AdS5 space, Phys. Rev. D 91 (2015) 045031, arXiv:1501.00184 [hep-th].
[49] M. Galante, G. Giribet, A. Goya, J. Oliva, Chemical potential driven phase transition of black holes in anti-de Sitter space, Phys. Rev. D 92 (2015) 104039, arXiv:1508.03780 [hep-th].
[50] S. Chakraborty, T. Padmanabhan, Evolution of spacetime arises due to the departure from holographic equipartition in all Lanczos–Lovelock theories of gravity, Phys. Rev. D 90 (2014) 124017, arXiv:1408.4679 [gr-qc].
[51] S. Chakraborty, T. Padmanabhan, Geometrical variables with direct thermodynamic significance in Lanczos–Lovelock gravity, Phys. Rev. D 90 (2014) 084021, arXiv:1408.4791 [gr-qc].
[52] S. Chakraborty, Lanczos–Lovelock gravity from a thermodynamic perspective, J. High Energy Phys. 1508 (2015) 029, arXiv:1505.07272 [gr-qc].
[53] S. Chakraborty, K. Parattu, T. Padmanabhan, A novel derivation of the boundary term for the action in Lanczos–Lovelock gravity, Gen. Relativ. Gravit. 49 (2017) 121, arXiv:1703.00624 [gr-qc].
[54] S. Chakraborty, Field equations for gravity: an alternative route, arXiv:1704.07366 [gr-qc].
[55] J. Oliva, S. Ray, A new cubic theory of gravity in five dimensions: black hole, Birkhoff's theorem and C-function, Class. Quantum Gravity 27 (2010) 225002, arXiv:1003.4773 [gr-qc].
[56] J. Oliva, S. Ray, Classification of six derivative Lagrangians of gravity and static spherically symmetric solutions, Phys. Rev. D 82 (2010) 124030, arXiv:1004.0737 [gr-qc].
[57] R.C. Myers, B. Robinson, Black holes in quasi-topological gravity, J. High Energy Phys. 1008 (2010) 067, arXiv:1003.5357 [hep-th].
[58] R.C. Myers, M.F. Paulos, A. Sinha, Holographic studies of quasi-topological gravity, J. High Energy Phys. 08 (2010) 035, arXiv:1004.2055 [hep-th].
[59] M. Chernicoff, O. Fierro, G. Giribet, J. Oliva, Black holes in quasi-topological gravity and conformal couplings, J. High Energy Phys. 1702 (2017) 010, arXiv:1612.00389 [hep-th].
[60] A. Cisterna, L. Guajardo, M. Hassaïne, J. Oliva, Quintic quasi-topological gravity, J. High Energy Phys. 1704 (2017) 066, arXiv:1702.04676 [hep-th].
[61] A. Ghodsi, F. Najafi, Ricci cubic gravity in d dimensions, gravitons and SAdS/Lifshitz black holes, arXiv:1702.06798 [hep-th].
[62] H. Dykaar, R.A. Hennigar, R.B. Mann, Hairy black holes in cubic quasi-topological gravity, J. High Energy Phys. 1705 (2017) 045, arXiv:1703.01633 [hep-th].
[63] Y.Z. Li, H.S. Liu, H. Lu, Quasi-topological Ricci polynomial gravities, arXiv:1708.07198 [hep-th].
[64] P. Bueno, P.A. Cano, Einsteinian cubic gravity, Phys. Rev. D 94 (2016) 104005, arXiv:1607.06463 [hep-th].
[65] R.A. Hennigar, R.B. Mann, Black holes in Einsteinian cubic gravity, Phys. Rev. D 95 (2017) 064055, arXiv:1610.06675 [hep-th].
[66] P. Bueno, P.A. Cano, Four-dimensional black holes in Einsteinian cubic gravity, Phys. Rev. D 94 (2016) 124051, arXiv:1610.08019 [hep-th].
[67] R.A. Hennigar, D. Kubizňák, R.B. Mann, Generalized quasi-topological gravity, Phys. Rev. D 95 (2017) 104042, arXiv:1703.01631 [hep-th].
[68] P. Bueno, P.A. Cano, On black holes in higher-derivative gravities, Class. Quantum Gravity 34 (2017) 175008, arXiv:1703.04625 [hep-th].
[69] J. Ahmed, R.A. Hennigar, R.B. Mann, M. Mir, Quintessential quartic quasi-topological quartet, J. High Energy Phys. 1705 (2017) 134, arXiv:1703.11007 [hep-th].

Graviton propagator, renormalization scale and black-hole like states

X. Calmet [a], R. Casadio [b], A.Yu. Kamenshchik [b,c,*], O.V. Teryaev [d,e]

[a] Department of Physics & Astronomy, University of Sussex, Falmer, Brighton, BN1 9QH, United Kingdom
[b] Dipartimento di Fisica e Astronomia, Università di Bologna, and INFN, Via Irnerio 46, 40126 Bologna, Italy
[c] L.D. Landau Institute for Theoretical Physics of the Russian Academy of Sciences, Kosygin str. 2, 119334 Moscow, Russia
[d] Bogoliubov Laboratory of Theoretical Physics, Joint Institute for Nuclear Research, 141980 Dubna, Russia
[e] Lomonosov Moscow State University, Leninskie Gory 1, 119991 Moscow, Russia

ARTICLE INFO

Editor: A. Ringwald

Keywords:
Gravitons
Renormalization
Black holes

ABSTRACT

We study the analytic structure of the resummed graviton propagator, inspired by the possible existence of black hole precursors in its spectrum. We find an infinite number of poles with positive mass, but both positive and negative effective width, and studied their asymptotic behaviour in the infinite sheet Riemann surface. We find that the stability of these precursors depend crucially on the value of the normalisation point scale.

1. Introduction

Propagators play a crucial role in both quantum mechanics (see e.g. [1]) and in quantum field theory (see e.g. [2,3]). As is well-known, the appearance of a pole in the free field propagator ($p^2 = m^2$) tells us that there exists a one-particle state with the corresponding mass m. The vacuum polarization loops summation for the photon propagator in quantum electrodynamics revealed the existence of a particular pole at a huge negative value of p^2 [4], which is called the "Landau pole". Because of gauge invariance, the existence of this pole implies the same pole in the effective charge of the electron. The latter can be removed by imposing causality and using some adequate analytic properties of propagators [5,6]. The generalisation of this procedure was later applied to quantum chromodynamics (QCD) [7], resulting in the successful description of various physical processes [8].

More recently, the resummed one-loop propagator of the graviton interacting with matter fields was obtained [9,10] (see also Appendix A). This propagator has a rather elegant, but involved form, namely

$$i\,D^{\alpha\beta}(p^2) = i\left(L^{\alpha\mu}L^{\beta\nu} + L^{\alpha\nu}L^{\beta\mu} - L^{\alpha\beta}L^{\mu\nu}\right)G(p^2)\,, \tag{1}$$

where

$$L^{\mu\nu}(p) = \eta^{\mu\nu} - \frac{p^\mu p^\nu}{p^2} \tag{2}$$

and

$$G^{-1}(p^2) = 2\,p^2\left[1 - \frac{N\,p^2}{120\pi\,m_{\rm P}^2}\ln\left(-\frac{p^2}{\mu^2}\right)\right]. \tag{3}$$

Here, $m_{\rm P}$ denotes the Planck mass, μ is the renormalization scale, $N = N_s + 3\,N_f + 12\,N_V$, where N_s, N_f, N_V are the number of scalar, fermion and vector fields, respectively. In the Standard Model, $N_s = 4$, $N_f = 45$, $N_V = 12$ and $N = 283$. The propagator (1) has a standard pole at $p^2 = 0$ and an infinite number of other poles, which are the zeros of the expression (3). It was suggested that these poles correspond to the appearance of a sort of precursors of quantum black holes in Refs. [11,12].

In this paper we shall study in detail the poles and discuss their possible physical interpretations. In particular, we reveal a multi-sheet structure of the corresponding Riemann surface, the role of the renormalisation point and some analogies with studies of the propagator in QCD.

2. Poles of the graviton propagator

We shall now proceed to study the structure of the graviton propagator that follows from the expression (3). We shall in partic-

* Corresponding author.
E-mail addresses: x.calmet@sussex.ac.uk (X. Calmet), Roberto.Casadio@bo.infn.it (R. Casadio), Alexander.Kamenshchik@bo.infn.it (A.Yu. Kamenshchik), teryaev@theor.jinr.ru (O.V. Teryaev).

ular derive expressions for the mass and width, and analyse their location in the Riemann surface.

2.1. Pole positions

It is convenient to rewrite the equation for the non-trivial zeros of (3) as

$$z \ln z = -A \,, \tag{4}$$

where the new variable z is defined as

$$z \equiv -\frac{p^2}{\mu^2} \tag{5}$$

and the positive constant

$$A = \frac{120\pi\, m_{\rm P}^2}{N\mu^2} \,. \tag{6}$$

Note that (4) is nothing but the well known Lambert equation [13]. Introducing as usual $z = \rho\, e^{i\theta}$, we can rewrite (4) as a pair of equations for the imaginary and real parts of the expression on the left-hand side, that is

$$\ln\rho\, \sin\theta + \theta\, \cos\theta = 0 \,, \tag{7}$$

$$\rho\, \ln\rho\, \cos\theta - \rho\theta\, \sin\theta = -A \,. \tag{8}$$

First of all, let us consider the particular case when $\theta = 0$. Then, Eq. (7) is satisfied automatically, while Eq. (8) takes the form

$$\rho\, \ln\rho = -A \,. \tag{9}$$

This equation has two solutions if $A < 1/e$, which merge at $A = 1/e$. Since $z = \mathrm{Re}(z) > 0$, the real part of the corresponding pole for p^2 is negative and both these solutions are tachyons. Obviously, they are stable because $\mathrm{Im}(p^2) = 0$.

If $\theta \neq n\pi$ (with n integer), it follows from Eq. (7) that

$$\ln\rho = -\frac{\theta}{\tan\theta} \,. \tag{10}$$

Substituting (10) into Eq. (8) yields

$$\rho = \frac{A \sin\theta}{\theta} \,. \tag{11}$$

Combining Eqs. (11) and (10), we obtain the equation for the phase θ,

$$f(\theta) \equiv \frac{\theta}{\sin\theta} \exp\left(-\frac{\theta}{\tan\theta}\right) = A \,, \tag{12}$$

where the function $f(\theta)$ is plotted for $0 < \theta < 2\pi$ in Fig. 1.

Obviously, we are interested only in solutions of Eq. (12) which correspond to a positive ρ from (11). Note, however, that solutions of Eq. (12) exist only if θ and $\sin\theta$ have the same sign, in which case the right-hand side of Eq. (11) is positive, as required. Therefore, Eq. (12) has solutions only in the intervals

$$2\pi n < \theta < (2n+1)\pi \,, \qquad n = 0, 1, 2, \ldots \tag{13}$$

and in the mirror symmetric intervals with negative values of θ. At the same time, no solutions exist in the intervals

$$(2n+1)\pi \le \theta \le (2n+2)\pi \,, \qquad n = 0, 1, 2, \ldots \tag{14}$$

and in the mirror intervals.

Let us consider the behaviour of the function $f(\theta)$ in Eq. (12) for $0 \le \theta < \pi$ (the interval (13) with $n = 0$). In this interval, $f(\theta)$ grows monotonically from the minimum $f(0) = 1/e$ to the particular value

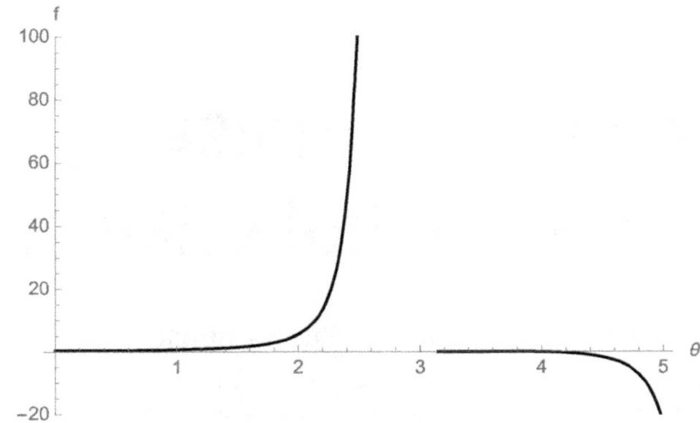

Fig. 1. Function f for $0 < \theta < 2\pi$. Solutions of Eq. (12) can only exist in $0 < \theta < \pi$ where $f > 1/e > 0$.

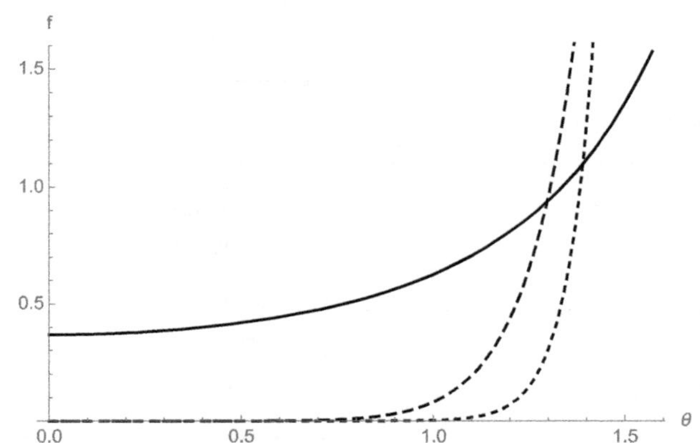

Fig. 2. Function f for $n = 0$ with $0 < \theta < \pi/2$ (solid line), for $n = 1$ with $2\pi < \theta < 5\pi/2$ (dashed line) and for $n = 2$ with $4\pi < \theta < 9\pi/2$ (dotted line). Note the origin is shifted to $2n\pi$.

$$f\left(\frac{\pi}{2}\right) = \frac{\pi}{2} \,, \tag{15}$$

where $\rho(\pi/2) = 1$, and then diverges for $\theta \to \pi$. That means that Eq. (12) has no roots if $A \le 1/e$ (excluding the two real roots of Eq. (9) already described above), has one root in the interval $0 \le \theta < \pi/2$ if $1/e \le A < \pi/2$, and one root in $\pi/2 \le \theta < \pi$ if $A \ge \pi/2$. It also implies that, for $A > \pi/2$, the equation for the phase has a unique root with $\pi/2 \le \theta < \pi$, and the corresponding z has a negative real part and a positive imaginary part.

As we are interested in the complete complex structure of the resummed graviton propagator, we shall consider the whole Riemann surface with all of its sheets. On the second sheet, the relevant interval is given by Eq. (13) with $n = 1$, that is $2\pi < \theta < 3\pi$. Here, the function (12) grows from 0 to $5\pi/2$ when the phase θ goes from 2π to $5\pi/2$, and then keeps on growing indefinitely. A comparison of $f(\theta)$ near its minimum at $2\pi n$ for $n = 0$, 1 and 2 is shown in Fig. 2, from which we see that the curve of $f(\theta)$ moves to the right and gets steeper for increasing n (this trend continues for larger values of n). It is worth remarking that this minimum is always $f(2\pi n) = 0$ except for the fundamental sheet $n = 0$, where $f(0) = 1/e$. In general, larger values of A result in larger values of the phases θ_n of the pole positions z_n, and the dependence of the phase θ_n on n will be analysed in more detail below.

2.2. Mass and width

The poles of propagators in the complex domain correspond to unstable particles and this relation is provided by the famous Breit–Wigner formula [14]. This formula may be presented in various ways [15,16], all of which practically coincide for non-relativistic particles. The covariant representation of the standard non-relativistic formula leads to the parametrization for the pole position given by

$$p^2 = (m - i\,\Gamma/2)^2 = m^2 - i\,\Gamma\, m - \Gamma^2/4\,, \tag{16}$$

which is suggested to be the preferable one (see Refs. [15,16]). We will hence use this expression for the interpretation of the poles of the graviton propagator. We would like to mention also the rather common expression

$$p^2 = m^2 - i\,\Gamma\, m\,, \tag{17}$$

which is obviously close to Eq. (16) for narrow resonances with $\Gamma \ll m$.

Note that the expression (16) is more suitable for investigating wide resonances. In particular, a negative real part of p^2 can be reconciled with positive m^2, which is not the case for (17). We shall dwell on this point in more detail. Let us try to find the parameters m and Γ corresponding to the poles found in the previous subsection. Comparing the real and imaginary parts of Eq. (16) with those of the position of a pole, we obtain

$$m^2 - \frac{\Gamma^2}{4} = -\mu^2\,\rho_n\,\cos\theta_n\,, \tag{18}$$

and

$$\Gamma = \frac{\mu^2\,\rho_n\,\sin\theta_n}{m}\,, \tag{19}$$

where the index n labels the sheet in Eq. (13). On substituting Eq. (19) into Eq. (18), we obtain a quadratic equation for m^2. The first solution is given by

$$m_1^2 = \mu^2\,\rho_n\,\sin^2\left(\frac{\theta_n}{2}\right)\,,$$
$$\Gamma_1^2 = 4\,\mu^2\,\rho_n\,\cos^2\left(\frac{\theta_n}{2}\right)\,, \tag{20}$$

with m_1^2 and Γ_1^2 positive definite and their ratio

$$\frac{\Gamma}{m} = 2\,\cot\left(\frac{\theta_n}{2}\right)\,. \tag{21}$$

The second solution may also be obtained from the first one by interchanging $m \leftrightarrow 4\,i\,\Gamma$ and reads

$$m_2^2 = -\mu^2\,\rho_n\,\cos^2\left(\frac{\theta_n}{2}\right)\,,$$
$$\Gamma_2^2 = -4\,\mu^2\,\rho_n\,\sin^2\left(\frac{\theta_n}{2}\right)\,, \tag{22}$$
$$\frac{\Gamma_2}{m_2} = -2\,\tan\left(\frac{\theta_n}{2}\right)\,,$$

with m_2^2 and Γ_2^2 negative definite. Obviously, the second solution (22) implies an imaginary mass m and an imaginary value of Γ. Imaginary m and Γ make the appeal to the Breit–Wigner type of the representation for the poles in the propagator meaningless, and we therefore completely discard Eq. (22) henceforth.

Let us then consider the solution (20) and further require that m is positive, so that the corresponding expression for the width is given by

$$\Gamma = \frac{\mu\,\sqrt{\rho_n}\,\sin\theta_n}{|\sin(\theta_n/2)|}\,. \tag{23}$$

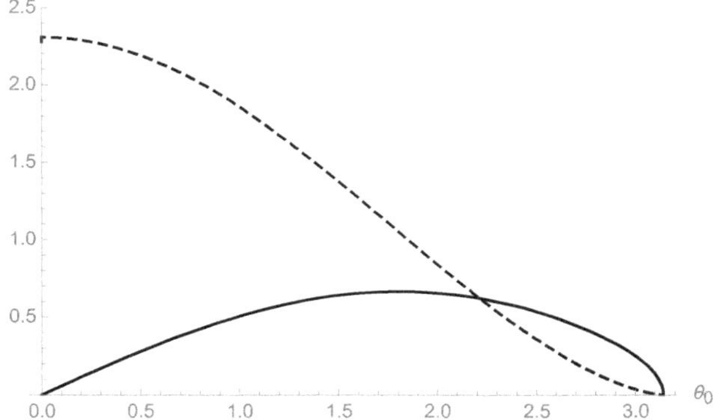

Fig. 3. Mass m (solid line) and width Γ (dashed line) in units of m_{P} versus the pole phase $0 < \theta_0 < \pi$.

Note that to any positive value of θ_n there corresponds a mirror negative value. Thus, we have pairs of poles with positive mass and both positive and negative values of the width Γ. In principle, it would also be possible to fix Γ positive and similarly get a pair of positive and negative masses. The appearance of such pairs of either Γ or m is simply related to the fact that Eq. (12) is even in θ.

Let us also remark that, had we chosen the representation (17) for the poles of the propagator, we would not be able to avoid the appearance of the negative mass squared and imaginary masses and widths for the poles with $\cos\theta_n > 0$. Thus, the choice of the representation (16) and the solution (20) looks justified in the case of the graviton propagator.

Now, using Eqs. (11) and (6), we can rewrite the expression for the mass as

$$m = m_{\mathrm{P}}\sqrt{\frac{120\,\pi}{N}\,\frac{\sin\theta_n}{\theta_n}}\,\left|\sin\left(\frac{\theta_n}{2}\right)\right|\,, \tag{24}$$

and the width as

$$\Gamma = m_{\mathrm{P}}\sqrt{\frac{120\,\pi}{N}\,\frac{\sin\theta_n}{\theta_n}}\,\frac{\sin\theta_n}{|\sin(\theta_n/2)|}\,, \tag{25}$$

and note that their ratio is again simply given by

$$\frac{\Gamma}{m} = 2\,\cot\left(\frac{\theta_n}{2}\right)\,. \tag{26}$$

Remarkably, these physical quantities depend only on the total number of fields N and the phase θ_n of the pole, but not on the renormalisation scale μ^2. We should however not forget that their very existence depends on the value of A which contains μ^2.

The mass and width are plotted for $0 < \theta_0 < \pi$ in Fig. 3, from which it appears that the Breit–Wigner approximation $\Gamma \ll m$ holds for θ_0 close to π. The specific case considered in Refs. [11, 12] is characterised by $m \simeq \Gamma$ and precisely occurs at $m \approx m_{\mathrm{P}}/2$. The mass and width are also plotted for the next sheet, with $2\pi < \theta_1 < 3\pi$ in Fig. 4, and a similar behaviour appears in higher sheets: the maximum value of the mass m and width Γ decrease with n, and the Breit–Wigner approximation holds for (relatively) large θ_n.

2.3. Riemann sheets

It is now interesting to look at the mass m and width Γ as functions of the renormalisation scale μ. This can be done by expressing μ as a function of the phase θ_n from Eq. (12) and then

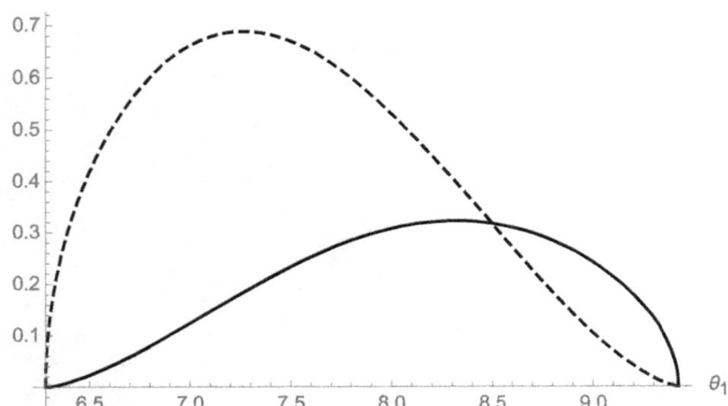

Fig. 4. Mass m (solid line) and width Γ (dashed line) in units of m_P versus the pole phase $2\pi < \theta_1 < 3\pi$.

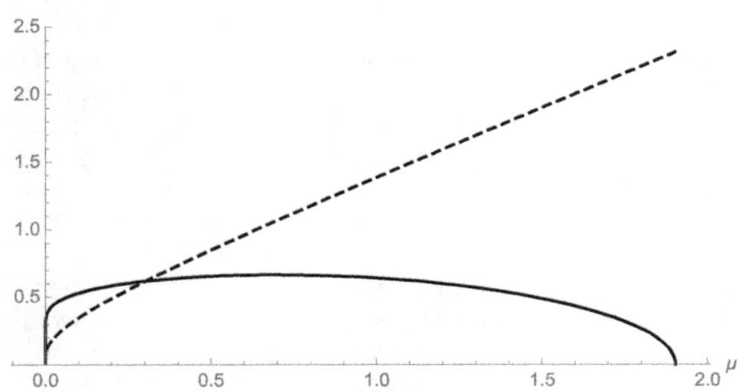

Fig. 5. Mass m (solid line) and width Γ (dashed line) versus the renormalisation scale μ (in units of m_P) for a pole in the fundamental sheet $n = 0$.

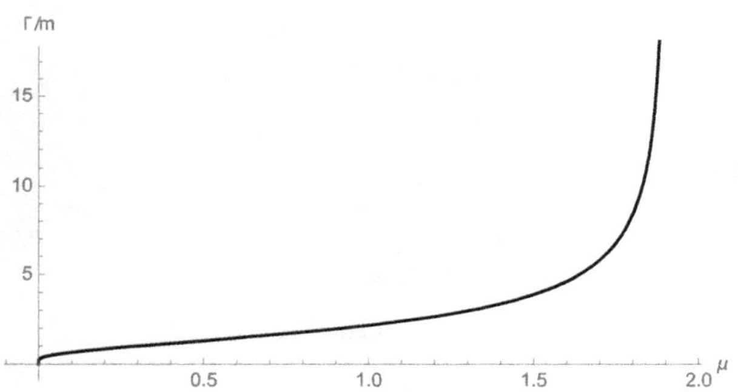

Fig. 6. Ratio Γ/m versus the renormalisation scale μ (in units of m_P) for a pole in the fundamental sheet $n = 0$.

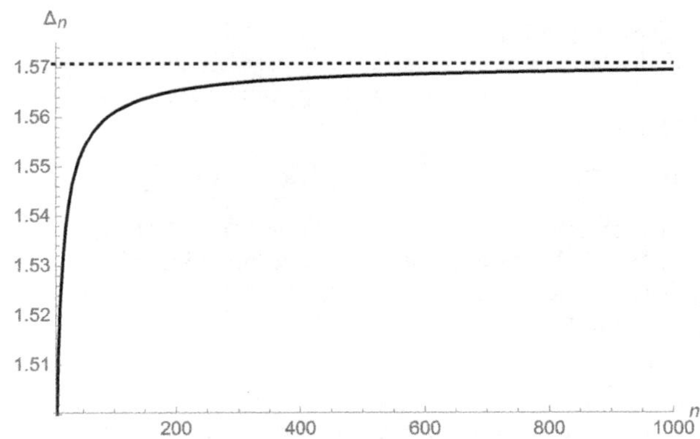

Fig. 7. Phase $\Delta_n \equiv \theta_n - 2\pi n$ for $\mu = m_P$ and $n = 1$ to $n = 1000$ (solid line). The dotted line is the asymptote $\pi/2$.

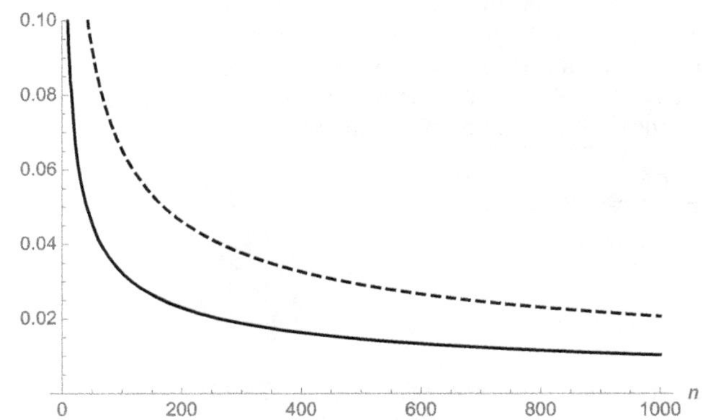

Fig. 8. Mass m (solid line) and width Γ (dashed line) for $\mu = m_P$ and $n = 1$ to $n = 1000$.

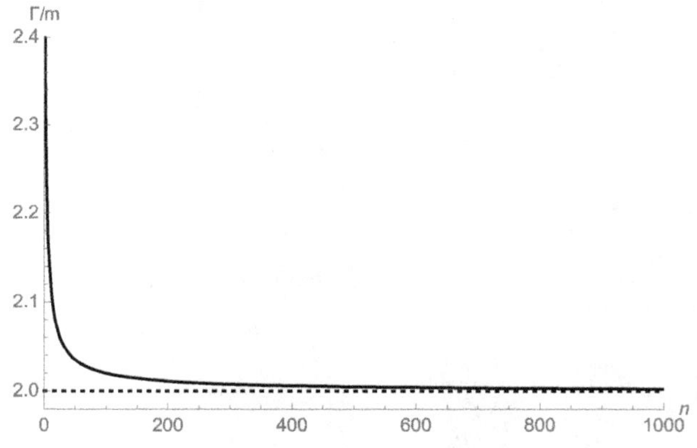

Fig. 9. Ratio Γ/m for $\mu = m_P$ and $n = 1$ to $n = 1000$. The asymptote is $\Gamma/m = 2$ (dotted line).

plotting m and Γ parametrically with respect to μ. For the fundamental sheet with $n = 0$, the mass m and the width Γ are shown in Fig. 5 and their ratio in Fig. 6. From these graphs, we see again that $m \simeq \Gamma \simeq \mu \simeq m_P$, in agreement with Refs. [11,12].

For describing the situation in sheets with $n > 0$, we find it convenient to fix the value of the renormalisation scale μ, hence A, and show the dependence of the phase θ_n, mass m and width Γ on n. For $\mu = m_P$, the difference $\Delta_n \equiv \theta_n - 2\pi n$ is plotted in Fig. 7, from which we see that Δ_n starts out smaller than $\pi/2$ and then asymptotes to $\pi/2$ monotonically from below. The corresponding mass m and width Γ are shown to decrease monotonically in Fig. 8, and their ratio (displayed in Fig. 9) asymptotes from above to $\Gamma/m = 2$, in agreement with Eq. (21). For $\mu = m_P/10$, the picture changes slightly. The shift Δ_n is initially larger than $\pi/2$ but decreases below it for increasing n, and eventually asymptotes to $\pi/2$ again from below (see Fig. 10). The mass and width in Fig. 11 behave qualitatively the same as before, but their ratio starts out smaller than $\Gamma/m = 2$, crosses over and then asymptotes to $\Gamma/m = 2$ again from above (see Fig. 12).

These behaviours can in fact be explained analytically from Eq. (12). Let us write $\Delta_n = \delta_n + \pi/2$, so that

$$\theta_n = 2\pi n + \pi/2 + \delta_n \equiv \bar{\theta}_n + \delta_n \, . \tag{27}$$

For $n \gg 1$, we can then see that $|\delta_n| \ll 1$. In fact, upon replacing (27) into Eq. (12), we obtain

$$\bar{\theta}_n \, e^{-\bar{\theta}_n \delta_n} \simeq A \, , \tag{28}$$

to leading order in δ_n, that is

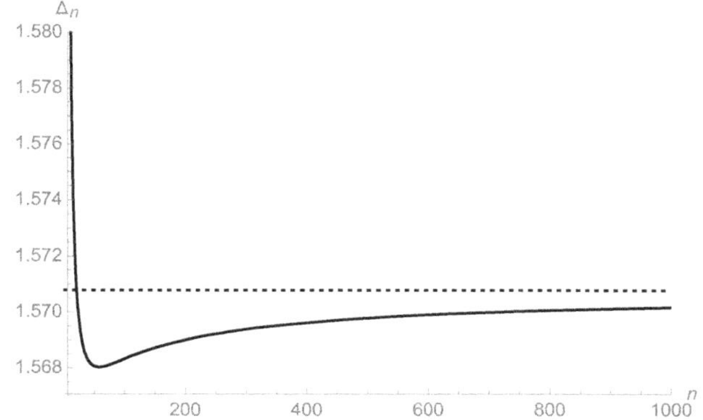

Fig. 10. Phase $\Delta_n \equiv \theta_n - 2\pi n$ for $\mu = m_P/10$ and $n = 1$ to $n = 1000$ (solid line). The dotted line is the asymptote $\pi/2$.

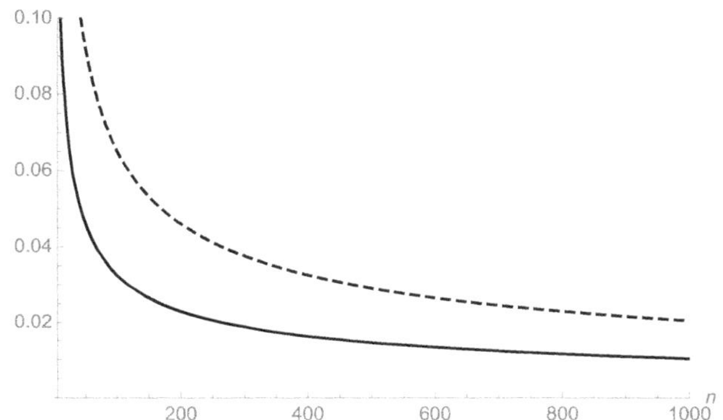

Fig. 11. Mass m (solid line) and width Γ (dashed line) for $\mu = m_P/10$ and $n = 1$ to $n = 1000$.

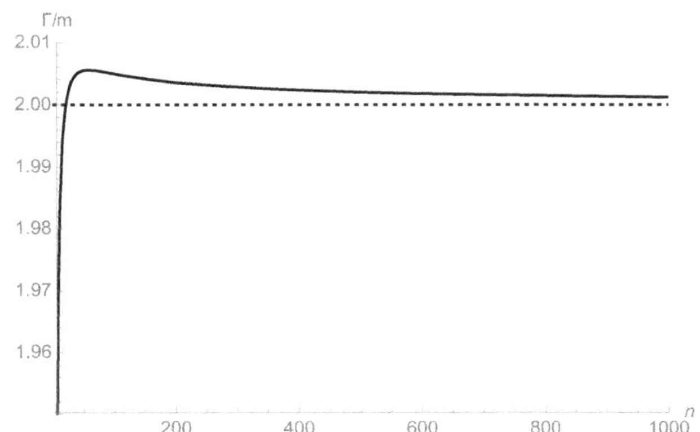

Fig. 12. Ratio Γ/m for $\mu = m_P/10$ and $n = 1$ to $n = 1000$. The asymptote is $\Gamma/m = 2$ (dotted line).

$$\delta_n \simeq \frac{1}{\bar{\theta}_n} \ln\left(\frac{A}{\bar{\theta}_n}\right) . \tag{29}$$

Since $\bar{\theta}_n$ will become necessarily larger than A, for sufficiently large n, the logarithm will become negative and $\delta_n < 0$ will asymptotically vanish. This clearly explains why Δ_n always asymptotes to $\pi/2$ from below. Moreover the ratio Γ/m asymptotes to 2 because of Eq. (21) and does so from above because $\delta_n < 0$ for sufficiently large n. Finally, let us note that for large n we have

$$m \sim \Gamma \sim \frac{1}{\sqrt{\bar{\theta}_n}} \sim \frac{1}{\sqrt{n}} . \tag{30}$$

3. Physical interpretation

In this paper we have studied the analytic structure of the resummed graviton propagator [9,10], inspired by the possible existence of black hole precursors [11,12]. Remarkably, this structure depends essentially on the value of the renormalisation scale parameter μ^2. It is instructive to compare this situation with the more studied cases of the QED and the QCD. In QED, the choice of the renormalisation point is conveniently performed in the classical limit, when the value of the charge is extracted from macroscopic measurements. At the same time it is possible to trade the dependence on the normalisation point μ^2 in favour of the parameter $\Lambda_{\rm QED}$, which is connected with the Landau pole [4], whose value is very large. In QCD, the analogous parameter $\Lambda_{\rm QCD}$ is typically of hadronic mass scale. Since QCD does not admit a classical limit, this parameter is the one used for describing experimental data [7,8].

Since gravity is not renormalisable, it must be treated as an effective theory with no parameter equivalent to Λ, and the dependence on the renormalisation scale μ^2 cannot be avoided. In order to have stable black hole-like quasi-particles in the spectrum, one would need widths $\Gamma \ll m$, which our analyses in turn showed would require a renormalisation scale $\mu^2 \ll m_P^2$. In any case, the number of such states would be finite, since the ratio Γ/m grows with the sheet number n towards the asymptotic value $\Gamma/m = 2$. On the other hand, the existence of such quasi-particle states with positive Γ implies the existence of equal mass states with negative Γ. One might speculate these states would be white hole precursors, although this physical interpretation is not completely clear to us. Perhaps highly unstable black and white hole-like states in higher Riemann sheets are not dangerous because of their highly virtual character and the cancellation of the corresponding imaginary parts in the amplitudes involving the graviton propagator. One should mention here that complex conjugated singularities do appear in the Gribov theory of quark confinement [17]. They appear in the solutions of the Schwinger–Dyson equation in QCD [18,19] and also in the extra-dimensional field theories [20].

It is perhaps more compelling to consider the lower Riemann sheets, where poles with positive mass come likewise with both positive and negative width. In order to deal with the states with negative Γ in lower Riemann sheets, one can consider (at least) two options. The first one is to define the contour of integration for computing the propagator in position space in such a way as to exclude these poles, in the spirit of Refs. [5,6]. In that case, the extra imaginary phase due to the black hole width might lead to observable effects, at least in principle, in particular, to single spin asymmetries [21]. Another option is to assume the renormalisation scale $\mu^2 \gtrsim m_P^2$, so that $|\Gamma| \gtrsim m$ and these states appear only as virtual particles. We conclude by saying that the connection between the black hole type states and the effective quantum field theory deserves further investigation. For instance, non-perturbative effects might come into play at scales of the order of m_P, which would require a different analysis like the one in Ref. [12].

Acknowledgements

We are indebted to D. Kazakov, E. Kolomeitsev, C. Roberts, G. Venturi and R. Yahibbaev for useful discussions. O.T. would like to thank INFN for kind hospitality during October 2016. X.C. is supported in part by the Science and Technology Facilities Council grant N. ST/J000477/1. R.C. and A.K. are partly supported by the INFN grant FLAG. A.K. was partially supported by the RFBR grant N. 17-02-01008. O.T. was partially supported by the RFBR grant N. 17-02-01108.

Appendix A. Gravitational action and propagator

As shown in [22], the resummed propagator (1) can be seen as coming from the variation with respect of the metric field $g_{\mu\nu}$ of the linearized version of the effective action

$$S = \int d^4x \sqrt{-g}\Bigg[\frac{\bar{m}_P^2}{2}\mathcal{R} + c_1\mathcal{R}^2 + c_2\mathcal{R}_{\mu\nu}\mathcal{R}^{\mu\nu} + b_1\mathcal{R}\log\left(\frac{\Box}{\mu^2}\right)\mathcal{R} + b_2\mathcal{R}_{\mu\nu}\log\left(\frac{\Box}{\mu^2}\right)\mathcal{R}^{\mu\nu} + b_3\mathcal{R}_{\mu\nu\rho\sigma}\log\left(\frac{\Box}{\mu^2}\right)\mathcal{R}^{\mu\nu\rho\sigma}\Bigg], \tag{A.1}$$

where $\mathcal{R}$, $\mathcal{R}^{\mu\nu}$ and $\mathcal{R}^{\mu\nu\rho\sigma}$ are respectively the Ricci scalar, Ricci tensor and Riemann tensor and $\bar{m}_P$ is the reduced Planck mass. The Wilson coefficients c_1 and c_2 are arbitrary within the effective field theory approach, and should be fixed by comparing with experimental data. On the other hand b_1, b_2 and b_3 are calculable from first principles and related to N.

The resulting complete propagator for the graviton then contains the function

$$G^{-1}(p^2) = 2p^2\left[1 - 16\pi c_2\frac{p^2}{\bar{m}_P^2} - \frac{N p^2}{120\pi \bar{m}_P^2}\ln\left(-\frac{p^2}{\mu^2}\right)\right]. \tag{A.2}$$

Clearly the position of the poles of (A.2) will depend on the value of c_2, which is arbitrary. For the self-healing mechanism to work, $c_2 = c_2(\mu)$ should be suppressed by $1/N$ in comparison to the coefficients b_i [23]. Let us also stress that Eq. (A.2) can be formally rewritten in the same form as Eq. (1), namely

$$G^{-1}(p^2) = 2p^2\left[1 - \frac{N p^2}{120\pi \bar{m}_P^2}\ln\left(-e^{1920\pi^2 c_2}\frac{p^2}{\mu^2}\right)\right], \tag{A.3}$$

so that the analytic structure does not change and our analysis still applies.

References

[1] B.R. Holstein, Topics in Advanced Quantum Mechanics, Dover, New York, 2014.
[2] N.N. Bogoliubov, D.V. Shirkov, Introduction to the Theory of Quantized Fields, John Wiley, New York, 1980.
[3] V.B. Berestetskii, E.M. Lifshitz, L.P. Pitaevskii, Quantum Electrodynamics, vol. 4, Elsevier Science, Amsterdam, 1982.
[4] L.D. Landau, A.A. Abrikosov, I.M. Khalatnikov, Dokl. Akad. Nauk SSSR 95 (1954) 773;
L.D. Landau, A.A. Abrikosov, I.M. Khalatnikov, Dokl. Akad. Nauk SSSR 95 (1954) 1177;
L.D. Landau, A.A. Abrikosov, I.M. Khalatnikov, Dokl. Akad. Nauk SSSR 96 (1954) 261.
[5] P.J. Redmond, Phys. Rev. 112 (1958) 1404.
[6] N.N. Bogolyubov, A.A. Logunov, D.V. Shirkov, Sov. Phys. JETP 37 (1960) 574.
[7] D.V. Shirkov, I.L. Solovtsov, Phys. Rev. Lett. 79 (1997) 1209;
D.V. Shirkov, I.L. Solovtsov, Theor. Math. Phys. 150 (2007) 132.
[8] A.P. Bakulev, S.V. Mikhailov, N.G. Stefanis, Phys. Rev. D 72 (2005) 074014;
M. Baldicchi, A.V. Nesterenko, G.M. Prosperi, D.V. Shirkov, C. Simolo, Phys. Rev. Lett. 99 (2007) 242001;
R.S. Pasechnik, D.V. Shirkov, O.V. Teryaev, Phys. Rev. D 78 (2008) 071902;
A.P. Bakulev, S.V. Mikhailov, N.G. Stefanis, J. High Energy Phys. 1006 (2010) 085;
V.L. Khandramai, R.S. Pasechnik, D.V. Shirkov, O.P. Solovtsova, O.V. Teryaev, Phys. Lett. B 706 (2012) 340.
[9] T. Han, S. Willenbrock, Phys. Lett. B 616 (2005) 215.
[10] U. Aydemir, M.M. Anber, J.F. Donoghue, Phys. Rev. D 86 (2012) 014025.
[11] X. Calmet, Mod. Phys. Lett. A 29 (2014) 1450204.
[12] X. Calmet, R. Casadio, Eur. Phys. J. C 75 (2015) 445.
[13] R.M. Corless, G.H. Gonnet, D.E.G. Hare, D.J. Jeffrey, D.E. Knuth, Adv. Comput. Math. 5 (1996) 329.
[14] G. Breit, E. Wigner, Phys. Rev. 49 (1936) 519.
[15] T. Bhattacharya, S. Willenbrock, Phys. Rev. D 47 (1993) 4022.
[16] A.R. Bohm, Y. Sato, Phys. Rev. D 71 (2005) 085018.
[17] V.N. Gribov, in: J. Nyiri (Ed.), The Gribov Theory of Quark Confinement, 2001, pp. 162–203, arXiv:hep-ph/9403218, arXiv:hep-ph/9404332, arXiv:hep-ph/9905285;
V.N. Gribov, Eur. Phys. J. C 10 (1999) 91.
[18] G. Krein, C.D. Roberts, A.G. Williams, Int. J. Mod. Phys. A 7 (1992) 5607;
C.J. Burden, C.D. Roberts, A.G. Williams, Phys. Lett. B 285 (1992) 347.
[19] P. Maris, Phys. Rev. D 52 (1995) 6087.
[20] D.I. Kazakov, G.S. Vartanov, J. High Energy Phys. 0706 (2007) 081.
[21] O.V. Teryaev, R.M. Yahibbaev, in preparation.
[22] X. Calmet, S. Capozziello, D. Pryer, in preparation.
[23] E. Tomboulis, Phys. Lett. B 70 (1977) 361.

6

Holographic equilibration under external dynamical electric field

M. Ali-Akbari [a,*], F. Charmchi [b]

[a] Department of Physics, Shahid Beheshti University G.C., Evin, Tehran 19839, Iran
[b] School of Particles and Accelerators, Institute for Research in Fundamental Sciences (IPM), P.O. Box 19395-5531, Tehran, Iran

ARTICLE INFO

Editor: N. Lambert

ABSTRACT

The holographic equilibration of a far-from-equilibrium strongly coupled gauge theory is investigated. In particular, the dynamics of a probe D7-brane in an AdS-Vaidya background is studied in the presence of an external time-dependent electric field. Defining the equilibration times t^c_{eq} and t^j_{eq}, at which condensation and current relax to their final equilibrated values, receptively, the smallness of transition time k_M or k_E is enough to observe a universal behaviour for re-scaled equilibration times $k_M k_E (t^c_{eq})^{-2}$ and $k_M k_E (t^j_{eq})^{-2}$. $k_M (k_E)$ is the time interval in which the temperature (electric field) increases from zero to finite value. Moreover, regardless of the values for k_M and k_E, t^c_{eq}/t^j_{eq} also behaves universally for large enough value of the ratio of the final electric field to final temperature. Then a simple discussion of the static case reveals that $t^c_{eq} \leq t^j_{eq}$. For an out-of-equilibrium process, our numerical results show that, apart from the cases for which k_E is small, the static time-ordering, that is $t^c_{eq} \leq t^j_{eq}$, persists.

1. Introduction and results

Describing time evolution of far-from-equilibrium processes is an attractive but difficult open problem, especially at strong coupling. An important consequence of the AdS/CFT correspondence, or more generally gauge-gravity duality, is that out-of-equilibrium dynamics in strongly coupled gauge theory is dual to a time-dependent problem in a classical gravity [1,2]. This new theoretical tool is applied to investigate various aspects of the far-from-equilibrium physics. Tracing the (instantaneous) thermalization [3] and isotropization [4] processes of quark-gluon plasma is one of the significant aspects which has attracted a lot of attention in recent times. Moreover, studying quantum quench, which is also a far-from-equilibrium situation, is another motivation to use the gauge-gravity duality [5]. Quantum quench is defined as the process of (rapidly) altering a physical coupling of a quantum system.

D3–D7 system is a reasonable candidate to explain different properties of QCD-like theories [6]. In fact, the fundamental matter (quark) is added to the gauge theory via introduction of N_f probe D7-branes in the background sourced by N_c colour D3-branes which is normally $AdS_5 \times S^5$ or AdS-black hole geometry [7]. By probe (limit) we mean the back-reaction of D7-branes on the background can be ignored. As a result, Dirac–Born–Infeld (DBI) action describes the dynamics of the D3–D7 system in the probe limit [12,23]. This model, apart from the static case, is applied to understand non-equilibrium processes which happen on the probe D7-brane such as black hole formation [8,9], time-dependent baryon injection [9], Schwinger effect [10] and time-dependent meson melting [11,14].

In the time evolution of non-equilibrium process, discussing various time-scales of relaxation is another interesting issue. In the context of gauge-gravity duality for a strongly coupled gauge theory with a holographic dual different time-scales have been introduced and studied in the literature. Furthermore time-ordering of these time-scales is also addressed in more recent papers such as [15,16]. Understanding how a non-equilibrium system relaxes to its final equilibrated state and what quantities are more effective during the relaxation are interesting problems to investigate.

In this paper, we consider a far-from-equilibrium strongly coupled gauge theory ($\mathcal{N} = 2$ super Yang–Mills) which is dual to a probe D7-brane in the AdS-Vaidya background. As is well-known, introducing the probe D7-brane in the static background decreases the number of supersymmetries from $\mathcal{N} = 4$ to $\mathcal{N} = 2$ (at zero temperature). The AdS-Vaidya background resembles thermalization in the $\mathcal{N} = 4$ gauge theory (gluon sector) which, in our system, is realized by the fields living on the probe brane. One can also define a time-dependent electric field on the probe D7-brane which is identified with the external time-dependent electric field

* Corresponding author.
E-mail addresses: m_aliakbari@sbu.ac.ir (M. Ali-Akbari), charmchi@ipm.ir (F. Charmchi).

in the gauge theory. The external electric field couples to the quarks and induces a time-dependent current on the probe brane. Therefore, we heat up the gauge theory (gluon sector), subject to an external time-dependent electric field. This system might resemble the early-time dynamics in the heavy ion collisions where very strong electric field is generated by the electric current induced by heavy ions passing by each other. We refer the reader for more details to [10] and references there in.

In our model, energy injection into the system is done by increasing external electric field and temperature or, equivalently, there are two ways where the system at hand deviates from its equilibrium. Comparing the systems considered in [11], [19] and our case in this paper, one can easily see that time-dependent temperature and electric field can be simultaneously studied in our model. In other words, we are able to investigate how different energy injections of the electric field and/or temperature affect the equilibration times of the system and to determine which one, between the electric field and temperature, plays more important role during the time evolution. More precisely, we study the effect of k_M, k_E, the final values of the temperature T_f and the electric field E_f on the equilibration times of the current, t^j_{eq}, and condensation, t^c_{eq}. As we shall see, the condensation c is defined by the expectation value of the $\bar{\psi}\psi$ operator where ψ represents the fermionic fields.

Our main findings can be summarized as follows:

- It was shown in [22] that in the static background for the massless quarks when the transition time k_E is small enough, $k_E t^j_{eq}$ behaves universally, meaning that its value is independent of the final value of the time-dependent electric field. In the system we consider here, there are two transition times k_E and k_M and thus one can question which dimensionless quantity behaves universally and, if any, whether both the transition times must be small to observe universal behaviour? Our numerical results indicate that $k_M k_E (t^c_{eq})^{-2}$ and $k_M k_E (t^j_{eq})^{-2}$ behave universally. Interestingly, in order to observe universality in our model, both k_M and k_E are *not* constrained to be small.
- Another significant question is whether, apart form $k_M k_E (t^c_{eq})^{-2}$ and $k_M k_E (t^j_{eq})^{-2}$, there are quantities that behave universally? We show that t^c_{eq}/t^j_{eq} for large enough value of the ratio of the final electric field to final temperature, i.e. E_f/T_f, *independent of the values of k_M and k_E*, also behaves universally. It means that the value of t^c_{eq}/t^j_{eq} is almost constant and independent of the final values of the temperature and electric field. Surprisingly, unlike the previous case, k_E or k_M can be small or large! To our knowledge, this new kind of universality has not been conjectured in the literature.
- Can we answer the question of there is any difference between transitions times? We find that, for small enough values of k_E, the static time-ordering $t^c_{eq} \leq t^j_{eq}$ can not be considered as a valid time-ordering or equivalently, the time-ordering is reversed. Note that the static time-ordering persists even for small enough values of k_M. From this point of view, *there is a remarkable difference between small values of k_E and k_M*. Notice that although the system is far from equilibrium, the static time-ordering is the correct one except for $k_E \ll 1$.

Needless to say, *the above results can only be observed in the system with two sources for the energy injection*. In the following we use the same numerical method and notations as the ones introduced in [11] and refer the reader to [11,14,17–19] for more details.

2. Model

The AdS-Vaidya background is considered as a toy model of describing thermalization process in the context of the gauge-gravity duality. This background specifies black hole formation in the gravity theory which corresponds to thermalization process in the dual gauge theory. The metric of AdS-Vaidya background is given by

$$ds^2 = G_{MN}dx^M dx^N = \frac{1}{z^2}\left[-F(V,z)dV^2 - 2dVdz + d\vec{x}^2\right] + d\varphi^2 + \cos^2\varphi\, d\Omega_3^2 + \sin^2\varphi\, d\theta^2, \tag{1}$$

where $F(V,z) = 1 - M(V)z^4$. We set the radius of the AdS spacetime to be one. The above metric is written in the Eddington–Finkelstein coordinates where the radial direction is represented by z. The boundary, where the gauge theory lives, is at $z = 0$. V shows the null direction where, at the boundary, V is equal to the time coordinate of the gauge theory, i.e. $t = V|_{z=0}$. The arbitrary function $M(V)$, related to the temperature of the gauge theory, represents the mass of the black hole which changes as time passes by until it reaches a constant value.

In order to introduce a dynamical electric field in the gauge theory, we need to add a probe brane to the bulk. The dynamics of the D7-brane is explained by the DBI action

$$S = -\tau_7 \int d^8\xi \sqrt{-\det(g_{ab} + 2\pi\alpha' F_{ab})}\,, \tag{2}$$

where tension of the D7-brane is denoted by τ_7 and a, b are the brane coordinates. $g_{ab} = G_{MN}\partial_a x^M \partial_b x^N$ is the induced metric and F_{ab} is the gauge field strength on the brane. The D7-brane is embedded along the six directions of the bulk metric, $\vec{x} = (x_1, x_2, x_3)$ and Ω_3. We choose the other two coordinates on the brane to be null coordinates u and v. In order to specify the D7-branes' embedding one must then specify the remaining space–time coordinates, V, z, φ and θ as functions of, in principle, all the world-volume coordinates. Translational symmetry in the $\vec{x}$-directions and rotational symmetry in the Ω_3-directions allow the rest of the bulk coordinates to depend only on u and v. Therefore, we choose the ansatz

$$V = V(u,v), \quad z = Z(u,v), \quad \varphi = \Phi(u,v), \quad \theta = 0. \tag{3}$$

Notice that because of the symmetry generated by ∂_θ in the background metric, without loss of generality, we set $\theta = 0$.

Since we are interested in studying the effect of the external time-dependent electric field on the equilibration time-scales, we also adopt an ansatz for the gauge filed as below ($x \equiv x_1$)

$$A_x(V,Z) = a_0(V) + a_x(V,Z) = \int_{V_0}^{V} E(V')dV' + a_x(V,Z), \tag{4}$$

and all other components of the gauge field are set to zero. The time-dependent electric field on the boundary is proportional to $E(V)$ and V_0 is an initial time reference which can be set, for instance, equal to zero or $-\infty$. Now by substituting (3) and (4) into the DBI action (2), it is easy, but lengthy, to find the following equations of motion

$$\begin{aligned} V_{,uv} =& \frac{Z^3}{2}A_{x,u}A_{x,v} + \frac{1}{2}\left(F_{,z} - \frac{5F}{Z}\right)V_{,u}V_{,v} \\ &+ \frac{3}{2}\tan(Z\Psi)\Big((Z\Psi)_{,u}V_{,v} + (Z\Psi)_{,v}V_{,u}\Big) \\ &+ \frac{3}{2}Z(Z\Psi)_{,u}(Z\Psi)_{,v}\,, \end{aligned} \tag{5a}$$

$$\begin{aligned}
Z_{,uv} =& -\frac{FZ^3}{2}A_{x,u}A_{x,v} - \frac{F_{,V}}{2}V_{,u}V_{,v} \\
&+\frac{1}{2}\left(\frac{5F}{Z} - F_{,Z}\right)\left(FV_{,u}V_{,v} + V_{,u}Z_{,v} + V_{,v}Z_{,u}\right) \\
&+\frac{3}{2}\tan(Z\Psi)\Big((Z\Psi)_{,u}V_{,v} + (Z\Psi)_{,v}V_{,u}\Big) \\
&-\frac{3F}{2}Z(Z\Psi)_{,u}(Z\Psi)_{,v} + \frac{5}{Z}Z_{,u}Z_{,v},
\end{aligned} \tag{5b}$$

$$\begin{aligned}
\Psi_{,uv} =& \frac{Z^2\Psi}{2}\left(F - \frac{3\tan(Z\Psi)}{Z\Psi}\right)A_{x,u}A_{x,v} \\
&+\frac{\Psi}{2Z}\left(F_{,Z} - \frac{5F}{Z} + \frac{3\tan(Z\Psi)}{Z^2\Psi}\right) \\
&\times\left(FV_{,u}V_{,v} + V_{,u}Z_{,v} + V_{,v}Z_{,u}\right) \\
&+\frac{\Psi F_{,V}}{2Z}V_{,u}V_{,v} + \frac{1}{2Z^2}\Big(1 - 3Z\Psi\tan(Z\Psi)\Big) \\
&\times\Big((Z\Psi)_{,u}Z_{,v} + (Z\Psi)_{,v}Z_{,u}\Big) \\
&+\frac{3}{2}\left(F\Psi + \frac{\tan(Z\Psi)}{Z}\right)(Z\Psi)_{,u}(Z\Psi)_{,v} - \frac{3\Psi}{Z^2}Z_{,u}Z_{,v},
\end{aligned} \tag{5c}$$

$$\begin{aligned}
A_{x,uv} =& \frac{1}{2Z}\Big(Z_{,u}A_{x,v} + Z_{,v}A_{x,u}\Big) \\
&+\frac{3}{2}\tan(Z\Psi)\Big((Z\Psi)_{,u}A_{x,v} + (Z\Psi)_{,v}A_{x,u}\Big),
\end{aligned} \tag{5d}$$

where $\Psi(u,v) \equiv \frac{\Phi(u,v)}{Z(u,v)}$. In fact, $\Phi(u,v)$ gives the shape of the brane [11,17,20] and, as we will see in (6a), based on the dictionary of the gauge-gravity duality, its asymptotic behaviour is related to the physical quantities in the gauge theory.

Quark in the fundamental representation of the gauge group with finite mass m can be added to the gauge theory by introducing probe D7-branes into the asymptotically $AdS_5 \times S^5$ background in the dual gravity theory [7]. As a matter of fact, by introducing D7-branes, a new term $m\bar{\psi}\psi$ has been added to the Lagrangian of the original gauge theory. The fermionic field ψ denotes the fundamental matter in the gauge theory. Furthermore, since an external electric field is also considered in our case, the new Lagrangian includes another term $j^x A_x$. According to the gauge-gravity duality, our observables in the gauge theory can be obtained by near boundary expansion of the corresponding fields on the gravity side. In the case at hand, the observables are time-dependent condensation $c \propto \langle\bar{\psi}\psi\rangle$ and current $j \propto \langle j^x\rangle$. These quantities are respectively related to the near boundary expansion of $\Psi(u,v)$ and $a_x(V,Z)$ as follows [19,20]

$$\Psi(V,Z) = m + \left(c(V) + \frac{m^3}{6}\right)Z^2 + \dots, \tag{6a}$$

$$a_x(V,Z) = \dot{a}_0(V)Z + \frac{1}{2}j(V)Z^2 + \dots, \tag{6b}$$

where $\dot{a}_0(V) = \frac{da_0(V)}{dV}$. By solving the equations of motion (5) numerically, one can find $\Psi(u,v)$, $Z(u,v)$ and $a_x(u,v)$. Then the time-dependent condensation and current can be obtained from (6a) and (6b), respectively.

In order to solve the above equations, we need to specify the boundary and initial conditions. These conditions are similar to the ones considered in [11,17,19]. Here we do not repeat the details of calculations and only state the final results. Therefore, we have

- **Boundary conditions at the AdS Boundary:** The AdS boundary has been set at $Z = 0$ where $u = v$. According to the dictionary of gauge-gravity duality, the shape of the probe brane and gauge field at the boundary give the mass of the quark and time-dependent electric field, i.e. $\Psi|_{u=v} = m$ and $A_x|_{u=v} = a_0(V)$, respectively. Moreover, by imposing the regularity condition on the equations of motion we get [11]

$$V_{,v}|_{u=v} = 2Z_{,u}|_{u=v}. \tag{7}$$

- **Initial conditions:** We assume that the energy injection into the system starts from $V \geq 0$ meaning that $V_0 = 0$ in (4). As a result, for $V < 0$, the shape of the brane can be found by solving (5) in the pure AdS background (i.e. $M(V) = 0$) with $A_x(V,Z) = 0$ for a given value of m. Thus, at the initial surface $v = 0$, we have [11]

$$\begin{aligned}
&V(u,0) = m^{-1}\big(\phi(u) - \sin\phi(u)\big) + V_{\text{ini}}, \\
&Z(u,0) = m^{-1}\sin\phi(u), \\
&\Psi(u,0) = \frac{m\phi(u)}{\sin\phi(u)}, \qquad A_x(u,0) = 0,
\end{aligned} \tag{8}$$

where $V_{\text{ini}} \leq 0$ is an integration constant and $\phi(u)$ is an arbitrary function corresponding to the residual coordinate freedom. We set V_{ini} to 0. An appropriate choice of the arbitrary function, introduced in the above equations, is $\phi(y) = y$. Our calculations in this paper are done by the same choice. For more details see [11].

Various types of energy injection are identified by the form of the functions chosen for the time-dependent temperature and electric field. These different forms have been investigated and classified in [13] and it seems that the final qualitative results are independent of the form of the functions. Here, in our case, the electric field (temperature) is turned on exactly at $V = 0$ and reaches its exact final maximum value, E_f (T_f), at some finite time. Therefore, the functions for $M(V)$ and $E(V)$ that we will work with are

$$M(V) = \begin{cases} 0 & V < 0, \\ \frac{M_f}{k_M}\left[V - \frac{k_M}{2\pi}\sin(\frac{2\pi V}{k_M})\right] & 0 \leqslant V \leqslant k_M, \\ M_f & V > k_M, \end{cases}$$

$$E(V) = \begin{cases} 0 & V < 0, \\ \frac{E_f}{k_E}\left[V - \frac{k_E}{2\pi}\sin(\frac{2\pi V}{k_E})\right] & 0 \leqslant V \leqslant k_E, \\ E_f & V > k_E, \end{cases} \tag{9}$$

where k_M (k_E) is the time interval at which the mass of the black hole (external electric field) increases from zero to M_f (E_f) which is constant. Note that the radius of the event horizon is $r_h = M_f^{\frac{1}{4}}$ and therefore $T_f = M_f^{\frac{1}{4}}/\pi$.

3. Numerical results

In order to determine suitable criteria for equilibration of the system, we define the following time-dependent equilibration parameters

$$\begin{aligned}
\epsilon_1(t) &= \frac{|c(t) - c_{eq}|}{c_{eq}}, \\
\epsilon_2(t) &= \frac{|j(t) - j_{eq}|}{j_{eq}},
\end{aligned} \tag{10}$$

where c_{eq} and j_{eq} are the final equilibrated values of the condensation and current, respectively. These values are found from our numerical calculations. Then equilibration time for condensation (current) is defined as the time at which $|\epsilon_1(t^c_{eq})| < 0.03$ ($|\epsilon_2(t^j_{eq})| < 0.03$) and stays below this limit afterwards.

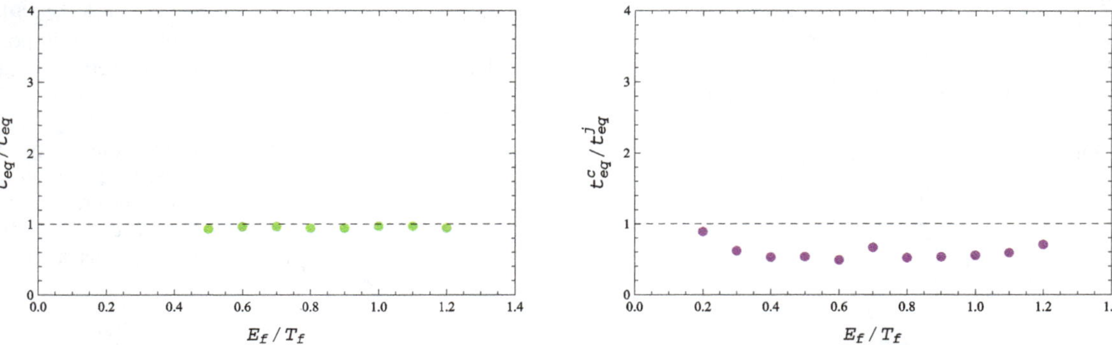

Fig. 1. Left: t^c_{eq}/t^j_{eq} in terms of E_f/T_f for $k_M = k_E = 5$ (slow–slow transition) and $T_f = 5$. Right: t^c_{eq}/t^j_{eq} in terms of E_f/T_f for $k_M = 5$, $k_E = 0.1$ (slow–fast transition) and $T_f = 5$.

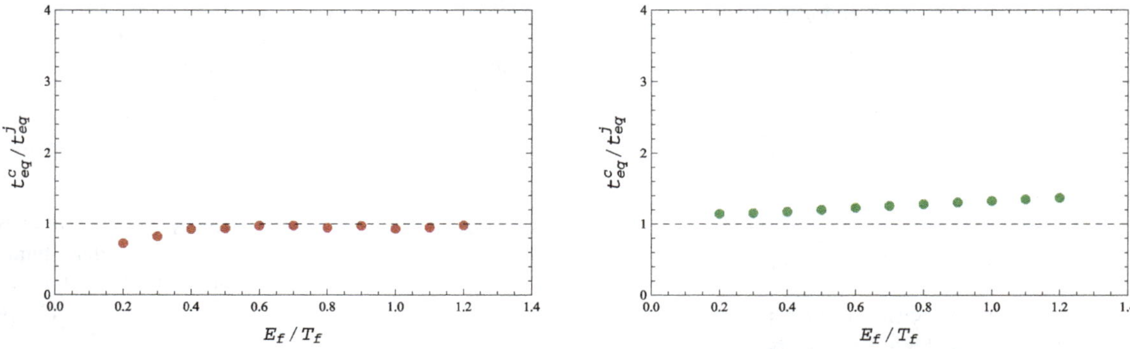

Fig. 2. Left: t^c_{eq}/t^j_{eq} in terms of E_f/T_f for $k_M = 0.1$, $k_E = 5$ (fast–slow transition) and $T_f = 5$. Right: t^c_{eq}/t^j_{eq} in terms of E_f/T_f for $k_M = 0.1$, $k_E = 0.1$ (fast–fast transition) and $T_f = 5$.

We continue our discussion for the case $m = 1$, $V_0 = 0$ without loss of generality and relevant values for the final temperature and electric field in such a way that the known black hole embedding can be produced as final states [24][6]. Then condensation $c(V)$ can be numerically found by solving the equations of motion (5). Since the shape of the D7-brane in an arbitrary background is described by $\Psi(V, Z)$ (or equivalently $\Phi(V, Z)$) [2], when the condensation reaches its final equilibrated value the shape of the brane is fixed and does not change any-more. However, from the static case, we know that the static current depends on the electric field as well as the shape of the brane in the probe limit. Therefore one may speculate that the equilibration time for the condensation is always smaller than or equal to the equilibration time for the current, that is $t^c_{eq} \le t^j_{eq}$. We will check this naive statement in the next section.

The parameters k_M and k_E are measures to show how fast the functions $M(V)$ and $E(V)$ can reach their maximum values. In other words, k_M and k_E represent the transition times from zero to finite value. It is obvious that for $k_M, k_E \ll 1$ ($k_M, k_E \gg 1$) the transition time is small (large) and hence indicates a fast (slow) growth in the electric field and mass of the black hole. Here we classify our numerical results [25] into four categories as follows

- *Slow–Slow transition*
  In Fig. 1(left), at fixed temperature, t^c_{eq}/t^j_{eq} is plotted in terms of E_f/T_f for $k_M = k_E = 5$. In this case the transition times are large enough to consider the system under study near equilibrium. As it is clearly seen from this figure, the equilibration times for the condensation and current are equal, in agreement with the static argument.
- *Slow–Fast transition*
  The case with $k_M = 5$ and $k_E = 0.1$ is plotted in Fig. 1(right). The equilibration time for condensation is smaller than the corresponding one for the current. It is comprehensible since smaller values of k_E leads to larger deviation from equilibrium [22].
- *Fast–Slow transition*
  Now let us consider $k_M = 0.1$ and $k_E = 5$ plotted in Fig. 2(left). In this case for small values of electric field the equilibration time for condensation is smaller than t^j_{eq}, that is $t^c_{eq} < t^j_{eq}$. For higher values, it is easy to see that $t^c_{eq} \simeq t^j_{eq}$. It is clearly seen that the ordering of time-scales is in agreement with our preliminary conclusion.
- *Fast–Fast transition*
  Up to now, all observations are consistent with our naive conclusion based on the static case, that is $t^c_{eq} \le t^j_{eq}$. Now, by considering $k_M = k_E = 0.1$, Fig. 2(right) indicates that time-ordering is reversed, i.e. $t^c_{eq} > t^j_{eq}$. This means that although the current on the D-brane reaches its final equilibrated value, the shape of the brane still changes! Thus, we observe a signal that indicates out-of-equilibrium situation. From this analysis one can conclude that *the shape of the probe brane is less important than the electric field in the equilibration time of the current.*

Now let us consider the slow–fast case in which E_f/T_f is larger than two, Fig. 3(left). Although in the Fig. 1(right) $t^c_{eq} < t^j_{eq}$, Fig. 3(left) shows that the ratio of the equilibration times is larger than two, that is $t^c_{eq} \gtrsim 2t^j_{eq}$, meaning that the time-ordering of the equilibration times has been reversed. Therefore, according to the Fig. 2(right) and Figs. 3, our numerical results indicate that in order to observe a reversed time-ordering a small value of k_E is essential. Then one may guess that the same behaviour can be also observed in Fig. 2(left) for large values of E_f/T_f with $k_M = 0.1$. But, up to $E_f/T_f \simeq 6$ for which our numerical data is reliable, the same behaviour is not observed. In other words, it seems that a small

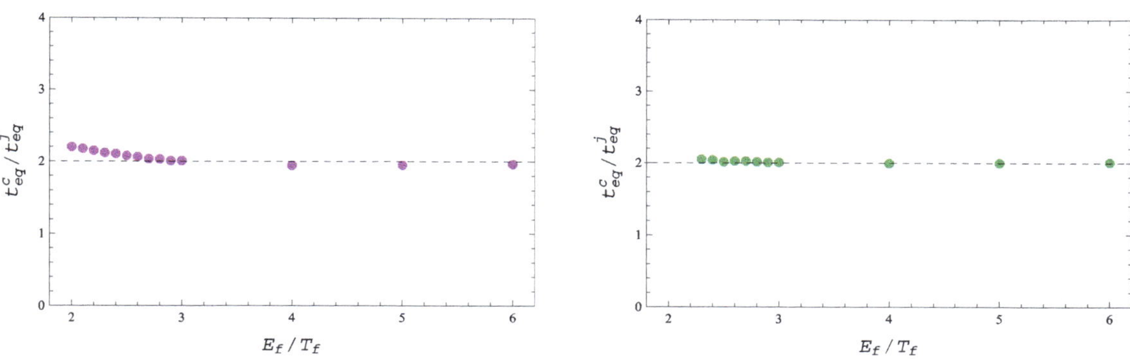

Fig. 3. Left: t^c_{eq}/t^j_{eq} in terms of $E_f/T_f > 2$ for $k_M = 5$, $k_E = 0.1$ (slow–fast transition) and $T_f = 5$. Right: t^c_{eq}/t^j_{eq} in terms of $E_f/T_f > 2$ for $k_M = 0.1$, $k_E = 0.1$ (fast–fast transition) and $T_f = 5$.

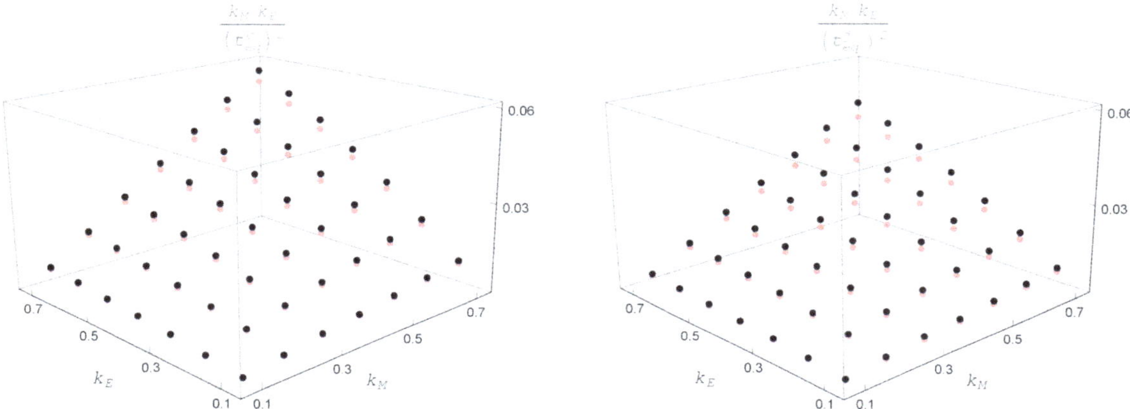

Fig. 4. The re-scaled equilibration times $k_M k_E (t^c_{eq})^{-2}$ (left) and $k_M k_E (t^j_{eq})^{-2}$ (right) verses k_M and k_E for $T_f = 1$ and $E_f = 1$ (black) and $E_f = 1.1$ (red). (For interpretation of the references to colour in this figure legend, the reader is referred to the web version of this article.)

value of k_E is necessary to reverse time-ordering of the equilibration times.

Surprisingly in the Figs. 3 for large value of E_f/T_f, the ratio of equilibration times converges to a constant value. But, since the convergence number does depend on the equilibration parameters $\epsilon_1(t)$ and $\epsilon_2(t)$, it does not have a physical interpretation. However, it is important to notice that this convergence is observed by different values of equilibration parameters. Thus this convergence can be considered as a universal behaviour. Because, for large E_f/T_f, the value of t^c_{eq}/t^j_{eq} is almost constant and independent of the final values of the temperature and electric field as it is clearly seen in Figs. 1(left), 2(left) and 3. Therefore we conclude that for large enough E_f/T_f the ratio of the equilibration times behaves universally.

For small values of the transition time, so-called fast quench, another universal behaviour is observed in the gauge theories with holographic dual [21]. For example, in [22], a probe D7-brane in the $AdS_5 \times S^5$ background subject to a dynamical external electric field, which is similar to our case, is considered and the calculations show that the re-scaled transition time $k_E(t^j_{eq})^{-1}$ behaves universally. In this case t^j_{eq} is equilibration time for the time-dependent current and universality means that $k_E(t^j_{eq})^{-1}$ is independent of the final value of the electric field. As the case under study in this paper is a generalization of the above system, we expect a universal behaviour as well. However, because there are two transition times k_M and k_E, we need to re-scale equilibration times for the condensation and current properly to observe universality. Therefore, in Fig. 4, we plot the new re-scaled equilibration times $k_M k_E (t^c_{eq})^{-2}$ and $k_M k_E (t^j_{eq})^{-2}$ in terms of k_M and k_E. Interestingly, at fixed and small value of k_M or k_E, a universal behaviour is observed meaning that the re-scaled equilibration times $k_M k_E (t^c_{eq})^{-2}$ and $k_M k_E (t^j_{eq})^{-2}$ are independent of the final values of the electric field and temperature. Therefore *the smallness of k_M or k_E is enough to observe a universal behaviour*. Obviously when both k_M and k_E are small the re-scaled equilibration times behave universally as well.

Acknowledgements

This work was financially supported by the following project of the Shahid Beheshti University, G.C.: 600-1475.

M. A. would like to thank School of Physics of Institute for research in fundamental sciences (IPM) for the research facilities and environment. The authors would like to thank H. Ebrahim and L. Shahkarami for useful discussions.

References

[1] J.M. Maldacena, The large N limit of superconformal field theories and supergravity, Int. J. Theor. Phys. 38 (1999) 1113, Adv. Theor. Math. Phys. 2 (1998) 231, arXiv:hep-th/9711200.

[2] J. Casalderrey-Solana, H. Liu, D. Mateos, K. Rajagopal, U.A. Wiedemann, Gauge/string duality, hot QCD and heavy ion collisions, arXiv:1101.0618 [hep-th].

[3] P.M. Chesler, L.G. Yaffe, Horizon formation and far-from-equilibrium isotropization in supersymmetric Yang–Mills plasma, Phys. Rev. Lett. 102 (2009) 211601, arXiv:0812.2053 [hep-th];
H. Ebrahim, M. Headrick, Instantaneous thermalization in holographic plasmas, arXiv:1010.5443 [hep-th];
V. Balasubramanian, et al., Holographic thermalization, Phys. Rev. D 84 (2011) 026010, arXiv:1103.2683 [hep-th];

P.M. Chesler, L.G. Yaffe, Boost invariant flow, black hole formation, and far-from-equilibrium dynamics in $N = 4$ supersymmetric Yang–Mills theory, Phys. Rev. D 82 (2010) 026006, arXiv:0906.4426 [hep-th];
D. Kaviani, D7-brane dynamics and thermalization in the Kuperstein–Sonnenschein model, arXiv:1608.02380 [hep-th].

[4] M.P. Heller, D. Mateos, W. van der Schee, M. Triana, Holographic isotropization linearized, JHEP 1309 (2013) 026, arXiv:1304.5172 [hep-th];
M.P. Heller, R.A. Janik, P. Witaszczyk, The characteristics of thermalization of boost-invariant plasma from holography, Phys. Rev. Lett. 108 (2012) 201602, arXiv:1103.3452 [hep-th].

[5] S.R. Das, Holographic quantum quench, J. Phys. Conf. Ser. 343 (2012) 012027, arXiv:1111.7275 [hep-th];
S.R. Das, T. Nishioka, T. Takayanagi, Probe branes, time-dependent couplings and thermalization in AdS/CFT, JHEP 1007 (2010) 071, arXiv:1005.3348 [hep-th];
A. Buchel, R.C. Myers, A. van Niekerk, Nonlocal probes of thermalization in holographic quenches with spectral methods, JHEP 1502 (2015) 017, arXiv:1410.6201 [hep-th].

[6] J. Erdmenger, N. Evans, I. Kirsch, E. Threlfall, Mesons in gauge/gravity duals—a review, Eur. Phys. J. A 35 (2008) 81, arXiv:0711.4467 [hep-th].

[7] A. Karch, E. Katz, Adding flavor to AdS/CFT, JHEP 0206 (2002) 043, http://dx.doi.org/10.1088/1126-6708/2002/06/043, arXiv:hep-th/0205236.

[8] S.R. Das, T. Nishioka, T. Takayanagi, Probe branes, time-dependent couplings and thermalization in AdS/CFT, JHEP 1007 (2010) 071, arXiv:1005.3348 [hep-th].

[9] K. Hashimoto, N. Iizuka, T. Oka, Rapid thermalization by baryon injection in gauge/gravity duality, Phys. Rev. D 84 (2011) 066005, arXiv:1012.4463 [hep-th];
M. Ali-Akbari, H. Ebrahim, JHEP 1303 (2013) 045, arXiv:1211.1637 [hep-th];
M. Ali-Akbari, H. Ebrahim, Meson thermalization in various dimensions, JHEP 1204 (2012) 145, arXiv:1203.3425 [hep-th].

[10] K. Hashimoto, T. Oka, A. Sonoda, Magnetic instability in AdS/CFT: Schwinger effect and Euler-Heisenberg Lagrangian of supersymmetric QCD, JHEP 1406 (2014) 085, arXiv:1403.6336 [hep-th].

[11] T. Ishii, S. Kinoshita, K. Murata, N. Tanahashi, Dynamical meson melting in holography, JHEP 1404 (2014) 099, arXiv:1401.5106 [hep-th].

[12] A. Karch, A. O'Bannon, Metallic AdS/CFT, JHEP 0709 (2007) 024, arXiv:0705.3870 [hep-th].

[13] S. Amiri-Sharifi, M. Ali-Akbari, A. Kishani-Farahani, N. Shafie, Double relaxation via AdS/CFT, Nucl. Phys. B 909 (2016) 778, http://dx.doi.org/10.1016/j.nuclphysb.2016.06.011, arXiv:1601.04281 [hep-th].

[14] M. Ali-Akbari, F. Charmchi, A. Davody, H. Ebrahim, L. Shahkarami, Evolution of Wilson loop in time-dependent $N = 4$ super Yang–Mills plasma, Phys. Rev. D 93 (8) (2016) 086005, arXiv:1510.00212 [hep-th].

[15] M. Ali-Akbari, F. Charmchi, H. Ebrahim, L. Shahkarami, Various time-scales of relaxation, Phys. Rev. D 94 (4) (2016) 046008, arXiv:1602.07903 [hep-th].

[16] M. Attems, J. Casalderrey-Solana, D. Mateos, I. Papadimitriou, D. Santos-Oliván, C.F. Sopuerta, M. Triana, M. Zilhão, Thermodynamics, transport and relaxation in non-conformal theories, JHEP 1610 (2016) 155, arXiv:1603.01254 [hep-th].

[17] M. Ali-Akbari, F. Charmchi, A. Davody, H. Ebrahim, L. Shahkarami, Time-dependent meson melting in an external magnetic field, Phys. Rev. D 91 (2015) 106008, arXiv:1503.04439 [hep-th].

[18] T. Ishii, K. Murata, Dynamical AdS strings across horizons, JHEP 1603 (2016) 035, arXiv:1512.08574 [hep-th].

[19] K. Hashimoto, S. Kinoshita, K. Murata, T. Oka, Electric field quench in AdS/CFT, JHEP 1409 (2014) 126, arXiv:1407.0798 [hep-th].

[20] C. Hoyos, T. Nishioka, A. O'Bannon, A chiral magnetic effect from AdS/CFT with flavor, JHEP 1110 (2011) 084, arXiv:1106.4030 [hep-th].

[21] A. Buchel, R.C. Myers, A. van Niekerk, Universality of abrupt holographic quenches, Phys. Rev. Lett. 111 (2013) 201602, arXiv:1307.4740 [hep-th];
S.R. Das, D.A. Galante, R.C. Myers, Quantum quenches in free field theory: universal scaling at any rate, JHEP 1605 (2016) 164, arXiv:1602.08547 [hep-th];
S.R. Das, D.A. Galante, R.C. Myers, Universality in fast quantum quenches, JHEP 1502 (2015) 167, arXiv:1411.7710 [hep-th].

[22] S. Amiri-Sharifi, H.R. Sepangi, M. Ali-Akbari, Electric field quench, equilibration and universal behavior, Phys. Rev. D 91 (2015) 126007, arXiv:1504.03559 [hep-th].

[23] Notice that the Chern–Simon action does not contribute for the case we consider, in this paper.

[24] It is important to notice that the final state depends on parameters of the injection. For a given value of m, for instance $m = 1$, depending on the values of E_f and T_f, the final state can be Minkowski or black hole embedding. Note that the final state is independent of the values of the transition times k_M and k_E. Since we would like to have a non-zero current on the probe brane, we have to choose the black hole embedding and it may happen in a range of numbers for final temperature and electric filed depending on the value of m. Moreover one can always choose $m = 1$ because of the following scaling symmetry of the equations of motion (5)

$$Z \to kZ,\, V \to kV,\, \Psi \to \Psi,\, /k, a_x \to ka_x,\, \vec{x} \to k\vec{x},\, M_f \to M_f/k^4,$$
$$k_{E,M} \to kk_{E,M}, m \to m/k, c \to c/k^3,\, j \to j/k,$$

where k is an arbitrary constant.

[25] We utilize finite difference method to solve equations of motion (5). This method has been firstly introduced in [11] to describe the dynamics of a probe D7-brane in the AdS-Vaidya background. We have applied a generalized version of the Mathematica code used in [11]. Since our Mathematica code with appropriate parameters, as a cross-check, reproduces the results of the mentioned papers, we trust our code and its numerical results.
It was discussed, for example in [22], that for slow (fast) quenches transition times $k_{E,M}$ can be considered in the range $3 \lesssim k_{E,M}$ $(0 < k_{E,M} \lesssim 0.2)$. We choose $k_{E,M} = 0.1$ and 5 for fast and slow quenches, respectively.

7

Holographic superconductivity from higher derivative theory

Jian-Pin Wu [a,*], Peng Liu [b]

[a] Institute of Gravitation and Cosmology, Department of Physics, School of Mathematics and Physics, Bohai University, Jinzhou 121013, China
[b] Department of Physics, Jinan University, Guangzhou 510632, China

ARTICLE INFO

Editor: M. Cvetič

ABSTRACT

We construct a 6 derivative holographic superconductor model in the 4-dimensional bulk spacetimes, in which the normal state describes a quantum critical (QC) phase. The phase diagram $(\gamma_1, \hat{T}_c)$ and the condensation as the function of temperature are worked out numerically. We observe that with the decrease of the coupling parameter γ_1, the critical temperature $\hat{T}_c$ decreases and the formation of charged scalar hair becomes harder. We also calculate the optical conductivity. An appealing characteristic is a wider extension of the superconducting energy gap, comparing with that of 4 derivative theory. It is expected that this phenomena can be observed in the real materials of high temperature superconductor. Also the Homes' law in our present models with 4 and 6 derivative corrections is explored. We find that in certain range of parameters γ and γ_1, the experimentally measured value of the universal constant C in Homes' law can be obtained.

1. Introduction

The AdS/CFT correspondence [1–4] provides a powerful tool to study the quantum critical (QC) dynamics, which are described by CFT and are strongly coupled systems without quasi-particles descriptions [5]. An interesting and important holographic QC dynamical system is that a probe Maxwell field coupled to the Weyl tensor $C_{\mu\nu\rho\sigma}$ in the Schwarzschild–AdS (SS–AdS) black brane background, which is a 4 derivative theory and has been fully studied in [6–15]. This system has zero charge density and can be understood as particle-hole symmetry. Of particular interest is the non-trivial optical conductivity due to the introduction of Weyl tensor, which is similar to the one in the superfluid-insulator quantum critical point (QCP) [6–8]. Further, higher derivative (HD) terms are introduced and the optical conductivity is studied in [15]. They find that an arbitrarily sharp Drude-like peak can be observed at low frequency in the optical conductivity and the bounds of conductivity found in [6,9] can be violated such that we have a zero DC conductivity at specific parameter. Especially, its behavior resembles quite closely that of the $O(N)$ $NL\sigma M$ in the large-N limit [16]. Therefore, the HD terms in SS–AdS black brane background provide possible route and alternative way to study the QC dynamics described by certain CFTs.

In this paper, we intend to study the holographic superconductor model in this holographic framework including HD terms. The holographic superconductor model [17–19] is an excellent example of the application of AdS/CFT in condensed matter theory (CMT), which provides valuable lessons to access the high temperature superconductor in CMT. In the original version of the holographic superconductor model [17–19], the superconducting energy gap is $\omega_g/T_c \approx 8$. This value is more than twice the one, which is 3.5, in the weakly coupled BCS theory, but roughly approximates that measured in high temperature superconductor materials [20]. Furthermore, by introducing 4 derivative term based on Weyl tensor, the holographic superconductor in the boundary theory dual to 5 dimensional AdS black brane is firstly constructed in [21]. This model exhibits an appealing characteristic of the extension of superconducting energy gap approximately varying from 6 to 10 [21]. Next, lots of Weyl holographic superconductor models, including p wave and different backgrounds, are constructed in [22–32]. In particular, the extension of superconducting energy gap is also observed in [22,31].[1] Here, we extend the analysis to include the 6 derivative term in the 4 dimensional bulk spacetime, in which the

* Corresponding author.
E-mail addresses: jianpinwu@mail.bnu.edu.cn (J.-P. Wu), phylp@jnu.edu.cn (P. Liu).

[1] In [23–30], they study the s or p wave superconducting condensation from 4 derivative term in 5 dimensional AdS geometry. But the computation of conductivity is absent. In [31], the Weyl holographic superconductor in the 4 dimensional Lifshitz black brane is explored and the extension of the superconducting energy gap is also observed. In addition, we would also like to point out that, the running of superconducting energy gap is also observed in other holographic superconductor model from higher derivative gravity, for example, the Gauss–Bonnet gravity [33,34]

normal state describes a QC phase and the electromagnetic (EM) self-duality is broken.

We also study the Homes' law over our model. Homes' law is an empirical law universally discovered in experiments of superconductors, which states that the product of the DC conductivity $\sigma_{DC}(T_c)$ and the critical temperature T_c has a linear relation to the superfluid density $\rho_S(T=0)$ at zero temperature. Holographic investigation of Homes' law can be found in [32,37–39]. Our results show that the constant of the Homes' law can be observed in certain range of parameter γ and γ_1 in our model, which can be extended by adding additional structures to study the universal realization of the Homes' law in holography.

2. Holographic framework

We shall construct a charged scalar hair black brane solution based SS–AdS black brane

$$ds^2 = \frac{L^2}{u^2}\Big(-f(u)dt^2 + dx^2 + dy^2\Big) + \frac{L^2}{u^2 f(u)}du^2\,,$$
$$f(u) = (1-u)p(u)\,, \qquad p(u) = u^2 + u + 1\,. \tag{1}$$

$u=0$ is the asymptotically AdS boundary while the horizon locates at $u=1$. The Hawking temperature of this system is $T = 3/4\pi L^2$. And then, we introduce the actions for gauge field A and charged complex scalar field Ψ

$$S_A = \int d^4x\sqrt{-g}\Big(-\frac{L^2}{8g_F^2}F_{\mu\nu}X^{\mu\nu\rho\sigma}F_{\rho\sigma}\Big)\,, \tag{2}$$

$$S_\Psi = \int d^4x\sqrt{-g}\Big(-|D_\mu\Psi|^2 - m^2|\Psi|^2\Big)\,. \tag{3}$$

In the action S_A, $F=dA$ is the curvature of gauge field A and the tensor X is an infinite family of HD terms as [15]

$$\begin{aligned} X_{\mu\nu}{}^{\rho\sigma} &= I_{\mu\nu}{}^{\rho\sigma} - 8\gamma_{1,1}L^2C_{\mu\nu}{}^{\rho\sigma} - 4L^4\gamma_{2,1}C^2I_{\mu\nu}{}^{\rho\sigma} \\ &-8L^4\gamma_{2,2}C_{\mu\nu}{}^{\alpha\beta}C_{\alpha\beta}{}^{\rho\sigma} - 4L^6\gamma_{3,1}C^3I_{\mu\nu}{}^{\rho\sigma} - 8L^6\gamma_{3,2}C^2C_{\mu\nu}{}^{\rho\sigma} \\ &-8L^6\gamma_{3,3}C_{\mu\nu}{}^{\alpha_1\beta_1}C_{\alpha_1\beta_1}{}^{\alpha_2\beta_2}C_{\alpha_2\beta_2}{}^{\rho\sigma} + \ldots\,, \end{aligned} \tag{4}$$

where $I_{\mu\nu}{}^{\rho\sigma} = \delta_\mu{}^\rho\delta_\nu{}^\sigma - \delta_\mu{}^\sigma\delta_\nu{}^\rho$ is an identity matrix and $C^n = C_{\mu\nu}{}^{\alpha_1\beta_1}C_{\alpha_1\beta_1}{}^{\alpha_2\beta_2}\ldots C_{\alpha_{n-1}\beta_{n-1}}{}^{\mu\nu}$. In the above equations (2) and (4), we have introduced the factor of L so that the coupling parameters g_F and $\gamma_{i,j}$ are dimensionless. But for later convenience, we shall work in units where $L=1$ in what follows. Notice that we shall set $g_F=1$ in the following numerical calculation. When $X_{\mu\nu}{}^{\rho\sigma} = I_{\mu\nu}{}^{\rho\sigma}$, the action S_A reduces to the standard version of Maxwell theory. For convenience, we denote $\gamma_{1,1}=\gamma$ and $\gamma_{2,i}=\gamma_i (i=1,2)$. In this paper, we mainly focus on the 6 derivative terms, i.e., γ_1 and γ_2 terms. But since the effect of γ_1 and γ_2 terms is similar, we only turn on γ_1 term through this paper. Note that when other parameters are turned off, γ_1 is constrained in the region $\gamma_1 \le 1/48$ in SS–AdS black brane background [15]. The upper bound of γ_1 is because of the requirement that the DC conductivity in the boundary theory is positive [15].

The action S_ψ supports a superconducting black brane [17]. Ψ is the charged complex scalar field with mass m and the charge q of the Maxwell field A, which can be written as $\Psi = \psi e^{i\theta}$ with ψ being a real scalar field and θ a Stückelberg field. $D_\mu = \partial_\mu - iqA_\mu$ is the covariant derivative. For convenience, we choose the gauge $\theta=0$ and then, the EOMs of gauge field and scalar field can be derived as

$$\nabla_\nu\big(X^{\mu\nu\rho\sigma}F_{\rho\sigma}\big) - 4q^2A^\mu\psi^2 = 0\,, \qquad \big[\nabla^2 - (m^2 + q^2A^2)\big]\psi = 0\,. \tag{5}$$

3. Condensation

In this section, we numerically construct a charged scalar hair black brane solution with 6 derivative term[2] and study its superconducting phase transition. The ansatz for the scalar field and gauge field is taken as

$$\psi = \psi(u)\,, \quad A = \mu(1-u)a(u)dt\,, \tag{6}$$

where μ is the chemical potential of the dual field theory. The background EOMs can be explicitly written as

$$\psi'' + \Big(\frac{f'}{f} - \frac{2}{u}\Big)\psi' - \frac{m^2}{u^2 f}\psi - q^2\frac{(u-1)^2\mu^2}{f^2}a^2\psi = 0\,, \tag{7}$$

$$a'' + \Big(\frac{X_3'}{X_3} + \frac{2}{u-1}\Big)a' + \Big(\frac{X_3'}{(u-1)X_3} - \frac{2q^2\psi^2}{u^2 f X_3}\Big)a = 0\,, \tag{8}$$

where the prime represents the derivative with respect to u and for convenience, the tensor $X_{\mu\nu}{}^{\rho\sigma}$ is denoted as $X_A{}^B = \{X_1(u), X_2(u), X_3(u), X_4(u), X_5(u), X_6(u)\}$ with $A, B \in \{tx, ty, tu, xy, xu, yu\}$. This system is characterized by the dimensionless quantities $\hat{T} \equiv T/\mu$. Without loss of generality, we set $m^2 = -2$ in what follows. It is easy to see that the asymptotical behavior of ψ at the boundary $u=0$ is

$$\psi = u\psi_1 + u^2\psi_2\,. \tag{9}$$

Here, we treat ψ_1 as the source and ψ_2 as the expectation value. And then, we set $\psi_1 = 0$ such that the condensate is not sourced.

We firstly work out the phase diagram of the critical temperature $\hat{T}_c$ for the formation of the superconducting phase as the function of the coupling parameter γ_1. It is convenient to estimate $\hat{T}_c$ by finding static normalizable mode of charged scalar field on the fixed background. This method has been described detailedly in [40,41]. Our result for the phase diagram $(\gamma_1, \hat{T}_c)$ is showed in the left plot in Fig. 1. The red points in this figure denote $\hat{T}_c$ for the representative γ_1, which is obtained by fully solving the coupled EOMs (7) as well as (8) and listed in Table 1. We see that the $\hat{T}_c$ by finding static normalizable mode is in agreement with that shown in Table 1. From the phase diagram $(\gamma_1, \hat{T}_c)$, we find that with the decrease of γ_1, $\hat{T}_c$ decreases. In particular, for the small γ_1, $\hat{T}_c$ seems to approach a fixed value. Therefore, it seems reasonable to infer that the 6 derivative term doesn't spoil the formation of charged scalar hair even if the absolute value of γ_1 ($\gamma_1 < 0$) is large enough. But note that the HD terms shall be introduced as a perturbative effect, so we shall restrict γ_1 in a small region, $|\gamma_1| \ll 1$. However, in order to see a more obvious effect from 6 derivative term, we also relax the restriction of γ_1 and study the effect of $\gamma_1 = -1$, in which the normal state has a sharp Drude peak and has been studied in [15].

Now, we solve the coupling EOMs (7) and (8) numerically and study the condensation behavior. The result of the condensation $\sqrt{\langle O_2\rangle}/T_c$ as a function of the temperature T/T_c is shown in the right plot in Fig. 1. We observe that with the decrease of γ_1, the

and the quasi-topological gravity [35,36]. But the value of ω_g/T_c is always greater than 8.

[2] As far as we know, the studies on 4 derivative holographic superconductor in the existing literatures except [31] are focus on the 5 dimensional bulk spacetime. Though the qualitative properties are expected to be similar between the 4 and 5 dimensional holographic superconductor with 4 derivative term, we also present the main properties in Appendix A so that the paper is self-contained.

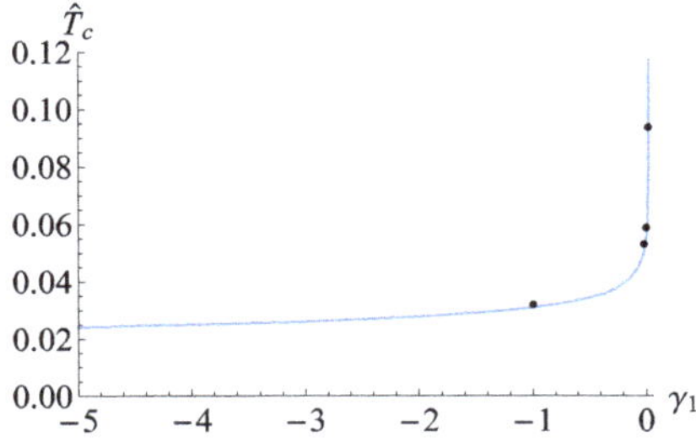

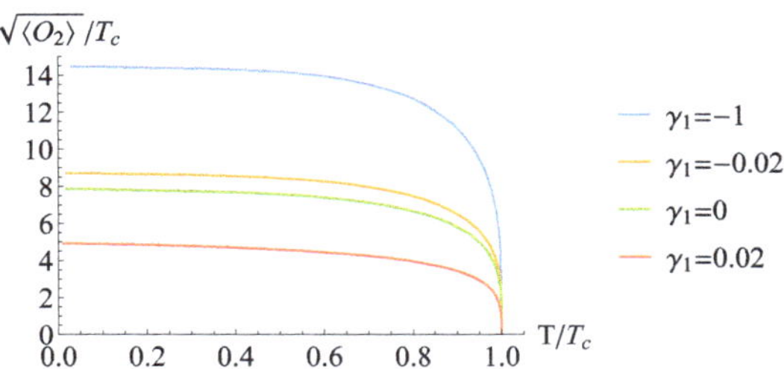

Fig. 1. Left plot: Phase diagram $(\gamma_1, \hat{T}_c)$ by finding static normalizable mode of charged scalar field on the fixed background. The red points denote $\hat{T}_c$ for the representative γ_1, which is obtained by fully solving the coupled EOMs (7) and (8) numerically and listed in Table 1. Right plot: The condensation as a function of temperature for representative γ_1. (For interpretation of the references to color in this figure, the reader is referred to the web version of this article.)

Table 1
The critical temperature $\hat{T}_c$ with different γ_1.

γ_1	−1	−0.02	0	0.02
$\hat{T}_c$	0.0319	0.0530	0.0587	0.0936

condensation becomes much larger. It implies a larger superconducting energy gap ω_g/T_c, which is also observed in the following study of optical conductivity. We also present the critical temperature $\hat{T}_c$ of the condensation phase transition with different γ_1 in Table 1. We see that when the γ_1 decreases, $\hat{T}_c$ goes down. The tendency is consistent with that shown in the left plot in Fig. 1.

4. Superconductivity

Given the charged scalar hair black brane solution from 6 derivative term, we can study the optical conductivity. To this end, we turn on the perturbations of the gauge field along x direction as $\delta a_x(t,u) \sim \delta a_x(u)e^{-i\omega t}$. And then the perturbative equation can be derived as

$$\delta a_x'' + \Big(\frac{X_5'}{X_5} + \frac{f'}{f}\Big)\delta a_x' + \Big(\frac{\omega^2}{f^2}\frac{X_1}{X_5} - \frac{2q^2\psi^2}{u^2 f X_5}\Big)\delta a_x = 0\,. \tag{10}$$

With the ingoing boundary condition at horizon, we numerically solve the above perturbative equation and read off the conductivity in terms of

$$\sigma(\omega) = \frac{\partial_u \delta a_x}{i\omega\delta a_x}\Big|_{u=0}\,. \tag{11}$$

The normal state from 6 derivative term has been studied in [15]. For $\gamma_1 = -1$, a sharp Drude peak emerges at small frequency in the conductivity. At mediate frequency, the optical conductivity exhibits a gap and then at large frequency, it saturates to the value $\sigma = 1$. While for $\gamma_1 = 1/48$, the DC conductivity is zero and the low frequency conductivity displays a hard-gap-like behavior. In addition, a pronounced peak emerges at mediate frequency. As pointed out in [15], these results are similar with that of the large-N $O(N)$ $NL\sigma M$ model and deserve further exploration [16].

Now, we study the properties of the conductivity when the charged scalar hair is formed. First, as the standard version of holographic superconductor model [17], a pole is observed at $\omega = 0$ in the imaginary part of the conductivity (right plot in Fig. 2), which corresponds to a delta function at $\omega = 0$ according to the Kramers–Kronig (KK) relation. It indicates that the superconductivity emerges. Subsequently, we mainly focus on two cases: one is for $\gamma_1 = -1$ and another for $\gamma_1 = 0.02$. For $\gamma_1 = -1$, we notice that after the superconducting phase has formed, the DC conductivity is still finite near the critical temperature (Fig. 2). It contributes to the normal component to the conductivity and means that our holographic model with $\gamma_1 = -1$ is a two-fluid model. With the decrease of the temperature, the DC conductivity goes down and finally vanishes at low temperature, which means that normal component of the electron fluid is decreasing to form the superfluid component. In addition, the soft gap at low frequency in normal state gradually becomes a superconducting gap with the decrease of the temperature. While for $\gamma_1 = 0.02$, we find that with the decrease of the temperature, the pronounced peak at the mediate frequency gradually decreases and a supercondcting energy gap comes into being.

One important property of our present model is the extension of the superconducting energy gap, which can be obtained by locating the minimal value of the imaginary part of the conductivity (see Fig. 2). The quantitative results are shown in Table 2.[3] We notice that comparing with the 4 derivative model (see [21] and also Appendix A) there is a wider extension of the energy gap, which ranges from 5.5 to 16.2 when γ_1 changes from 0.02 to −1. We expect that the extension of the energy gap in our holographic model has a corresponding partner in high temperature superconductor and we pursuit it in future.

5. Homes' law

In previous sections and the appendix we investigate the superconductivity in presence of 4 derivative and 6 derivative terms. Next we intend to study the Homes' law [42,43], a universal relation observed in laboratory, in our model. For a large class of superconductivity materials there exists an empirical law, i.e., Homes' law, regardless of the details of the materials,

$$\rho_s(\hat{T}=0) = C\sigma_{DC}(\hat{T}_c)\hat{T}_c\,, \tag{12}$$

where $\sigma_{DC}(\hat{T}_c)$ is the DC conductivity at a temperature slightly above the critical temperature $\hat{T}_c$, $\rho_s(\hat{T}=0)$ the density of the condensation and C a constant.

The universal constant $C \simeq 4.4$ for ab-plane high temperature superconductors and clean BCS superconductors, while for c-axis high temperature materials and BCS superconductors in dirty limit $C \simeq 8.1$. The most recent results show that for organic superconductors the $C = 4 \pm 2.1$ [44].[4] Theoretically, Zaanen proposed an explanation for the Homes' law of high temperature superconductors in terms of the Planckian dissipation [45]. The superfluid density $\rho_s = 4\pi n_S e^2/m_e$, which has dimension $[t]^{-2}$. Meanwhile, DC conductivity $\sigma_{DC} = 4\pi n_N e^2\tau/m_e$, which has dimension $[t]^{-1}$ and τ is a timescale related to the dissipation. To balance the Homes relation in dimension, the critical temperature T_c has to be converted to the inverse of a timescale. This timescale is naturally

[3] We adopted the convention that the unit of the system is the chemical potential here as [41] but not the charge density as most of the present literatures. So the superconducting energy gap is $\omega_g/T_c \simeq 9$ but not $\omega_g/\tilde{T}_c \simeq 8$.

[4] Note that all the data of C in here has been converted from the original experimental data into our holographic framework.

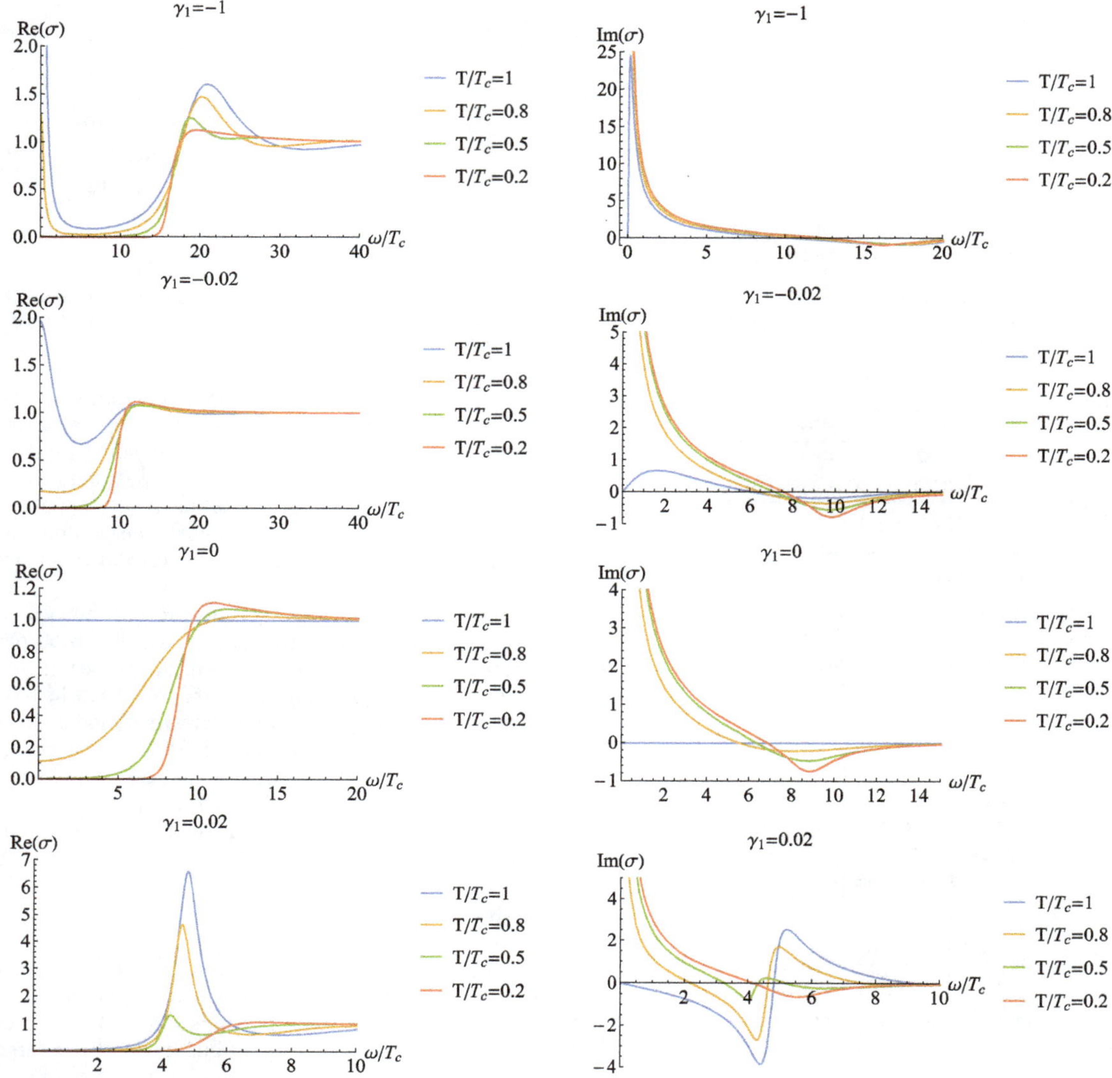

Fig. 2. Real and imaginary parts of conductivity as a function of the frequency with different γ_1. (For interpretation of the references to color in this figure, the reader is referred to the web version of this article.)

Table 2
The superconducting energy gap ω_g/T_c with different γ_1.

γ_1	−1	−0.02	0	0.02
ω_g/T_c	16.253	9.886	8.880	5.529

deduced from the temperature through the energy-time uncertainty relation $\tau \sim \hbar/k_B T$. Therefore the Homes law is equivalent to that the time scale of the dissipation is as short as permitted by quantum physics, namely the Planckian dissipation [45].

It is desirable to test if the universal Homes' law would also emerge in holographic method. Earlier literatures concerning Homes' relation in holographic framework include [32,37–39]. In [38,39] the holographic models with helical structures and Q-lattices structures have been studied respectively, and Homes' relation has been found solid in narrow range of parameters. In [32] the axion model with Weyl correction have been studied, and it is found that a generalized Homes' relation holds in a relatively large range of parameters,

$$\rho_s(\hat{T}=0) = C\sigma_{DC}(\hat{T}_c)\hat{T}_c + D\,, \tag{13}$$

where D is a constant. Next we investigate Homes' relation in 4 derivative and 6 derivative theories respectively.

The required ingredient to study Homes' law is $\rho_S(\hat{T}=0)$, $\hat{T}_c$ and $\sigma_{DC}(\hat{T}_c)$. The computation of T_c and $\sigma_{DC}(\hat{T}_c)$ is straightforward, while the superfluid density $\rho_S(\hat{T}=0)$ is identified as the coefficient of the pole in the imaginary part of the optical conductivity,

$$\mathrm{Im}\,\sigma(\omega) = \frac{2\rho_S}{\pi\omega}. \tag{14}$$

The constant C vs. γ (γ_1) is shown in Fig. 3. It is interesting to notice that in both 4 derivative and 6 derivative theories, there exists a range of γ (γ_1) where C falls between the experimental results of Homes' relation. We expect the Homes' relation could be realized on these models when other structures, such as axions, lattices and dilatons are turned on.

6. Conclusion and discussion

In this paper, we construct a 6 derivative holographic superconductor model in the 4-dimensional bulk spacetimes, in which the

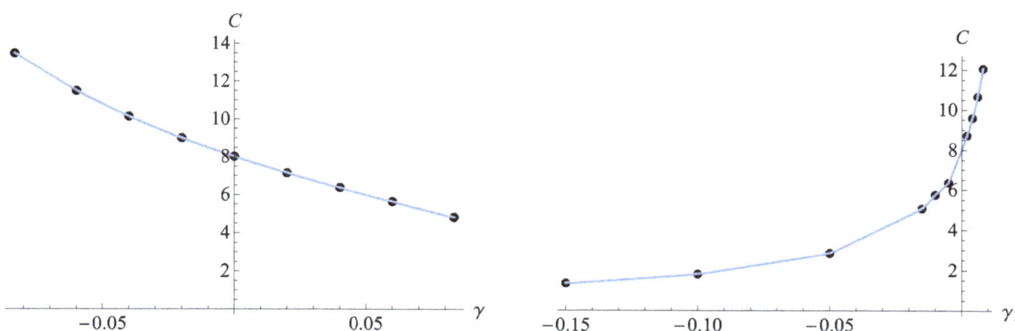

Fig. 3. Left: Homes' coefficient C vs. γ in range $\gamma \in [-1/12, 1/12]$ for 4 derivative correction. Right: Homes' coefficient C vs. γ_1 in range $\gamma_1 \in [-0.15, 0.008]$ for 6 derivative correction.

normal state is a QC phase. We find that with the decrease of the coupling parameter γ_1, the critical temperature $\hat{T}_c$ decreases and the formation of charged scalar hair becomes harder. But in terms of the present data we have obtained and the varying tendency of $\hat{T}_c$ with γ_1, the 6 derivative term doesn't seem to spoil the formation of the charged scalar hair and superconductivity. And then, we study the optical conductivity. One appealing property of our present model is the extension of the superconducting energy gap. Comparing with that of 4 derivative one, a wider extension of the energy gap, which ranges from 5.5 to 16.2 when γ_1 changes from 0.02 to -1, can be obtained. We expect that this phenomena can be observed in the real materials of high temperature superconductor and we shall further explore it in future. We also study the empirical and universal Homes' law in models with HD corrections. The experimental results of Homes' law can be satisfied in certain range of parameters both in 4 derivative and 6 derivative theories.

There are some interesting topics worthy of further investigation in future.

- In our previous works [46–48], the spatial linear dependent axions are incorporated into HD theory to study the optical conductivity [46,47] and to implement metal-insulator transition [48]. We can incorporate more interesting structures, including axions [49], Q-lattice [50], helical lattice [51], and the periodic lattice structure [52,53], into the HD theory to study the superconductivity, in particular the running of the energy gap, and the Homes' law in holography.
- We can construct the EM dual theory of the present model with HD terms and study its properties. Note that by introducing axion and dilaton fields, the EM duality for holographic p-wave superconductor has been explored in [54].
- It is valuable to study the transports and the quasi-normal modes in the superconducting state with HD terms at full momentum and energy spaces, which can provide far deeper insights into this holographic system than that at the zero momentum [11,55].

We shall address these topics in future.

Acknowledgements

This work is supported by the Natural Science Foundation of China under Grant Nos. 11775036, 11305018, and by the Natural Science Foundation of Liaoning Province under Grant No. 201602013.

Appendix A. Holographic superconductivity from 4 derivatives

In this Appendix, we present the main results of the holographic superconductivity from 4 derivative term in 4 dimensional AdS geometry and give some comments. Fig. 4 exhibits the condensation $\sqrt{\langle O_2 \rangle}/T_c$ as a function of temperature for various values of γ. We see that with the decrease of γ, the condensation is formed harder, which is consistent with the case of 5 dimensional background [21].

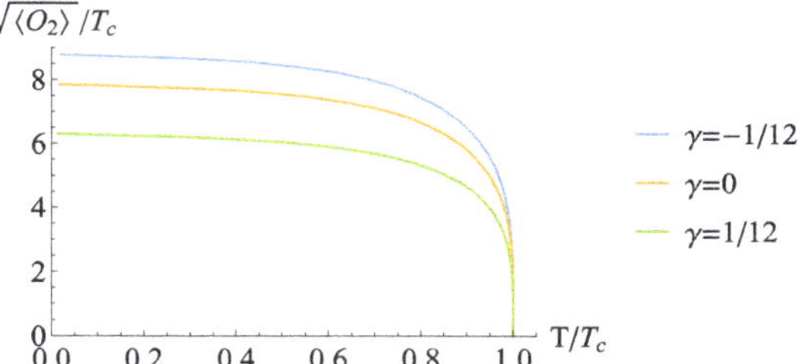

Fig. 4. The condensation from 4 derivative theory as a function of temperature for different γ. (For interpretation of the references to color in this figure, the reader is referred to the web version of this article.)

Table 3
The energy gap ω_g/T_c with different γ.

γ	$-1/12$	0	$1/12$
ω_g/T_c	9.886	8.880	7.205

Fig. 5 shows the real and imaginary parts of the conductivity as a function frequency for $\gamma = \pm 1/12$, respectively. The main properties are presented as follows.

- *Superconductivity.* The imaginary part of the conductivity exhibits a pole at $\omega = 0$ (right plot in Fig. 5) when the temperature is below the critical temperature. It means that there is a delta function in the real part of the conductivity at $\omega = 0$ and the superconductivity emerges.
- *DC conductivity due to normal fluid.* The DC conductivity after the superconducting phase transition happens is still finite (left plot in Fig. 5), which implies our model with 4 derivative correction is a two-fluid model as the original version of holographic superconductor model [17–19]. But with the decrease of the temperature, the DC conductivity decreases and disappears at extremal low temperature, which implies that the normal component of the electron fluid is decreasing to form the superfluid component and disappears at extremal low temperature.
- *Extension of energy gap.* From Fig. 5, we see that there is a running of the energy gap when we tune the coupling parameter γ, which has been observed in the holographic superconductor model with 4 derivative correction in 5 dimensional AdS background [21]. The quantitative results are listed in Table 3.

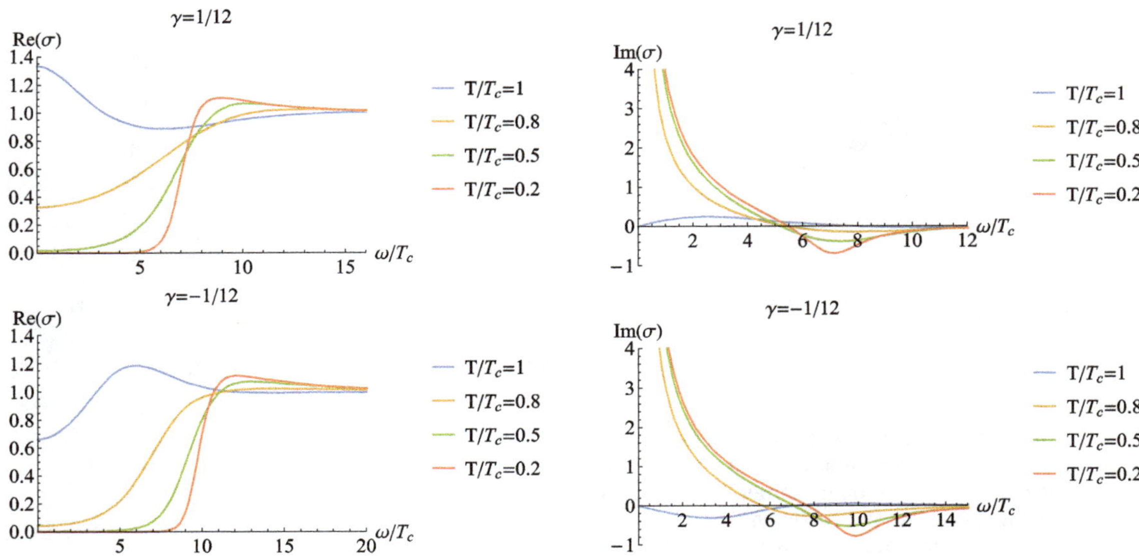

Fig. 5. Real and imaginary parts of conductivity as a function of the frequency for different γ. (For interpretation of the references to color in this figure, the reader is referred to the web version of this article.)

References

[1] J.M. Maldacena, Int. J. Theor. Phys. 38 (1999) 1113, Adv. Theor. Math. Phys. 2 (1998) 231.
[2] S.S. Gubser, I.R. Klebanov, A.M. Polyakov, Phys. Lett. B 428 (1998) 105.
[3] E. Witten, Adv. Theor. Math. Phys. 2 (1998) 253.
[4] O. Aharony, S.S. Gubser, J.M. Maldacena, H. Ooguri, Y. Oz, Phys. Rep. 323 (2000) 183.
[5] S. Sachdev, Quantum Phase Transitions, 2nd edition, Cambridge University Press, England, 2011.
[6] R.C. Myers, S. Sachdev, A. Singh, Phys. Rev. D 83 (2011) 066017.
[7] S. Sachdev, Annu. Rev. Condens. Matter Phys. 3 (2012) 9.
[8] S.A. Hartnoll, A. Lucas, S. Sachdev, arXiv:1612.07324 [hep-th].
[9] A. Ritz, J. Ward, Phys. Rev. D 79 (2009) 066003.
[10] W. Witczak-Krempa, S. Sachdev, Phys. Rev. B 86 (2012) 235115.
[11] W. Witczak-Krempa, S. Sachdev, Phys. Rev. B 87 (2013) 155149.
[12] W. Witczak-Krempa, E.S. Sørensen, S. Sachdev, Nat. Phys. 10 (2014) 361.
[13] E. Katz, S. Sachdev, E.S. Sørensen, W. Witczak-Krempa, Phys. Rev. B 90 (24) (2014) 245109.
[14] S. Bai, D.W. Pang, Int. J. Mod. Phys. A 29 (2014) 1450061.
[15] W. Witczak-Krempa, Phys. Rev. B 89 (16) (2014) 161114.
[16] K. Damle, S. Sachdev, Phys. Rev. B 56 (14) (1997) 8714.
[17] S.A. Hartnoll, C.P. Herzog, G.T. Horowitz, Phys. Rev. Lett. 101 (2008) 031601.
[18] S.A. Hartnoll, C.P. Herzog, G.T. Horowitz, J. High Energy Phys. 0812 (2008) 015.
[19] G.T. Horowitz, Lect. Notes Phys. 828 (2011) 313.
[20] K.K. Gomes, A.N. Pasupathy, A. Pushp, S. Ono, Y. Ando, A. Yazdani, Nature 447 (2007) 569.
[21] J.P. Wu, Y. Cao, X.M. Kuang, W.J. Li, Phys. Lett. B 697 (2011) 153.
[22] D.Z. Ma, Y. Cao, J.P. Wu, Phys. Lett. B 704 (2011) 604.
[23] D. Momeni, M.R. Setare, Mod. Phys. Lett. A 26 (2011) 2889.
[24] D. Momeni, N. Majd, R. Myrzakulov, Europhys. Lett. 97 (2012) 61001.
[25] D. Momeni, M.R. Setare, R. Myrzakulov, Int. J. Mod. Phys. A 27 (2012) 1250128.
[26] D. Roychowdhury, Phys. Rev. D 86 (2012) 106009.
[27] Z. Zhao, Q. Pan, J. Jing, Phys. Lett. B 719 (2013) 440.
[28] D. Momeni, R. Myrzakulov, M. Raza, Int. J. Mod. Phys. A 28 (2013) 1350096.
[29] D. Momeni, M. Raza, R. Myrzakulov, Int. J. Geom. Methods Mod. Phys. 13 (2016) 1550131.
[30] L. Zhang, Q. Pan, J. Jing, Phys. Lett. B 743 (2015) 104.
[31] S.A.H. Mansoori, B. Mirza, A. Mokhtari, F.L. Dezaki, Z. Sherkatghanad, J. High Energy Phys. 1607 (2016) 111.
[32] Y. Ling, X. Zheng, Nucl. Phys. B 917 (2017) 1.
[33] R. Gregory, S. Kanno, J. Soda, J. High Energy Phys. 0910 (2009) 010.
[34] Q. Pan, B. Wang, E. Papantonopoulos, J. Oliveira, A.B. Pavan, Phys. Rev. D 81 (2010) 106007.
[35] X.M. Kuang, W.J. Li, Y. Ling, J. High Energy Phys. 1012 (2010) 069.
[36] X.M. Kuang, W.J. Li, Y. Ling, Class. Quantum Gravity 29 (2012) 085015.
[37] J. Erdmenger, P. Kerner, S. Muller, J. High Energy Phys. 1210 (2012) 021.
[38] J. Erdmenger, B. Herwerth, S. Klug, R. Meyer, K. Schalm, J. High Energy Phys. 1505 (2015) 094.
[39] K.Y. Kim, C. Niu, J. High Energy Phys. 1610 (2016) 144.
[40] G.T. Horowitz, J.E. Santos, J. High Energy Phys. 1306 (2013) 087.
[41] Y. Ling, P. Liu, C. Niu, J.P. Wu, Z.Y. Xian, J. High Energy Phys. 1502 (2015) 059.
[42] C.C. Homes, S.V. Dordevic, M. Strongin, D.A. Bonn, R. Liang, W.N. Hardy, S. Komiya, Y. Ando, G. Yu, N. Kaneko, et al., Nature 430 (Jul. 2004) 539–541.
[43] C.C. Homes, S.V. Dordevic, T. Valla, M. Strongin, Phys. Rev. B 72 (Oct. 2005).
[44] S.V. Dordevic, D.N. Basov, C.C. Homes, Sci. Rep. 3 (Apr. 2013).
[45] J. Zaanen, Superconductivity: why the temperature is high, Nature 430 (Jul. 2004) 512–513.
[46] J.P. Wu, arXiv:1609.04729 [hep-th].
[47] G. Fu, J.P. Wu, B. Xu, J. Liu, Phys. Lett. B 769 (2017) 569.
[48] Y. Ling, P. Liu, J.P. Wu, Z. Zhou, Phys. Lett. B 766 (2017) 41.
[49] T. Andrade, B. Withers, J. High Energy Phys. 1405 (2014) 101.
[50] A. Donos, J.P. Gauntlett, J. High Energy Phys. 1404 (2014) 040.
[51] A. Donos, S.A. Hartnoll, Nat. Phys. 9 (2013) 649.
[52] G.T. Horowitz, J.E. Santos, D. Tong, J. High Energy Phys. 1207 (2012) 168.
[53] Y. Ling, C. Niu, J.P. Wu, Z.Y. Xian, J. High Energy Phys. 1311 (2013) 006.
[54] A. Gorsky, E. Gubankova, R. Meyer, A. Zayakin, arXiv:1706.05770 [hep-th].
[55] I. Amado, M. Kaminski, K. Landsteiner, J. High Energy Phys. 0905 (2009) 021.

8

Imprint of quantum gravity in the dimension and fabric of spacetime

Giovanni Amelino-Camelia [a,b], Gianluca Calcagni [c,*], Michele Ronco [a,b]

[a] *Dipartimento di Fisica, Università di Roma "La Sapienza", P.le A. Moro 2, 00185 Roma, Italy*
[b] *INFN, Sez. Roma1, P.le A. Moro 2, 00185 Roma, Italy*
[c] *Instituto de Estructura de la Materia, CSIC, Serrano 121, 28006 Madrid, Spain*

ARTICLE INFO

Editor: M. Cvetič

ABSTRACT

We here conjecture that two much-studied aspects of quantum gravity, *dimensional flow* and *spacetime fuzziness*, might be deeply connected. We illustrate the mechanism, providing first evidence in support of our conjecture, by working within the framework of multifractional theories, whose key assumption is an anomalous scaling of the spacetime dimension in the ultraviolet and a slow change of the dimension in the infrared. This sole ingredient is enough to produce a scale-dependent deformation of the integration measure with also a fuzzy spacetime structure. We also compare the multifractional correction to lengths with the types of Planckian uncertainty for distance and time measurements that was reported in studies combining quantum mechanics and general relativity heuristically. This allows us to fix two free parameters of the theory and leads, in one of the scenarios we contemplate, to a value of the ultraviolet dimension which had already found support in other quantum-gravity analyses. We also formalize a picture such that fuzziness originates from a fundamental discrete scale invariance at short scales and corresponds to a stochastic spacetime geometry.

1. Introduction and main goal

The landscape of quantum gravity (QG) looks like a variegated compound of approaches that start from different conceptual premises and use different mathematical formalisms (see, e.g., Refs. [1–21]). Rather surprisingly, despite this heterogeneity, over the past few years a generic prediction has emerged: *dimensional flow* [22–39], i.e., a change of spacetime dimension with the scale of the observer. In all QG models, the dimensionality of spacetime exhibits a dependence on the scale, changing (or "flowing") from the topological dimension D in the infrared (IR) to a different value in the ultraviolet (UV). So far, there has been no deep explanation for this universal property. Understanding its origin is just as important as looking for its physical characterization, needed to relate the flow of dimensions to physical observables.

We here put forward and motivate the conjecture that dimensional flow is directly related to the presence of limitations on the measurability of distances close to the Planck length $\ell_{\rm Pl} = \sqrt{G\hbar/c^3}$, a feature (*spacetime fuzziness*) which has been of interest for QG research for decades [40–47]. More precisely, we shall provide preliminary "theoretical evidence" in support of a connection between the number of spacetime dimensions in the UV and the form of the uncertainty on spacetime distances. Important from our perspective is the fact that such a connection might set the stage for a role for dimensional flow in QG phenomenology [48]. Indeed, it has been shown that, in some cases, spacetime fuzziness could be investigated in ongoing and forthcoming experiments, even if the fuzziness is introduced at the Planck scale. This was first explored in analyses of the interferometers used for gravity-wave searches [48–50], and more recently is focusing mainly on the implications of fuzziness for the formation of halo structures in the images of distant quasars [48,51].

2. Example: multifractional theories

We provide preliminary support for our conjecture within the context of multifractional theories [25,52] fully reviewed in [53]. These are a class of field theories of matter and gravity where spacetime is "anomalous" and changes properties with the probed scale, in a way similar to a multifractal. While in other quantum gravities dimensional flow is a derived property not required *a priori*, here it is part of the definition of the framework. Thanks to their peculiar properties, these field theories living on a multifractal spacetime reproduce a wealth of phenomena found in QG. In

* Corresponding author.
E-mail addresses: giovanni.amelino-camelia@roma1.infn.it (G. Amelino-Camelia), calcagni@iem.cfmac.csic.es (G. Calcagni), michele.ronco@roma1.infn.it (M. Ronco).

particular, the running of dimensions is produced by an integration measure of the type $d^D q(x) := dq^0(x^0) dq^1(x^1) \cdots dq^{D-1}(x^{D-1}) = \partial_0 q^0 dx^0 \partial_1 q^1 dx^1 \cdots \partial_{D-1} q^{D-1} dx^{D-1}$. The factorizable form is assumed for technical reasons [53] not especially important here, while the specific form of the distributions $q^\mu(x^\mu)$ is obtained by requiring that dimensional flow is slow at large scales. This assumption (spacetime dimension almost constant in the IR), true in all quantum gravities without known exception, is at the core of a result we will invoke often later, the second flow-equation theorem [52] (a "first" version holds for nonfactorizable measures). An approximation of the full measure, which is physically nonrestrictive but will be refined later, is the binomial space-isotropic profile

$$q^\mu(x^\mu) \simeq (x^\mu - \bar{x}^\mu) + \frac{\ell_*}{\alpha_\mu} \left| \frac{x^\mu - \bar{x}^\mu}{\ell_*} \right|^{\alpha_\mu}, \tag{1}$$

where the index μ is not summed over and takes values $0, 1, 2, \ldots, D-1$. For simplicity, we assume $\alpha_\mu = \delta_{0\mu}\alpha_0 + (1 - \delta_{0\mu})\alpha$, i.e., the exponents $\alpha_{\mu \neq 0}$ associated with spatial directions have all the same value α; moreover, we also enforce $0 < \alpha_0, \alpha < 1$, to avoid negative dimensions and obtain the correct IR limit [53]. Note that (1) is uniquely determined parametrically as soon as dimensional flow is switched on and is slow (almost constant spacetime dimension) in the IR [52]. This means that different models of quantum gravity can predict different values of the parameters α_μ and ℓ_* (plus other parameters that appear in the full expression at mesoscopic scales [52]), but the general form of the measure as a parametric profile are the same and given by (1). The only ambiguity left undecided by the second flow-equation theorem is a shift in the coordinates, represented by the given point $\bar{x}^\mu$. This shift ambiguity is a puzzling aspect from the viewpoint of interpretation, since it is a sort of preferred point in the universe. However, our results will neutralize this feature and embed it into a more amenable physical interpretation. We will comment on this shortly.

Depending on the symmetries of the Lagrangian, there are four possible multifractional theories, classified according to the derivative operators appearing in kinetic terms. Here we will concentrate on two theories with the same asymptotic expression for lengths, with so-called q- and fractional derivatives. For the purposes of this paper, suffice it to say that q-derivatives are defined as $\partial_{q^\mu} = (dq^\mu/dx^\mu)^{-1}\partial_\mu$. Details on fractional derivatives are discussed in [53].

To get the Hausdorff dimensions d_H of spacetime, one computes the volume $\mathcal{V}$ of a D-cube with size edge ℓ, leading to the result that, if $\alpha_0 = \alpha$ (as fixed by the arguments below), then $\mathcal{V} = \int_\mathrm{cube} d^D q(x) \simeq \ell_*^D[(\ell/\ell_*)^D + (\ell/\ell_*)^{D\alpha}]$. Thus, we have $d_\mathrm{H} \simeq D\alpha$ in the UV ($\ell < \ell_*$). Here we have neglected mesoscopic contributions to $\mathcal{V}$, which are not relevant to get the number of dimensions in the far UV [54]. For the two multifractional theories considered here, it is not difficult to prove that, in the UV, the spectral dimension (the scaling of the return probability $\mathcal{P} \sim \ell^{-d_\mathrm{S}}$ measuring how likely it is to find a test particle in a neighborhood of its actual position when probing spacetime with an apparatus with resolution $1/\ell$) coincides with the Hausdorff dimension, $d_\mathrm{S} \simeq D\alpha \simeq d_\mathrm{H}$, for $\alpha_0 = \alpha$ [53]. Both α and ℓ_* are free parameters of the theory with the only requirement that ℓ_* must be small enough to comply with experimental constraints [53]. As said above, the measure $q^\mu(x^\mu)$ is fixed by the second flow-equation theorem [52], but there remains an ambiguity related to the choice of a preferred frame, which amounts to the choice of $\bar{x}^\mu$ in Eq. (1). In fact, physical observables have to be compared in the picture with x^μ coordinates representing clocks and rods that do not adapt to the scale. This poses the so-called presentation problem [29,53], which consists in the choice of the physical frame where Eq. (1) is defined and observables are calculated.

3. Connecting dimensional flow and fuzziness: first glimpse

As announced, we shall use multifractional theories as a testing ground for our conjecture. We shall seek a connection between dimensional flow in multifractional theories and the limitations on the measurability of spacetime distances obtained by many authors heuristically combining aspects of quantum mechanics (QM) and general relativity (GR) [41,44–47]. It is noteworthy that the presence of these distance-measurement uncertainties, though originally discussed exclusively with heuristic reasoning, has found confirmation in concrete QG theories in recent years (see, e.g., Refs. [1, 2]), each of which realizes the corresponding UV features in very different ways [22,23,25,53]. The observations we here report can also be viewed as an explanation of why one gets a correct intuition about distance fuzziness even just resorting to the qualitative interplay of QM and GR. The link is provided by the fact that limitations on geometric measurements are intimately related to dimensional flow. As a byproduct of our analysis, we will also give a physical interpretation for the ambiguities of multifractional theories and select two sets of preferred values for α and ℓ_*. Remarkably, in one of these cases, we obtain $\alpha = 1/2$ and, consequently, $d_\mathrm{H} \simeq d_\mathrm{S} \simeq 2$ in the UV, a value that has already been singled out for independent reasons in many QG studies (see Refs. [11,16,17, 22,23,25,30,33–36] and references therein).

We focus on the $(1+1)$-dimensional theory with q-derivatives, a context where the analysis progresses more simply but without loss of any characteristic feature. Using Eq. (1), the reader can easily realize that the spatial distance between two points A and B is

$$L := \int_{x_\mathrm{A}}^{x_\mathrm{B}} dq^1 = \ell + \frac{1}{\alpha}\frac{\ell_*}{\ell}\left(\left| \frac{x_\mathrm{B} - \bar{x}}{\ell_*} \right|^\alpha - \left| \frac{x_\mathrm{A} - \bar{x}}{\ell_*} \right|^\alpha \right), \tag{2}$$

with $\ell = x_\mathrm{B} - x_\mathrm{A}$. Thus, different presentations (i.e., different values of $\bar{x}$ [29,53]) give different results for the distance, although they do not change the anomalous scaling, which is solely governed by α. Up to now, this has been regarded as a freedom of the model, but we here suggest that the presentation ambiguity should be viewed as a manifestation of spacetime fuzziness.

Four presentation choices have been identified as special among the others [29], but the second flow-equation theorem [52] selects only two of these: the initial-point presentation, where $\bar{x} = x_\mathrm{A}$, and the final-point presentation, where $\bar{x} = x_\mathrm{B}$. In both cases, Eq. (2) simplifies in such a way that the difference between L and the value ℓ that would be measured in an ordinary space is [29]

$$\delta L_\alpha \simeq \pm \frac{\ell_*}{\alpha}\left(\frac{\ell}{\ell_*}\right)^\alpha, \tag{3}$$

approximately valid in any space dimensions, where the plus sign is for the initial-point presentation and the minus is for the final-point presentation.

Strikingly, the multifractional contribution to distances (3) is of the same type of the lower bound on distances found by heuristically combining QM and GR arguments [41,44–47]. In particular, in Ref. [47], one of us proposed an argument leading to a minimal length uncertainty $\delta L \sim \sqrt{\ell_\mathrm{Pl}^2 \ell / s}$, where s is a length scale characterizing the measuring apparatus. Using a somewhat different line of reasoning, the authors of Ref. [46] suggested instead fluctuations of magnitude $\sim (\ell_\mathrm{Pl}^2 \ell)^{\frac{1}{3}}$. Both of these well-studied scenarios for distance fuzziness match quantitatively the multifractional contribution to distances (3) upon adopting

$$\alpha = \frac{1}{2}, \qquad \ell_* = \frac{\ell_\mathrm{Pl}^2}{s}, \tag{4}$$

in agreement with Ref. [47], or

$$\alpha = \frac{1}{3}\,, \qquad \ell_* = \ell_{\rm Pl}\,, \tag{5}$$

in agreement with Ref. [46].

This leads us to advocate a novel interpretation of (3), such that it gives an intrinsic uncertainty on the measurement of spacetime distances. According to this interpretation, the initial-point presentation generates a positive fluctuation $+\delta L_\alpha$, while the final-point presentation produces a negative fluctuation $-\delta L_\alpha$, with the possibilities $\alpha = 1/2$ and $\alpha = 1/3$ being favored by the connection with [46,47] we are starting to build up.

The value (4) has been already recognized as special for several theoretical reasons [53]. In particular, it gives the aforementioned result $d_S \simeq 2$ in the UV. What is more, the length scale ℓ_* turns out to be related to the Planck length. In the case (4), we have $\ell_* = \ell_{\rm Pl}^2/s < \ell_{\rm Pl}$, where s is the observation scale. Thus, the dependence on the scales at which the measurement is being performed becomes explicit. This is exactly what is expected to happen in multifractal geometry and, in particular, in multifractional theories, where the results of measurements depend on the observation scale [53]. In the case (5), ℓ_* coincides with $\ell_{\rm Pl}$. In both cases, the relation of ℓ_* with $\ell_{\rm Pl}$ exposes the possibility of encoding highly nontrivial *quantum* features within multifractional theories. A similar line of thought applies also to the time direction, which leads us to entertain the concrete possibility that the binomial measure should be isotropic in space and time, i.e.,

$$\alpha_0 = \alpha\,. \tag{6}$$

It is intriguing that, in the illustrative example for our main claim, a connection is established between a multifractional theory with a built-in dimensional flow (a feature usually derived, rather than assumed, in top-down approaches to QG) and uncertainties on distance measurements motivated by heuristic bottom-up approaches, combining just QM and GR principles without adding any hypothetical QG ingredient. We are thereby conjecturing that the connection between the form of dimensional flow and the form of spacetime fuzziness should have *wider applicability*. However, also within the limits of our example some additional consistency checks are appropriate. A more in-depth analysis is needed in order to establish satisfactorily that, in multifractional models, both dimensional flow and spacetime fuzziness are obtained without introducing internal contradictions or external ingredients.

4. Core of the connection: stochastic spacetime emerges

From the multifractional perspective, the reinterpretation we are proposing is not arbitrary. In Ref. [29], it was observed that the theory with fractional derivatives describes spacetimes with a microscopic stochastic structure, i.e., a nowhere-differentiable geometry where location of events ("points" in space) cannot be determined with arbitrary accuracy and particle trajectories are nonsmooth. The presentation label $\bar{x}^\mu$ prescribes how integrals on stochastic spacetime variables can be performed, as in the Itô–Stratonovich dilemma in random processes. Inspired by this, instead of defining as many physically inequivalent theories (but with the same anomalous scaling) as the number of presentations, and to choose one presentation among the others, one can take "all presentations at the same time." In this case, the measures $\{q^\mu(x-\bar{x}^\mu) : \bar{x}^\mu \in \mathbb{R}^D\}$ would not correspond to a class of (in)finitely many theories labeled by $\bar{x}^\mu$ all with the same anomalous scaling: they would be *one* measure corresponding to *one* theory with an intrinsic microscopic uncertainty. This *stochastic view* holds only in the multifractional theory with fractional derivatives and also in the case with q-derivatives, which is an approximation of the former [53].

A direct and rigorous way to understand where stochasticity may come from in classical multifractional spacetimes is the following. A connection between a fractal and a stochastic structure has been advanced long ago by Nottale in his *scale relativity* (e.g., Ref. [55]). There, assuming that spacetime is fractal, the expression for a length was found to be $L = \ell + \zeta\ell_*(\ell/\ell_*)^\alpha$, where one discriminates between a deterministic differentiable part ℓ (the length on usual space) and a stochastic, nowhere-differentiable part ζ. The latter is a wildly fluctuating random variable such that $\langle\zeta\rangle = 0$ and $\langle\zeta^2\rangle = \mp 1$, depending on whether the distance is time- or spacelike. Since both scale relativity and multifractional spacetimes rely on a fractal geometry, it is not surprising that they lead to similar descriptions of lengths. However, the original fractal-spacetime formulation of multifractional theories [54] has been made much more solid thanks to a fundamental principle (slow IR dimensional flow) [52] that reproduces the measure dictated by fractal geometry and fixes some of the free parameters of scale relativity. In particular, not only is the stochastic random variable ζ of Nottale's "fractal" length L present in a more general multifractional length if we go beyond the approximation (1) of a binomial measure, but it is also fixed by the second flow-equation theorem, in contrast with the *ad hoc* variable ζ in scale relativity. In fact, considering the second-order truncation of the full measure determined by the flow-equation theorem [52], we have (index μ omitted everywhere)

$$q(x) = x + \frac{\ell_*}{\alpha}\left|\frac{x}{\ell_*}\right|^\alpha F_\omega(x)\,, \tag{7}$$

where $F_\omega(x) = F_\omega(\lambda_\omega x)$ is a complex modulation factor encoding a fundamentally discrete spacetime symmetry $x \to \lambda_\omega x$ in the far UV (λ_ω is fixed). This symmetry arises as a consequence of the theorem, it is not imposed by hand, and Eq. (7) is the generalization of (1) (with $\bar{x} = 0$ for simplicity; presentation does not affect the argument here) to higher orders in the flow equation [52]. Requiring the measure to be real-valued, one has [52–54]

$$F_\omega(x) = \sum_{n=0}^{+\infty} F_n(x)\,, \qquad \omega_n = \omega n\,, \tag{8a}$$

$$F_n(x) := A_n \cos\left(\omega_n \ln\left|\frac{x}{\ell_\infty}\right|\right) + B_n \sin\left(\omega_n \ln\left|\frac{x}{\ell_\infty}\right|\right), \tag{8b}$$

where A_n and B_n are constant amplitudes and $\ell_{\rm Pl} \sim \ell_\infty \lesssim \ell_*$. The coordinate dilation of the discrete scale invariance is governed by the frequency ω, $\lambda_\omega = \exp(-2\pi/\omega)$. The log-oscillating structure is determined by the flow-equation theorem [52], while the simple but crucial linear relation $\omega_n = \omega n$ is determined by discrete scale invariance, the trade mark of iterative (also called deterministic) fractals [54]. For phenomenological reasons, the modulation factor (8) is usually approximated by only two frequencies, the zero mode $n = 0$ [$F_0(x) = A_0 = \text{const}$] and the $n = 1$ mode. This approximation is quite effective in capturing the physical imprint of the logarithmic oscillations in particle-physics and cosmological observables [53], but here we prefer to retain the full structure (8). Defining $y := \ln|x/\ell_\infty|$ and taking the average $\langle f(y)\rangle := (2\pi)^{-1}\int_0^{2\pi} dy\, f(y)$, we get

$$\langle F_\omega\rangle = A_0\,, \qquad \langle F_\omega^2\rangle = A_0^2 + \sum_{n>0}\frac{A_n^2 + B_n^2}{2}\,. \tag{9}$$

Thus, *if we drop the zero mode* and set $A_0 = 0$, the profile $\tilde{F}_\omega(x) := \sum_{n>0} F_n(x)$ reproduces Nottale's fractal coordinates upon the identification $\zeta = \tilde{F}_\omega$. Since the sign and magnitude of the multiscale correction to lengths are modulated by log oscillations, the latter solve the presentation problem by making the presentation choice irrelevant. Moreover, for certain n-dependences of the amplitudes A_n and B_n (corresponding to introducing ergodic mixing phases in the oscillations), Eq. (8) is a Weierstrass-type nowhere-differentiable function [56]. Non-differentiability is a key property of random distributions. As a result, we reach the neat conclusion that, in multifractional theories, the "stochastic fluctuations" of the geometry are provided by the logarithmic oscillatory modulation of the measure.

5. Extending to other quantum gravities

The logic so far has been to obtain fuzziness, an effect of quantum mechanics that can be found in quantum gravity, as a byproduct of the multifractal structure of spacetime. Assuming that spacetime has dimensional flow is sufficient to obtain some quantum-gravity effects tightly related with the quantization paradigm. In this respect, an intrinsic fractal structure of spacetime can, in some loose sense, "replace" quantum mechanics, inasmuch as it is responsible for uncertainty in time and distance measurements. In particular, this explains why classical multifractional theories can encode these features efficiently. An added advantage gained by this perspective is that, as ensured by the flow-equation theorem, the choice of such a fractal structure at short distances is unambiguous. However, by definition quantum mechanics cannot be dispensed with in quantum gravity, and fuzziness and dimensional flow are, at least in part, its consequences. Therefore, the consequence of the present results for quantum gravity is not a change in the main paradigm (gravity is still quantized, in each model by a different fashion) but, rather, the unification of two concepts, dimensional flow and fuzziness, so far considered as separate entities.

In closing, let us offer some additional comments on how our observations might shed light on why the flow of dimensions in the UV is a universal property of QG approaches. Our findings indicate the possibility that dimensional flow is linked to distance fuzziness, whose form can be inferred from arguments combining QM and GR, without knowledge of the detailed features of one or another QG model. In this respect, spacetime fuzziness could be viewed in analogy with the Hawking temperature for black holes, also derived from semiquantitative model-independent arguments combining QM and GR.

Multifractional theories are particularly manageable for what concerns the structures that one needs to investigate in order to test our conjecture. Of course, this is the reason why we chose them as the example for this first exploratory study. The test may be harder in other formalisms of quantum gravity, but we hope that the encouraging results reported here will energize efforts in that direction. All the main elements of our arguments are already in place in some of the major proposals in the literature. In particular, string theory and nonlocal quantum gravity both realize dimensional flow [19,26] and coarse-grain (in the case of the string) or eliminate UV divergences (see [57] for nonlocal quantum gravity). Asymptotically-safe quantum gravity and the discrete-geometry, mutually related frameworks of loop quantum gravity, spin foams and group field theory all have dimensional flow [9,11–13,27,28,30,31] and implement fuzziness by the presence of minimal lengths, areas or resolutions [4,58,59]. Maybe also causal dynamical triangulations [9] realize fuzziness, as indicated by modified-dispersion-relation arguments [39]. However, although a relation between anomalous dimensions and fuzzy features certainly seems to exist in these cases, as well as in general in QG [35], so far it has not been understood beyond the merely technical level. Revisiting those theories in search of a physical connection similar to that found here may help to clarify some of their formal aspects and even give new tools by which to extract testable phenomenology.

6. Phenomenology

We conclude with the implications of our conjecture for phenomenology. If indeed our conjecture was confirmed, then the phenomenology would be empowered by the possibility of combining experimental bounds on dimensional flow and experimental bounds on fuzziness. For example, for multifractional theories the established bounds on dimensional flow [53] acquire the added significance of bounds on the minimal resolution $1/\ell_*$ achievable. In turn, from Eq. (3) we can infer constraints on time–space isotropic d_S (or d_H) using bounds on fuzziness [48–51]. In fact, neglecting an $\mathcal{O}(1)$ numerical factor, Eq. (3) yields spacetime fuzziness of the form $\sigma \sim (\ell_*)^{1-\alpha}\ell^{\alpha}$. For models in which this form of fuzziness admits phenomenological description in terms of distance fluctuations (which one would naturally expect, but needs to be checked in each specific model [48]), one would then expect to find [48,49] a strain noise $\sigma^2 = \int d\nu S^2(\nu)$ with spectral density $S(\nu) \propto c^{\alpha}(\ell_*)^{1-\alpha} \times \nu^{-\frac{1+2\alpha}{2}}$ (ν here denoting the frequency), and this form of strain noise can be meaningfully constrained, even for very small ℓ_*, using modern gravity-wave interferometers, such as LIGO and VIRGO [49,50,60]. Since $\alpha = d_{S,H}^{UV}/D$ (see above), we find for the UV dimension $d_{S,H}^{UV} \propto D \log(S\sqrt{\nu}/\ell_*)/\log(c/\nu\ell_*)$, and for a first order-of-magnitude estimate we can take as reference the LIGO sensitivity level of $S \sim 10^{-20}\,\mathrm{m\,Hz}^{-1/2}$ at $\nu \sim 10^3$ Hz. This allows to establish meaningful constraints even for "Planckian values" of ℓ_*: for example for $\ell_* \simeq \ell_{\rm Pl}$ at 10^3 Hz one would expect fuzziness noise at the level of $10^{-20}\,\mathrm{m\,Hz}^{-1/2}$ for $d_{S,H}^{UV} \sim 1.7$. So this is a rare case for quantum-gravity research where experimental sensitivities are at a level comparable to where we are with theoretical understanding, since most arguments point to $1.5 \lesssim d_S^{UV} \lesssim 2.5$.

Acknowledgements

G.C. is under a Ramón y Cajal contract and is supported by the I+D grant FIS2014-54800-C2-2-P. M.R. thanks Instituto de Estructura de la Materia (CSIC) for the hospitality during the first stage of elaboration of this work. The contribution of G.C. and M.R. is based upon work from COST Action CA15117, supported by COST (European Cooperation in Science and Technology).

References

[1] D. Oriti (Ed.), Approaches to Quantum Gravity, Cambridge University Press, Cambridge, U.K., 2009.

[2] G.F.R. Ellis, J. Murugan, A. Weltman (Eds.), Foundations of Space and Time, Cambridge University Press, Cambridge, U.K., 2012.

[3] B. Zwiebach, A First Course in String Theory, Cambridge University Press, Cambridge, U.K., 2009.

[4] C. Rovelli, Quantum Gravity, Cambridge University Press, Cambridge, U.K., 2007.

[5] T. Thiemann, Modern Canonical Quantum General Relativity, Cambridge University Press, Cambridge, U.K., 2007;
T. Thiemann, Introduction to modern canonical quantum general relativity, arXiv:gr-qc/0110034.

[6] A. Perez, The spin-foam approach to quantum gravity, Living Rev. Relativ. 16 (2013) 3.

[7] S. Gielen, L. Sindoni, Quantum cosmology from group field theory condensates: a review, SIGMA 12 (2016) 082, arXiv:1602.08104.

[8] G. Amelino-Camelia, S. Majid, Waves on noncommutative space–time and gamma-ray bursts, Int. J. Mod. Phys. A 15 (2000) 4301, arXiv:hep-th/9907110.

[9] J. Ambjørn, A. Görlich, J. Jurkiewicz, R. Loll, Nonperturbative quantum gravity, Phys. Rep. 519 (2012) 127, arXiv:1203.3591.

[10] F. Dowker, Introduction to causal sets and their phenomenology, Gen. Relativ. Gravit. 45 (2013) 1651.

[11] O. Lauscher, M. Reuter, Fractal spacetime structure in asymptotically safe gravity, J. High Energy Phys. 10 (2005) 050, arXiv:hep-th/0508202.
[12] M. Niedermaier, M. Reuter, The asymptotic safety scenario in quantum gravity, Living Rev. Relativ. 9 (2006) 5.
[13] M. Reuter, F. Saueressig, Asymptotic safety, fractals, and cosmology, Lect. Notes Phys. 863 (2013) 185, arXiv:1205.5431.
[14] P. Aschieri, M. Dimitrijevic, P. Kulish, F. Lizzi, J. Wess, Noncommutative Spacetimes, Springer, Berlin, Germany, 2009.
[15] A.P. Balachandran, A. Ibort, G. Marmo, M. Martone, Quantum fields on noncommutative spacetimes: theory and phenomenology, SIGMA 6 (2010) 052, arXiv:1003.4356.
[16] P. Hořava, Spectral dimension of the universe in quantum gravity at a Lifshitz point, Phys. Rev. Lett. 102 (2009) 161301, arXiv:0902.3657.
[17] G. Calcagni, A. Eichhorn, F. Saueressig, Probing the quantum nature of spacetime by diffusion, Phys. Rev. D 87 (2013) 124028, arXiv:1304.7247.
[18] E.T. Tomboulis, Super-renormalizable gauge and gravitational theories, arXiv: hep-th/9702146.
[19] L. Modesto, Super-renormalizable quantum gravity, Phys. Rev. D 86 (2012) 044005, arXiv:1107.2403.
[20] T. Biswas, E. Gerwick, T. Koivisto, A. Mazumdar, Towards singularity- and ghost-free theories of gravity, Phys. Rev. Lett. 108 (2012) 031101, arXiv:1110.5249.
[21] G. Calcagni, L. Modesto, Nonlocal quantum gravity and M-theory, Phys. Rev. D 91 (2015) 124059, arXiv:1404.2137.
[22] G. 't Hooft, Dimensional reduction in quantum gravity, in: A. Ali, J. Ellis, S. Randjbar-Daemi (Eds.), Salamfestschrift, World Scientific, Singapore, 1993, arXiv:gr-qc/9310026.
[23] S. Carlip, Spontaneous dimensional reduction in short-distance quantum gravity?, AIP Conf. Proc. 1196 (2009) 72, arXiv:0909.3329.
[24] L. Modesto, P. Nicolini, Spectral dimension of a quantum universe, Phys. Rev. D 81 (2010) 104040, arXiv:0912.0220.
[25] G. Calcagni, Fractal universe and quantum gravity, Phys. Rev. Lett. 104 (2010) 251301, arXiv:0912.3142.
[26] G. Calcagni, L. Modesto, Nonlocality in string theory, J. Phys. A 47 (2014) 355402, arXiv:1310.4957.
[27] G. Calcagni, D. Oriti, J. Thürigen, Spectral dimension of quantum geometries, Class. Quantum Gravity 31 (2014) 135014, arXiv:1311.3340.
[28] G. Calcagni, D. Oriti, J. Thürigen, Dimensional flow in discrete quantum geometries, Phys. Rev. D 91 (2015) 084047, arXiv:1412.8390.
[29] G. Calcagni, ABC of multi-fractal spacetimes and fractional sea turtles, Eur. Phys. J. C 76 (2016) 181, arXiv:1602.01470.
[30] L. Modesto, Fractal structure of loop quantum gravity, Class. Quantum Gravity 26 (2009) 242002, arXiv:0812.2214.
[31] M. Ronco, On the UV dimensions of loop quantum gravity, Adv. High Energy Phys. 2016 (2016) 9897051, arXiv:1605.05979.
[32] D. Benedetti, Fractal properties of quantum spacetime, Phys. Rev. Lett. 102 (2009) 111303, arXiv:0811.1396.
[33] J. Ambjorn, J. Jurkiewicz, R. Loll, Spectral dimension of the universe, Phys. Rev. Lett. 95 (2005) 171301, arXiv:hep-th/0505113.
[34] G. Amelino-Camelia, M. Arzano, G. Gubitosi, J. Magueijo, Dimensional reduction in the sky, Phys. Rev. D 87 (2013) 123532, arXiv:1305.3153.
[35] T. Padmanabhan, S. Chakraborty, D. Kothawala, Spacetime with zero point length is two-dimensional at the Planck scale, Gen. Relativ. Gravit. 48 (2016) 55, arXiv:1507.05669.
[36] A. Eichhorn, Spectral dimension in causal set quantum gravity, Class. Quantum Gravity 31 (2014) 125007, arXiv:1311.2530.
[37] D.N. Coumbe, Hypothesis on the nature of time, Phys. Rev. D 91 (2015) 124040, arXiv:1502.04320.
[38] A. Belenchia, D.M.T. Benincasa, A. Marcianò, L. Modesto, Spectral dimension from nonlocal dynamics on causal sets, Phys. Rev. D 93 (2016) 044017, arXiv: 1507.00330.
[39] D.N. Coumbe, Quantum gravity without vacuum dispersion, Int. J. Mod. Phys. D 26 (2017) 1750119, arXiv:1512.02519.
[40] G. Veneziano, A stringy nature needs just two constants, Europhys. Lett. 2 (1986) 199.
[41] T. Padmanabhan, Limitations on the operational definition of spacetime events and quantum gravity, Class. Quantum Gravity 4 (1987), L107.
[42] K. Konishi, G. Paffuti, P. Provero, Minimum physical length and the generalized uncertainty principle in string theory, Phys. Lett. B 234 (1990) 276.
[43] J. Ellis, N.E. Mavromatos, D.V. Nanopoulos, String theory modifies quantum mechanics, Phys. Lett. B 293 (1992) 37, arXiv:hep-th/9207103.
[44] L.J. Garay, Quantum gravity and minimum length, Int. J. Mod. Phys. A 10 (1995) 145, arXiv:gr-qc/9403008.
[45] D.V. Ahluwalia, Quantum measurements, gravitation, and locality, Phys. Lett. B 339 (1994) 301, arXiv:gr-qc/9308007.
[46] Y.J. Ng, H. Van Dam, Limit to space–time measurement, Mod. Phys. Lett. A 9 (1994) 335.
[47] G. Amelino-Camelia, Limits on the measurability of space–time distances in the semiclassical approximation of quantum gravity, Mod. Phys. Lett. A 9 (1994) 3415, arXiv:gr-qc/9603014.
[48] G. Amelino-Camelia, Quantum-spacetime phenomenology, Living Rev. Relativ. 16 (2013) 5, arXiv:0806.0339.
[49] G. Amelino-Camelia, An interferometric gravitational wave detector as a quantum gravity apparatus, Nature 398 (1998) 216, arXiv:gr-qc/9808029.
[50] Y.J. Ng, H. Van Dam, Measuring the foaminess of space–time with gravity-wave interferometers, Found. Phys. 30 (2000) 795, arXiv:gr-qc/9906003.
[51] W.A. Christiansen, Y.J. Ng, H. Van Dam, Probing spacetime foam with extra-galactic sources, Phys. Rev. Lett. 96 (2006) 051301, arXiv:gr-qc/0508121.
[52] G. Calcagni, Multiscale spacetimes from first principles, Phys. Rev. D 95 (2017) 064057, arXiv:1609.02776.
[53] G. Calcagni, Multifractional theories: an unconventional review, J. High Energy Phys. 03 (2017) 138, arXiv:1612.05632.
[54] G. Calcagni, Geometry and field theory in multi-fractional spacetime, J. High Energy Phys. 01 (2012) 065, arXiv:1107.5041.
[55] L. Nottale, Scale relativity and fractal space–time: theory and applications, Found. Sci. 15 (2010) 101, arXiv:0812.3857.
[56] S. Gluzman, D. Sornette, Log-periodic route to fractal functions, Phys. Rev. E 65 (2002) 036142, arXiv:cond-mat/0106316.
[57] L. Modesto, L. Rachwał, Super-renormalizable and finite gravitational theories, Nucl. Phys. B 889 (2014) 228, arXiv:1407.8036.
[58] M. Reuter, J.-M. Schwindt, A minimal length from the cutoff modes in asymptotically safe quantum gravity, J. High Energy Phys. 01 (2006) 070, arXiv: hep-th/0511021.
[59] M. Reuter, J.-M. Schwindt, Scale-dependent metric and causal structures in quantum Einstein gravity, J. High Energy Phys. 01 (2007) 049, arXiv:hep-th/0611294.
[60] D.V. Martynov, et al., Sensitivity of the advanced LIGO detectors at the beginning of gravitational wave astronomy, Phys. Rev. D 93 (2016) 112004, arXiv:1604.00439.

Integrability and chemical potential in the (3 + 1)-dimensional Skyrme model

P.D. Alvarez [a], F. Canfora [b], N. Dimakis [c], A. Paliathanasis [c,d,*]

[a] Universidad Técnica Federico Santa María, Santiago, Chile
[b] Centro de Estudios Científicos (CECS), Casilla 1469, Valdivia, Chile
[c] Instituto de Ciencias Fisicas y Matematicas, Universidad Austral de Chile, Valdivia, Chile
[d] Institute of Systems Science, Durban University of Technology, PO Box 1334, Durban 4000, South Africa

ARTICLE INFO

Editor: M. Cvetič

ABSTRACT

Using a remarkable mapping from the original $(3+1)$dimensional Skyrme model to the Sine-Gordon model, we construct the first analytic examples of Skyrmions as well as of Skyrmions–anti-Skyrmions bound states within a finite box in $3+1$ dimensional flat space-time. An analytic upper bound on the number of these Skyrmions–anti-Skyrmions bound states is derived. We compute the critical isospin chemical potential beyond which these Skyrmions cease to exist. With these tools, we also construct topologically protected time-crystals: time-periodic configurations whose time-dependence is protected by their non-trivial winding number. These are striking realizations of the ideas of Shapere and Wilczek. The critical isospin chemical potential for these time-crystals is determined.

1. Introduction

One of the most beautiful examples of the interplay between topology and field theory is provided by the Skyrme theory [1]. Its importance in nuclear and particle physics (for instance see [2–7]) arises from its relations with low energy QCD [8] and it is the prototype of a non-integrable model. The non-trivial role of topology appears since the Skyrme model supports solitons, called *Skyrmions*, which are topologically stable and represent Baryonic degrees of freedom (see [9] [10] [11] [8] [12] [13] [14] [15] and references therein). The identification of the Baryon number in particle physics with the third homotopy class of the Skyrmion [8] showed that the original Skyrme intuition was correct.

Moreover, the ideas of Skyrme have nontrivial applications not only in particles and nuclear physics but also in many other areas of physics. In [16] it was found that skyrmion lattices exist in semiconductors. What is more, in an antiferromagnetic spinor Bose–Einstein condensate a two-dimensional skyrmion was observed [17] to be stable on a short time. Applications of Skyrme model are also encountered in gravitational physics, for instance in cosmology [18,19] or black holes physics where it has been found that the "no hair" conjecture can be violated [20–25]. A further field of research in which the role of Skyrmions is extremely relevant is the analysis of magnetic materials (a nice review being [26]).

The original Skyrme model is very far from being integrable and only very few explicit analytic results are known. In particular, there is no explicit solution with non-vanishing Baryon number on flat space-times. Consequently, the Skyrme phase diagrams (which could provide very valuable informations on nuclear matter) are very difficult to approach with analytical methods.

By using the original spherical hedgehog ansatz of Skyrme, Klebanov proposed a phenomenological approach to the analysis of Skyrmions at finite density a long time ago [27]. By following his point of view and using the technique of [28] [29], a non-vanishing isospin chemical potential was introduced in [30] [31]. However, both finite volume effects and isospin chemical potential break spherical symmetry, and this fact makes difficult to apply the spherical hedgehog ansatz in these cases. Without an appropriate ansatz (which is both topologically non-trivial and non-spherically symmetric) it becomes very hard to derive analytic results on how the Isospin chemical potential as well as finite-volume effects affect the behavior of Skyrmions.

In recent years, a generalized hedgehog ansatz was proposed, which provides with a strategy to construct a non-spherical hedgehog-like ansatz with the right properties (see [32–34], [35],

* Corresponding author.
E-mail addresses: pedro.alvarezn@usm.cl (P.D. Alvarez), canfora@cecs.cl (F. Canfora), nsdimakis@gmail.com (N. Dimakis), anpaliat@phys.uoa.gr (A. Paliathanasis).

[36], [37] and references therein) both in the Skyrme and in the Yang–Mills–Higgs cases. By using this technique, we construct analytic and topologically non-trivial solutions of the Skyrme model without spherical symmetry living within a finite box in flat space-times. We also construct analytic Skyrmions–anti-Skyrmions bound states. We derive a bound on the number of Skyrmions–anti-Skyrmions bound states in terms of the coupling constant. The isospin chemical potential can be included keeping, at the same time, the nice properties of the ansatz. The critical isospin chemical potential, beyond which the Skyrmion living in the box ceases to exist, can be explicitly determined.

Remarkably, the generalized hedgehog ansatz allowed us to construct novel types of topological configurations of the Skyrme model that can be defined as *topologically protected time crystals*, see below for more information.

The idea of time crystal was introduced by Wilczek and Shapere in [38] [39] [40], based on the following observations. *Spontaneous Symmetry Breaking* is a general property of nature that manifests itself in many different situations. It refers to situations where the observed configurations of a given system possess less symmetries than the corresponding action. They proposed the following very interesting and intriguing question: *is it possible to break spontaneously time translation symmetry?*

Powerful no-go theorems [41] [42] ruled out the original proposals but they inspired a huge number of physicists to open new research lines (a nice review is [43]). New types of time crystals in condensed matter physics have been proposed and realized in laboratories [44] [45] [46] [47] [48] [49] (an up-to-date list of references can be found in [43]). However, no example in nuclear and particles physics has been considered so far.

Here we show explicitly that, when finite volume effects are taken into account, the Skyrme model predicts the existence of a new type of time-crystal. These are exact time-periodic configurations of the $(3+1)$-dimensional Skyrme model that cannot be deformed continuously to the trivial vacuum as they possess a non-trivial winding number extending along the time-direction (which is a sort of Lorentzian version of the Euclidean instanton number). Due to topological reasons, these time crystals can only decay into other time-periodic configurations: correspondingly, the time-periodicity is topologically protected. Hence the name *topologically protected time crystals*.

This paper is organized as follows: in section 2 we introduce the Skyrme action. In section 3, we discuss the sine-Gordon mapping and the effects of the chemical potential. In section 4, we describe the topologically protected time-crystals. In section 5, we draw some concluding ideas.

2. The Skyrme model

We consider the $SU(2)$ Skyrme model in four dimensions. The action of the system is

$$S=\frac{K}{2}\int d^4x\sqrt{-g}\left[\frac{1}{2}\mathrm{Tr}\left(R^{\mu}R_{\mu}\right)+\frac{\lambda}{16}\mathrm{Tr}\left(G_{\mu\nu}G^{\mu\nu}\right)\right], \qquad (1)$$

$$R_{\mu}=U^{-1}\nabla_{\mu}U\ ,\quad G_{\mu\nu}=\left[R_{\mu},R_{\nu}\right]\ , \qquad (2)$$

$$U\in SU(2)\ ,\quad R_{\mu}=R_{\mu}^{j}t_{j}\ ,\quad t_{j}=i\sigma_{j}\ , \qquad (3)$$

where $\sqrt{-g}$ is the (square root of minus) the determinant of the metric, the positive parameters K and λ are fixed experimentally and σ_j are the Pauli matrices. In our conventions $c=\hbar=1$, the space-time signature is $(-,+,+,+)$ and Greek indices run over space-time. The stress-energy tensor is

$$T_{\mu\nu}=-\frac{K}{2}\mathrm{Tr}\left[R_{\mu}R_{\nu}-\frac{1}{2}g_{\mu\nu}R^{\alpha}R_{\alpha}\right.$$
$$\left.\times\frac{\lambda}{4}\left(g^{\alpha\beta}G_{\mu\alpha}G_{\nu\beta}-\frac{g_{\mu\nu}}{4}G_{\sigma\rho}G^{\sigma\rho}\right)\right],$$

and the matter field equations are

$$\nabla^{\mu}\left(R_{\mu}+\frac{\lambda}{4}\left[R^{\nu},G_{\mu\nu}\right]\right)=0. \qquad (4)$$

We adopt a standard parametrization of the $SU(2)$-valued scalar $U(x^{\mu})$

$$U^{\pm1}(x^{\mu})=Y^{0}(x^{\mu})\mathbf{I}\pm Y^{i}(x^{\mu})t_{i}\ ,\quad \left(Y^{0}\right)^{2}+Y^{i}Y_{i}=1\ , \qquad (5)$$

where $\mathbf{I}$ is the 2×2 identity and

$$Y^{0}=\cos C\ ,\quad Y^{i}=n^{i}\cdot\sin C\ , \qquad (6)$$

$$n^{1}=\sin F\cos G\ ,\quad n^{2}=\sin F\sin G\ ,n^{3}=\cos F\ . \qquad (7)$$

The Skyrme field possesses a non-trivial topological charge which, mathematically, is a suitable homotopy class or winding number: its explicit expression as an integral over a suitable three-dimensional hypersurface Σ is

$$W=-\frac{1}{24\pi^{2}}\int_{\Sigma}\epsilon^{ijk}Tr\left(U^{-1}\partial_{i}U\right)\left(U^{-1}\partial_{j}U\right)\left(U^{-1}\partial_{k}U\right)$$
$$=-\frac{1}{24\pi^{2}}\int_{\Sigma}\rho_{B}\ , \qquad (8)$$

where the baryon density is defined by $\rho_B=12\sin^2C\sin F\ dC\wedge dF\wedge dG$. A necessary condition to have a non-vanishing baryon density is $dC\wedge dF\wedge dG\neq0$.

When, in the above integral, the three-dimensional hypersurface Σ is space-like then the topological charge is interpreted as Baryon number. However, due to the fact that ρ_B does not depend on the metric, there are two further options: Σ can be time-like or light-like. The last two possibilities have not been explored so far in the literature. In fact they are extremely interesting as whenever $W\neq0$ (whether Σ is space-like, time-like or light-like) the corresponding Skyrme configuration has a non-trivial homotopy and, consequently, cannot be deformed continuously into the trivial vacuum $U=\mathbf{I}$. The cases in which Σ is time-like and $W\neq0$ correspond to topologically protected time crystals as it will be explained below. We will only consider an ansatz in which $\rho_B\neq0$.

The natural generalization of the hedgehog ansatz introduced in [36] in the cases in which the metric is flat reads

$$G=\frac{\gamma+\phi}{2}\ ,\quad \tan F=\frac{\tan H}{\cos A}\ ,\quad \tan C=\frac{\sqrt{1+\tan^{2}F}}{\tan A}\ , \qquad (9)$$

where

$$A=\frac{\gamma-\phi}{2}\ ,\quad H=H(r,z)\ . \qquad (10)$$

It can be verified directly that, the topological density ρ_B is non-vanishing. From the standard parametrization of $SU(2)$ ([50]) it follows that

$$0\leq\gamma\leq4\pi\ ,\quad 0\leq\phi\leq2\pi\ , \qquad (11)$$

while the boundary condition for H will be discussed below.

3. Sine-Gordon and Skyrmions

Let us consider the following flat metric

$$ds^{2}=-dz^{2}+\ell^{2}\left(dr^{2}+d\gamma^{2}+d\phi^{2}\right)\ , \qquad (12)$$

(in this section z is the time variable). The length ℓ represents the size of the box where the Skyrmion lives. The coordinates r, γ and ϕ are angular coordinates; the domain of γ and ϕ is given by (11), while for r we choose the finite interval $0 \leq r \leq 2\pi$.

The full Skyrme field equations (4) with the generalized hedgehog ansatz in Eqs. (6), (7), (9) and (10) *reduce to just one scalar differential equation for the profile H*

$$\Box H - \frac{\lambda}{8\ell^2(\lambda + 2\ell^2)} \sin(4H) = 0 \,, \tag{13}$$

where $\Box$ is the two-dimensional D' Alambert operator.

The energy of the configuration is given by

$$E = \int \ell^3 T^{00} dr d\gamma d\phi \,, \tag{14}$$

where

$$T_{00} = \frac{K}{64\ell^4}\Big[16(\lambda + 2\ell^2)\left((\partial_r H)^2 + \ell^2(\partial_z H)^2\right) + \lambda(1 - \cos(4H)) + 16\ell^2\Big] \,. \tag{15}$$

The topological Baryon charge B and charge density ρ_B become respectively

$$B = -\frac{1}{24\pi^2}\int_{t=const} \rho_B \,, \quad \rho_B = 3\sin(2H) dH d\gamma d\phi \,. \tag{16}$$

If we replace the topologically non-trivial ansatz in Eqs. (6), (7), (9) and (10) into the original action (1) we obtain an effective action given by

$$\mathcal{L}(H) = 16\ell^2(\lambda + 2\ell^2)\nabla_\mu H \nabla^\mu H - \lambda\cos(4H), \tag{17}$$

which reproduces equation of motion (13). The boundary conditions for the function H are

$$H(0) = 0 \,, \; H(2\pi) = \pm\frac{\pi}{2} \,, \tag{18}$$

which corresponds to $B = \pm 1$ and

$$H(0) - H(2\pi) = 0 \,, \tag{19}$$

which corresponds to $B = 0$. The sector $B = 0$ is relevant in the construction of Skyrmion anti-Skyrmion bound states.

Taking into account that the Skyrme model in $(3+1)$ dimensions is the prototype of non-integrable systems, the above results in Eqs. (13), (15) and (17) are quite remarkable since they show that the full $(3+1)$-dimensional Skyrme field equations, energy density and effective action in a topologically non-trivial sector (as $\rho_B \neq 0$) can be reduced to the corresponding quantities of the $(1+1)$-dimensional sine-Gordon model. The latter is a well known example of integrable models, see [51] for a detailed review. In particular, it is trivial to construct kink-like solutions of Eq. (13) satisfying the boundary conditions in Eq. (18) (see [51] and references therein) and which (due to Eq. (16)) represent analytic (anti)Skyrmions living in the finite flat box defined above.

Since Eqs. (13), (15) and (17) allow to use all the available results in Sine-Gordon theory to analyze the $(3+1)$-dimensional Skyrme model at finite density, it is useful to follow the conventions of [52]. The effective action for a rescaled the Skyrmion profile Φ is

$$S = \ell^3 \int d\gamma d\phi \int dt dr \mathcal{L}(\Phi), \quad H = \frac{\ell}{(\lambda + 2\ell^2)^{1/2}}\Phi \,,$$

where ℓ^3 comes from the square root of the determinant of the metric. Thus the effective Lagrangian for Φ reads

$$\mathcal{L}(\Phi) = -\frac{1}{2}\nabla^\mu\Phi\nabla_\mu\Phi + \frac{\alpha}{\beta^2}(\cos(\beta\Phi) - 1) \,, \tag{20}$$

$$\alpha = \frac{\lambda}{2\ell^2(\lambda + 2\ell^2)}, \quad \beta = \frac{4\ell}{\sqrt{\lambda + 2\ell^2}} \,. \tag{21}$$

The effective sine-Gordon coupling constant that appears from the Skyrme model always satisfies the Coleman bound $\beta^2 < 8\pi$.

The mapping presented above allows to construct analytic Skyrmion–anti-Skyrmion bound states. Namely, the breather-like solutions of Eq. (13) satisfying the boundary conditions in Eq. (19) (which correspond to kink anti-kink bound states) correspond[1] to *analytic Skyrmion–anti-Skyrmion bound states.* To the best of authors knowledge, this is the first analytic construction of Skyrmions–anti-Skyrmions bound states in the original $(3+1)$dimensional Skyrme model. In particular the number n_B of bound states satisfies $n_B \leq \frac{8\pi}{\beta^2} - 1$. Note that already Skyrme and Perring [53] used sine-Gordon in $1+1$ dimensions as a "toy model" for the $3+1$ dimensional Skyrme model. What is remarkable about the present treatment is that we found a nontrivial topological sector of the full Skyrme model in which they are *exactly equivalent.*

The semi-classical quantization in the present sector of the Skyrme model can be analyzed following [14] [13]. One first has to identify the (classical) low energy modes and then it is necessary to quantize such modes. In the present case, the task is simplified by one of the results mentioned above: namely, not only the Skyrme field equations with the generalized hedgehog ansatz in Eqs. (6), (7), (9) and (10) reduce to the sine-Gordon equation but also the full Skyrme action reduces to the corresponding sine-Gordon action in $1+1$ dimensions with the coupling constants defined in Eq. (21). Thus, the *principle of symmetric criticality* [54] applies in the present case. Consequently, the low energy semi-classical fluctuations of the Skyrme model in the sector described by Eqs. (6), (7), (9) and (10) are described by the reduced action itself (which is nothing but the sine-Gordon action). Thus, all the known semi-classical results on the sine-Gordon theory hold.

3.0.1. An interesting function

Here we consider an interesting function Δ of the Skyrmions with charge ± 1 defined above which encodes the information about how close they can get to saturate the Skyrme-BPS bound (which, as already emphasized, cannot be saturated on flat space-times). Nevertheless, it is interesting to analyze the following function defined[2] as

$$\Delta = E - 12\sqrt{2}\pi^2 |B| = E - 12\sqrt{2}\pi^2 \tag{22}$$

where E is the energy of the (anti)Skyrmion defined above and B is its baryon charge. This relation is nothing but the Bogomol'ny bound, as can be found in [55], expressed in our conventions. It is worth to emphasize that in the usual case of the spherical Skyrmion found by Skyrme himself, the energy exceeds the topological charge by 23%.

In this case, the above difference Δ for static configurations $H = H(r)$ can be evaluated explicitly in terms of elliptic integral as follows. From Eq. (15) one gets the following expression for the energy-density

[1] Indeed, under the present mapping from the $(3+1)$-dimensional Skyrme model into the $(1+1)$-dimensional Sine-Gordon model, the (anti)Skyrmion is mapped into the (anti)kink. Consequently, kink-antikink bound states correspond to Skyrmion–antiSkyrmion bound states.

[2] Only in this subsection, we will adopt the convention that $K = 2$ and $\lambda = 1$ which means, roughly (see page 25 of [11], taking into account that the authors use the opposite convention for the space-time metric with respect to our), that we are measuring lengths in fm and energy in MeV.

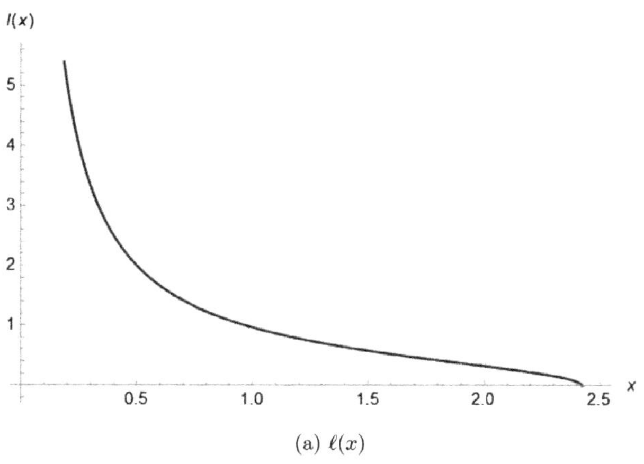

(a) $\ell(x)$

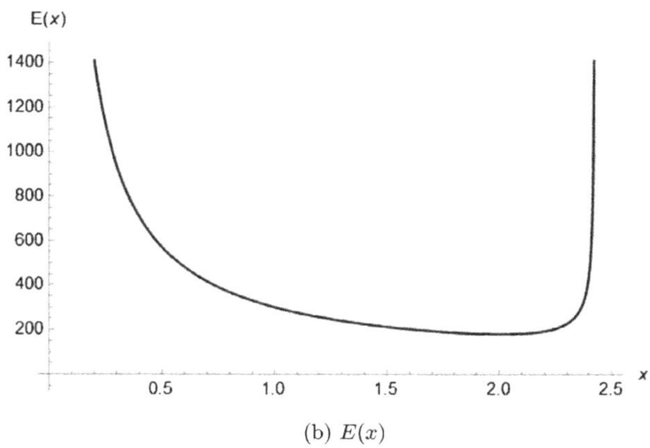

(b) $E(x)$

Fig. 1. The size of the box ℓ and the energy E as functions of $x = \left(\frac{2}{32I_0-1}\right)^{1/2}$.

$$T_{00} = \frac{K}{64\,\ell^4}\left[16(\lambda + 2\,\ell^2)(\partial_r H)^2 + \lambda\,(1-\cos(4H)) + 16\,\ell^2\right]. \tag{23}$$

The field equations Eq. (13) can be reduced to

$$\left(1+2\,\ell^2\right)\frac{(H')^2}{2} + \frac{1}{32}\cos(4H) = I_0 \Rightarrow \tag{24}$$

$$\frac{dH}{dr} = \pm\sqrt{Q(H)} = \pm\frac{1}{\sqrt{1+2\,\ell^2}}\left(2I_0 - \frac{\cos 4H}{16}\right)^{1/2}. \tag{25}$$

In the above equations I_0 is an integration constant satisfying the following condition:

$$\int_0^{\pi/2} \frac{dH}{Q(H)^{1/2}} = \pm 2\pi\,, \tag{26}$$

arising from the requirement to have Baryon charge ± 1 (in this subsection, from now on we will consider the + sign). The above condition fixes the integration constant I_0 as a function of ℓ:

$$\int_0^{\pi/2} \frac{dH}{Q(H)^{1/2}} = 2\pi \Rightarrow I_0 = I_0(\ell)\,. \tag{27}$$

An explicit expression however cannot be extracted since,

$$\int_0^{\pi/2} \frac{dH}{Q(H)^{1/2}} = 2\sqrt{2}(1+2\ell^2)xK\left(-x^2\right) \tag{28}$$

where $x^2 = \frac{2}{32I_0-1} > 0$ and $K(x)$ is the complete elliptic integral of the first kind.

We can evaluate the energy using Eq. (26)

$$E = \pi^2 \int_0^{\pi/2} \frac{dH}{Q(H)^{1/2}}\left(4\frac{(1+2\,\ell^2)}{\ell}Q(H) + \frac{(1-\cos(4H))}{4\,\ell} + 4\ell\right) = E(\ell)\,. \tag{29}$$

Thus, the natural question is: how far is the energy of the Skyrmions from saturating the topological bound? The answer depends on the size of the box ℓ. From Eqs. (26) and (29) one can write the energy $E(\ell)$ as combinations of elliptic integrals.

Although relation (27) cannot provide us with a closed form relation for $I_0(\ell)$, we can invert it and express ℓ as a function of I_0 or of x. In this manner we can also express the energy with respect to x. Thus we obtain the two graphs that we see in Fig. 1. From the first graph we can see that ℓ and x are in inverse proportion with each other. At the same time there is a cutoff at a finite value of x, this is owed to taking a finite bound in the integration of the r variable in (26). As can be seen, by combining (27) and (28) to solve algebraically for ℓ; for x above a certain value, ℓ becomes imaginary. The extension of integration over the full real line should be seen as a pushing of this bound to infinity. It can be checked arithmetically that the most economic energy configuration corresponds to a size of the box of $\ell \simeq 0.32$ where $E \simeq 181.312$ which is approximately 8.25% above the lower bound $12\sqrt{2}\pi^2$ as seen in (22).

These results have a quite natural interpretation. When the size of the box is small (as the order of fm) one should expect strong deviations from spherical symmetry as extended objects feel strongly the presence of boundaries. Indeed, the present non-spherical skyrmion is closer to the topological bound than the usual spherical skyrmion (which exceeds the topological bound by about 23%). On the other hand, when the box size increases, it should be expected that the presence of the box itself becomes less relevant. The above plot shows that this is the case since the energy of the present nonspherical skyrmion grows very rapidly (well above the topological bound) as the size of the box increases. Thus, when the box is large enough only small deviations from the spherical skyrmion should be expected. Consequently, the type of Skyrmions analyzed here is expected to be favored at small volumes (or high densities).

3.1. Inclusion of chemical potential

The effects of the isospin chemical potential can be taken into account by using the following covariant derivative (see [28] [29])

$$D_\mu = \nabla_\mu + \bar{\mu}[t_3,\ ]\delta_{\mu 0}\,. \tag{30}$$

Thus R_μ becomes $\bar{R}_\mu = U^{-1}D_\mu U$ and the equations of motion read

$$D^\mu\left(\bar{R}_\mu + \frac{\lambda}{4}[\bar{R}^\nu, \bar{G}_{\mu\nu}]\right) = 0\,, \tag{31}$$

where $\bar{G}_{\mu\nu} = [\bar{R}_\mu, \bar{R}_\nu]$. For static configurations $H(r,z) = H(r)$, the good property of the hedgehog ansatz is not lost: *the full Skyrme field equations with isospin chemical potential in Eq. (31) reduce to just one scalar ODE for* H

$$\left(\lambda + 2\,\ell^2 - 8\,\lambda\,\ell^2\bar{\mu}^2\sin^2(H(r))\right)H''(r) - 4\lambda\ell^2\bar{\mu}^2\sin(2H(r))H'^2$$

$$+\lambda\left(\bar{\mu}^2\ell^2-\frac{1}{8}\right)\sin(4H(r))+4\bar{\mu}^2\ell^4\sin(2H(r))=0. \quad (32)$$

This is a quite non-trivial technical achievement in itself (see, for instance, [30] [31]). Moreover, the above differential equation can be reduced to

$$Y(H)\frac{(H')^2}{2}+V(H)=E_0\,, \quad (33)$$

where

$$Y(H)=\lambda+2\ell^2-8\lambda\ell^2\bar{\mu}^2\sin^2(H),$$
$$V(H)=-\frac{\lambda}{4}\left(\bar{\mu}^2\ell^2-\frac{1}{8}\right)\cos(4H)-2\bar{\mu}^2\ell^4\cos(2H)\,.$$

E_0 is an integration constant to be determined imposing the physical boundary condition:

$$\int_0^{\pi/2}\frac{[Y(H)]^{1/2}}{[E_0-V(H)]^{1/2}}dH=\sqrt{2}2\pi\,. \quad (34)$$

Thus, we can compute the critical isospin chemical potential $\overline{\mu}_c$ as the one for which the above boundary condition cannot be satisfied anymore as Y can be negative[3] when $\bar{\mu}\geq\overline{\mu}_c$:

$$(\overline{\mu}_c)^2=\frac{\lambda+2\ell^2}{8\lambda\ell^2}\,.$$

4. Time crystals

Obviously, not any time-periodic solution of the Skyrme model is a time-crystal. For instance, the Skyrmion–anti-Skyrmion bound state constructed above are time-periodic. However they are not topologically protected since, if one 'pays' the corresponding binding energies, they decay into the trivial vacuum.

Here we adopt the flat space line element

$$ds^2=-d\gamma^2+\ell^2\left(dz^2+dr^2+d\phi^2\right)\,, \quad (35)$$

where γ plays the role of time. We have to make the following modification to ansatz (9), (10)

$$A=\frac{\omega\gamma-\phi}{2},\quad G=\frac{\omega\gamma+\phi}{2}, \quad (36)$$

where $0\leq\omega\gamma\leq4\pi$ and the frequency ω is necessary to keep A and G dimensionless.

The adoption of line-element (35) means that in this case the profile H depends on two space-like coordinates. The Skyrme configurations U defined in Eqs. (6), (7), (9), (10) and (36) are necessarily time-periodic. The full Skyrme field equations (4) reduce in this case to

$$\Delta H-\frac{\lambda\omega^2}{4\left(\ell^2(\lambda\omega^2-4)-\lambda\right)}\sin(4H)=0\,, \quad (37)$$

$$\omega^2\neq\omega_c^2=\frac{\lambda+4\ell^2}{\ell^2\lambda}\,, \quad (38)$$

where Δ is the two-dimensional Laplacian in z and r. Eq. (37) is the Euclidean sine-Gordon equation.[4] Exact solutions of Eq. (37)[5] can easily be constructed taking, for instance, $H=H(r)$.

To construct a time crystal configuration, we firstly need to find stable kinks satisfying Eq. (37). As in the previous section, it is useful to obtain the reduced action $\mathcal{L}$ corresponding to the Eq. (37),

$$\mathcal{L}(H)=\nabla_\mu H\nabla^\mu H+\frac{\lambda\omega^2}{4\left(\ell^2(\lambda\omega^2-4)-\lambda\right)}\sin^2(2H)\,. \quad (39)$$

This allows to use well known results on quantization of sine-Gordon theory also in this sector, for instance, a sine-Gordon kink with $H=H(r)$.

Secondly, the configuration has to have a non-trivial winding number. The topological density is given by $\rho_B=3\sin(2H)dH\wedge d(\omega\gamma)\wedge d\phi$, and thus the winding number can be evaluated to

$$W=-\frac{\omega}{8\pi^2}\int_{z=\text{const}}\sin(2H)dHd\gamma d\phi=\pm1. \quad (40)$$

This is one of the main results of the paper. We have shown that there are smooth time-periodic regular configurations of the Skyrme model living at finite volume possessing a non-trivial winding number along a three-dimensional time-like surface. Since the winding number is invariant under any continuous deformation, these configurations can only decay into other configurations which are also time-periodic (as for static configurations the above winding number vanishes). Thus, the time periodicity of these configurations is topologically protected. These are classical *topologically protected* time-crystals in the sense of [39]. Interestingly enough, the principle of symmetric criticality [54] can be applied to time-crystal as well since, with the above time-crystal ansatz not only the Skyrme field equations reduce to the sine-Gordon system but also the full Skyrme action reduces to the corresponding sine-Gordon action. Thus, the low energy semi-classical fluctuations around the time-crystals constructed here are described by the semi-classical analysis of sine-Gordon theory. Hence, well-known results on sine-Gordon theory suggest that such time-crystals should also be present at semi-classical level.

4.1. The chemical potential

We can introduce the chemical potential as in the previous section. The full Skyrme field equations with isospin chemical potential reduce to just one scalar partial differential equation for the profile H,

$$\begin{aligned}&\left(\lambda+4\ell^2-\lambda\ell^2\omega^2+8\lambda\ell^2\bar{\mu}(\omega-2\bar{\mu})\sin^2(H)\right)\Delta H\\&+4\lambda\ell^2\bar{\mu}(\omega-2\bar{\mu})\sin(2H)\nabla_\mu H\nabla^\mu H\\&+4\ell^2\bar{\mu}(2\bar{\mu}-\omega)\sin(2H)+\frac{\lambda}{4}(\omega-2\bar{\mu})^2\sin(4H)=0\,.\end{aligned} \quad (41)$$

In this case the critical chemical potential $\overline{\mu}^*$ corresponding to time crystal can be found easily as in the previous section (see the comments below Eqs. (32) and (33)). In particular, let us consider

[3] When $Y(H)$ becomes negative, the numerator in the left hand side of Eq. (34) develops an imaginary part which cannot be compensated by the denominator.

[4] It is worth to note that there is a critical value ω_c for the frequency ω of the time crystal (defined in Eq. (38)) at which Eq. (37) becomes degenerate. On the other hand, in the case of the Skyrmions described in the previous section the theory is defined for all values of the parameters of the model.

[5] Previous literature on the analogies between sine-Gordon and Skyme models can be found in [56–58] and references therein. As it has been emphasized previously, sine-Gordon theory was believed to be just a "toy model" for the $3+1$ dimensional Skyrme model. In fact, we proved that in a nontrivial topological sector they exactly coincide.

a kink-like solution of Eq. (41) in which the profile only depends on one coordinate $H = H(r)$ and satisfying the boundary condition in Eq. (18). Then, Eq. (41) reads

$$Y_1(H)\frac{(H')^2}{2} + V_1(H) = E_0 , \quad (42)$$

where

$$Y_1(H) = \lambda + \ell^2\left[4 - \lambda\omega^2 + 8\lambda\bar{\mu}(\omega - 2\bar{\mu})\sin^2(H)\right],$$

$$V_1(H) = -\frac{\lambda}{16}(\omega - 2\bar{\mu})^2\cos(4H) - 2\ell^2\bar{\mu}(2\bar{\mu} - \omega)\cos(2H).$$

Thus, the critical chemical potential $\overline{\mu}^*$ can be determined by requiring

$$\lambda + \ell^2\left[4 - \lambda\omega^2 + 8\lambda\overline{\mu}^*(\omega - 2\overline{\mu}^*)\right] \leq 0 , \quad (43)$$

since, when this happens, the boundary condition in Eq. (18) cannot be satisfied anymore. It is also interesting to note that high values (compared to λ) of the time-crystal frequency ω^2 decrease the critical chemical potential: hence, low values of ω^2 are favored from the thermodynamical point of view.

There is a further special value for the chemical potential (which does not coincide with the one defined in Eq. (43)) for time-crystal configurations. Indeed, for $\overline{\mu} = \omega/2$, the non-linear partial differential equation Eq. (41) reduces to the linear $\triangle H = 0$, which, obviously, possesses more symmetries than the generic one for $\overline{\mu} < \omega/2$. Since this value $\overline{\mu} = \omega/2$ of the isospin chemical potential corresponds to a symmetry enhancement of the field equations, it is natural to wonder whether it is related to some phase transition of the system. We hope to come back on this interesting issue in a future publication.

5. Conclusions

We constructed the first analytic examples of Skyrmions as well as of Skyrmions–anti-Skyrmions bound states on flat spaces at finite volume. We have derived an analytic upper bound for the number of Skyrmion–anti-Skyrmion bound states in terms of the parameters of the model. The critical isospin chemical potential can also be computed. With the same formalism, one can build topologically protected time-crystals: these are exact configurations of the Skyrme model whose time-dependence is topologically protected by the non-vanishing winding number. We computed the corresponding critical isospin chemical potential and determined a possible experimental signature of these time-crystals.

The present construction answers positively to the question posed in [39] on the existence of a classical time crystal in systems possessing non-vanishing topological charges. In fact, as these classical configurations are topologically protected, the presence of quantum fluctuations cannot destroy them. Using the results presented above it is easy to show that these configurations are also present in the semi-classical quantization of the model. The reason why the powerful no-go theorems in [41] [42] do not apply in the present case is that, in these theorems (both explicitly and implicitly) it is assumed that the ground state of the theory is static. On the other hand, in theories with non-Abelian internal symmetries each non-trivial topological sector has its own ground state.

For instance, in the case of non-Abelian gauge theories admitting BPS monopoles, the ground state in the sector with unit non-Abelian magnetic charge is the well-known BPS monopole which cannot be deformed continuously to the trivial vacuum. Such a ground state is not invariant under the full symmetry group of the trivial ground state $|0>$. In particular, it is not invariant under spatial rotations (unless they are compensated by internal rotations). This fact, is the origin of the "spin from isospin effect" discovered in the seventies. In our case, in the sectors we have named time-crystal, the ground state is time-periodic and consequently the theorems in [41] and [42] do not apply, as we discussed above.

In the present paper, we focused on the $SU(2)$ Skyrme model, but our results can be generalized to any theory with $SU(N)$ internal symmetry.

Acknowledgements

This work has been funded by the FONDECYT grants 1160137, 1161150, 3150016 and 3160121. The Centro de Estudios Científicos (CECs) is funded by the Chilean Government through the Centers of Excellence Base Financing Program of Conicyt.

References

[1] T. Skyrme, Proc. R. Soc. Lond. A 260 (1961) 127;
T. Skyrme, Proc. R. Soc. Lond. A 262 (1961) 237;
T. Skyrme, Nucl. Phys. 31 (1962) 556.
[2] T.S. Walhout, Nucl. Phys. A 531 (1991) 596.
[3] C. Adam, M. Haberichter, A. Wereszczynski, Phys. Rev. C 92 (2015) 055807.
[4] J.-i Fukuda, S. Žumer, Nat. Commun. 2 (2011) 246.
[5] H. Stefan, et al., Nat. Phys. 7 (2011) 713.
[6] D. Fostar, S. Krusch, Nucl. Phys. B 897 (2015) 697.
[7] M. Gillard, Nucl. Phys. B 895 (2015) 272.
[8] E. Witten, Nucl. Phys. B 223 (1983) 422;
E. Witten, Nucl. Phys. B 223 (1983) 433.
[9] D. Finkelstein, J. Rubinstein, J. Math. Phys. 9 (1968) 1762–1779.
[10] N. Manton, P. Sutcliffe, Topological Solitons, Cambridge University Press, Cambridge, 2007.
[11] V.G. Makhanov, Y.P. Rybakov, V.I. Sanyuk, The Skyrme Model, Springer-Verlag, 1993.
[12] D. Giulini, Mod. Phys. Lett. A 8 (1993) 1917–1924.
[13] A.P. Balachandran, A. Barducci, F. Lizzi, V.G.J. Rodgers, A. Stern, Phys. Rev. Lett. 52 (1984) 887.
[14] G.S. Adkins, C.R. Nappi, E. Witten, Nucl. Phys. B 228 (1983) 552–566.
[15] E. Guadagnini, Nucl. Phys. B 236 (1984) 35–47.
[16] W. Münzer, et al., Phys. Rev. B 81 (2010) 041203.
[17] J.-y. Choi, W.J. Kwon, Y.-i. Shin, Phys. Rev. Lett. 108 (2012) 035301.
[18] K. Benson, M. Bucher, Nucl. Phys. B 406 (1993) 355.
[19] L. Parisi, N. Radicella, G. Vilakis, Phys. Rev. D 91 (2015) 063533.
[20] S. Droz, M. Heusler, N. Straumann, Phys. Lett. B 268 (1991) 371.
[21] H. Luckock, I. Moss, Phys. Lett. B 176 (1986) 341.
[22] N. Shiki, N. Sawado, Phys. Rev. D 71 (2005) 104031.
[23] T. Ioannidou, B. Kleihaus, J. Kunz, Phys. Lett. B 643 (2006) 213.
[24] S.B. Gudnason, M. Nitta, N. Sawado, J. High Energy Phys. 1609 (2016) 055.
[25] G. Dvali, A. Gußmann, Nucl. Phys. B 913 (2016) 1001.
[26] S. Seki, M. Mochizuki, Skyrmions in Magnetic Materials, Springer, 2016.
[27] I. Klebanov, Nucl. Phys. B 262 (1985) 133.
[28] A. Actor, Phys. Lett. B 157 (1985) 53.
[29] H.A. Weldon, Phys. Rev. D 26 (1982) 1394.
[30] M. Loewe, S. Mendizabal, J.C. Rojas, Phys. Lett. B 632 (2006) 512.
[31] J.A. Ponciano, N.N. Scoccola, Phys. Lett. B 659 (2008) 551.
[32] F. Canfora, H. Maeda, Phys. Rev. D 87 (2013) 084049.
[33] F. Canfora, Phys. Rev. D 88 (2013) 065028.
[34] F. Canfora, F. Correa, J. Zanelli, Phys. Rev. D 90 (2014) 085002.
[35] S. Chen, Y. Li, Y. Yang, Phys. Rev. D 89 (2014) 025007.
[36] E. Ayon-Beato, F. Canfora, J. Zanelli, Phys. Lett. B 752 (2016) 201–205.
[37] F. Canfora, G. Tallarita, Nucl. Phys. B 921 (2017) 394.
[38] F. Wilczek, Phys. Rev. Lett. 109 (2012) 160401.
[39] A. Shapere, F. Wilczek, Phys. Rev. Lett. 109 (2012) 160402.
[40] F. Wilczek, Phys. Rev. Lett. 111 (2013) 250402.
[41] P. Bruno, Phys. Rev. Lett. 111 (2013) 070402;
P. Bruno, Phys. Rev. Lett. 110 (2013) 118901;
P. Bruno, Phys. Rev. Lett. 111 (2013) 029301.
[42] H. Watanabe, M. Oshikawa, Phys. Rev. Lett. 114 (2015) 251603.
[43] K. Sacha, J. Zakrzewski, Time crystals: a review, arXiv:1704.03735.
[44] K. Sacha, Phys. Rev. A 91 (2015) 033617.
[45] S. Choi, J. Choi, R. Landig, G. Kucsko, H. Zhou, J. Isoya, F. Jelezko, S. Onoda, H. Sumiya, V. Khemani, C. von Keyserlingk, N.Y. Yao, E. Demler, M.D. Lukin, Nature 543 (7644) (2017) 221–225, letter.

[46] J. Zhang, P.W. Hess, A. Kyprianidis, P. Becker, A. Lee, J. Smith, G. Pagano, I.-D. Potirniche, A.C. Potter, A. Vishwanath, N.Y. Yao, C. Monroe, Nature 543 (7644) (2017) 217–220, letter.
[47] N.Y. Yao, A.C. Potter, I.-D. Potirniche, A. Vishwanath, Phys. Rev. Lett. 118 (2017) 030401.
[48] D.V. Else, B. Bauer, C. Nayak, Phys. Rev. Lett. 117 (2016) 090402;
D.V. Else, B. Bauer, C. Nayak, Phys. Rev. Lett. 118 (2017) 030401.
[49] V. Khemani, A. Lazarides, R. Moessner, S.L. Sondhi, Phys. Rev. Lett. 116 (2016) 250401.
[50] Y.M. Shnir, Magnetic Monopoles, Springer-Verlag, Berlin, Heidelberg, 2005, p. 500.
[51] Jesús Cuevas-Maraver, Panayotis G. Kevrekidis, Floyd Williams (Eds.), The Sine-Gordon Model and Its Applications: from Pendula and Josephson Junctions to Gravity and High-Energy Physics, Springer, 2014.
[52] S. Coleman, Phys. Rev. D 11 (1975) 2088.
[53] J.K. Perring, T.H.R. Skyrme, Nucl. Phys. 31 (1962) 550–555.
[54] R.S. Palais, Commun. Math. Phys. 69 (1979) 19–30.
[55] I. Zahed, G.E. Brown, Phys. Rep. 142 (1 & 2) (1986) 1–102.
[56] S.B. Gudnason, M. Nitta, Phys. Rev. D 90 (2014) 085007.
[57] M. Nitta, Nucl. Phys. B 895 (2015) 288.
[58] M. Eto, M. Nitta, Phys. Rev. D 91 (2015) 085044.

10

Invariant vacuum

Salvador Robles-Pérez [a,b,*]

[a] Estación Ecológica de Biocosmología, Pedro de Alvarado 14, 06411 Medellín, Spain
[b] Instituto de Física Fundamental, CSIC, Serrano 121, 28006 Madrid, Spain

A R T I C L E   I N F O

Editor: M. Cvetič

A B S T R A C T

We apply the Lewis–Riesenfeld invariant method for the harmonic oscillator with time dependent mass and frequency to the modes of a charged scalar field that propagates in a curved, homogeneous and isotropic spacetime. We recover the Bunch–Davies vacuum in the case of a flat DeSitter spacetime, the equivalent one in the case of a closed DeSitter spacetime and the invariant vacuum in a curved spacetime that evolves adiabatically. In the three cases, it is computed the thermodynamical magnitudes of entanglement between the modes of the particles and antiparticles of the invariant vacuum, and the modification of the Friedmann equation caused by the existence of the energy density of entanglement. The amplitude of the vacuum fluctuations are also computed.

1. Introduction

All the machinery of a quantum field theory is ultimately rooted on the definition of the vacuum state. Once this is defined a Fock space can be generated from the number eigenstates of the corresponding representation and the general quantum state of the field can be written as a vector of such space. The field can then be interpreted as composed of many particles propagating along the spacetime.

However, the definition of the vacuum state and the associated definition of particle cannot be always unambiguously stated in a curved spacetime. The most appropriate definition of the vacuum state in a local region of the spacetime may not correspond to the vacuum state in another local region, and that may lead to the creation of particles [1–6]. The question is then which vacuum state has to be selected from the set of possible vacuum states, with a twofold consideration: which quantum representation can determine the appropriate boundary condition for the field; and, which one can represent the observable particles.

A customary approach [7,8] is to define the vacuum state in an "IN" and "OUT" regions that asymptotically behave like Minkowski spacetime, where the vacuum state is therefore well defined. The corresponding "IN" vacuum is assumed to supply the initial boundary condition for the field and the "OUT" vacuum is expected to define the kind of measurable particles. Generally, the result is that the initial vacuum state turns out to be full of particles of the "OUT" representation. A problem with this approach is that it is not always possible to find in a curved spacetime two asymptotically flat regions where to define these vacuum states. That might wrongly induce us to think that a well defined vacuum state cannot be then given.

In this paper we shall adopt a different point of view. On the one hand, one would expect that the appropriate boundary condition for a cosmological field should be global, i.e. not tied to a local initial state, and such that the field should remain in the same state along the entire evolution of the field if no external force is present. In that case the state of the field should be invariant under time evolution. Furthermore, in cosmology there is no external element to the universe[1] so in particular, one would expect the field to stay in the ground state or the state of minimal excitation of some invariant representation.

In most cases of interest the wave equation of the field modes in a curved spacetime turns out to be the wave equation of a harmonic oscillator with time dependent mass and frequency. Then, we can apply the method of the invariants of the harmonic oscillator, developed by Lewis–Riesenfield [9,10] and others [11–16], to find the invariant representation of the field modes. The important property of the invariant representation is that the associated number operator turns out to be a constant of motion. It means that once the field is in a given quantum superposition of the number eigenstates of the invariant representation it remains in

* Correspondence to: Instituto de Física Fundamental, CSIC, Serrano 121, 28006 Madrid, Spain.
E-mail address: salvarp@imaff.cfmac.csic.es.

[1] We are not considering a multiverse scenario here. If that would be the case the same would apply to the multiverse as a whole instead of a single universe.

the same state along the entire evolution of the field. In particular, if the field is in the vacuum state of the invariant representation at a given moment of time it will remain in the same vacuum state along the entire evolution of the field.

Then, we shall assume that the field is in the vacuum state of the invariant representation. Furthermore, instead of imposing an *initial* condition on the state of the field at some given time t_0, we shall impose the *boundary* condition that the largest modes of the field must be the positive frequency modes of a field that propagates in a Minkowski spacetime. This is a boundary condition that is ultimately rooted in the equivalence principle of the theory of relativity. For a sufficiently closed neighborhood, the spacetime looks always like a flat spacetime and, therefore, the largest modes of the field must not feel the curvature of the spacetime. This boundary condition will fix the invariant representation to be used and, thus, it will fix the invariant vacuum state.

In terms of the invariant representation the invariant vacuum state will then represent the ground state along the entire evolution of the field. However, in terms of the number states of any other representation the vacuum state of the invariant representation may contain particles. Let us notice that the concept of particle is a local concept that is based on the definition of the particle detector and, thus, the number of detected particles is an observer-dependent quantity. In particular, for an observer that is making measurements in a local region of the spacetime, the most appropriate representation of the vacuum seems to be the vacuum of instantaneous Hamiltonian diagonalization [8], which represents the state of minimal excitation at a given moment of time. More concretely, an actual detector will only detect particles with wavelength smaller than the characteristic length of the detector. We shall then show that such a detector will in practice detect no particles in a small local region of the spacetime because, as a consequence of the boundary condition, the field modes remain there in the vacuum state along the entire evolution of the field. However, on cosmological grounds, the invariant vacuum turns out to be full of particle-antiparticle pairs of the diagonal representation, which are created in entangled states. We can then analyze the quantum state of each component of the entangled pair and their evolution separately.

The paper is outlined as follows. In Sect. 2 we briefly review the customary procedure of canonical quantization of a charged scalar field. In Sect. 3 we obtain the invariant representation of the associated Hamiltonian and define the invariant vacuum state. In Sect. 4 we apply the results to the case of a DeSitter spacetime and in Sect. 5 the same is done for a homogeneous and isotropic spacetime that evolves adiabatically. Finally, we summarize and draw some conclusions in Sect. 6.

2. Field quantization

Let us briefly summarize the standard procedure of canonical quantization for a charged scalar field, $\phi(x) = \phi(\mathbf{x}, t)$, by starting from the action integral

$$S = \int dt d^3\mathbf{x}\, \mathcal{L} = \int dt\, L, \tag{1}$$

with the Lagrangian density $\mathcal{L}$ given by [5,7,17,18]

$$\mathcal{L}(x) = \sqrt{-g}\left(g^{\mu\nu}\partial_\mu\phi\partial_\nu\phi^* - \left(m^2 + \xi R(x)\right)\phi(x)\phi^*(x)\right), \tag{2}$$

where m is the mass of the field and $g_{\mu\nu}$ is the metric tensor, with $g \equiv \det(g_{\mu\nu})$. The coupling between the scalar field and the gravitational field is represented by the term $\xi R\phi^2$, where $R(x)$ is the Ricci scalar. The value $\xi = 0$ corresponds to the so-called minimal coupling and the value $\xi = \frac{1}{6}$ corresponds to the conformal coupling. Unless otherwise indicated, we shall assume minimal coupling ($\xi = 0$) but a similar procedure can be followed with any other value of ξ. The variational principle of the action (1) yields the field equation

$$\left(\Box_x + m^2 + \xi R(x)\right)\phi(x) = 0, \tag{3}$$

where the d'Alembertian operator $\Box_x$ is given by [7]

$$\Box_x\phi = g^{\mu\nu}\nabla_\mu\nabla_\nu\phi = \frac{1}{\sqrt{-g}}\partial_\mu\left(\sqrt{-g}g^{\mu\nu}\partial_\nu\phi\right). \tag{4}$$

In particular, let us consider a homogeneous and isotropic spacetime with metric element given by

$$ds^2 = dt^2 - a^2\, dl^2, \tag{5}$$

where, $a = a(t)$ is the scale factor and $dl^2 = h_{ij}dx^i dx^j$, is the metric element of the three dimensional space with the constant curvature $\kappa = 0, \pm1$. It is customary to work in conformal time η, and to scale the scalar field according to, $\phi = a^{-1}\chi$. In that case, the modes of the field χ satisfy the wave equation of a harmonic oscillator with constant mass and time dependent frequency. However, we shall work in cosmic time t and retain the charged scalar field $\phi(\mathbf{x}, t)$ for at least for three reasons: i) the scaling is unnecessary for obtaining the invariant representation of the scalar field $\phi(x)$; ii) unlike in the wave equation of χ, the frequency of the wave equation of ϕ is always real, so we shall avoid imaginary values of the frequency of the modes; and, iii) the invariant representation of any two field variables is the same provided that they are related by a canonical transformation, i.e. the invariant representation of the field $\chi(x)$ is also the invariant representation of the field $\phi(x)$, so the vacuum state of the invariant representation is the same for both fields.

The isotropy of the spacetime described by the metric (5) allows us to expand the field in Fourier modes

$$\phi(\mathbf{x}, t) = \int d\mu(k)\psi_{\mathbf{k}}(\mathbf{x})\phi_{\mathbf{k}}(t), \tag{6}$$

where $\psi_{\mathbf{k}}$ are the eigenfunctions of the three-dimensional Laplacian,

$$\Delta^{(3)}\psi_{\mathbf{k}}(\mathbf{x}) = -(k^2 - \kappa)\psi_{\mathbf{k}}(\mathbf{x}), \tag{7}$$

and, $k = |\mathbf{k}|$ with $\mathbf{k} = \{k_x, k_y, k_z\}$ with $-\infty < k_i < \infty$ in the flat case, or just k in $\mathbf{k} = \{k, l, m\}$ with $0 < k < \infty$, $l = 0, 1, 2, \ldots$ in the open case, $k = 1, 2, \ldots$ and $l = 0, 1, \ldots, k-1$ in the closed case, with $-l \le m \le l$ in both cases, and $d\mu(k)$ is the measure of the Fourier space (see Refs. [4,5,7] for the details). With (6) and (7), integrating by parts and using the orthogonality properties of the functions $\psi_{\mathbf{k}}(\mathbf{x})$ [7], the Lagrangian in (1) turns out to be

$$L = \int d\mu(k) M(t)\left\{\dot\phi_{\mathbf{k}}\dot\phi^*{}_{\mathbf{k}} - \omega_k^2(t)\,\phi_{\mathbf{k}}\phi_{\mathbf{k}}^*\right\}, \tag{8}$$

where, $M(t) = a^3(t)$,

$$\omega_k^2(t) = \frac{k^2 - \kappa}{a^2} + m^2 + \xi R. \tag{9}$$

The Lagrangian (8) is the Lagrangian of a set of harmonic oscillators with time dependent mass and frequency. Let us now proceed to quantize the field modes by writing [5,7,8]

$$\phi_{\mathbf{k}}(t) = \frac{1}{\sqrt{2}}\left(v_k(t)a_{\mathbf{k}} + (-1)^{\kappa m}v_k^*(t)b_{-\mathbf{k}}^\dagger\right), \tag{10}$$

where, $-\mathbf{k} = \{-k_x, -k_y, -k_z\}$, in the flat case and, $-\mathbf{k} = \{k, l, -m\}$ in the open and closed cases and, $\psi_{\mathbf{k}}^* = (-1)^{\kappa m}\psi_{-\mathbf{k}}$, for $\kappa = 0, \pm1$.

In (10), $a_{\mathbf{k}}^{\dagger}$ and $a_{\mathbf{k}}$ are constant operators that describe the creation and annihilation operators of particles and $b_{\mathbf{k}}^{\dagger}$ and $b_{\mathbf{k}}$ are those for antiparticles. They obey the standard commutation relations

$$[a_{\mathbf{k}}, a_{\mathbf{k}'}^{\dagger}] = \delta(\mathbf{k} - \mathbf{k}'), [a_{\mathbf{k}}, a_{\mathbf{k}'}] = [a_{\mathbf{k}}^{\dagger}, a_{\mathbf{k}'}^{\dagger}] = 0, \quad (11)$$

$$[b_{\mathbf{k}}, b_{\mathbf{k}'}^{\dagger}] = \delta(\mathbf{k} - \mathbf{k}'), [b_{\mathbf{k}}, b_{\mathbf{k}'}] = [b_{\mathbf{k}}^{\dagger}, b_{\mathbf{k}'}^{\dagger}] = 0, \quad (12)$$

and define the vacuum state, $|0_a 0_b\rangle = |0\rangle_a |0\rangle_b$, as usual by the relation

$$a_{\mathbf{k}}|0\rangle_a = 0 \ , \ b_{\mathbf{k}}|0\rangle_b = 0, \quad (13)$$

for all $\mathbf{k}$. The field amplitudes, $v_k(t)$ in (10), satisfy then

$$\ddot{v}_k + \frac{\dot{M}}{M}\dot{v}_k + \omega_k^2(\eta) v_k = 0. \quad (14)$$

Because the time dependence of the mass and frequency of the harmonic oscillator (14) the vacuum state defined at t_0 contains particles and antiparticles at any other moment of time t_1. Therefore, it does not represent the no particle state along the evolution of the scalar field.

3. Invariant vacuum state

3.1. Classical description

There is a quantum representation that can describe a non-particle state along the entire evolution of the scalar field. It is given by the invariant representation. We shall briefly sketch the general procedure developed in Refs. [9–16]. Particularly, we shall closely follow the formulation given in Refs. [11,14]. Let us therefore consider the following point transformation

$$\zeta_{\mathbf{k}} = \frac{1}{\sigma}\phi_{\mathbf{k}}, \quad (15)$$

where $\sigma \equiv \sigma_k(t)$ is an auxiliary real function that satisfies the non-linear equation

$$\ddot{\sigma} + \frac{\dot{M}}{M}\dot{\sigma} + \omega_k^2\sigma = \frac{k^2}{M^2\sigma^3}, \quad (16)$$

with the frequency ω_k being given by (9). Let us here notice that a solution of (16) can be generally given by[2]

$$\sigma = \sqrt{\sigma_1^2 + \sigma_2^2}, \quad (17)$$

where σ_1 and σ_2 are two real independent solutions of

$$\ddot{\sigma}_{1,2} + \frac{\dot{M}}{M}\dot{\sigma}_{1,2} + \omega_k^2\sigma_{1,2} = 0, \quad (18)$$

with the normalization condition, $\sigma_1\dot{\sigma}_2 - \sigma_2\dot{\sigma}_1 = \frac{k}{M}$. Let us also perform the following change of time variable, $t \to \tau_k$, given by

$$d\tau_k = \frac{1}{M\sigma_k^2}dt. \quad (19)$$

Then, the action (1) with the Lagrangian (8) transforms into

$$S = \int d\mu(k) S_{\mathbf{k}}, \quad (20)$$

where

$$S_{\mathbf{k}} = \int d\tau_k \left\{ \frac{d\zeta_{\mathbf{k}}}{d\tau_k}\frac{d\zeta_{\mathbf{k}}^*}{d\tau_k} - k^2\zeta_{\mathbf{k}}^*\zeta_{\mathbf{k}} \right\}, \quad (21)$$

is the action of a harmonic oscillator with constant frequency k. The action (20) is the sum of the actions of a set of uncoupled harmonic oscillators, each one evolving however with a different time variable, τ_k. The momenta conjugated to $\zeta_{\mathbf{k}}$ and $\zeta_{\mathbf{k}}^*$ are

$$\tilde{\pi}_{\mathbf{k}} = \frac{d\zeta_{\mathbf{k}}^*}{d\tau_k} \ , \ \tilde{\pi}_{\mathbf{k}}^* = \frac{d\zeta_{\mathbf{k}}}{d\tau_k}, \quad (22)$$

and the corresponding Hamiltonian reads

$$\tilde{H}_{\mathbf{k}} = \tilde{\pi}_{\mathbf{k}}\tilde{\pi}_{\mathbf{k}}^* + k^2\zeta_{\mathbf{k}}\zeta_{\mathbf{k}}^*. \quad (23)$$

The wave equation for the field $\zeta_{\mathbf{k}}(\tau_k)$ is

$$\frac{d^2\zeta_{\mathbf{k}}}{d\tau_k^2} + k^2\zeta_{\mathbf{k}}^2 = 0, \quad (24)$$

with normalized solutions given by

$$\zeta_{\mathbf{k}} = \frac{1}{\sqrt{k}}e^{-ik\tau_k}, \quad (25)$$

which is positive frequency with respect to τ_k. Then, the corresponding solutions of the original field modes are

$$\phi_{\mathbf{k}} = \frac{\sigma}{\sqrt{k}}e^{-ik\tau_k} = \frac{\sigma}{\sqrt{k}}e^{-ik\int\frac{1}{M\sigma^2}dt}. \quad (26)$$

The invariant value of the field ϕ relies then in the computation of the auxiliary function σ. In order to fix the value of σ we must impose a boundary condition. For this, one has to realize that in terms of the rescaled field, $\chi = a\phi$, and in conformal time, $\eta = \int\frac{dt}{a}$, the wave equation (14) becomes, in the limit of large modes of the field, the customary equation of a harmonic oscillator with unit mass and constant frequency k, i.e.

$$\chi_k'' + k^2\chi_k = 0, \quad (27)$$

where the prime denotes derivative with respect to the conformal time. The positive frequency solutions of (27) are

$$\chi_k(\eta) = \frac{1}{\sqrt{k}}e^{-ik|\eta|}. \quad (28)$$

Then, in order for the field modes (26) to be the modes associated to the positive frequency solutions (28) we have to impose the boundary condition

$$\sigma = a^{-1} \ , \ (\Rightarrow \tau_k = \eta), \quad (29)$$

in the limit of large modes, $k \gg 1$. The normalization condition given after (18) and the boundary condition (29) fix the invariant representation to be used and, thus, they fix the invariant vacuum state.

Let us finally point out that the transformation

$$(\phi_{\mathbf{k}}, \phi_{\mathbf{k}}^*, p_{\phi_{\mathbf{k}}}, p_{\phi_{\mathbf{k}}}^*; t) \to (\zeta_{\mathbf{k}}, \zeta_{\mathbf{k}}^*, \pi_{\mathbf{k}}, \pi_{\mathbf{k}}^*; \tau_k), \quad (30)$$

is a canonical transformation given by

$$\zeta_{\mathbf{k}} = \frac{1}{\sigma}\phi_{\mathbf{k}} \ , \ \pi_{\mathbf{k}} = \sigma p_{\phi_{\mathbf{k}}} - M\dot{\sigma}\phi_{\mathbf{k}}^*, \quad (31)$$

$$\zeta_{\mathbf{k}}^* = \frac{1}{\sigma}\phi_{\mathbf{k}}^* \ , \ \pi_{\mathbf{k}}^* = \sigma p_{\phi_{\mathbf{k}}}^* - M\dot{\sigma}\phi_{\mathbf{k}}, \quad (32)$$

which is generated by the following generating function [14]

$$F_2(\phi_{\mathbf{k}}, \phi_{\mathbf{k}}^*, \pi_{\mathbf{k}}, \pi_{\mathbf{k}}^*) = \frac{1}{\sigma}\left(\phi_{\mathbf{k}}\pi_{\mathbf{k}} + \phi_{\mathbf{k}}^*\pi_{\mathbf{k}}^*\right) + \frac{M\dot{\sigma}}{\sigma}\phi_{\mathbf{k}}\phi_{\mathbf{k}}^*, \quad (33)$$

through the relations [14]

[2] For more general solutions of (16) see Ref. [11].

$$p_{\phi_{\mathbf{k}}} = \frac{\partial}{\partial \phi_{\mathbf{k}}} F_2 \, , \; p^*_{\phi_{\mathbf{k}}} = \frac{\partial}{\partial \phi^*_{\mathbf{k}}} F_2 \tag{34}$$

$$\zeta_{\mathbf{k}} = \frac{\partial}{\partial \pi_{\mathbf{k}}} F_2 \, , \; \zeta^*_{\mathbf{k}} = \frac{\partial}{\partial \pi^*_{\mathbf{k}}} F_2 \tag{35}$$

$$H(\zeta_{\mathbf{k}}, \pi_{\mathbf{k}}, \zeta^*_{\mathbf{k}}, \pi^*_{\mathbf{k}}) \dot{t}_k = H(\phi_{\mathbf{k}}, p_{\phi_{\mathbf{k}}}, \phi^*_{\mathbf{k}}, p^*_{\phi_{\mathbf{k}}}; t) + \partial_t F_2 \tag{36}$$

3.2. Invariant creation and annihilation operators

The Hamiltonian (23) is the Hamiltonian of a harmonic oscillator with unit mass and constant frequency given by k. Thus, the creation and annihilation operators defined in terms of the field $\zeta_{\mathbf{k}}$ are invariant under the time evolution. Therefore, the annihilation operator of the invariant representation defines a vacuum state that is stable along the entire evolution of the scalar field. The invariant representation of particles and antiparticles, $\tilde{a}_{\mathbf{k}}$, $\tilde{a}^\dagger_{\mathbf{k}}$ and $\tilde{b}_{-\mathbf{k}}$, $\tilde{b}^\dagger_{-\mathbf{k}}$, respectively, is defined in terms of the invariant field and its conjugated momenta as usual, by

$$\zeta_{\mathbf{k}} = \frac{1}{\sqrt{2k}} (\tilde{a}_{\mathbf{k}} + \tilde{b}^\dagger_{-\mathbf{k}}) \, , \; \pi^*_{\mathbf{k}} = -i \sqrt{\frac{k}{2}} (\tilde{a}_{\mathbf{k}} - \tilde{b}^\dagger_{-\mathbf{k}}), \tag{37}$$

$$\zeta^*_{\mathbf{k}} = \frac{1}{\sqrt{2k}} (\tilde{b}_{-\mathbf{k}} + \tilde{a}^\dagger_{\mathbf{k}}) \, , \; \pi_{\mathbf{k}} = -i \sqrt{\frac{k}{2}} (\tilde{b}_{-\mathbf{k}} - \tilde{a}^\dagger_{\mathbf{k}}), \tag{38}$$

in terms of which the Hamiltonian (23) reads

$$H_{\mathbf{k}} = k \left(\tilde{a}^\dagger_{\mathbf{k}} \tilde{a}_{\mathbf{k}} + \tilde{b}^\dagger_{-\mathbf{k}} \tilde{b}_{-\mathbf{k}} + 1 \right). \tag{39}$$

Using (31)–(32) and the inverse relation of (37)–(38) we can express the invariant representation in terms of the original field modes $\phi_{\mathbf{k}}$ and the conjugated momentum $p_{\mathbf{k}}$. It yields (see the analogy with the invariant representation given in Refs. [10,19])

$$\tilde{a}_{\mathbf{k}} = \sqrt{\frac{k}{2}} \left(\frac{1}{\sigma} \phi_{\mathbf{k}} + \frac{i}{k} (\sigma p^*_{\phi_{\mathbf{k}}} - M \dot{\sigma} \phi_{\mathbf{k}}) \right), \tag{40}$$

$$\tilde{b}_{-\mathbf{k}} = \sqrt{\frac{k}{2}} \left(\frac{1}{\sigma} \phi^*_{\mathbf{k}} + \frac{i}{k} (\sigma p_{\phi_{\mathbf{k}}} - M \dot{\sigma} \phi^*_{\mathbf{k}}) \right). \tag{41}$$

The important property of the invariant representation is that the eigenstates of the number operators of particles and antiparticles, $\tilde{N}^a_{\mathbf{k}} \equiv \tilde{a}^\dagger_{\mathbf{k}} \tilde{a}_{\mathbf{k}}$ and $\tilde{N}^b_{\mathbf{k}} \equiv \tilde{b}^\dagger_{\mathbf{k}} \tilde{b}_{\mathbf{k}}$, respectively, are stable along the entire evolution of the scalar field, because

$$\frac{d\tilde{N}_{\mathbf{k}}}{dt} = \dot{\tau} \frac{d\tilde{N}_{\mathbf{k}}}{d\tau} = -i\dot{\tau} [\tilde{N}_{\mathbf{k}}, H_{\mathbf{k}}] = 0. \tag{42}$$

It means that once the field is in a given eigenstate of the invariant number operator, or more generally in a quantum superposition of number eigenstates, it remains in the same state along the entire evolution of the spacetime. In particular, the vacuum state of the invariant representation, defined as $|\tilde{0}_a \tilde{0}_b\rangle = |\tilde{0}\rangle_a |\tilde{0}\rangle_b$, with

$$\tilde{a}_{\mathbf{k}} |\tilde{0}\rangle_a = 0 \, , \; \tilde{b}_{\mathbf{k}} |\tilde{0}\rangle_b = 0, \; \forall \, \mathbf{k}, \tag{43}$$

describes the no particle state along the entire evolution of the field irrespective of whether there is or not an asymptotically flat region of the spacetime. It is therefore a stable definition for the vacuum state and an appropriate representation to provide a global, observer-independent boundary condition for the state of the field.

3.3. Relation with the diagonal representation

At a given moment of time t_0, however, and for small changes around t_0 the representation that describes instantaneously the ground state of the Hamiltonian is the diagonal representation, $c_{\mathbf{k}}$, $c^\dagger_{\mathbf{k}}$ and $d_{-\mathbf{k}}$, $d^\dagger_{-\mathbf{k}}$, defined as

$$\phi_{\mathbf{k}} = \frac{1}{\sqrt{2M\omega_k}} \left(c_{\mathbf{k}} + d^\dagger_{-\mathbf{k}} \right), \tag{44}$$

$$p^*_{\phi_{\mathbf{k}}} = -i \sqrt{\frac{M\omega_k}{2}} \left(c_{\mathbf{k}} - d^\dagger_{-\mathbf{k}} \right). \tag{45}$$

The instantaneous diagonal representation of the Hamiltonian at a given moment of time cannot define a stable vacuum state of the field because it entails the continuous generation of particles detected by a local particle detector at any other moment of time [6]. This can easily be seen by noting that, because the time dependence of M and ω_k in (44)–(45), two different representations, $c_0 \equiv c_{\mathbf{k}}(t_0)$ and $d_0 \equiv d_{-\mathbf{k}}(t_0)$, and, $c_1 \equiv c_{\mathbf{k}}(t_1)$ and $d_1 \equiv d_{-\mathbf{k}}(t_1)$, at two given moments of time t_0 and t_1 are related by the Bogolyubov transformation

$$c_1 = \mu_0 c_0 - \nu^*_0 d^\dagger_0, \tag{46}$$

$$d_1 = \mu_0 d_0 - \nu^*_0 c^\dagger_0, \tag{47}$$

where

$$\mu_0 = \frac{1}{2} \left(\sqrt{\frac{M_1 \omega_1}{M_0 \omega_0}} + \sqrt{\frac{M_0 \omega_0}{M_1 \omega_1}} \right), \tag{48}$$

$$\nu_0 = = \frac{1}{2} \left(\sqrt{\frac{M_1 \omega_1}{M_0 \omega_0}} - \sqrt{\frac{M_0 \omega_0}{M_1 \omega_1}} \right), \tag{49}$$

with, $|\mu_0|^2 - |\nu_0|^2 = 1$, and, $M_{0,1} \equiv M(t_{0,1})$ and $\omega_{0,1} \equiv \omega_k(t_{0,1})$. In the limit of large modes the particles measured in a local region of the space at time t_1 would then be given, in an expanding universe, by

$$N(t_1) = |\nu_0|^2 \approx \frac{M_1 \omega_1}{4 M_0 \omega_0} \sim \left(\frac{a(t_1)}{a(t_0)} \right)^2 , \forall k \gg 1. \tag{50}$$

It means that a local particle detector would detect a large amount of particles in a large expanding universe like ours. It does not seem to be therefore a consistent boundary condition to impose that the field has to be in the vacuum state of the diagonal representation at a given initial time t_0.

A more appropriate boundary condition seems to be imposing that the field is in the vacuum state[3] of the invariant representation. First, because of (42), the invariant vacuum state represents the no-particle state along the entire evolution of the field. Then, in terms of the invariant representation there is no particle production at all for all time [7]. However, we shall assume that the measurable particles are given, in a local region, by the number states of the instantaneous diagonal representation. Even though, we shall now show that a local detector will in practice detect no particles within a small region of the spacetime. Let us first notice that the invariant representation (40)–(41) can be related to the diagonal representation (44)–(45) through the Bogolyubov transformation

$$\tilde{a}_{\mathbf{k}} = \mu(t) \, c_{\mathbf{k}} - \nu^*(t) \, d^\dagger_{-\mathbf{k}}, \tag{51}$$

$$\tilde{b}_{-\mathbf{k}} = \mu(t) \, d_{-\mathbf{k}} - \nu^*(t) \, c^\dagger_{\mathbf{k}}, \tag{52}$$

where

[3] Or generally speaking, in a linear combination of number states.

$$\mu(t)=\frac{1}{2}\left(\sigma\sqrt{\frac{M\omega_k}{k}}+\frac{1}{\sigma}\sqrt{\frac{k}{M\omega_k}}-i\dot{\sigma}\sqrt{\frac{M}{\omega_k k}}\right), \tag{53}$$

$$\nu(t)=\frac{1}{2}\left(\sigma\sqrt{\frac{M\omega_k}{k}}-\frac{1}{\sigma}\sqrt{\frac{k}{M\omega_k}}-i\dot{\sigma}\sqrt{\frac{M}{\omega_k k}}\right), \tag{54}$$

with, $|\mu|^2-|\nu|^2=1$ for all time. In the limit of a large value of the mode k, i.e. within a small volume of the space, $\omega_k\sim\frac{k}{a}$ and $\sigma\sim a^{-1}$ (see (29)), and thus

$$N_k=|\nu|^2\to\frac{\dot{a}^2}{4k^2}\sim\left(\frac{\lambda_{\text{ph}}}{H^{-1}}\right)^2, \tag{55}$$

where, $\lambda_{\text{ph}}=\frac{2\pi a}{k}$, is the physical wavelength of the mode and, $H^{-1}=\frac{a}{\dot{a}}$, is the curvature radius at a given time. It can be easily seen from (55) that for sub-horizon modes, $\lambda_{\text{ph}}\ll H^{-1}$, the field does not feel the curvature of the spacetime and these modes remain in the vacuum state. A local particle detector of a practical length scale would measure then no particle at all within a small region of the space, irrespective of the moment of time. On cosmological grounds, however, there is a significant production of modes[4] but this is not surprising in an expanding universe whose evolution is determined, according to the Friedmann equation, by the matter content of the universe. The energy of the spacetime is negative and it balances the energy of the matter fields so the total energy is zero (see, for instance, Ref. [20]). Therefore, in an expanding universe the energy of the field is not conserved and it grows as the universe expands.

3.4. Thermodynamical magnitudes of entanglement

Let us now assume that the field is in the vacuum state of the invariant representation. It seems to be an appropriate boundary condition because it means that the field will remain in the same vacuum state along the entire evolution of the field, with a quantum state described by the density matrix

$$\rho=|\tilde{0}_a\tilde{0}_b\rangle\langle\tilde{0}_a\tilde{0}_b|. \tag{56}$$

Using the Bogolyubov transformation (51)–(52) the vacuum state of the invariant representation can be written as [8]

$$|\tilde{0}_a\tilde{0}_b\rangle=\prod_{\mathbf{k}}\frac{1}{|\mu|}\left(\sum_{n=0}^{\infty}\left(\frac{\nu}{\mu}\right)^n|\,n_{c,\mathbf{k}}n_{d,-\mathbf{k}}\rangle\right), \tag{57}$$

where, $\mu\equiv\mu_k$ and $\nu\equiv\nu_k$, and

$$|n_{c,\mathbf{k}}\rangle=\frac{(c^\dagger_{\mathbf{k}})^n}{\sqrt{n!}}|0_{c,\mathbf{k}}\rangle\,,\ |n_{d,-\mathbf{k}}\rangle=\frac{(d^\dagger_{-\mathbf{k}})^n}{\sqrt{n!}}|0_{d,-\mathbf{k}}\rangle, \tag{58}$$

are the number states of the diagonal representation (44)–(45). It means that the vacuum state of the invariant representation is full of particle-antiparticle pairs created with opposite momenta in entangled states. Let us consider just the quantum state of the particles. The reduced density matrix that represents the quantum state of the particles alone can be obtained by tracing out from the density matrix (56) the degrees of freedom of the antiparticles. It typically yields [19–21]

$$\rho_c=\text{Tr}_d\rho=\prod_{\mathbf{k}}\frac{1}{Z_k}\sum_n e^{-\frac{1}{T_k}(n+\frac{1}{2})}|\,n_{c,\mathbf{k}}\rangle\langle n_{c,\mathbf{k}}|, \tag{59}$$

where, $Z_k^{-1}=2\sinh\frac{1}{2T_k}$, with a specific temperature of entanglement [21] given by

$$T_k\equiv T_k(t)=\frac{1}{\ln\frac{|\mu(t)|^2}{|\nu(t)|^2}}=\frac{1}{\ln\left(1+|\nu(t)|^{-2}\right)}. \tag{60}$$

The temperature of entanglement is a measure of the entanglement between the particles and antiparticles of the charged scalar field. Therefore, it is also a measure of the effects of the curvature of the spacetime. For a large value of the mode k, $T_k\to 0$, and there is thus no entanglement, as it is expected because the largest modes (or at the shortest distances) do not feel the curvature of the spacetime.

One can even define the thermodynamical magnitudes of entanglement associated to the quasi thermal state (59). They are given, for each mode, by [22]

$$E_k(t)=\frac{\omega_k}{2}\operatorname{cotanh}\frac{1}{2T_k}=\omega_k\left(N_k+\frac{1}{2}\right), \tag{61}$$

$$Q_k(t)=\frac{\omega_k}{2}\operatorname{cotanh}\frac{1}{2T_k}-\omega_k T_k\ln\sinh\frac{1}{2T_k}, \tag{62}$$

$$W_k(t)=\omega_k T_k\ln\sinh\frac{1}{2T_k}, \tag{63}$$

where, $N_k\equiv|\nu|^2$. The first principle of thermodynamics, $E_k(t)=Q_k(t)+W_k(t)$, is satisfied for all modes k individually, and the energy densities that correspond to E_n, Q_n, and W_n, are given by

$$\varepsilon_n=\frac{E_n}{V}\,,\ q_n=\frac{Q_n}{V}\,,\ w_n=\frac{W_n}{V}, \tag{64}$$

with, $V=a^3(t)$. The entropy of entanglement [22,23] can also be easily obtained from the von Neumann formula

$$S(\rho)=-\text{Tr}\,(\rho\ln\rho)\,, \tag{65}$$

with ρ given by (59). It yields [22]

$$S_{\text{ent}}(a)=|\mu|^2\ln|\mu|^2-|\nu|^2\ln|\nu|^2. \tag{66}$$

4. DeSitter spacetime

4.1. Flat DeSitter spacetime

Let us now consider a flat DeSitter spacetime described by the metric element (5) and a scale factor given by

$$a(t)=\frac{1}{H}e^{Ht}, \tag{67}$$

where, $-\infty<t<\infty$, and $\Lambda\equiv H^2$ is the cosmological constant. The invariant representation is given by (40)–(41) with the function σ being given by (17) with σ_1 and σ_2 satisfying

$$\ddot{\sigma}_{1,2}+3H\dot{\sigma}_{1,2}+\left(H^2k^2e^{-2Ht}+m^2\right)\sigma_{1,2}=0. \tag{68}$$

The two solutions that make σ in (17) satisfying the boundary condition (29) are,

$$\sigma_1(t)=H\sqrt{\frac{\pi k}{2}}e^{-\frac{3H}{2}t}\mathcal{J}_\mu(ke^{-Ht}), \tag{69}$$

$$\sigma_2(t)=H\sqrt{\frac{\pi k}{2}}e^{-\frac{3H}{2}t}\mathcal{Y}_\mu(ke^{-Ht}), \tag{70}$$

where

$$\mu=\sqrt{\frac{9}{4}-\frac{m^2}{H^2}}. \tag{71}$$

[4] These modes would be like *global* particles in the sense that their associated wavelength are of order of the curvature radius.

Let us notice that with the value of σ given by (17) with σ_1 and σ_2 given by (69)–(70), the invariant field modes (26) are nothing more than the modes associated to the Bunch–Davies vacuum, as it was expected. By using the properties of the Bessel functions [24], one can easily check that

$$\sigma = \sqrt{\sigma_1^2 + \sigma_2^2} = \frac{1}{a}\sqrt{\frac{\pi k}{2Ha}}|\mathcal{H}_\mu^{(2)}(x)|, \tag{72}$$

where $\mathcal{H}_\mu^{(2)}(x)$ is the Hankel function of second kind and order μ, with

$$x = ke^{-Ht} = \frac{k}{Ha} = k|\eta|, \tag{73}$$

where η is the conformal time and, $\theta \equiv -k\tau_k$, is the phase of the Hankel function satisfying (see (9.2.21) of Ref. [24])

$$|\mathcal{H}_\mu^{(2)}(x)|^2 \frac{d\theta}{dx} = \frac{2}{\pi x}. \tag{74}$$

Therefore,

$$\phi_{\mathbf{k}} = \frac{\sigma}{\sqrt{k}} e^{-ik|\tau_k|} = \frac{1}{a}\sqrt{\frac{\pi|\eta|}{2}}\mathcal{H}_\mu^{(2)}(k|\eta|). \tag{75}$$

It means that the Bunch–Davies vacuum corresponds to the invariant vacuum in the sense of the Lewis–Riesenfeld formulation too, as it was expected [25]. But it also means that we can find the values of σ and τ_k in (16) and (19), respectively, by computing the modulus and phase of the invariant wave function of the field propagating in a more general, curved spacetime.

In terms of the diagonal representation, the number of particles of the field in the flat DeSitter spacetime,

$$N_a = |\nu|^2 \approx \frac{9H^2}{16k_{\rm ph}^2} \sim \left(\frac{L_{\rm ph}}{H^{-1}}\right)^2, \tag{76}$$

on a given physical scale, $k_{\rm ph} = k/a$, does not depend on time (in the limit $k \gg 1$), and it is negligible for a practical detector of human length scale. For large values of the scale factor the energy density of the particles is given by

$$\varepsilon \approx \int_0^{k_{\rm m}} dk\, k^2 \varepsilon_k \propto \frac{9H^2 k_{\rm m}^2}{16a^2}. \tag{77}$$

It is therefore a function that decreases exponentially in time for an evolved universe like ours.

Finally, the amplitude of fluctuations of the field can be easily obtained from

$$\delta\phi_{\mathbf{k}} = \frac{k^{\frac{3}{2}}}{2\pi}\Delta\phi_{\mathbf{k}}, \tag{78}$$

with

$$(\Delta\phi_{\mathbf{k}})^2 = \langle|\phi_{\mathbf{k}}|^2\rangle - |\langle\phi_{\mathbf{k}}\rangle|^2 = \frac{\sigma^2}{2k}, \tag{79}$$

where the expected values are computed in the vacuum state of the invariant representation and $\phi_{\mathbf{k}}$ can be obtained from the inverse relation of (40)–(41). It gives the standard expressions for the spectrum of fluctuations [8]

$$\delta\phi(k_{\rm ph}) = \frac{H}{4\sqrt{\pi}}\left(\frac{k_{\rm ph}}{H}\right)^{\frac{3}{2}}\left(\mathcal{J}_\mu^2(\frac{k_{\rm ph}}{H}) + \mathcal{Y}_\mu^2(\frac{k_{\rm ph}}{H})\right)^{\frac{1}{2}}. \tag{80}$$

4.2. Closed DeSitter spacetime

The relation between the Lewis–Riesenfeld formalism of the invariant modes and the customary formulation of the invariant wave function is not restricted to the flat DeSitter spacetime and it is indeed quite general. Let us notice that σ and τ_k are nothing more the modulus and the phase of the wave function (26), and that the equations (16) and (19) are, respectively, the real and the complex parts of the wave equation (14) for the modes (26), i.e. inserting (26) in (14) one obtains (16) and (19). Then, the customary solutions of the modes of a scalar field in a curved spacetime are recovered here by taking the appropriate solutions of σ_1 and σ_2. On the other hand, the values of σ and τ_k can be obtained by computing the modulus and phase of the normalized solutions of the wave equation of the scalar field.

In the case of a closed DeSitter spacetime with metric element (5) and a scale factor given by

$$a(t) = \frac{1}{H}\cosh Ht, \tag{81}$$

where, $-\infty < t < \infty$, the functions $\sigma_1(t)$ and $\sigma_2(t)$ satisfy (18) with the frequency ω_k given by

$$\omega_k^2 = \frac{H^2(k^2-1)}{\cosh^2 Ht} + m^2. \tag{82}$$

The solutions of (18) with the frequency (82) can be written in terms of the hypergeometric functions [18,26,27], or equivalently in terms of Legendre functions [7,26,28]. As it is pointed out in Ref. [7], if one follows the procedures used in Ref. [26] (see also Refs. [27,29]) and defines the IN and OUT vacuum states by taking the positive frequency solutions of (18) in the asymptotic limits, $t \to \pm\infty$, one obtains an infinite particle production, irrespective of the value of k. It means that one should measure an infinite number of particles even in a small region of the space as the universe expands.

On the contrary, we are here imposing the boundary condition that the field is in the vacuum state of the invariant representation, in terms of which there is no particle production at all at any time because it represents the no particle state along the entire evolution of the field. Following Ref. [18] (see also, Ref. [27]) we can express the solutions of (18) in terms of the hypergeometric function as

$$\begin{aligned}\chi_k = \frac{1}{k!}&\sqrt{\Gamma(k+\frac{1}{2}-\mu)\Gamma(k+\frac{1}{2}+\mu)}e^{-ik\eta}\\ &\times F(\frac{1}{2}-\mu, \frac{1}{2}+\mu; 1+k; \frac{1-i\tan\eta}{2}),\end{aligned} \tag{83}$$

where, $-\frac{\pi}{2} < \eta < \frac{\pi}{2}$, is the conformal time and, $F(a,b;c;z) = {}_2F_1(a,b;c;z)$, is the hypergeometric function. By taking into account the expansion of (83) in powers of k, given by [18]

$$\chi_k = \frac{1}{\sqrt{k}}e^{-ik\eta}\left(1 + \mathcal{O}(k^{-1})\right), \tag{84}$$

one can easily check that the modes (83) reduce to (28) in the limit of large modes ($k \to \infty$), so the modes (83) already satisfy our boundary condition. However, in order to give an explicit expression of σ it is more convenient to rewrite the modes (83) as [7,28]

$$\chi_k(\eta) = N_k \cos^{\frac{1}{2}}\eta\left(P^\mu_{k-\frac{1}{2}}(\sin\eta) - \frac{2i}{\pi}Q^\mu_{k-\frac{1}{2}}(\sin\eta)\right), \tag{85}$$

where $P_\nu^\mu(x)$ and $Q_\nu^\mu(x)$ are the associated Legendre functions of first and second kind, respectively, of degree ν and order μ, being μ given by (71), and

$$N_k = \left(\frac{\pi \, \Gamma(k + \frac{1}{2} - \mu)}{2 \, \Gamma(k + \frac{1}{2} + \mu)} \right)^{\frac{1}{2}} e^{i \frac{\mu\pi}{2}}. \tag{86}$$

Then, the appropriately normalized solutions of σ_1 and σ_2 in (18) are given by

$$\sigma_1 = \frac{|N_k|\sqrt{k}}{a} \sin^{\frac{1}{2}} \eta P^{\mu}_{k-\frac{1}{2}}(-\cos\eta), \tag{87}$$

$$\sigma_2 = -\frac{2|N_k|\sqrt{k}}{\pi \, a} \sin^{\frac{1}{2}} \eta Q^{\mu}_{k-\frac{1}{2}}(-\cos\eta), \tag{88}$$

so that (see (26)),

$$\sigma = \sqrt{\sigma_1^2 + \sigma_2^2} = \frac{\sqrt{k}}{a}|\chi_{\mathbf{k}}|. \tag{89}$$

It can also be checked that

$$\tau_k = \frac{1}{k} \arctan \frac{2 \, Q^{\mu}_{k-\frac{1}{2}}(-\cos\eta)}{\pi \, P^{\mu}_{k-\frac{1}{2}}(-\cos\eta)} - \frac{\mu\pi}{2}, \tag{90}$$

satisfies (19).

The value of ν in (54) turns out to be then

$$\nu(t) \approx -i(\mu - \frac{3}{2})\frac{\dot{a}}{k} = -i(\mu - \frac{3}{2})\frac{\sqrt{H^2a^2 - 1}}{k} \to 0, \tag{91}$$

in the limit, $k \to \infty$, for all time t, so again the vacuum state of the invariant representation (40)–(41), with σ given by (17) with σ_1 and σ_2 given by (87)–(88), defines a stable adiabatic vacuum state. The energy density behaves similar to (77) for large values of the scale factor, as it was expected.

The amplitude of fluctuations (78) gives now,

$$\delta\phi(k_{\rm ph}) \approx \frac{H}{4\sqrt{\pi}} \left(\frac{k_{\rm ph}}{H} \right)^{\frac{3}{2}} \sqrt{\frac{\Gamma(k + \frac{1}{2} - \mu)}{\Gamma(k + \frac{1}{2} + \mu)}} \times \sqrt{\left(P^{\mu}_{k-\frac{1}{2}}(z) \right)^2 + \frac{4}{\pi^2} \left(Q^{\mu}_{k-\frac{1}{2}}(z) \right)^2}, \tag{92}$$

where,

$$z \equiv \tanh Ht = \left(1 - H^{-2}a^{-2} \right)^{\frac{1}{2}}. \tag{93}$$

5. Adiabatic solutions

The general solution of the function σ in (16) is given by (17) with σ_1 and σ_2 satisfying (18). Let us not consider the following two WKB solutions of (18) satisfying the given boundary condition

$$\sigma_1(t) = \sqrt{\frac{k}{M\omega_k}} \cos S \, , \; \sigma_2(t) = \sqrt{\frac{k}{M\omega_k}} \sin S, \tag{94}$$

where, $M(t) = a^3(t)$, $\omega_k(t)$ is given by (9), and

$$S(t) = \int^t \omega(t')dt'. \tag{95}$$

Thus,

$$\sigma \equiv \sigma_k = \sqrt{\frac{k}{M\omega_k}} \, , \; \tau_k(t) = \frac{1}{k} \int^t \omega(t')dt', \tag{96}$$

which satisfy the asymptotic conditions

$$\sigma \to \frac{1}{a} \, , \; \tau_k \to \eta, \tag{97}$$

in the limit, $\frac{k}{a} \to \infty$, for which, $\omega_k \to \frac{k}{a}$ and $\zeta_{\mathbf{k}}(\tau_k) \to \chi_{\mathbf{k}}(\eta)$. The function σ given by (96) satisfies (16) provided that

$$\sigma \left(\frac{\dot{M}^2}{4M^2} - \frac{\ddot{M}}{2M} + \frac{3\dot{\omega}^2}{4\omega^2} - \frac{\ddot{\omega}}{2\omega} \right) \to 0, \tag{98}$$

in some appropriate limit. In the case of minimal coupling this is accomplished whenever

$$-\frac{\sigma}{2} \left(\frac{(3\alpha + 2)^2 - 6\alpha^2}{(1 + \alpha)^2} \frac{\dot{a}^2}{2a^2} + \frac{3\alpha + 2}{1 + \alpha} \frac{\ddot{a}}{a} \right) \to 0, \tag{99}$$

where

$$\alpha \equiv \frac{m^2a^2}{k^2}. \tag{100}$$

The limit (99) is satisfied for large values of the physical modes, $\frac{k}{a} \gg 1$, for which $\sigma \to \frac{1}{a}$ and $\alpha \to 0$, provided that

$$\frac{1}{a} \left(\frac{\dot{a}^2}{a^2} + \frac{\ddot{a}}{a} \right) \to 0. \tag{101}$$

It is also satisfied for large values of the scale factor, $\frac{k}{a} \ll 1$, for which $\sigma \to \sqrt{\frac{k}{a^3m}}$ and $\alpha \gg 1$, whenever

$$\sqrt{\frac{k}{ma}} \frac{1}{a} \left(\frac{9\dot{a}^2}{4a^2} + \frac{3\ddot{a}}{2a} \right) \to 0. \tag{102}$$

Therefore, the adiabatic solution (96) is valid for many cases of interest, including those for which

$$\frac{\dot{a}^2}{a^3} \to 0 \text{ and, } \frac{\ddot{a}}{a^2} \to 0. \tag{103}$$

In those cases, the value of ν in (54) can be approximated by

$$\nu(t) = \frac{i}{4\omega} \left(\frac{\dot{M}}{M} + \frac{\dot{\omega}}{\omega} \right) = \frac{i}{4\omega} \left(\frac{2 + 3\alpha}{1 + \alpha} \right) \frac{\dot{a}}{a} \to \frac{i}{2} \frac{\dot{a}}{k}, \tag{104}$$

in the limit of large modes, which is similar to that given in (76) and in (91). The energy density[5] associated to the mode k is given, in the limit $m \ll \frac{k}{a}$, by

$$\varepsilon_k = \frac{\dot{a}^2}{4a^4} \frac{1}{k}, \tag{105}$$

so the energy density of entanglement is given by

$$\varepsilon \propto \int_0^{k_{\rm m}} dk \, k^2 \varepsilon_k \approx \frac{k_{\rm m}^2}{4a^2} \frac{\dot{a}^2}{a^2}, \tag{106}$$

where an ultraviolet cut-off, $k_{\rm m}$, has been introduced. The dynamics of the background spacetime turns out to be then modified by the existence of the energy of entanglement between the particles and antiparticles of the invariant vacuum. The modified Friedmann equation would read, in the region $a \gg k_{\rm m}$,

$$\left(\frac{\dot{a}}{a} \right)^2 \approx \rho_0 \left(1 + \frac{k_{\rm m}^2}{4a^2} \right). \tag{107}$$

For instance, let us consider a flat deSitter universe for which $\rho_0 = \Lambda \equiv H^2$. The Friedmann equation would be modified by the

[5] Above the zero point energy density.

existence of the entanglement between the modes of the particles-antiparticle pairs, and the scale factor would end up evolving like

$$a(t) \approx \frac{k_{\rm m}}{2} \sinh H\Delta t, \tag{108}$$

instead of the customary exponential expansion (67). Thus, the entanglement between the modes of the scalar field would produce a departure from the evolution of the initial flat deSitter spacetime that might be observable.

Finally, let us notice that the amplitude of fluctuations (78) turns out to be given, in the case of the adiabatic solution (96), by

$$\delta\phi(k_{\rm ph}) \propto \frac{k^{\frac{3}{2}}}{2\pi M^{\frac{1}{2}}\omega^{\frac{1}{2}}} \approx \begin{cases} \frac{k_{\rm ph}}{2\pi}, & k_{\rm ph} \gg m, \\ \frac{m}{2\pi}\left(\frac{k_{\rm ph}}{m}\right)^{\frac{3}{2}}, & k_{\rm ph} \ll m, \end{cases} \tag{109}$$

which are scale independent in both sub-curvature and super-curvature scales. Besides, for short-wavelength modes the spectrum is in agreement with the spectrum of fluctuations in Minkowski spacetime [8] and thus the field modes are not significantly affected on sub-curvature scales, as it was expected.

6. Conclusions

We have applied the method originally developed by Lewis and Riesenfeld and further developed by others for obtaining the invariant representation of the field modes of a charged scalar field. Then, we have assumed that the field is in the vacuum state of the invariant representation, in terms of which there is no particle production at all at any time, because it represents the ground state along the entire evolution of the field. In order to fix the vacuum state we have further imposed that the largest modes of the field must not feel the curvature of the spacetime, which is a boundary condition ultimately rooted in the equivalence principle of the theory of relativity.

We have assumed however that the observable modes of the field are those described by the instantaneously diagonal representation of the Hamiltonian at a given moment of time, when the observer is performing the measurement. In a small local region of the space, any practical particle detector would measure in practice no particles at all. However, on cosmological grounds it turns out that the vacuum state of the invariant representation is full of particle-antiparticle pairs of the diagonal representation, which are created with opposite momentum in entangled states. The quantum state of each single component is given by a quasi-thermal state with a specific temperature of entanglement that measures the rate of entanglement between the component of the created pair.

We have computed the thermodynamical magnitudes of entanglement and represented the energy density of entanglement in the case of a DeSitter spacetime. It is large for the early phases of the universes and becomes very small for an evolved universe like ours.

We have also computed the vacuum state of the invariant representation for a charged scalar field that propagates in a homogeneous and isotropic spacetime that evolves adiabatically. The energy density of the particles of the field modifies the Friedmann equation producing a departure from the unperturbed evolution that could be detected, at least in principle.

We have computed the amplitude of the vacuum fluctuations. In the case of a DeSitter spacetime the amplitude of fluctuations are the expected one. For a general spacetime that evolves adiabatically, they become scale independent for both sub-curvatures and super-curvatures scales.

This work supplies us with a new point of view for the evolution of matter and radiation fields in curved spacetime that can help us to make new further developments, particularly in the context of the thermodynamics of entanglement in curved spacetime backgrounds.

References

[1] L. Parker, Phys. Rev. Lett. 21 (1968) 562.
[2] L. Parker, Phys. Rev. 183 (1969) 1057.
[3] L. Parker, J. Navarro-Salas, arXiv:1702.07132v1, 2017.
[4] A.A. Grib, B.A. Levitskii, V.M. Mostepanenko, Theor. Math. Phys. 19 (1974) 349.
[5] S.G. Mamaev, V.M. Mostepanenko, A.A. Starobinskii, Sov. Phys. JETP 43 (1976) 823.
[6] S.A. Fulling, Gen. Relativ. Gravit. 10 (1979) 807.
[7] N.D. Birrell, P.C.W. Davies, Quantum Fields in Curved Space, Cambridge University Press, Cambridge, UK, 1982.
[8] V.F. Mukhanov, S. Winitzki, Quantum Effects in Gravity, Cambridge University Press, Cambridge, UK, 2007.
[9] H.R. Lewis, J. Math. Phys. 9 (1968) 1976.
[10] H.R. Lewis, W.B. Riesenfeld, J. Math. Phys. 10 (1969) 1458.
[11] P.G.L. Leach, J. Phys. A 16 (1983) 3261.
[12] I.A. Pedrosa, Phys. Rev. D 36 (1987) 1279.
[13] C.M.A. Dantas, et al., Phys. Rev. A 45 (1992) 1320.
[14] H. Kanasugui, H. Okada, Prog. Theor. Phys. 93 (1995) 949.
[15] D.-Y. Song, Phys. Rev. A 62 (2000) 014103.
[16] D.G. Vergel, E.J.S. Villaseñor, Ann. Phys. 324 (2009) 1360.
[17] K.A. Bronnikov, E.A. Tagirov, Gravit. Cosmol. 10 (2004) 249.
[18] N.A. Chernikov, E.A. Tagirov, Ann. Inst. Henri Poincaré A 9 (1968) 109.
[19] S. Robles-Pérez, P.F. González-Díaz, Phys. Rev. D 81 (2010) 083529, arXiv: 1005.2147v1.
[20] S.J. Robles-Pérez, Quantum cosmology of a conformal multiverse, Phys. Rev. D 96 (2017) 063511, https://doi.org/10.1103/PhysRevD.96.063511.
[21] S. Robles-Pérez, A. Balcerzak, M.P. Dabrowski, M. Kramer, Phys. Rev. D 95 (2017) 083505.
[22] S. Robles-Pérez, P.F. González-Díaz, J. Exp. Theor. Phys. 118 (2014) 34, arXiv:1111.4128v1.
[23] R. Horodecki, P. Horodecki, M. Horodecki, K. Horodecki, Rev. Mod. Phys. 81 (2009) 865.
[24] M. Abramovitz, I.A. Stegun (Eds.), Handbook of Mathematical Functions, NBS, 1972.
[25] J.K. Kim, S.P. Kim, J. Phys. A 32 (1999) 2711, arXiv:quant-ph/9806096.
[26] M. Gutzwiller, Helv. Phys. Acta 29 (1956) 313.
[27] E. Mottola, Phys. Rev. D 31 (1985) 754.
[28] E.A. Tagirov, Ann. Phys. 76 (1973) 561.
[29] J.S. Dowker, R. Crichley, Phys. Rev. D 13 (1976) 224.

11

Isotropic Lifshitz point in the $O(N)$ theory

Dario Zappalà

INFN, Sezione di Catania, Via Santa Sofia 64, 95123 Catania, Italy

ARTICLE INFO

Editor: A. Ringwald

Keywords:
Functional Renormalization Group
Lifshitz point
1/N expansion

ABSTRACT

The presence of an isotropic tricritical Lifshitz point for the $O(N)$ scalar theory is investigated at large N in the improved Local Potential Approximation (LPA′) by means of the Functional Renormalization Group equations. At leading order, the non-trivial Lifshitz point is observed if the number of dimensions d is taken between $d = 4$ and $d = 8$, and the eigenvalue spectrum of the associated eigendirections is derived. At order $1/N$ of the LPA′ the anomalous dimension η_N is computed and it is found to vanish both in $d = 4$ and $d = 8$. The dependence of our findings on the infrared regulator is discussed.

1. Introduction

The description of a tricritical Lifshitz point by a Landau–Ginzburg ϕ^4 model, where the derivatives of the field with respect to the coordinates of a m-dimensional subset of a d-dimensional space and those of the complementary $(d-m)$-dimensional subspace possess different scaling laws, was first presented in [1]. More specifically, in [1] the kinetic term with square gradient of the field, $O(\partial^2)$, is kept finite only for the second subset of coordinates, while the corresponding term of the m-dimensional subset is suppressed, so that the term with four powers of the gradient, $O(\partial^4)$, becomes the leading kinetic term of the m-dimensional subspace and this induces drastic changes in the scaling properties of the theory.

The Lifshitz points, which are related to the coexistence on the phase diagram of three phases, one with vanishing order parameter, another with finite constant order parameter and the third characterized by a modulated order parameter with finite wave vector, find application in various fields such as magnetic systems as well as polymer mixtures or high T_C superconductors (for reviews see [2,3]), but, recently, also in different contexts such as Lorentz symmetry violation, [4,5], or emergent gravity theories [6–10]. In addition, an oscillating phase has been predicted for a very wide class of systems [11–15], and it is conceivable to expect that a Lifshitz point could be associated to these modulated phases. In this sense, a more complete understanding of the properties of the Lifshitz point is certainly desirable.

E-mail address: dario.zappala@ct.infn.it.

Rather than considering the general case with $0 < m < d$, where the different scaling properties in the two separate subspaces lead to a peculiar critical behaviour that involves two different anomalous dimensions and correlation lengths, we shall focus on the isotropic case with $m = d$. In fact, if $m < d$, due to the different behaviour of the two sets of coordinates, the isotropy of the problem is lost while, when $m = d$, all the space coordinates have the same critical behaviour and spatial isotropy is preserved. Clearly, in this latter case the critical scaling remains different from the standard one because, as explained before, the kinetic term in the action is quartic, rather than quadratic, in the field derivatives.

The critical properties of the Lifshitz point were studied in the ε-expansion [1] as well as in the $O(1/N)$ expansion [16]. The isotropic case $m = d$ was considered within an expansion around $d = 8$ and $\varepsilon = 8 - d$ [17] while, recently, a numerical Monte-Carlo study indicated a possible disappearance of the Lifshitz point, when fluctuations are properly included [18].

Furthermore, another non-perturbative technique already employed to study this problem is the Functional Renormalization Group (FRG) [19–21] which consists of a set of differential flow equations either for various operators entering the effective action of the theory, or for one or more n-point Green functions derived from the effective action. Fixed points correspond to stationary points of these equations and the critical exponents, that classify relevant, marginal and irrelevant operators, are extracted by determining the eigenvalue spectrum of the linear reduction of the differential equations around the fixed point solutions. Coming to the Lifshitz point, the FRG was applied to study this problem for a one component scalar theory, $N = 1$, [22], and for the $N = 3$ theory, [23], both in the uniaxial $(m = 1)$ case. Finally, the isotropic case ($m=d$) with $N=1$ was considered in [24] and, in this last case, the Proper Time version [25–27] of the FRG, which can be formally derived in the framework of the background field flows [28,29], was used because it proved to be quite accurate and suitable for the numerical analysis of the critical properties of a theory at a fixed point [27,30–32] and, in addition, the Proper Time flow equation of the $O(\partial^4)$ operator (coupled to the potential and to the $O(\partial^2)$ operator equations), that is necessary to treat a Lifshitz point, had been already derived in [32].

The numerical analysis performed in [24] for the $N=1$ theory, shows at the lowest order (in the Local Potential Approximation – LPA), i.e. by considering the fixed potential equation only, that a non-trivial solution exists when the number of spatial dimensions is $4<d<8$, and for $d\geq 8$ the solution merges with the trivial, gaussian fixed point, while for $d\leq 4$ the asymptotic structure of the differential equation changes and no discrete set of non-trivial solution is available. Then, when going beyond the LPA and including the differential equations for the $O(\partial^2)$ and $O(\partial^4)$ operators, a solution was observed in the range $5.5<d<8$, but the numerical analysis for smaller d becomes too demanding and it was not possible to establish whether the Lifshitz point survives down to $d=4$ or, rather, the fluctuations associated with higher derivatives terms, $O(\partial^2)$ and $O(\partial^4)$, effectively destroy the critical behaviour when d approaches 4.

In this letter we consider another aspect of the problem and analyze the existence of a Lifshitz point for a scalar $O(N)$-symmetric theory, in order to find out whether the critical behaviour survives to the presence of the strong infrared fluctuations due to the transverse modes. To this aim, a numerical analysis would require the resolution of a very large number of differential equations that would probably present the same kind of problems observed for the simpler $N=1$ case.

Therefore, we follow a different approach. We start by considering the procedure developed in [33,34], where the flow equation for the effective action is projected onto a set of flow equations for the n-point Green functions which is to be truncated at some specific n and, for our purposes, we retain the three equations for the potential and the longitudinal and transverse two-point functions. Then, we treat these equations in the framework of the $1/N$ expansion to extract the corresponding $O(1/N)$ contribution to the anomalous dimension η.

Actually, this combined procedure neither amounts to a full resolution of the flow equations for the two point functions derived in [33,34], nor to a complete $O(1/N)$ computation, however it allows us to go one step beyond the leading order of the $1/N$ expansion (at which the full eigenvalue spectrum of the Lifshitz point is determined) and establish both the survival of the Lifshitz point and the first non-vanishing contribution to η at criticality. In particular we find that the expression derived for η according to this procedure reduces to the result that is obtained in the minimal improvement of the LPA [39], also known as LPA′.

2. Flow equations

In order to write down the fixed point equation, we start from the full FRG equation [21]: ($\partial_t \equiv k\partial_k$):

$$\partial_t \Gamma_k[\phi] = \frac{1}{2}\int_q \partial_t R_k(q)\left[\Gamma_k^{(2)}[q,-q;\phi]+R_k(q)\right]^{-1} \tag{1}$$

$\Gamma_k[\phi]$ being the running effective action at scale k, and $R_k(q)$ a suitable regulator that suppress the modes with $q\ll k$ and allows to integrate those with $q\gg k$. The specific choice of the regulator $R_k(q)$ is discussed below.

Rather than introducing the running parameters by means of an explicit form of the effective action, we proceed by displaying the second functional derivative of the effective action $\Gamma^{(2)}_{ab}(p;\phi)\equiv \delta^2\Gamma_k/(\delta\phi_a(p)\delta\phi_b(-p))$ that, according to the $O(N)$ symmetry, has the general form ($\rho\equiv\phi_a\phi_a/2$):

$$\Gamma^{(2)}_{ab}(p,\phi)=\Gamma_A(p,\rho)\delta_{ab}+\phi_a\phi_b\Gamma_B(p,\rho) \tag{2}$$

Then, we parametrize Γ_A and Γ_B in terms of the potential V and of the renormalization functions Z_A, Z_B, W_A, W_B, i.e. the coefficients of the quadratic and quartic powers of the momentum p:

$$\Gamma_A(p^2,\rho)=W_A(\rho)\,p^4+Z_A(\rho)p^2+V' \tag{3}$$

$$\Gamma_B(p^2,\rho)=N\,W_B(\rho)\,p^4+N\,Z_B(\rho)p^2+V'' \tag{4}$$

where prime indicates the derivative with respect to ρ and N is the number of field components.

The factor N appearing in front of W_B and Z_B is due to a specific rescaling of the potential and of the field with respect to the standard definitions, $V\to N\,V$ and $\phi_a\to\sqrt{N}\phi_a$, which is made in order to derive a fixed point equation that is directly arranged in a $1/N$ expanded structure. Clearly, this rescaling has no effect on V', while it changes $V''\to V''/N$ as well as the factor $\phi_a\phi_b\to N\,\phi_a\phi_b$ in Eq. (2), these last two transformations being responsible for the factor N appearing in the definition of Γ_B in Eq. (4). Therefore, from Eqs. (3) and (4), it is easy to expect the parameters W_B, Z_B to be $1/N$ suppressed with respect to W_A, Z_A, as it will be checked below.

Then, if we separate the longitudinal (L) and transverse (T) components in the inverse of $\Gamma^{(2)}_{ab}(p,\phi)$, that is the propagator of the theory $G_{ab}(p,\phi)$, according to:

$$G_{ab}(p,\phi)=\left(\delta_{ab}-\frac{\phi_a\phi_b}{2\rho}\right)G_T(p,\rho)+\frac{\phi_a\phi_b}{2\rho}G_L(p,\rho) \tag{5}$$

one finds

$$G_T^{-1}(p,\rho)=\Gamma_A(p,\rho) \tag{6}$$

and

$$G_L^{-1}(p,\rho)=\Gamma_A(p,\rho)+2\rho\Gamma_B(p,\rho)\,. \tag{7}$$

It is understood that the field dependent parameters V, W_A, W_B, Z_A, Z_B also depend on the running scale k and, with these settings, we can rely on the derivation of the flow equations carried out in [34]. We define the integrals

$$J_n^{\alpha\beta}(p,\rho)=\int_q \partial_t R_k(q)\widetilde{G}_\alpha^{n-1}(q,\rho)\widetilde{G}_\beta(p+q,\rho)\,, \tag{8}$$

$$I_n^{\alpha\beta}(\rho)=J_n^{\alpha\beta}(0,\rho), \tag{9}$$

where $n\geq 1$, $\int_q\equiv\int\frac{d^dq}{(2\pi)^d}$, α and β stand either for L or T, and

$$(\widetilde{G}_\alpha^n(q,\rho))^{-1}\equiv(G_\alpha^n(q,\rho))^{-1}+R_k(q)\,. \tag{10}$$

Then, by following [34] (see also [35]), we get the flow equation for the potential V:

$$\partial_t V(\rho)=\frac{1}{2}\left\{I_1^{TT}(\rho)+\frac{1}{N}\left[I_1^{LL}(\rho)-I_1^{TT}(\rho)\right]\right\} \tag{11}$$

and for the two-point functions, properly subtracted of the zero-momentum contribution:

$$\partial_t\left[\Gamma_X(p^2,\rho)-\Gamma_X(0,\rho)\right]=F_X(p^2,\rho)-F_X(0,\rho) \tag{12}$$

where X stands either for A or B, and

$$F_A(p^2,\rho)=-\frac{1}{2}I_2^{TT}\,\Gamma_A'+\frac{1}{N}\Bigg[2\rho\left(J_3^{LT}\,\Gamma_A'^{\,2}+J_3^{TL}\,\Gamma_B{}^2\right)$$
$$-I_2^{LL}\left(\frac{\Gamma_A'}{2}+\rho\Gamma_A''\right)-I_2^{TT}\left(\Gamma_B-\frac{\Gamma_A'}{2}\right)\Bigg], \qquad (13)$$

$$F_B(p^2,\rho)=J_3^{TT}\,\Gamma_B{}^2-\frac{1}{2}I_2^{TT}\,\Gamma_B'+O\left(\frac{1}{N}\right) \qquad (14)$$

Eqs. (12), (13), (14) can be reduced to flow equations either for W_X or Z_X, by selecting in F_X the terms proportional respectively to p^4 or p^2. Then, it is evident from Eqs. (4), (12) and (14) that, in order to avoid any inconsistency in the $1/N$ expansion, W_B and Z_B must be $O(1/N)$ so that $\Gamma_B\sim O(1)$. Accordingly, we are allowed to neglect $O(1/N)$ corrections in Eq. (14), as they contribute to W_B and Z_B to order $O(1/N^2)$.

Let us now consider the regulator $R_k(q)$. A particularly useful regulator, that has the advantage of reducing the integrals to simple structures which can be analytically solved in most cases, was introduced in [36] and has the form:

$$R_k^\theta(q)=(k^2-q^2)\,\widehat{Z}_k\,\theta(k^2-q^2) \qquad (15)$$

where θ is the Heaviside step function and a k-dependent (but field independent) normalization factor $\widehat{Z}_k$ is included. For the present problem the regulator in Eq. (15) should be modified into $R_k^\theta(q)=(k^4-q^4)\,\widehat{W}_k\,\theta(k^2-q^2)$ with $\widehat{W}_k$ taken equal to W_A, evaluated at a particular value of ρ: $\widehat{W}_k=W_A(\overline{\rho})$, with $\overline{\rho}$ to be specified. However, due to the presence of the Heaviside function, the second and higher derivatives of $R_k^\theta(q)$ with respect to q^4, generate a singular behaviour of the integrals involved in this analysis. Therefore it is preferable to replace $R_k^\theta(q)$ with a smooth, one-parameter (α) regulator:

$$R_k(q)=\frac{\widehat{W}_k}{2}\left[(k^4-q^4)+\sqrt{(k^4-q^4)^2+(2\alpha)^{-2}}\right] \qquad (16)$$

In fact, $R_k(q)$ in Eq. (16) approaches $R_k^\theta(q)$ in the limit $\alpha^{-1}\to 0$, and, for values of the dimensionless parameter $k^4\alpha\sim 10^3$ or larger, $R_k(q)$ (and its first derivative) can be practically replaced by $R_k^\theta(q)$ in the resolution of the integrals but, on the other hand, all its derivatives are regular so that it does not generate any singularity as long as α is kept finite, i.e. $\alpha^{-1}\neq 0$.

Incidentally, as the vanishing of the regulator at $k=0$, $R_{k=0}(q)=0$, is a necessary requirement of the flow equations, then α^{-1} must be a function of the scale k that vanishes at $k=0$. This can be easily achieved e.g. by taking $(2\alpha)^{-1}=\lambda\Lambda^4\tanh[(k/\Lambda)^\mu]$, where Λ is a fixed mass scale and λ and μ two small dimensionless parameters that can be adjusted to set the size of $(2\alpha)^{-1}$ and of its derivatives. Due to the dependence of α on k, the fixed point equations do contain additional terms proportional to $k[\partial(2\alpha)^{-1}/\partial k]=2\mu\,(k/\Lambda)^\mu\,(2\alpha)^{-1}\sinh^{-1}[2\,(k/\Lambda)^\mu]$, but one easily realizes that even the largest contributions (proportional to α) encountered in the following calculations, when multiplied by this factor, for sufficiently small values of μ turn out to be systematically suppressed with respect to the other terms appearing in the fixed point equations. Therefore, we neglect the contributions proportional to $k[\partial(2\alpha)^{-1}/\partial k]$ and simply treat α as a free parameter.

However, as discussed below, even the regulator in Eq. (16) is not sufficient to get rid of all potentially large (divergent in the limit $\alpha^{-1}\to 0$) terms and therefore at some point we find convenient to analyze our equations by adopting the smoother exponential regulator

$$R_k^b(q)=\frac{b\,\widehat{W}_k q^4}{e^{q^4/k^4}-1} \qquad (17)$$

where b is a dimensionless adjustable parameter.

3. Leading order of the $1/N$ expansion

As anticipated, Eqs. (11), (12), (13) and (14), are already arranged in a $1/N$ expansion structure and we can straightforwardly extract the leading ($1/N=0$) flow equations for the suitably rescaled parameters, and also the associated fixed point equations, which are obtained by requiring the rescaled parameters to be t-independent. The rescaled parameters, relevant for our analysis, are $\varrho=k^{-d+4-\eta}\rho$, $v=k^{-d}V$, $w^A=k^\eta W_A$, $w^B=k^{d-4+2\eta}W_B$, $z^A=k^{\eta-2}Z_A$, $z^B=k^{d-6+2\eta}Z_B$, where the scaling dimensions, i.e. the exponents in the powers of the scale k, are given in [23,24], and the fixed point equations for w^A and z^A at $1/N=0$ are:

$$-\eta_0 w_0^A+(d-4+\eta_0)\varrho w_0^{A'}=-\frac{1}{2}I_2^{TT}\,w_0^{A'} \qquad (18)$$

$$(2-\eta_0)z_0^A+(d-4+\eta_0)\varrho z_0^{A'}=-\frac{1}{2}I_2^{TT}\,z_0^{A'}\ . \qquad (19)$$

In Eqs. (18) and (19) the prime indicates derivation with respect to ϱ and the subscript 0 indicates the lowest order of the $1/N$ expansion. It is easy to check that a field independent w_0^A (and therefore $w_0^{A'}=0$) together with $\eta_0=z_0^A=0$ is a solution of this set of equations. Therefore, we can take $w_0^A=1$ to set the overall normalization of the effective action.

Then, we turn to the fixed point equation for the potential, Eq. (11), and, after setting $\widehat{W}_k=1$ in Eq. (16), the integral I_1^{TT} can be solved and Eq. (11) conveniently written as:

$$\left[(x+f(x))^2\,d_+-1\right]f(x)=\left[(x+f(x))^2\,d_--1\right]xf_x(x) \qquad (20)$$

with the following definitions $x=\sqrt{2\varrho}$; $f(x)=dv/dx$; $f_x(x)=df/dx$; $d_\pm=(d\pm 4)/(2\tau)$ and finally $\tau=2/[(4\pi)^{d/2}\Gamma(1+d/2)]$ is the factor coming from the resolution of the integral I_1^{TT}.

Eq. (20) can be easily attacked numerically, but all the essential features can be deduced by simple inspection. In fact we immediately see that the constant function $f_G(x)=0$ is a solution of Eq. (20), that plays the same role of the gaussian fixed point for the standard scaling. In addition, we observe that a viable non-trivial Lifshitz solution $f_L(x)$ must vanish at the origin $f_L(0)=0$ due to the symmetry of the problem and, in addition, another zero of f_L must occur at

$$\bar{x}^2=\frac{2\tau}{(d-4)} \qquad (21)$$

i.e. $f_L(\bar{x})=0$ with non-vanishing derivative $f_{Lx}(\bar{x})\neq 0$. By expanding Eq. (20) around $\bar{x}$, one finds from the linear terms:

$$f_{Lx}(\bar{x})=\frac{8-d}{d-4}\ . \qquad (22)$$

Eq. (21) loses meaning when $d\leq 4$, while the vanishing of $f_{Lx}(\bar{x})$ from Eq. (22) at $d=8$ indicates a flattening of the solution f_L onto the trivial solution f_G. The latter result accords with the numerical analysis of [24] with $N=1$, which indicates that the two solutions merge at $d=8$ and only the trivial solution survives for $d\geq 8$. Therefore we limit the study of Eq. (20) to the range $4\leq d\leq 8$.

With the information collected above, we are able to determine the eigenvalues λ_L of the flow equation, linearized around the fixed point solution. To this aim we follow the procedure originally worked out in [37,38] for the standard Wilson–Fisher (WF) fixed point, (see also [31]) and, by writing the t-dependent function $f(t,x) = f_L(x) + e^{\lambda t} h(x)$, as the sum of the fixed point solution $f_L(x)$ and a perturbation $h(x)$, we get the following linear (in $h(x)$) equation:

$$\frac{\lambda h}{\tau} = \left[d_+ - (x+f_L)^{-2}\right] h - \left[d_- - (x+f_L)^{-2}\right] x h_x \\ +2\,(x+f_L)^{-3}\left(f_L h - x f_{Lx} h\right) \tag{23}$$

The function h is supposed to be regular at any finite x and can be expanded around $\bar{x}$:

$$h(x) = \sum_{i=n}^{\infty} a_i (x-\bar{x})^i \tag{24}$$

where the lowest power n must be a non-negative integer, $n \geq 0$. At $x = \bar{x}$ the coefficient of $x h_x$ in square brackets vanishes and therefore $h_x(x)/h(x)$ is either singular at $x = \bar{x}$ (with a simple pole singularity) or finite, the former case corresponding to $n > 0$ and the latter to $n = 0$ in Eq. (24). In both cases, after dividing both members of Eq. (23) by h, one can make the replacement $h_x(x)/h(x) = n/(x-\bar{x})$ in order to compute the linear corrections in the expansion of Eq. (23) around the point $\bar{x}$. This expansion, with the help of Eq. (22), yields the following eigenvalue spectrum (we recall $4 \leq d \leq 8$):

$$\lambda_L = d - 4 - 4n \tag{25}$$

parameterized by the non-negative integer $n \geq 0$. By following the same procedure, one derives from Eq. (23) the eigenvalues associated to f_G (again with integer $n \geq 0$):

$$\lambda_G = 4 - (d-4)\,n \tag{26}$$

In particular one can determine those values of n that correspond to relevant (positive) eigenvalues, namely $0 \leq n < (d-4)/4$ from Eq. (25), and $0 \leq n < 4/(d-4)$ from Eq. (26). In addition, we observe that in $d = 8$ the two spectra in Eq. (25) and (26) are equal, as the two fixed point solutions become coincident.

In conclusion, the solutions found at $1/N = 0$ with this particular scaling, clearly resemble those obtained with standard scaling where, aside from the constant gaussian solution with eigenvalue spectrum $\lambda_g = 2 - (d-2)n$, one has the WF fixed point with $\lambda_{WF} = d - 2 - 2n$. One clearly sees that the difference, at this order, is only in the range spanned by d which, in this case, goes from $d = 2$ to $d = 4$, while, in the analysis of the tricritical Lifshitz point, from $d = 4$ to $d = 8$. In fact, even the number of relevant directions is the same in the two cases, once the proper change in d is taken into account.

4. $1/N$ corrections

At the leading order $1/N = 0$, the equations for the momentum dependent parts admit the elementary field-independent solutions $w_0^A = 1$ and $z_0^A = 0$, together with $\eta_0 = 0$, while the equations for w^B and z^B at this order decouple from the other equations and one is left with the fixed point equation for the potential only.

For the next step, we consider the potential expansion $v = v_0 + v_N/N + O(1/N^2)$ and the analogous expansions for η, w^A, z^A, w^B, z^B, and insert them into the fixed point equations in order to analyze the $1/N$ corrections. We start by observing that Eq. (11) for v_N involves the $(1/N)$ corrections of all the above variables (we recall here that $w_0^B = z_0^B = 0$, but the first non-vanishing terms of the expansion of w^B and z^B, which are $O(1/N)$, contribute to the leading order $(1/N = 0)$ longitudinal propagator G_L, because of the factor N in Eq. (4)). Therefore, a full determination of the $1/N$ corrections requires the resolution of five coupled equations.

However, it is possible to determine η_N without solving the whole set of equations. To this aim, a direct inspection of equations (12), (13), (14) shows that the vanishing field-independent solution $w_N^A = z_N^A = w_N^B = z_N^B = 0$ is not allowed because of the non-vanishing coefficients of the integrals J_3 in Eqs. (13), (14), respectively ${\Gamma'_A}^2$ and ${\Gamma_B}^2$, which are finite and field dependent due to their dependence on V' and V'', as shown in Eqs. (3), (4).

Nevertheless, at least for one particular value of the field $\varrho = \bar{\varrho}$, we can extend at $1/N$ the normalization of the propagators, already fixed by the leading order solution $w_0^A = 1$; $z_0^A = w_0^B = z_0^B = 0$. This immediately implies $w_N^A(\bar{\varrho}) = z_N^A(\bar{\varrho}) = w_N^B(\bar{\varrho}) = z_N^B(\bar{\varrho}) = 0$ and it is natural to take $\bar{\varrho}$ as the point where the derivative of the leading order potential vanishes, i.e. $\bar{\varrho} = \bar{x}^2/2$, with $\bar{x}$ defined in Eq. (21). Finally, we extract from Eq. (13) the two equations for w_N^A, z_N^A, directly computed at $\bar{\varrho}$:

$$-\eta_N - (d-4)\,\bar{\varrho}\, w_N^{A\prime}(\bar{\varrho}) = \\ -\frac{1}{2}(I_2^{TT})_0\, w_N^{A\prime}(\bar{\varrho}) + 2\bar{\varrho}\, v_0''(\bar{\varrho})^2 \left(J_3^{LT}\big|_{p^4} + J_3^{TL}\big|_{p^4}\right)_0 \tag{27}$$

$$-(d-4)\,\bar{\varrho}\, z_N^{A\prime}(\bar{\varrho}) = \\ -\frac{1}{2}(I_2^{TT})_0\, z_N^{A\prime}(\bar{\varrho}) + 2\bar{\varrho}\, v_0''(\bar{\varrho})^2 \left(J_3^{LT}\big|_{p^2} + J_3^{TL}\big|_{p^2}\right)_0 \tag{28}$$

where the subscript 0 of the various integrals indicates that they must be computed by using the leading order $(1/N = 0)$ solution of the various parameters, while the subscript p^4 in Eq. (27) and p^2 in Eq. (28) of the integrals J_3, indicates that only the coefficient of that particular power of the momentum p in the expansion of the addressed integral is to be retained. We find that the $1/N$ correction to the anomalous dimension η_N does not appear in Eq. (28), but it is directly obtained from Eq. (27), if one neglects the terms proportional to $w_N^{A\prime}(\bar{\varrho})$. At the same time, the $O(1/N)$ corrections to the fixed point potential in Eq. (11) are under control and easily computable by numerical integration.

As anticipated, we notice that the procedure adopted to compute η_N essentially coincides with the scheme introduced in [39] which leads to the improved Local Potential Approximation LPA′. This can be straightforwardly checked by replacing Eq. (27) with the equation obtained by repeating the previous steps for the case of the anomalous dimension η_{WF} at the WF fixed point, which gives $\eta_{WF} = -2\bar{\varrho}\, v_0''(\bar{\varrho})^2\,(J_3^{LT}|_{p^2} + J_3^{TL}|_{p^2})_0$. In this case the expansion is to be taken to order p^2 and the regulator in (15) can be safely chosen, because it does not generate any singularity. The corresponding integrals can be analytically computed, as shown in [39], and one finds $(J_3^{LT}|_{p^2} + J_3^{TL}|_{p^2})_0 = -\tau/(1 + 2\bar{\varrho}\, v_0''(\bar{\varrho}))^2$ and, therefore, $\eta_{WF} = 2\tau\, \bar{\varrho}\, v_0''(\bar{\varrho})^2/(1 + 2\bar{\varrho}\, v_0''(\bar{\varrho}))^2$. This is exactly the expression of the anomalous dimension which is used in the LPA′ [21,39,40].

In order to test the reliability of this procedure, we can go one step further and replace in η_{WF}, the particular value of $\bar{\varrho}$ and $v_0''(\bar{\varrho})$) that are obtained from the leading order analysis $(1/N = 0)$ for the WF fixed point. Then, instead of Eq. (21), one has $\bar{x}^2 = \tau/(d-2)$ and Eq. (22) becomes $f_{WF x}(\bar{x}) = 2\bar{\varrho}\, v_0''(\bar{\varrho}) = (4-d)/(d-2)$ (these changes are due to the different scaling of the various quantities in the two cases and also to the different dimension of the regulator R_k that, when derived with respect to the scale k, $\partial_t R_k$, produces a different factor). Thus, one finds the following $1/N$ correction to the anomalous dimension at the WF fixed point:

$$\eta_{WF} = \frac{(d-2)(4-d)^2}{4} \tag{29}$$

that is to be compared to the full result directly obtained in the $1/N$ expansion, [41] ($\epsilon \equiv 4-d$ and Γ indicates the Gamma function):

$$\eta = \frac{4\epsilon}{(4-\epsilon)\pi} \frac{\sin(\pi\epsilon/2)\,\Gamma(2-\epsilon)}{\Gamma(1-\epsilon/2)\,\Gamma(2-\epsilon/2)} . \tag{30}$$

Remarkably, Eqs. (29) and (30) have the same behaviour both for $d=2+\delta$ (with $\delta \gtrsim 0$), i.e. $\eta_{WF}=\eta=\delta$, and for $d \lesssim 4$ (with $\epsilon \gtrsim 0$), i.e. $\eta_{WF}=\eta=\epsilon^2/2$. Instead, in $d=3$, where the difference between Eq. (29) and Eq. (30) is largest, one finds $\eta_{WF}=1/4$ and $\eta = 8/(3\pi^2) \simeq 1/(3.7)$. We take this small discrepancy as the measure of the reliability of the LPA′ here considered even in the case of the Lifshitz critical behaviour.

Going back to the Lifshitz fixed point problem, we have to compute η_N from Eq. (27) by neglecting the terms proportional to $w_N^{A'}(\bar{\varrho})$. However, as anticipated, this time a strong dependence on the regulator is observed. In particular, the parameter α introduced in Eq. (16) explicitly shows up in the resolution of the integrals, because $(\partial^2 R_k(q^4)/\partial q^8)_{q^4=k^4} = \alpha$. Namely, we get

$$\eta_N = \frac{4\tau\,\bar{\varrho}\,v_0''(\bar{\varrho})^2}{D^2} \left[4\bar{\alpha} - \frac{24\bar{\alpha}+d+8}{(d+2)\,D} + \frac{6}{(d+2)\,D^2} \right] \tag{31}$$

where we introduced the dimensionless parameter $\bar{\alpha} = k^4\alpha$ and $D = (1+2\bar{\varrho}\,v_0''(\bar{\varrho}))$. Then, with the help of Eqs. (21) and (22) one gets the analogous of Eq. (29) for the Lifshitz case, with no need to solve the fixed point equation for the $O(1/N)$ corrections to the potential, or the wave function renormalizations:

$$\eta_N = \frac{(d-4)\,(8-d)^2}{16} \left\{ 4\bar{\alpha} - \frac{(d-4)}{4(d+2)}\,(24\bar{\alpha}+d+8) + \frac{3(d-4)^2}{8(d+2)} \right\} \tag{32}$$

We observe the explicit dependence on the parameter $\bar{\alpha}$ in Eq. (32) and it is evident that the alternative use of the Heaviside cutoff R_k^{θ}, associated to the limit $1/\bar{\alpha} \to 0$ would produce a singular behaviour of η_N. Instead, for finite values of $\bar{\alpha}$, η_N is finite in the whole range $4<d<8$.

However the $\bar{\alpha}$-dependence in Eq. (32) has strong drawbacks: for instance in $d=6$, one can take $\bar{\alpha}$ sufficiently large that the term η_N/N in the $1/N$ expansion of the anomalous dimension is so big, even with $N >> 1$, that the expansion itself become questionable. The only two cases in which the $\bar{\alpha}$-dependence becomes irrelevant are the two limits of η_N for $d \to 4^+$ and for $d \to 8^-$, that vanish for any fixed value of $\bar{\alpha}$, due to the factor in front of the curly bracket in the right hand side of Eq. (32).

In order to collect further indications on the effect of the regulator in the computation of η_N, we solve the integrals in Eq. (27) with the exponential regulator defined in Eq. (17), which has the advantage of being essentially smoother than the one in Eq. (16) and free of additional dimensionful parameters but, on the other hand, no analytical expression for η_N can be derived.

Therefore, we report in Table 1 the values of η_N obtained with different values of the parameter b of the regulator in Eq. (17) for two values of the dimension d, namely $d=4.1$ and $d=5$. In both cases η_N shows the same qualitative behaviour, by reaching a maximum value (indicated by a star) around $b \simeq 10^{-2}$, and then systematically decreasing down to large negative values. Unlike the result in Eq. (32) that shows a linear dependence on $\bar{\alpha}$, in this case we can invoke the minimal sensitivity criterion to select the maximal values as estimates of the anomalous dimension η_N.

Table 1
η_N as obtained for different values of b in Eq. (17).

$d=4.1$		$d=5$	
b	η_N	b	η_N
10^{-4}	0.081	10^{-4}	0.137
10^{-3}	0.086	10^{-3}	0.230
10^{-2}	0.089	10^{-2}	0.313
$1.5\ 10^{-2}$	*0.0894	$1.2\ 10^{-2}$	*0.3144
10^{-1}	0.085	10^{-1}	0.147
$2\ 10^{-1}$	0.080	$2\ 10^{-1}$	0.023
20	−0.036	20	−0.720

Table 2
η_N as obtained with d approaching 4 and 8.

$b=0.015$		$b=1$	
d	$\eta_N/(d-4)$	d	$-8\,\eta_N/(8-d)^2$
4.1	0.8939	7.9	0.9762
4.05	0.9454	7.95	0.9880
4.01	0.9888	7.99	0.9976
4.005	0.9944	7.995	0.9988
4.001	0.9989	7.999	0.9997

However when d is increased to 6 or to larger values, a more cumbersome picture shows up. In fact, already at $d=6$ the simple b-dependence of Table 1 is lost and one finds three different extrema in η_N when b grows (two maxima with a minimum in between), namely $\eta_N = 0.154, -0.282, -0.214$, respectively for $b = 6\,10^{-3}, 0.41, 2.5$. We notice that $\eta_N < 0$ both at the minimum and at the second maximum and the same pattern of three extrema is observed for larger d.

We conclude the analysis with the regulator in Eq. (17), by showing in Table 2 the behaviour of η_N when d approaches the two extremal values $d=4$ and $d=8$. In the former case, b is obviously selected by the presence of a single maximum in η_N, while in the latter case we took $b=1$ that corresponds to a rather stable (for $d \simeq 8$), negative value of η_N, very close to its second maximum (note that for the remaining two extrema, the effect shown in Table 2 is not observed). Remarkably, Table 2 shows that the power law behaviour already seen in Eq. (29) for the WF fixed point, and in Eq. (32) for the Lifshitz fixed point with the other regulator, is in fact recovered in this case both when $d \to 4^+$ and when $d \to 8^-$.

5. Discussion

We investigated the existence of the isotropic tricritical Lifshitz point for the $O(N)$ theory in the $1/N$ expansion and explicitly computed the associated anomalous dimension in the LPA′. More specifically, instead of directly implementing the LPA′ to the Lifshitz case, our analysis started from a set of coupled flow equations for the potential and the two point functions derived in [33, 34], which were then evaluated at the next to leading order in the $1/N$ expansion and under further assumptions. This procedure produced an equation for the anomalous dimension η that turned out to be equivalent to the equation for η derived in the LPA′.

It must be remarked that η_N determined in the LPA′ does not include the full $O(1/N)$ corrections to the anomalous dimension and, therefore, an indication of the difference between the two determinations was obtained in the case of the WF fixed point, where the maximum discrepancy amounts to about 8% at $d=3$ while, close to the extremal values, $d=2$ and $d=4$, the two calculations coincide.

We find that, already at leading order, the non-trivial Lifshitz point is observed only between $4<d<8$. At order $1/N$ of the LPA′, the presence of the Lifshitz point is confirmed and the anomalous dimension η_N vanishes both at $d=4$ and $d=8$. This is in agreement with the conjecture that these two values respectively represent the lower and upper critical dimension for the Lifshitz point of the $O(N)$ theory.

In particular, in [24], it is argued for the Lifshitz point of the $N = 1$ theory, that the lower critical dimension could be associated to the large field behaviour of the fixed potential, corresponding to the particular value of d below which the potential does no longer diverge as a power law for large values of the field ϕ but, instead, a continuous set of solutions (constant at large ϕ) of the fixed potential equation is found. This value of d is related to the change of sign of the scaling dimension of ϕ, that for the case considered is $D_\phi = (d - 4 + \eta)/2$ and therefore, if the anomalous dimension vanishes or is neglected, $d = 4$ is the requested value. For the Lifshitz point of the $O(N)$ theory where, as shown above, $\eta_N = 0$ in $d = 4$, it is natural to accept it as the lower critical dimension. Needless to say, this argument is the restatement of what occurs for the scaling of the $O(N)$ theory at the lower critical dimension of the WF fixed point, $d = 2$.

In addition, one can focus on the leading order potential equation, Eq. (20), directly in $d = 4$. Actually, this equation can be solved analytically and, as for the case with $d < 4$, one ends up with a continuous set of solutions, parameterized by one real parameter.

Finally we comment on the dependence on the regulator of the result obtained for η_N. While we checked that the regulator (15) is well behaved for the computation of η_{WF} in the WF case, even its smoothened version in Eq. (16) produces potentially dangerous terms in the Lifshitz point case; terms that become irrelevant only in the limits $d \to 4^+$ and $d \to 8^-$. Then, the use of the smoother regulator (17) on the one hand confirms the behaviour of η_N in the region close to $d = 4$ and $d = 8$ but on the other hand, still produces the undesired effect, at least only for more than six dimensions, of generating multiple spurious extrema in η_N, regarded as a function of b. Therefore, away from the extremal points $d = 4$ and $d = 8$, no firm statement can be made on η_N in the LPA′. Conceivably, this is due to the modified two point functions (3), (4) with a leading $O(p^4)$ term, which provide the major difference between the Lifshitz point and the standard fixed point case where, conversely, the LPA′ provides reliable results.

We conclude by observing that the tricritical Lifshitz point which, when looking at the eigenvalue spectrum at the leading order of the $1/N$ expansion, could appear as a trivial duplicate of the WF fixed point with a suitable redefinition of the scaling dimensions of the various operators, does actually show original features. In fact, not only rather different properties of the anomalous dimension (with respect to the WF case) show up at order $1/N$, but it must also be noticed that, as soon as the wave function renormalizations are explicitly included in the fixed point equations, the coefficient of $(\partial\phi)^2$, Z, has positive scaling dimension $2 - \eta$, which indicates the existence of a relevant direction that has no correspondence at the WF critical point. On the other hand the similarities in the two cases could be a hint that the structure observed around $d = 2$, such as the presence of multi-critical solutions [40,42–45], or the relation with phase transitions of different nature [46,47], could have a counterpart in the Lifshitz scaling around $d = 4$.

Acknowledgements

This work has been carried out within the INFN project QFT-HEP.

References

[1] R.M. Hornreich, M. Luban, S. Shtrikman, Phys. Rev. Lett. 35 (1975) 1678–1681.
[2] W. Selke, Phys. Rep. 170 (1988) 213–264.
[3] H.W. Diehl, Acta Phys. Slovaca 52 (2002) 271–283.
[4] J. Alexandre, Int. J. Mod. Phys. A 26 (2011) 4523, arXiv:1109.5629 [hep-ph].
[5] K. Kikuchi, Prog. Theor. Phys. 127 (2012) 409–431, arXiv:1111.6075 [hep-th].
[6] P. Horava, Phys. Rev. D 79 (2009) 084008, arXiv:0901.3775 [hep-th].
[7] D. Benedetti, F. Guarnieri, J. High Energy Phys. 1403 (2014) 078, arXiv:1311.6253 [hep-th].
[8] G. D'Odorico, F. Saueressig, M. Schutten, Phys. Rev. Lett. 113 (2014) 171101, arXiv:1406.4366 [gr-qc].
[9] X. Bekaert, M. Grigoriev, Nucl. Phys. B 876 (2013) 667–714, arXiv:1305.0162 [hep-th].
[10] G. Cognola, R. Myrzakulov, L. Sebastiani, S. Vagnozzi, S. Zerbini, Class. Quantum Gravity 33 (2016) 225014, arXiv:1601.00102 [gr-qc].
[11] R. Casalbuoni, G. Nardulli, Rev. Mod. Phys. 76 (2004) 263, arXiv:hep-ph/0305069.
[12] T. Mizushima, K. Machida, M. Ichioka, Phys. Rev. Lett. 94 (2005) 060404, arXiv:cond-mat/0409417.
[13] P. Castorina, M. Grasso, M. Oertel, M. Urban, D. Zappalà, Phys. Rev. A 72 (2005) 025601, arXiv:cond-mat/0504391.
[14] R. Anglani, R. Casalbuoni, M. Ciminale, N. Ippolito, R. Gatto, M. Mannarelli, M. Ruggieri, Rev. Mod. Phys. 86 (2014) 509–561, arXiv:1302.4264 [hep-ph].
[15] M. Buballa, S. Carignano, Prog. Part. Nucl. Phys. 81 (2015) 39–96, arXiv:1406.1367 [hep-ph].
[16] M.A. Shpot, H.W. Diehl, Y.M. Pis'mak, J. Phys. A 41 (2008) 135003, arXiv:0802.2434 [cond-mat.stat-mech].
[17] H.W. Diehl, M.A. Shpot, J. Phys. A 35 (2002) 6249–6260, arXiv:cond-mat/0204267.
[18] M. Mueller, F. Schmid, Advanced Computer Simulation Approaches for Soft Matter Sciences II, Adv. Polym. Sci., vol. 185, 2005, p. 1, arXiv:cond-mat/0501076.
[19] C. Wetterich, Phys. Lett. B 301 (1993) 90.
[20] T.R. Morris, Int. J. Mod. Phys. A 9 (1994) 2411, arXiv:hep-ph/9308265.
[21] J. Berges, N. Tetradis, C. Wetterich, Phys. Rep. 363 (2002) 223–386, arXiv:hep-ph/0005122.
[22] C. Bervillier, Phys. Lett. A 331 (2004) 110–116, arXiv:hep-th/0405027.
[23] K. Essafi, J.P. Kownacki, D. Mouhanna, Europhys. Lett. 98 (2012) 51002, arXiv:1202.5946 [cond-mat.stat-mech].
[24] A. Bonanno, D. Zappalà, Nucl. Phys. B 893 (2015) 501–511, arXiv:1412.7046 [hep-th].
[25] S.-B. Liao, Phys. Rev. D 53 (1996) 2020, arXiv:hep-th/9501124.
[26] O. Bohr, B.-J. Schaefer, J. Wambach, Int. J. Mod. Phys. A 16 (2001) 3823–3852, arXiv:hep-ph/0007098.
[27] A. Bonanno, D. Zappalà, Phys. Lett. B 504 (2001) 181–187, arXiv:hep-th/0010095.
[28] D.F. Litim, J.M. Pawlowski, Phys. Rev. D 66 (2002) 025030, arXiv:hep-th/0202188.
[29] D.F. Litim, J.M. Pawlowski, Phys. Lett. B 546 (2002) 279–286, arXiv:hep-th/0208216.
[30] D. Zappalà, Phys. Lett. A 290 (2001) 35–40, arXiv:quant-ph/0108019.
[31] M. Mazza, D. Zappalà, Phys. Rev. D 64 (2001) 105013, arXiv:hep-th/0106230.
[32] D.F. Litim, D. Zappalà, Phys. Rev. D 83 (2011) 085009, arXiv:1009.1948 [hep-th].
[33] J.-P. Blaizot, R. Mendez Galain, N. Wschebor, Phys. Lett. B 632 (2006) 571, arXiv:hep-th/0503103.
[34] F. Benitez, J.-P. Blaizot, H. Chate, B. Delamotte, R. Mendez-Galain, N. Wschebor, Phys. Rev. E 85 (2012) 026707, arXiv:1110.2665 [cond-mat.stat-mech].
[35] D. Zappalà, Phys. Rev. D 86 (2012) 125003, arXiv:1206.2480 [hep-th].
[36] D.F. Litim, Phys. Rev. D 64 (2001) 105007, arXiv:hep-th/0103195.
[37] M. D'Attanasio, T.R. Morris, Phys. Lett. B 409 (1997) 363, arXiv:hep-th/9704094.
[38] T.R. Morris, M.D. Turner, Nucl. Phys. B 509 (1988) 637, arXiv:hep-th/9704202.
[39] C. Wetterich, Z. Phys. C 57 (1993) 451–470.
[40] A. Codello, G. D'Odorico, Phys. Rev. Lett. 110 (2013) 141601, arXiv:1210.4037 [hep-th].
[41] J. Zinn-Justin, Quantum Field Theory and Critical Phenomena, Clarendon Press, Oxford, 2002.
[42] T.R. Morris, Phys. Lett. B 345 (1995) 139–148, arXiv:hep-th/9410141.
[43] A. Codello, J. Phys. A 45 (2012) 465006, arXiv:1204.3877 [hep-th].
[44] A. Codello, N. Defenu, G. D'Odorico, Phys. Rev. D 91 (2015) 105003, arXiv:1410.3308 [hep-th].
[45] N. Defenu, P. Mati, I.G. Marian, I. Nandori, A. Trombettoni, J. High Energy Phys. 1505 (2015) 141, arXiv:1410.7024 [hep-th].
[46] G. Von Gersdorff, C. Wetterich, Phys. Rev. B 64 (2001) 054513, arXiv:hep-th/0008114.
[47] I. Nandori, S. Nagy, K. Sailer, A. Trombettoni, Phys. Rev. D 80 (2009) 025008, arXiv:0903.5524 [hep-th].

12

Knotted solutions for linear and nonlinear theories: Electromagnetism and fluid dynamics

Daniel W.F. Alves [a], Carlos Hoyos [b], Horatiu Nastase [a,*], Jacob Sonnenschein [c]

[a] Instituto de Física Teórica, UNESP-Universidade Estadual Paulista, Rua Dr. Bento T. Ferraz 271, Bl. II, São Paulo 01140-070, SP, Brazil
[b] Department of Physics, Universidad de Oviedo, Calle Federico García Lorca 18, 33007, Oviedo, Spain
[c] School of Physics and Astronomy, The Raymond and Beverly Sackler Faculty of Exact Sciences, Tel Aviv University, Ramat Aviv 69978, Israel

ARTICLE INFO

Editor: N. Lambert

ABSTRACT

We examine knotted solutions, the most simple of which is the "Hopfion", from the point of view of relations between electromagnetism and ideal fluid dynamics. A map between fluid dynamics and electromagnetism works for initial conditions or for linear perturbations, allowing us to find new knotted fluid solutions. Knotted solutions are also found to be solutions of nonlinear generalizations of electromagnetism, and of quantum-corrected actions for electromagnetism coupled to other modes. For null configurations, electromagnetism can be described as a null pressureless fluid, for which we can find solutions from the knotted solutions of electromagnetism. We also map them to solutions of Euler's equations, obtained from a type of nonrelativistic reduction of the relativistic fluid equations.

1. Introduction

Solutions with knotted topological structures play an important role in various areas of physics, but in this paper we will concern ourselves with two, electromagnetism and fluid dynamics. Despite the simplicity of the theory, the basic knotted solution of free Maxwell electromagnetism was only found in [1,2], following an earlier work in [3]. There are null solutions, $\vec{E}^2 - \vec{B}^2 = 0$, as well as generically non-null solutions in Rañada's construction, both of which are explicitly time dependent. In [4], it was shown that we can construct more general null solutions obtaining (m, n) knotted structures, and in [5], new knotted solutions were found using conformal $SO(4,2)$ transformations with complex parameters on known ones. For a review of this subject, and more complete references, see [6].

On the other hand, the theory of knots was actually developed in the 19th century based on knotted fluid lines, whose topological robustness was already discovered by Lord Kelvin, following the work of Helmholtz in 1858. The abstract study of knots and their evolution [7] is a fertile subject, for reviews see [8–10] and the book [11]. Remarkably, however, explicit theoretical solutions of fluid equations were very scarce, whereas experimental creation of knots waited until a few years ago [12] (see also [13] for some numerical construction). Moffat [14] finally defined a "helicity" $\mathcal{H}_v$ for the fluid flow similar, as we will see, to a magnetic helicity for electromagnetism, and wrote some explicit solutions with $\mathcal{H}_v \neq 0$. The properties of these were studied in [15,16]. More solutions were found in [17–19]. A map of magnetohydrodynamics to just fluid dynamics was used extensively (for instance, [20–22]), and one also was able to show that in magnetohydrodynamics a defined "velocity of lines of force" $\vec{v}_p = (\vec{E} \times \vec{H})/H^2$ [23] can be measured, and in some cases ("frozen field condition") coincides with the velocity of the fluid transporting it (see for instance [22,24], the last also considering transporting the electromagnetic Hopfion).

In this letter, we will use connections between electromagnetism and ideal fluid dynamics to find both new knotted solutions in electromagnetism, as well as new (time dependent) knotted solutions in fluid dynamics, that we believe have not been written explicitly before.

2. Knots in electromagnetism

In this section we review electromagnetic knotted solutions and some of their properties. Using the Riemann–Sielberstein (RS) vector $\vec{F} = \vec{E} + i\vec{B}$ $(c = 1)$, the source-free Maxwell's equations are

$$\vec{\nabla} \times \vec{F} = i\frac{\partial}{\partial t}\vec{F}; \quad \vec{\nabla} \cdot \vec{F} = 0 . \qquad (1)$$

* Corresponding author.
E-mail addresses: dwfalves@ift.unesp.br (D.W.F. Alves), hoyoscarlos@uniovi.es (C. Hoyos), nastase@ift.unesp.br (H. Nastase), cobi@post.tau.ac.il (J. Sonnenschein).

Electromagnetic duality is manifest as a rotation $\vec{F} \to e^{i\phi}\vec{F}$. We will define electric and magnetic potentials as $\vec{F} = \vec{B}_e + i\vec{B}_m = \vec{\nabla} \times (\vec{A}_e + i\vec{A}_m)$. To characterize the non-trivial topology of electromagnetic fields, a common set of observables are the helicities, that give a measure of the mean value of the linking number of the electromagnetic field lines. The helicities $\mathcal{H}_{ab}$, $a, b = e, m$ are defined as

$$\mathcal{H}_{ab} = \int d^3x \vec{A}_a \cdot \vec{B}_b \,. \tag{2}$$

Calculating their time derivatives, we find that $\mathcal{H}_{ee}$ and $\mathcal{H}_{mm}$ are conserved if $\vec{E} \cdot \vec{B} = 0$ and $\mathcal{H}_{em}$ and $\mathcal{H}_{me}$ are conserved if $\vec{E}^2 - \vec{B}^2 = 0$. One can find solutions that have conserved helicities and contain "knotted" structures for the electric and magnetic fields, characterized by Hopf or winding number invariants of the field structures.

2.1. Knotted solutions in Bateman's construction

One way to obtain knotted solutions is in Bateman's construction, imposing an ansatz

$$\vec{F} = \vec{\nabla}\alpha \times \vec{\nabla}\beta \,, \tag{3}$$

with $\alpha, \beta \in \mathbb{C}$. One Maxwell equation, $\vec{\nabla} \cdot \vec{F} = 0$, is automatically satisfied, and the other imposes the constraint $\vec{F}^2 = 0 \Rightarrow \vec{E}^2 - \vec{B}^2 = 0$ and $\vec{E} \cdot \vec{B} = 0$, so these are *null* solutions. Using this construction one can show that there is a "Hopfion" solution

$$\alpha = \frac{r^2 - t^2 - 1 + 2iz}{r^2 - t^2 + 1 + 2it}; \quad \beta = \frac{2(x - iy)}{r^2 - t^2 + 1 + 2it}. \tag{4}$$

Where (t, x, y, z) are the spacetime coordinates and $r^2 = x^2 + y^2 + z^2$. The Hopfion has $\mathcal{H}_{em} = \mathcal{H}_{me} = 0$ and nonzero helicities $\mathcal{H}_{mm} = \mathcal{H}_{ee}$, but their value depends on the amplitude of the electromagnetic field $|\vec{E}|$. An invariant that only depends on the topology can be constructed by introducing a map to unit quaternions $q = (\alpha + \beta j)/\sqrt{|\alpha|^2 + |\beta|^2} \in \mathbb{H}$ [5]. Using $\mathbb{R}^3 \cup \{\infty\} \cong S^3$ and $\mathbb{H}_{|q|^2=1} \cong SU(2) \cong S^3$, at any fixed time the "Hopfion" solution maps $S^3 \to S^3$ with unit winding number $w = 1$. By replacing α with α^m and β with β^n in (3), we find a (m, n) knot solution with winding number $w = mn$.

2.2. Knotted solutions in Rañada's construction

There are more general, *non-null* ($\vec{E}^2 - \vec{B}^2 \neq 0$), knotted solutions. Consider the ansatz for $F_{\mu\nu}$ and $*F_{\mu\nu}$ of the type

$$F_{\mu\nu} = \frac{\sqrt{a}}{2\pi i}\frac{1}{(1 + \bar{\phi}\phi)^2}(\partial_\mu\bar{\phi}\partial_\nu\phi - \partial_\nu\bar{\phi}\partial_\mu\phi) \,,$$

$$*F_{\mu\nu} = \frac{\sqrt{a}}{2\pi i}\frac{1}{(1 + \bar{\theta}\theta)^2}(\partial_\mu\bar{\theta}\partial_\nu\theta - \partial_\nu\bar{\theta}\partial_\mu\theta). \tag{5}$$

The condition $*F_{\mu\nu} = \frac{1}{2}\epsilon_{\mu\nu\rho\sigma}F^{\rho\sigma}$, determines the equations for ϕ, θ. The solutions then solve Maxwell's equations, and by construction $F_{\mu\nu} * F^{\mu\nu} \propto \vec{E} \cdot \vec{B} = 0$. Moreover, the 2-form F decomposes as $F = dq \wedge dp$ (Clebsch representation), where p, q are *real* functions, and in particular

$$\vec{B} = \vec{\nabla}p \times \vec{\nabla}q \,. \tag{6}$$

The relation between p, q and ϕ is

$$p = \frac{1}{1 + |\phi|^2}; \quad q = \frac{arg(\phi)}{2\pi} \,, \tag{7}$$

and there is a similar one for $*F = du \wedge dv$, with u, v in terms of θ. If p and q are *single-valued and well-defined in the whole of space*, then the magnetic helicity $\mathcal{H}_{mm}$ is zero. If not, we have

$$\vec{A} = p\vec{\nabla}q + \vec{\nabla}\chi \,, \tag{8}$$

where χ is such that $\vec{A}$ is well defined. The magnetic helicity is

$$\mathcal{H}_{mm} = \int d^3x \vec{\nabla}\chi \cdot (\vec{\nabla}p \times \vec{\nabla}q). \tag{9}$$

Defining the function

$$\phi_H(x, y, z) = \frac{2(x + iz)}{2z + i(r^2 - 1)}, \tag{10}$$

the "Hopfion" solution in Rañada's construction is such that, at $t = 0$,

$$\phi(x, y, z) = \phi_H(z, -y, x), \quad \theta(x, y, z) = \phi_H(x, z, -y). \tag{11}$$

The solution has $\mathcal{H}_{ee} = \mathcal{H}_{mm} \neq 0$ and $\mathcal{H}_{em} = \mathcal{H}_{me} = 0$. A more general set of solutions with non-zero helicities [25] are, at $t = 0$,

$$\phi = \frac{(x + iy)^{(n)}}{(z + i(r^2 - 1)/2)^{(m)}} \quad \theta = \frac{(y + iz)^{(l)}}{(z + i(r^2 - 1)/2)^{(s)}} \,, \tag{12}$$

where the (m) index means we leave the modulus intact, but we raise the phase to the m-th power.

2.3. Hopf index

Using $\mathbb{R}^3 \cup \{\infty\} \cong S^3$ and $\mathbb{C} \cup \{\infty\} \cong S^2$, at any fixed time ϕ can be seen as a map $S^3 \to S^2$, and the magnetic helicity is its Hopf index $\mathcal{H}(\phi)$. Indeed, we can define an area 2-form on S^2,

$$\omega = \frac{1}{2\pi i}\frac{d\phi^* \wedge d\phi}{(1 + |\phi|^2)^2} \,, \tag{13}$$

whose pullback onto S^3 is a 2-form $\mathcal{F}$ whose components are the spatial components in (5) (taking $a = 1/4$) $\mathcal{F}_{ij} = F_{ij}$. The Hopf index then equals the magnetic helicity

$$\mathcal{H}(\phi) = \int_{S^3} \mathcal{A} \wedge \mathcal{F} = \int d^3x \vec{A} \cdot \vec{B} = \mathcal{H}_{mm} \,. \tag{14}$$

A Hopf index $\mathcal{H}(\phi_H) = 1$ is found for instance for (10).

3. Solutions to nonlinear theories

We now show that the knotted solutions in Bateman's construction are also solutions of any nonlinear electromagnetism theory that reduces to the Maxwell case for small fields. This is true in general for any null configurations (for recent discussions see e.g. [26,27]). Since $\vec{F}^2 = 0$, both the two possible Lorentz invariants constructed out of $\vec{E}$ and $\vec{B}$ vanish on the solutions. Let us define

$$L \equiv \frac{F_{\mu\nu}F^{\mu\nu}}{2b^2} = \frac{1}{b^2}(\vec{B}^2 - \vec{E}^2);$$

$$P \equiv \frac{1}{8b^2}\epsilon^{\mu\nu\rho\sigma}F_{\mu\nu}F_{\rho\sigma} = \frac{1}{b^2}\vec{E} \cdot \vec{B} \,, \tag{15}$$

where b is a dimension 2 constant. We assume that fields vary on distances much larger than $b^{-1/2}$, so that we can ignore possible derivatives on $F_{\mu\nu}$. Then, nonlinear generalizations of electromagnetism are described by actions of the form

$$\mathcal{L} = b^2\left[-\frac{L}{2} + \sum_{n\geq 2}\sum_{m\geq 0} c_{n,m}L^n P^m\right] \,, \tag{16}$$

i.e., which reduces to Maxwell at small fields, and has nonlinear corrections written solely in terms of L and P. This includes actions like the Born–Infeld Lagrangian, $\mathcal{L} = -b^2[\sqrt{1+L-P^2}-1]$ [28], introduced to get rid of diverging electric fields in electromagnetism, as well as actions obtained by integrating out other fields, like in the case of the one-loop Euler–Heisenberg Lagrangian for QED, where the fermions have been integrated out, $\mathcal{L} = b^2\left[-\frac{L}{2} + \beta b^2[L^2+7P^2]\right]$, where $b = m^2$ and $\beta = 2\alpha^2/45$ (m is the fermion mass and α the fine structure constant). This form also applies for *any higher loop* integration of the coupling to *any field* (see [29] for a review). Therefore knotted solutions are valid in the full quantum theory after integrating out the other fields. Moreover, this is not only true for usual quantum fields, but also string modes in string theory can be integrated out to give the same result. Indeed, for electromagnetism confined to a D-brane in string theory, the integrating out of higher modes (in $\alpha' = l_s^2$, the string scale) results in the BI action with $b = 1/\alpha'$. Note that in this case, the action is α' exact at leading order in g_s, and there are no $\partial_\mu F_{\nu\rho}$ terms.

In terms of quantities analogous to the ones of electromagnetism in a medium (see [30]),

$$\vec{H} \equiv -\frac{\partial \mathcal{L}}{\partial \vec{B}}, \quad \vec{D} \equiv +\frac{\partial \mathcal{L}}{\partial \vec{E}}, \tag{17}$$

the equations of motion look formally the same as Maxwell's in a medium,

$$\begin{aligned} &\vec{\nabla}\times\vec{E} + \partial_t\vec{B} = 0, \ \vec{\nabla}\cdot\vec{B} = 0 \\ &\vec{\nabla}\times\vec{H} - \partial_t\vec{D} = 0, \ \vec{\nabla}\cdot\vec{D} = 0 . \end{aligned} \tag{18}$$

For solutions such that $L = P = 0$, $\vec{H}$ reduces to $\vec{B}$ and $\vec{D}$ to $\vec{E}$, so we have Maxwell's equations in vacuum and thus knotted solutions found using Bateman's construction in Maxwell theory are solutions of the nonlinear theory as well.

4. Mapping electromagnetic to fluid knots

An ideal fluid, for an adiabatic flow with potential per unit mass π is governed by the Euler's and continuity equations,

$$\begin{aligned} &\partial_t\vec{v} + (\vec{v}\cdot\vec{\nabla})\vec{v} = -\frac{1}{\rho}\vec{\nabla}p - \vec{\nabla}\pi, \\ &\partial_t\rho + \vec{\nabla}\cdot(\rho\vec{v}) = 0. \end{aligned} \tag{19}$$

The right hand side of the Euler's equations (19) equals $-\vec{\nabla}h$ ($\delta h = \frac{\delta p}{\rho} + \delta\pi$), where h is the enthalpy per unit mass. The continuity equation can also be rewritten as

$$\partial_t h + \vec{v}\cdot\vec{\nabla}h + c_s^2\vec{\nabla}\cdot\vec{v} = 0 , \tag{20}$$

where $c_s = \sqrt{\partial p/\partial\rho}$ is the sound speed.

For incompressible fluids $\vec{\nabla}\cdot\vec{v} = 0$ and compressible barotropic fluids $p = p(\rho)$, there is also a conserved helicity. From the Euler's equations (19), one finds

$$\partial_t(\vec{v}\cdot\vec{\omega}) - \vec{\nabla}\cdot\left[\vec{\omega}\left(\frac{\vec{v}^2}{2} - \int\frac{dp}{\rho} - \pi\right) - \vec{v}(\vec{v}\cdot\vec{\omega})\right] = 0 , \tag{21}$$

where $\vec{\omega} = \vec{\nabla}\times\vec{v}$ is the vorticity, which means that we have the conserved *fluid helicity*, the integral of the velocity Chern–Simons term,

$$\mathcal{H}_v = \int d^3x\,\vec{v}\cdot\vec{\omega} = \int d^3x\,\vec{v}\cdot(\vec{\nabla}\times\vec{v}) . \tag{22}$$

Knotted fluid solutions are solutions for which there is a linking of $\vec{v}(t,\vec{x})$ at fixed time t, i.e. nonzero fluid helicity.

We see the analogy with electromagnetism [31,32]: $\vec{v}$ is the analog of $\vec{A}$, so $\vec{B}$ is the analog of the vorticity $\vec{\omega}$, and $\mathcal{H}_v$ is the analog of $\mathcal{H}_{mm}$. One can define Clebsch variables λ and μ for the fluid, as in [33], giving the velocity field

$$\vec{v} = \lambda\vec{\nabla}\mu + \vec{\nabla}\Phi , \tag{23}$$

where Φ is the fluid potential, just like the decomposition for $\vec{A}$ in (8) in the Rañada construction. Note that this is not the much-used map from magneto-hydrodynamics (fluid coupled to electromagnetism) to hydrodynamics, where $\vec{v}$ is mapped to $\vec{v}$, but $\vec{\omega}$ is mapped to $\vec{B}$, and one restricts the configurations to the ones with $\vec{B} = \vec{\omega} = \vec{\nabla}\times\vec{v}$. Instead, we can define a full map, from fluid to electromagnetism, by

$$\vec{B} = \vec{\omega} = \vec{\nabla}\times\vec{v}, \quad \vec{E} = -\partial_t\vec{v} - \vec{\nabla}h , \tag{24}$$

which means that really $\vec{A} = \vec{v}$; $A_t = h$, though we have no gauge invariance now, since $\vec{v}$ is physical (observable). Two of Maxwell's equations $\vec{\nabla}\cdot\vec{B} = \vec{\nabla}\times\vec{E} + \partial_t\vec{B} = 0$ are automatically satisfied, but from the Euler's equations (19) we find the condition

$$\vec{E} = (\vec{v}\cdot\vec{\nabla})\vec{v} = \vec{\omega}\times\vec{v} + \vec{\nabla}\left(\frac{\vec{v}^2}{2}\right) = \vec{B}\times\vec{A} + \vec{\nabla}\left(\frac{\vec{A}^2}{2}\right) , \tag{25}$$

which does not hold for knotted solutions of Maxwell's equations. At the linearized level however, it implies $\vec{E} = 0$, i.e., pure magnetism, and the continuity equation, in electromagnetic variables

$$\partial_t A_t + (\vec{A}\cdot\vec{\nabla})A_t + c_s^2\vec{\nabla}\cdot\vec{A} = 0 , \tag{26}$$

becomes the Lorenz gauge condition, identifying c_s with the speed of light.

Rather than mapping the full solution, we will use the map between electromagnetism and fluid variables at a fixed time

$$\vec{v}(t=0,\vec{x}) = \vec{A}(t=0,\vec{x}), \ h(t=0,\vec{x}) = h_0. \tag{27}$$

Where h_0 is an arbitrary constant. In this case the knotted solutions supply initial conditions for Euler's and continuity equations. The time-dependent solutions have non-zero helicity and correspond to a fluid configurations of non-trivial topology.

In [33], knotted solutions for an incompressible fluid were (implicitly) found by giving an initial condition of the form (23), where $\lambda = \cos\vartheta(\vec{x})$, $\mu = \varphi(\vec{x})$, with (ϑ, φ) the polar and azimuthal angles of a S^2. Φ would be determined by the incompressibility condition, but finding an explicit solution is the main obstacle to obtaining an analytic expression for the velocity. We will avoid this issue by considering more general cases of compressible barotropic fluids, so the only constraint on Φ is that the velocity should be a smooth function of the spatial coordinates. Knotted solutions in Rañada's construction can be mapped for instance using the Clebsch decompositions (8) and (23), making $\lambda = p$, $\mu = q$ and $\Phi = \chi$, where p, q are given by (7) and ϕ can be taken to be (11) or (12). In terms of ϕ, the velocity is

$$\vec{v} = \frac{1}{4\pi i(1+|\phi|^2)}\left(\frac{\vec{\nabla}\phi}{\phi} - \frac{\vec{\nabla}\phi^*}{\phi^*}\right) + \vec{\nabla}\Phi . \tag{28}$$

In addition, we present here an additional set of initial conditions for knotted solutions. Consider the stereographic projection of S^3 on $\mathbb{R}^3$:

$$X_i = \frac{2x_i}{1+r^2}, \quad X_4 = \frac{1-r^2}{1+r^2}. \tag{29}$$

We define the complex coordinates $Z_1 = X_1 + iX_2$, $Z_2 = X_3 + iX_4$, that then we use them to define a ϕ

$$\phi = \frac{Z_1^n}{Z_2^m}, \; n, m \geq 1, \; n, m \in \mathbb{Z}. \tag{30}$$

A non-singular velocity is obtained for

$$\Phi = -\frac{n}{2\pi} \arctan\left(\frac{y}{x}\right). \tag{31}$$

The helicity of these configurations is $\mathcal{H}_v = -nm$.

5. Electromagnetism as a fluid and its knotted solutions

Any gapless quantum system is expected to have an effective fluid description at low energies. For a relativistic system this means that the energy–momentum tensor can be put in the form

$$T_{\mu\nu} = \rho u_\mu u_\nu + p(\eta_{\mu\nu} + u_\mu u_\nu) + \pi_{\mu\nu}, \tag{32}$$

where u_μ is the 4-velocity of the fluid, ρ is the energy density and p the pressure. $\pi_{\mu\nu}$ depends on derivatives of u_μ, ρ and p. From $\partial_\mu T^{\mu\nu} = 0$ we obtain the relativistic fluid equations. This program can be applied to any quantum system as in [34–37]. For a perfect fluid $\pi_{\mu\nu} = 0$ we obtain in the non-relativistic limit $u^\mu \simeq (1, \vec{v})$, $|\vec{v}|^2 \ll 1$, $p \ll \rho$ the continuity and Euler's equations. However, in electromagnetism we can also find another, more unusual, fluid, a "null pressureless fluid", or "null dust", with $p = 0$ and $u^\mu u_\mu = 0$.

As has been observed in [38], this is the case for null configurations with $\vec{E}^2 - \vec{B}^2 = 0$ and $\vec{E} \cdot \vec{B} = 0$. In this case the electromagnetic energy–momentum tensor takes the form $T_{\mu\nu} = \rho u_\mu u_\nu$, with $u_\mu = (1, \vec{v})$ ($\vec{v}^2 = 1$) and

$$\rho = \frac{1}{2}(\vec{E}^2 + \vec{B}^2); \quad \vec{v} = \frac{1}{\rho} \vec{E} \times \vec{B}. \tag{33}$$

The Hopfion solution determined by (3) and (4) is null, so we can map it to fluid dynamics. We find an energy density

$$\rho = \frac{16\left((t-z)^2 + x^2 + y^2 + 1\right)^2}{\left(\left(r^2 - t^2 + 1\right)^2 + 4t^2\right)^3}. \tag{34}$$

The velocity field is

$$v_x = \frac{2(y + x(t-z))}{1 + x^2 + y^2 + (t-z)^2}, \; v_y = \frac{-2(x - y(t-z))}{1 + x^2 + y^2 + (t-z)^2}, \tag{35}$$

and $v_z^2 = 1 - v_x^2 - v_y^2$. The topological structure can be seen in Fig. 1. Although the fluid helicity $\mathcal{H}_v$ diverges due to non-vanishing contributions at spatial infinity, these can be subtracted to produce a finite nonzero result. The velocity of the null fluid can be rewritten in terms of the RS vector $\vec{F}$ and its complex conjugate $\vec{F}^*$, as

$$\vec{v} = 2i\vec{F} \times \vec{F}^* / \vec{F} \cdot \vec{F}^*. \tag{36}$$

This implies that (m, n) knots, and more general null solutions found by the replacement in (3) $\alpha \to f(\alpha)$, $\beta \to g(\beta)$, have the same velocity as the Hopfion. On the other hand, the energy density ρ is different for each configuration.

5.1. Maps to non-relativistic fluids

Switching to lightcone coordinates, $x^\pm = t \pm z$, the velocity of the Hopfion solution (35), is independent of x^+, in such a way that it can be mapped to a solution of a non-relativistic 2 + 1-dimensional system, with $\tau = x^-$ playing the role of time coordinate. Defining β^a by $\beta^a = \frac{v^a}{1 - v^z}$, $a = x, y$, for any configuration satisfying $\partial_+ v^a = 0$, the relativistic fluid equations become the

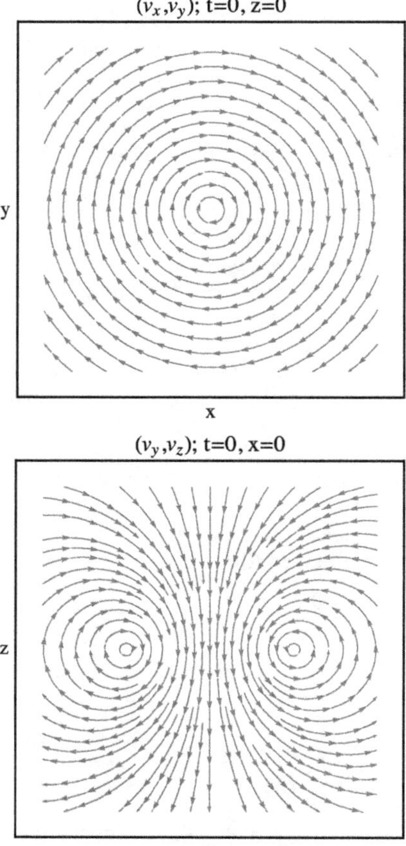

Fig. 1. Orthogonal sections of the velocity field for the Hopfion solution, on the (x, y) plane (top) and (y, z) plane (bottom). Using rotational symmetry in the (x, y) directions the linked torus structure is apparent.

2 + 1-dimensional Euler's equations of a compressible fluid with velocity β^a and constant pressure[1]:

$$\partial_\tau \beta^a + \beta^b \partial_b \beta^a = 0. \tag{37}$$

In this description, the Hopfion solution is a spherical bouncing shock

$$\beta^a = \epsilon^{ab} \partial_b \tilde{\psi} + \tau \partial^a \tilde{\psi}, \; \tilde{\psi} = \log(x^2 + y^2 - 1 - \tau^2). \tag{38}$$

If $\partial_+ v^a = 0$, it is also possible to establish a map to an *incompressible* fluid. The relativistic fluid equations are

$$(1 - v^z)\partial_\tau v^a + v^b \partial_b v^a = 0. \tag{39}$$

At any fixed τ we can make an identification of the first term with the pressure in Euler's equations

$$\partial^a p \equiv (1 - v^z)\partial_\tau v^a. \tag{40}$$

Then v^a, p map to *steady state* (time-independent) solutions of Euler's equations, with τ seen as a parameter. The Hopfion solution can be written as

$$v^a = \epsilon^{ab} \partial_b \psi + \tau \partial^a \psi, \; \psi = \log(1 + x^2 + y^2 + \tau^2). \tag{41}$$

At $\tau = 0$ the velocity satisfies the incompressibility condition $\partial_a v^a = 0$. In this case, an explicit solution to (40) can be found

[1] This map should not be confused with other non-relativistic limits of electromagnetism where the speed of light is taken to infinity [39].

$$p = p_\infty - \frac{2}{1 + x^2 + y^2}. \tag{42}$$

This solution is a smooth vortex, it can be obtained by the stereographic projection of a constant vorticity configuration on the sphere (see e.g. [18]). This map can also be used to give initial conditions, at $\tau = 0$, for the null fluid, given a steady state (v_S^a, p_S) solution in $2+1$ dimensions, by

$$v^a(\tau = 0) = v_S^a, \quad \partial_\tau v^a(\tau = 0) = \frac{\partial^a p_S}{1 - v_S^2}. \tag{43}$$

Acknowledgements

We would like to thank Manuel Arrayás, Ori Ganor and Nilanjan Sircar for useful comments and discussions. This work was supported in part by a center of excellence supported by the Israel Science Foundation (grant number 1989/14), and by the US-Israel bi-national fund (BSF) grant number 2012383 and the Germany Israel bi-national fund GIF grant number I-244-303.7-2013. The work of HN is supported in part by CNPq grant 304006/2016-5 and FAPESP grant 2014/18634-9. HN would also like to thank the ICTP-SAIFR for their support through FAPESP grant 2016/01343-7. C.H. is supported by the Ramon y Cajal fellowship RYC-2012-10370, the Asturian grant FC-15-GRUPIN14-108 and the Spanish national grant MINECO-16-FPA2015-63667-P. D.F.W.A. is supported by CNPq grant 146086/2015-5.

References

[1] A.F. Ranada, A topological theory of the electromagnetic field, Lett. Math. Phys. 18 (1989) 97.
[2] A.F. Ranada, Knotted solutions of the maxwell equations in vacuum, J. Phys. A, Math. Gen. 23 (16) (1990) L815.
[3] A. Trautman, Solutions of the Maxwell and Yang–Mills equations associated with Hopf fibrings, Int. J. Theor. Phys. 16 (1977) 561.
[4] H. Kedia, I. Bialynicki-Birula, D. Peralta-Salas, W.T.M. Irvine, Tying knots in light fields, Phys. Rev. Lett. 111 (2013) 150404, arXiv:1302.0342 [math-ph].
[5] C. Hoyos, N. Sircar, J. Sonnenschein, New knotted solutions of Maxwell's equations, J. Phys. A 48 (25) (2015) 255204, arXiv:1502.01382 [hep-th].
[6] M. Arrayás, D. Bouwmeester, J. Trueba, Knots in electromagnetism, Phys. Rep. 667 (2017) 1–61.
[7] T. Kambe, T. Takao, Motion of distorted vortex rings, J. Phys. Soc. Jpn. 31 (2) (1971) 591–599.
[8] R.L. Ricca, M.A. Berger, Topological ideas and fluid mechanics, Phys. Today 49 (12) (1996) 28–34.
[9] R.L. Ricca, Applications of knot theory in fluid mechanics, Banach Cent. Publ. 42 (1998) 321–346.
[10] R.L. Ricca, New developments in topological fluid mechanics, Nuovo Cimento Soc. Ital. Fis. C 32 (1) (2009) 185.
[11] V.I. Arnold, B.A. Khesin, Topological Methods in Hydrodynamics, vol. 125, Springer Science & Business Media, 1999.
[12] D. Kleckner, W.T. Irvine, Creation and dynamics of knotted vortices, Nat. Phys. 9 (4) (2013) 253–258.
[13] D. Proment, M. Onorato, C.F. Barenghi, Vortex knots in a Bose–Einstein condensate, Phys. Rev. E 85 (3) (2012) 036306.
[14] H.K. Moffatt, The degree of knottedness of tangled vortex lines, J. Fluid Mech. 35 (01) (1969) 117–129.
[15] H. Moffatt, A. Tsinober, Helicity in laminar and turbulent flow, Annu. Rev. Fluid Mech. 24 (1) (1992) 281–312.
[16] H. Moffatt, R.L. Ricca, Helicity and the Calugareanu invariant, Proc. R. Soc. Lond. A Math. Phys. Eng. Sci. 439 (1992) 411–429.
[17] A. Morgulis, V.I. Yudovich, G.M. Zaslavsky, Compressible helical flows, Commun. Pure Appl. Math. 48 (5) (1995) 571–582.
[18] D.G. Crowdy, Stuart vortices on a sphere, J. Fluid Mech. 498 (2004) 381402.
[19] E. Ifidon, E. Oghre, Vortical structures on spherical surfaces, J. Niger. Math. Soc. 34 (2) (2015) 216–226.
[20] A.L. Kholodenko, Optical knots and contact geometry I. From Arnold inequality to Ranada's dyons, Anal. Math. Phys. 6 (2) (2016) 163–198, arXiv:1402.1793 [math-ph].
[21] A.L. Kholodenko, Optical knots and contact geometry II. From Ranada dyons to transverse and cosmetic knots, Ann. Phys. 371 (2016) 77–124, arXiv:1406.6108 [math-ph].
[22] A.H. Boozer, Mathematics and Maxwell's equations, Plasma Phys. Control. Fusion 52 (12) (2010) 124002.
[23] W.A. Newcomb, Motion of magnetic lines of force, Ann. Phys. 3 (Apr. 1958) 347–385.
[24] W.T.M. Irvine, Linked and knotted beams of light, conservation of helicity and the flow of null electromagnetic fields, J. Phys. A, Math. Theor. 43 (38) (2010) 385203.
[25] M. Arrayas, J.L. Trueba, A class of non-null toroidal electromagnetic fields and its relation to the model of electromagnetic knots, J. Phys. A, Math. Theor. 48 (2015) 025203, arXiv:1106.1122 [hep-th].
[26] M. Ortaggio, V. Pravda, Class. Quantum Gravity 33 (11) (2016) 115010, arXiv: 1506.04538 [gr-qc].
[27] E. Goulart, Nonlinear electrodynamics is skilled with knots, Europhys. Lett. 115 (1) (2016) 10004, arXiv:1602.05071 [gr-qc].
[28] M. Born, L. Infeld, Foundations of the new field theory, Proc. Roy. Soc. Lond. A 144 (1934) 425–451.
[29] G.V. Dunne, Heisenberg–Euler effective agrangians: basics and extensions, in: M. Shifman, A. Vainshtein, J. Wheater (Eds.), From Fields to Strings: Circumnavigating Theoretical Physics. Ian Kogan Memorial Collection (3 volume set), 2004, pp. 445–522, arXiv:hep-th/0406216.
[30] H. Nastase, J. Sonnenschein, More on Heisenberg's model for high energy nucleon–nucleon scattering, Phys. Rev. D 92 (2015) 105028, arXiv:1504.01328 [hep-th].
[31] H. Marmanis, Analogy between the Navier–Stokes equations and Maxwell's equations: application to turbulence, Phys. Fluids 10 (1998) 1428–1437.
[32] S. Sridhar, Turbulent transport of a tracer: an electromagnetic formulation, Phys. Rev. E 58 (1998) 522–525.
[33] E. Kuznetsov, A. Mikhailov, On the topological meaning of canonical Clebsch variables, Phys. Lett. A 77 (1) (1980) 37–38.
[34] H. Nastase, DBI scalar field theory for QGP hydrodynamics, Phys. Rev. D 94 (2) (2016) 025014, arXiv:1512.05257 [hep-th].
[35] S. Endlich, A. Nicolis, R. Rattazzi, J. Wang, The quantum mechanics of perfect fluids, J. High Energy Phys. 04 (2011) 102, arXiv:1011.6396 [hep-th].
[36] S. Dubovsky, L. Hui, A. Nicolis, D.T. Son, Effective field theory for hydrodynamics: thermodynamics, and the derivative expansion, Phys. Rev. D 85 (2012) 085029, arXiv:1107.0731 [hep-th].
[37] L. Berezhiani, J. Khoury, J. Wang, A Universe without dark energy: cosmic acceleration from dark matter–baryon interactions, arXiv:1612.00453 [hep-th].
[38] I. Bialynicki-Birula, Field theory of photon dust, Acta Phys. Pol. B 23 (1992) 553–559.
[39] G. Rousseaux, Forty years of Galilean Electromagnetism (1973–2013), Eur. Phys. J. Plus 128 (2013) 81.

Logarithmic corrections to entropy of magnetically charged AdS_4 black holes

Imtak Jeon [a], Shailesh Lal [b,*]

[a] Harish-Chandra Research Institute, Chhatnag Road, Jhusi, Allahabad 211019, India
[b] LPTHE – UMR 7589, UPMC Paris 06, Sorbonne Universités, Paris 75005, France

ARTICLE INFO

Editor: M. Cvetič

Keywords:
String theory
AdS/CFT correspondence
Black holes
Quantum entropy
AdS black hole
Logarithmic corrections

ABSTRACT

Logarithmic terms are quantum corrections to black hole entropy determined completely from classical data, thus providing a strong check for candidate theories of quantum gravity purely from physics in the infrared. We compute these terms in the entropy associated to the horizon of a magnetically charged extremal black hole in $AdS_4 \times S^7$ using the quantum entropy function and discuss the possibility of matching against recently derived microscopic expressions.

1. Introduction

Providing a microscopic interpretation to the Bekenstein Hawking formula in the context of certain classes of supersymmetric extremal black holes in flat space has been a main success of string theory as a theory of quantum gravity [1–6]. The expression for the microscopic entropy obtained by explicit enumeration and counting of black hole microstates in these cases contains the area law as the leading formula, but also contains higher-derivative and quantum corrections to it. We refer the reader to [7] for a review of these developments as well as more exhaustive references. Importantly, since extremal black holes expectedly possess an AdS_2 factor in the near horizon geometry, one may use the AdS_2/CFT_1 correspondence to provide an alternative, but equivalent, definition of the quantum entropy of extremal black holes in string theory. This proposal is known as the quantum entropy function, and for extremal black holes carrying charges $\vec{q} \equiv q_i$ [8,9],

$$d_{hor}(\vec{q}) \equiv \left\langle \exp\left[i \oint q_i d\theta \mathcal{A}^i_\theta \right] \right\rangle^{finite}_{AdS_2}, \qquad (1)$$

where d_{hor} is the full quantum degeneracy associated with the black hole horizon, and $\mathcal{A}^i_\theta$ is the component of the ith gauge field along the boundary of the AdS_2. In this picture the entropy associated to the horizon degrees of freedom of an extremal blackhole is essentially the free energy corresponding to the partition function (1). The superscript 'finite' reminds us that the quantity on the right hand side of (1) is naively divergent due to the infinite volume of AdS_2 but this divergence may be regulated in accordance with general principles of the AdS/CFT correspondence and a cutoff-insensitive finite part extracted, which is then identified to d_{hor} [8–10]. The path integral is carried out over all fields that asymptote to the black hole near horizon geometry. In the context of supersymmetric extremal black holes in flat space, this has been evaluated using saddle point techniques [10–14,16,17] as well as supersymmetric localization [18–24] and the answer matched with microscopic results wherever available. Importantly, even in the cases where the full microscopic formula is unavailable, this quantity may be evaluated at least using saddle-point methods to gain some insight into the full microscopic formula [13,17].

In contrast, the situation for entropy computations for extremal black holes in AdS space is relatively in its infancy. While the Wald entropy [25] associated to the horizon may be computed for such black holes using for example the entropy function formalism, the quantum corrections to it are still unknown.[1] In this situation, it

* Corresponding author.
E-mail address: shailesh.hri@gmail.com (S. Lal).

[1] Recently the quantum entropy of the topological black hole in AdS_4 has been computed using supersymmetric localization [26], building on the analysis of [27].

would clearly be of interest to evaluate (1) for such black holes to obtain the set of quantum corrections to the Wald formula. Such computations are further motivated by the recent proposal for a *microscopic* computation of a CFT_3 index argued to capture the quantum entropy of an AdS_4 extremal black hole [29].

In this note we will focus on the computation of (1) in the semi-classical approximation where the event horizon of the black hole has a large length scale a associated with it. In the case of extremal black holes in flat space, this large length scale arises when the charges of the black hole are taken to be large. In the present case of AdS black holes, this corresponds to taking the rank N of the gauge group in the dual CFT to be large, while the charges themselves are not scaled.

One saddle-point of the path integral (1) is the black hole's near-horizon geometry itself. By evaluating the on-shell action on this field configuration, it is possible to show that [8,9]

$$d_{hor} \simeq e^{S_{\text{Wald}}} \quad \Rightarrow S_{BH} = \ln d_{hor} \simeq S_{\text{Wald}}. \tag{2}$$

Hence the quantum entropy function produces as the leading contribution, Wald's formula for the entropy of the horizon. In this note, we will consider subleading corrections of the form $\ln a$ to this formula, i.e.

$$S_{BH} = S_{\text{Wald}} + c \ln a + \ldots, \tag{3}$$

where c is a coefficient which depends on the details of the quantum gravity that the black hole is embedded in. For example, the same four-dimensional black hole which is a quarter-BPS black hole in $\mathcal{N} = 4$ supergravity may be embedded as a one-eighth BPS black hole in $\mathcal{N} = 8$ supergravity and while the leading Bekenstein–Hawking answer is the same for the black hole, the log terms computed in both theories are different [12], and match with the microscopic computations carried out respectively in $\mathcal{N} = 4$ and $\mathcal{N} = 8$ string compactifications. This matching is an important test of the consistency of the quantum entropy function proposal.

The main reason why the log term is an important contribution to the microscopic formula is that it is a genuinely quantum correction to the Bekenstein Hawking formula, but is determined completely from one-loop fluctuations of massless fields of two-derivate supergravity, which essentially constitute the IR data of the black hole.[2] To see this, let us recall some elements of a scaling argument presented in [15]. Consider the ℓ-loop free energy for a theory defined on a D dimensional background with a length scale a associated with it. A typical Feynman graph contributing to this quantity would scale as (see [15] for details)

$$\ell_P^{(D-2)(\ell-1)} a^{-(D-2)(\ell-1)} \int^{a/\sqrt{\epsilon}} d^{D\ell}\tilde{k}\, \tilde{k}^{2-2\ell} F\left(\tilde{k}\right), \tag{4}$$

where $\tilde{k} = ka$ and $\{k\}$ are the loop momenta, and F is a function which approaches 1 at large values of its arguments. By focusing on the regime where all loop momenta are of the same order, and working at large $\tilde{k}$, we see that a $\ln a$ term arises from the $\tilde{k}^{2\ell-2-D\ell}$ term in the $\frac{1}{\tilde{k}}$ expansion of F at large $\tilde{k}$, and the full a dependence of this term is

$$\left(\frac{1}{a}\right)^{(D-2)(\ell-1)} \ln a, \tag{5}$$

which is highly suppressed in the large a limit unless $\ell = 1$. This verifies the above claim that only one-loop fluctuations give rise to the log term. Further, by considering different scaling regimes where various subsets of loop momenta scale to be much larger than the rest, one may verify the fact that only the two derivative sector of massless fields contributes to the log term. However, we do not do so here and instead refer the reader to [15] for those details.

Therefore, as argued earlier, the log term may be regarded as an IR probe of the microscopic theory, in the sense that any putative microscopic description of the black hole must correctly reproduce not only the leading Bekenstein Hawking area law, but also the log correction to it.

In this note we shall compute the log term for a class of magnetically charged extremal black holes which asymptote to $\text{AdS}_4 \times S^7$ for which a complete expression for the microscopic entropy has recently been computed via the computation of a topologically twisted index in ABJM theory [29]. We omit details of the microscopic formula, referring the reader to [29] as well as the companion papers [30,31]. Further, the only features of the near horizon geometry which shall be relevant to us are that it is $\text{AdS}_2 \times S^2 \times S^7$ where the S^7 is bundled over the S^2 and that the AdS_2, S^2 and S^7 have a common length scale a associated to them. That is, the metric over the full 11 dimensional near horizon geometry can be brought to the form $g_{\mu\nu} = a^2 g^{(0)}_{\mu\nu}$ where the metric $g^{(0)}$ and the coordinates are a independent.

Further the S^2 has $SO(3)$ isometry, while the S^7 has $U(1)^4$ isometry. These inputs suffice for the macroscopic computation of the log term as a prediction for the microscopic formula of [29]. We shall finally discuss a few aspects of the proposed match.[3] Details of the full black hole solution are reviewed in [29].

2. The log term from the quantum entropy function

We will now describe how the log term may be extracted from the path integral (1). To do so, it is useful to phrase the problem more generally. In particular, we consider a theory in D dimensions, with a dynamical field Φ, admitting a saddle-point which is a background with length scale a.

$$\mathcal{Z}[\Phi] = \int \mathcal{D}\Phi\, e^{-\frac{1}{\hbar}S[\Phi]}. \tag{6}$$

Then, as argued from Equation (4), to extract the term in the free energy proportional to $\ln a$, it is sufficient to concentrate on the one-loop partition function of the theory. The techniques for this analysis are well-known and we refer the reader to [14,38,16,17] for accounts of how these computations are carried out. Firstly, the one-loop partition function is then given by

$$\mathcal{Z}_{1-\ell} = \det{}' \mathcal{O}^{-\frac{1}{2}} \cdot (\mathcal{Z}_{\text{zero}})^{n_0}, \tag{7}$$

where $\det' \mathcal{O}$ is the determinant of $\mathcal{O}$ evaluated over its non-zero modes, n_0 is the number of zero modes of $\mathcal{O}$, and $\mathcal{Z}_{\text{zero}}$ is the residual zero-mode integral. Therefore

$$\ln \mathcal{Z}_{1-\ell} = -\frac{1}{2} \ln \det{}' \mathcal{O} + n_0 \ln \mathcal{Z}_{\text{zero}}. \tag{8}$$

The final result is that the coefficient of the $\ln a$ term in the free energy computed about this saddle-point depends on $K(t;0)$,

to find a complete match with the corresponding microscopic expression. We thank Jun Nian for helpful correspondence regarding this. We also note that the quantum entropy of AdS_4 hyperbolic black holes was studied in [28].

[2] The term 'massless' has to be carefully defined on curved manifolds. The more precise statement is that the eigenvalues of the kinetic operator should scale as $\frac{1}{a^2}$, which is how the eigenvalues of the kinetic operator over a massless scalar, $-\Delta$ would scale.

[3] While this draft was being readied for submission, we learned of [32] where this comparison has been carried out in a numerical scheme to find a mismatch.

which is the t^0 coefficient of the heat kernel expansion of the kinetic operator $\mathcal{O}$ about this saddle-point, and β_Φ, which determines how the zero mode contribution $\mathcal{Z}_{\text{zero}}$ to the path integral scales with a.

$$\mathcal{Z}_{\text{zero}} = a^{\beta_\Phi} \hat{\mathcal{Z}}_{\text{zero}}, \tag{9}$$

where $\hat{\mathcal{Z}}_{\text{zero}}$ does not scale with a. We eventually obtain the formula

$$\ln \mathcal{Z}_{1-\ell} = K(t;0) \ln a + (\beta_\Phi - 1) n_0 \ln a. \tag{10}$$

It is well known that in odd-dimensional spacetimes, $K(0;t) = 0$ and hence we have the formula

$$\ln \mathcal{Z}_{1-\ell} = (\beta_\Phi - 1) n_0 \ln a. \tag{11}$$

The extension to the case of multiple fields $\{\Phi\}$ having zero modes is apparent.

$$\ln \mathcal{Z}_{1-\ell} = \sum_{\phi \in \{\Phi\}} (\beta_\phi - 1) n_0^\phi \ln a, \tag{12}$$

where n_0^ϕ is the number of zero modes of the kinetic operator over the field ϕ. The specific values for β_Φ that will be relevant to us are for the vector, the graviton, the three-form and the gravitino. These are given by

$$\beta_v = \frac{D-2}{2}, \quad \beta_m = \frac{D}{2}, \quad \beta_f = D-1, \quad \beta_C = \frac{D}{2} - 3. \tag{13}$$

Here v denotes the vector field, m the metric or the graviton, f the gravitino, and the corresponding β values have been listed in Equation (2.37) of [14]. C denotes the three-form field and its β value may be determined exactly in the same manner as the previous fields [14,38]. We start with the expression for the normalization of the field C_{MNP} in D-dimensions.

$$\int \mathcal{D}C_{MNP}\, e^{\left[-\int d^dx \sqrt{g} g^{MU} g^{NV} g^{PW} C_{MNP} C_{UVW}\right]} = 1. \tag{14}$$

Here the metric scales as $g_{MN} = a^2 g^{(0)}_{MN}$ where $g^{(0)}$ does not scale with a. Therefore we have

$$\int \mathcal{D}C_{MNP}\, e^{\left[-a^{D-6} \int d^dx \sqrt{g^{(0)}} g^{(0)MU} g^{(0)NV} g^{(0)PW} C_{MNP} C_{UVW}\right]} = 1. \tag{15}$$

Hence the correctly normalized integration measure is

$$\prod_{x,(MNP)} d\left(a^{\frac{D}{2}-3} C_{MNP}\right). \tag{16}$$

From this we can obtain that the zero mode of C_{MNP} corresponds to

$$\beta_C = \frac{D}{2} - 3. \tag{17}$$

3. Counting the number of zero modes

From the above discussion it is clear that if we are to extract the contribution to the (logarithm of the) quantum entropy function which scales as $\ln a$, where a is the length scale associated with the radii of AdS_2, S^2 and S^7, it is enough to compute the zero modes of the fields appearing in the path integral (1). Zero modes can in principle appear in the massless spectrum of AdS_2 fields obtained from fields of the 11 dimensional supergravity reduced on to $S^2 \times S^7$. In this section we will enumerate these zero modes and compute their contribution to the quantum entropy of magnetically charged AdS_4 black holes. In isolating these zero modes, a special role is played by the so-called *discrete modes* of the Laplacian for spin-1 and spin-2 fields, and the Dirac operator for spin-$\frac{3}{2}$ fields on AdS_2 explicitly enumerated in [33,34] respectively, and counting the total number of zero modes is essentially equivalent to counting the total number of discrete modes of these fields. These modes have also been listed in [12,14,17]. An important observation for us is that it turns out that naively the number of zero modes for each of these fields turns out to be infinite. However, this divergence turns out to be essentially equivalent to the volume divergence of the free energy and may be regulated in the same way. Two slightly different, but equivalent, procedures for doing this are available in [11,12,14] and [16,17] and we shall use those results for the regularized number of zero modes in the computations that follow.

3.1. Bosonic zero modes

The bosonic fields are the graviton h_{MN} and the 3-form C_{MNP}. Quantization of the graviton gives rise to a ghost vector field but this has no zero modes. Quantization of the 3-form C gives rise to a ghost 2-form B with Grassmann odd statistics, and a ghost-for-ghost 1-form A with Grassmann even statistics. We use the following conventions: M is an 11-d vector index, μ is an AdS_2 index, a is an S^2 index and i is an S^7 index. Finally α is either a or i.

Consider the metric zero modes first. The graviton h_{MN} decomposes into the AdS_2 graviton $h_{\mu\nu}$, 3 massless AdS_2 vectors $h_{\mu a}$, and 4 massless AdS_2 vectors $h_{\mu i}$, along with the AdS_2 scalars g_{ia}, g_{ij} and g_{ab}. In counting the number of massless vectors, we used the fact that these are given by

$$h_{\mu\alpha} = v_\mu k_\alpha, \tag{18}$$

where k_α is a Killing vector along an internal direction. Since the internal space has $SU(2) \otimes U(1)^{\otimes 4}$ isometry, there are $3 + 4 = 7$ Killing vectors. Each massless vector field on AdS_2 contributes

$$n_0^v = -1, \tag{19}$$

hence there are -7 zero modes from the $h_{\mu\alpha}$. Also there are -3 zero modes from the AdS_2 metric $h_{\mu\nu}$. Therefore total number of metric zero modes is

$$n_0^m = -7 - 3 = -10. \tag{20}$$

Then the contribution to the log term from the 11d metric zero modes is

$$\delta Z|_{\text{metric}} = (\beta_m - 1) n_0^m \ln a = -45 \ln a. \tag{21}$$

We next consider the 3-form field C_{MNP}. Its quantization was carried out in [35–37] and is reviewed in Section 3.1 of [38]. We will just require the following result. The quantization of a p-form requires p generalized ghost fields which are $(p-j)$-forms, where j runs from 1 to p. Further the logarithmic contribution of all these fields to the free energy may be packaged into the expression

$$\Delta F = \sum_{j=0}^{p} (-1)^j \left(\beta_{p-j} - j - 1\right) n^0_{\mathcal{O}_{p-j}} \ln a, \tag{22}$$

where $\mathcal{O}$ is the kinetic operator. For $j = 1, \ldots, p$ which are the generalized ghost fields, this is just the Hodge Laplacian. For $j = 0$, which is the physical field, this can have couplings to background fluxes as well. We now consider the reduction of C_{MNP} onto AdS_2. This firstly leads to $C_{\mu\nu a} C_{\mu\nu i}$ which are two-forms on AdS_2 which are Hodge duals of scalars. These have no zero modes. Next, we consider a 3-form which is the wedge product of a 1-form along AdS_2 with a harmonic 2-form along $S^2 \times S^7$.

$$C^{(3)} = C^{(AdS_2)}_{(1)} \wedge C^{(S^2 \times S^7)}_{(2)}. \tag{23}$$

The number of such harmonic forms is given by the second Betti number b_2 of the manifold, which is 1 in this case, as may be readily seen from multiplying the Poincaré polynomials of S^2 and S^7. Hence there is a single massless vector fields along AdS_2 from the 11 dimensional 3-form field. Finally we have the scalars C_{abi}, C_{aij}, C_{ijk} which don't have zero modes. Hence the 3-form contributes

$$\delta Z|_C = (\beta_C - 1)\, n_0^C \ln a = -\frac{3}{2}\ln a \tag{24}$$

to the log term. We next consider the ghost B_{MN} which arises from the quantization of C. This decomposes into the following massless fields on AdS_2. First we have $B_{\mu\nu}$ which contributes no zero modes. Next we may obtain a massless 1-form on AdS_2 from this field by decomposing B_{MN} into a wedge product of an AdS_2 1-form with a harmonic 1-form on $S^2\times S^7$. The number of such harmonic 1-forms is the first Betti number of $S^2\times S^7$ which is zero. Finally we have the scalars B_{ab}, B_{ai} and B_{ij}, which contribute no zero modes. Therefore the contribution to the log term is

$$\delta Z|_B = -\,(\beta_B - 2)\, n_0^B \ln a = 0. \tag{25}$$

The overall minus sign is on account of Grassmann odd statistics of this field. Finally we have the ghost-for-ghost field A_M from the quantization of B. This leads to one massless vector field on AdS_2 and therefore

$$\delta Z|_A = (\beta_A - 3)\, n_0^A \ln a = -\frac{3}{2}\ln a. \tag{26}$$

We therefore add (21), (24), (25) and (26) to obtain

$$\delta Z = \left(-45 - \frac{3}{2} - \frac{3}{2}\right)\ln a = -48\ln a. \tag{27}$$

The residual scalar generalized ghost has no zero modes.

3.2. Fermionic zero modes

To count fermion zero modes, we need to compute the regularized number of discrete modes $\xi_\mu^{(k)+}$ and $\hat{\xi}_\mu^{(k)+}$, $k=1,2,\ldots$, on AdS_2. The relevant computations are available in [12,14,17] and we only mention final results. Firstly, it may be shown that the regularized number of modes of both ξ and $\hat{\xi}$ is given by -1. Further, these modes should be tensored with the spinors associated with directions transverse to AdS_2. This will give rise to additional multiplicity factors. To determine the multiplicity, we note first that the near horizon geometry of the black hole has a superconformal symmetry $su(1,1|1)$ with fermionic generators G_n^α where $\alpha=\pm$ and $n\in\mathbb{Z}+\frac{1}{2}$. The gravitino zero modes we consider are associated with the generators G_n^α where $|n|\geq\frac{3}{2}$. In particular, we identify G_n^α with $n\geq\frac{3}{2}$ with ξ^k where $n=k+\frac{1}{2}$ and G_n^α with $n\leq-\frac{3}{2}$ with ξ^k where $n=-k-\frac{1}{2}$. Hence, there is an overall multiplicity factor of 2 coming from α taking values $+$ and $-$. Therefore the number of fermion zero modes is

$$n_0^f = (-1-1)\times 2 = -4. \tag{28}$$

Then the contribution to the log term from the fermionic zero modes is

$$\delta Z|_{\text{fermions}} = -\left(\beta_f - 1\right) n_0^f \ln a = +36\ln a. \tag{29}$$

The overall minus is on account of Grassmann odd statistics. Adding (27) and (29) we see that the log term is given by

$$\Delta F = (-48+36)\ln a = -12\ln a. \tag{30}$$

4. On the comparison with microscopics

We have so far computed the log term for magnetically charged AdS_4 extremal black holes using the quantum entropy function. In this section we shall very briefly discuss how in principle a match with the proposed microscopic answer of [29] may be carried out. At first glance one may expect that the large N expansion of the logarithm of the topologically twisted index in the CFT computed in [29] may be matched term by term with the large a expansion of the logarithm of the quantum entropy function. We note however, that several steps should in principle be necessary to carry out before a match can be meaningfully proposed.

Firstly, the index computed by [29] measures the black hole entropy in the grand canonical ensemble as it employs the AdS_4/CFT_3 correspondence. In contrast, the natural boundary conditions for the quantum entropy function (1) pick out the *microcanonical* ensemble [9]. While the choice of ensemble is irrelevant in the strict large N limit, it is typically important when finite N effects are taken into account. Indeed examples exist where the choice of ensemble is explicitly shown to affect the value of the log term [15]. Therefore, as a first step, we expect that one would need to go to the microcanonical ensemble when computing the CFT answer to match with the quantum entropy function.

Second, it is not obvious how the naive large N scalings of the log of the index are to be reproduced by the quantum entropy function. In particular, it appears from the analysis of [29] that $\ln Z$ scales as

$$\ln Z \sim N^{3/2} + \mathcal{O}(N\ln N). \tag{31}$$

The $N^{3/2}$ term precisely matches with the Bekenstein Hawking entropy of the black hole, but the possible subleading term of order $N\ln N$ is particularly surprising from the point of view of the general scalings of contributions to the quantum entropy as expected from the quantum entropy function. In particular, on going through the scaling analysis of Equation (4), it is apparent that we do not expect a term of the form $N\ln N$, and that this term should drop from the CFT_3 answer if a match with the quantum entropy function is to be possible.

Remarkably, it seems that a numerical estimate of the large N behavior of the index produces the pattern one would naturally expect from the quantum entropy function scalings, including an absence of the $N\ln N$ term, and the presence of the $\ln N$ term, albeit with a mismatching coefficient [32]. This remarkable feature should certainly be better understood by carrying out a systematic large N expansion of the CFT index while accounting for the choice of ensemble as we have indicated above. It would be interesting to return to these questions in the future.

Acknowledgements

We would like to thank Rajesh Gupta, Nick Halmagyi, Euihun Joung, Tomoki Nosaka, Chiung Hwang, Leopoldo Pando Zayas, Ashoke Sen, Piljin Yi and Alberto Zaffaroni for helpful discussions. SL thanks the Korea Institute of Advanced Study and Kyung Hee University for hospitality while this work was carried out. SL's work is supported by a Marie Sklodowska Curie 2014 Individual Fellowship.

References

[1] A. Strominger, C. Vafa, Microscopic origin of the Bekenstein–Hawking entropy, Phys. Lett. B 379 (1996) 99, arXiv:hep-th/9601029.

[2] R. Dijkgraaf, E.P. Verlinde, H.L. Verlinde, Counting dyons in N=4 string theory, Nucl. Phys. B 484 (1997) 543, http://dx.doi.org/10.1016/S0550-3213(96)00640-2, arXiv:hep-th/9607026.

[3] B. Pioline, BPS black hole degeneracies and minimal automorphic representations, J. High Energy Phys. 0508 (2005) 071, http://dx.doi.org/10.1088/1126-6708/2005/08/071, arXiv:hep-th/0506228.
[4] D. Shih, A. Strominger, X. Yin, Counting dyons in N=8 string theory, J. High Energy Phys. 0606 (2006) 037, http://dx.doi.org/10.1088/1126-6708/2006/06/037, arXiv:hep-th/0506151.
[5] J.R. David, A. Sen, CHL dyons and statistical entropy function from D1–D5 system, J. High Energy Phys. 0611 (2006) 072, http://dx.doi.org/10.1088/1126-6708/2006/11/072, arXiv:hep-th/0605210.
[6] A. Sen, N=8 dyon partition function and walls of marginal stability, J. High Energy Phys. 0807 (2008) 118, http://dx.doi.org/10.1088/1126-6708/2008/07/118, arXiv:0803.1014 [hep-th].
[7] A. Sen, Black hole entropy function, attractors and precision counting of microstates, Gen. Relativ. Gravit. 40 (2008) 2249, http://dx.doi.org/10.1007/s10714-008-0626-4, arXiv:0708.1270 [hep-th].
[8] A. Sen, Entropy function and AdS(2) / CFT(1) correspondence, J. High Energy Phys. 0811 (2008) 075, http://dx.doi.org/10.1088/1126-6708/2008/11/075, arXiv:0805.0095 [hep-th].
[9] A. Sen, Quantum entropy function from AdS(2)/CFT(1) correspondence, Int. J. Mod. Phys. A 24 (2009) 4225, http://dx.doi.org/10.1142/S0217751X09045893, arXiv:0809.3304 [hep-th].
[10] A. Sen, Arithmetic of quantum entropy function, J. High Energy Phys. 0908 (2009) 068, http://dx.doi.org/10.1088/1126-6708/2009/08/068, arXiv:0903.1477 [hep-th].
[11] S. Banerjee, R.K. Gupta, A. Sen, Logarithmic corrections to extremal black hole entropy from quantum entropy function, J. High Energy Phys. 1103 (2011) 147, arXiv:1005.3044 [hep-th].
[12] S. Banerjee, R.K. Gupta, I. Mandal, A. Sen, Logarithmic corrections to N=4 and N=8 black hole entropy: a one loop test of quantum gravity, J. High Energy Phys. 1111 (2011) 143, arXiv:1106.0080 [hep-th].
[13] A. Sen, Logarithmic corrections to N=2 black hole entropy: an infrared window into the microstates, Gen. Relativ. Gravit. 44 (5) (2012) 1207, http://dx.doi.org/10.1007/s10714-012-1336-5, arXiv:1108.3842 [hep-th].
[14] A. Sen, Logarithmic corrections to rotating extremal black hole entropy in four and five dimensions, Gen. Relativ. Gravit. 44 (2012) 1947, http://dx.doi.org/10.1007/s10714-012-1373-0, arXiv:1109.3706 [hep-th].
[15] A. Sen, Logarithmic corrections to Schwarzschild and other non-extremal black hole entropy in different dimensions, J. High Energy Phys. 1304 (2013) 156, http://dx.doi.org/10.1007/JHEP04(2013)156, arXiv:1205.0971 [hep-th].
[16] R.K. Gupta, S. Lal, S. Thakur, Heat kernels on the AdS(2) cone and logarithmic corrections to extremal black hole entropy, J. High Energy Phys. 1403 (2014) 043, http://dx.doi.org/10.1007/JHEP03(2014)043, arXiv:1311.6286 [hep-th].
[17] R.K. Gupta, S. Lal, S. Thakur, Logarithmic corrections to extremal black hole entropy in $\mathcal{N}=2$, 4 and 8 supergravity, J. High Energy Phys. 1411 (2014) 072, http://dx.doi.org/10.1007/JHEP11(2014)072, arXiv:1402.2441 [hep-th].
[18] A. Dabholkar, J. Gomes, S. Murthy, Quantum black holes, localization and the topological string, J. High Energy Phys. 1106 (2011) 019, http://dx.doi.org/10.1007/JHEP06(2011)019, arXiv:1012.0265 [hep-th].
[19] A. Dabholkar, J. Gomes, S. Murthy, Localization & exact holography, J. High Energy Phys. 1304 (2013) 062, http://dx.doi.org/10.1007/JHEP04(2013)062, arXiv:1111.1161 [hep-th].
[20] R.K. Gupta, S. Murthy, All solutions of the localization equations for N=2 quantum black hole entropy, J. High Energy Phys. 1302 (2013) 141, http://dx.doi.org/10.1007/JHEP02(2013)141, arXiv:1208.6221 [hep-th].
[21] A. Dabholkar, J. Gomes, S. Murthy, Nonperturbative black hole entropy and Kloosterman sums, J. High Energy Phys. 1503 (2015) 074, http://dx.doi.org/10.1007/JHEP03(2015)074, arXiv:1404.0033 [hep-th].
[22] S. Murthy, V. Reys, Functional determinants, index theorems, and exact quantum black hole entropy, J. High Energy Phys. 1512 (2015) 028, http://dx.doi.org/10.1007/JHEP12(2015)028, arXiv:1504.01400 [hep-th].
[23] R.K. Gupta, Y. Ito, I. Jeon, Supersymmetric localization for BPS black hole entropy: 1-loop partition function from vector multiplets, J. High Energy Phys. 1511 (2015) 197, http://dx.doi.org/10.1007/JHEP11(2015)197, arXiv:1504.01700 [hep-th].
[24] S. Murthy, V. Reys, Single-centered black hole microstate degeneracies from instantons in supergravity, J. High Energy Phys. 1604 (2016) 052, http://dx.doi.org/10.1007/JHEP04(2016)052, arXiv:1512.01553 [hep-th].
[25] R.M. Wald, Black hole entropy is the Noether charge, Phys. Rev. D 48 (1993) 3427, arXiv:gr-qc/9307038.
[26] J. Nian, X. Zhang, Entanglement entropy of ABJM theory and entropy of topological black hole, J. High Energy Phys. 1707 (2017) 096, http://dx.doi.org/10.1007/JHEP07(2017)096, arXiv:1705.01896 [hep-th].
[27] A. Dabholkar, N. Drukker, J. Gomes, Localization in supergravity and quantum AdS_4/CFT_3 holography, J. High Energy Phys. 1410 (2014) 90, http://dx.doi.org/10.1007/JHEP10(2014)090, arXiv:1406.0505 [hep-th].
[28] A. Cabo-Bizet, V.I. Giraldo-Rivera, L.A. Pando Zayas, Microstate counting of AdS_4 hyperbolic black hole entropy via the topologically twisted index, J. High Energy Phys. 1708 (2017) 023, http://dx.doi.org/10.1007/JHEP08(2017)023, arXiv:1701.07893 [hep-th].
[29] F. Benini, K. Hristov, A. Zaffaroni, Black hole microstates in AdS_4 from supersymmetric localization, J. High Energy Phys. 1605 (2016) 054, http://dx.doi.org/10.1007/JHEP05(2016)054, arXiv:1511.04085 [hep-th].
[30] F. Benini, A. Zaffaroni, A topologically twisted index for three-dimensional supersymmetric theories, J. High Energy Phys. 1507 (2015) 127, http://dx.doi.org/10.1007/JHEP07(2015)127, arXiv:1504.03698 [hep-th].
[31] F. Benini, K. Hristov, A. Zaffaroni, Exact microstate counting for dyonic black holes in AdS4, Phys. Lett. B 771 (2017) 462, http://dx.doi.org/10.1016/j.physletb.2017.05.076, arXiv:1608.07294 [hep-th].
[32] J.T. Liu, L.A. Pando Zayas, V. Rathee, W. Zhao, Toward microstate counting beyond large N in localization and the dual one-loop quantum supergravity, arXiv:1707.04197 [hep-th].
[33] R. Camporesi, A. Higuchi, Spectral functions and zeta functions in hyperbolic spaces, J. Math. Phys. 35 (1994) 4217, http://dx.doi.org/10.1063/1.530850.
[34] R. Camporesi, A. Higuchi, On the eigenfunctions of the Dirac operator on spheres and real hyperbolic spaces, J. Geom. Phys. 20 (1996) 1, http://dx.doi.org/10.1016/0393-0440(95)00042-9, arXiv:gr-qc/9505009.
[35] W. Siegel, Hidden ghosts, Phys. Lett. B 93 (1980) 170, http://dx.doi.org/10.1016/0370-2693(80)90119-7.
[36] J. Thierry-Mieg, BRS structure of the antisymmetric tensor gauge theories, Nucl. Phys. B 335 (1990) 334, http://dx.doi.org/10.1016/0550-3213(90)90497-2.
[37] E.J. Copeland, D.J. Toms, Quantized antisymmetric tensor fields and selfconsistent dimensional reduction in higher dimensional space-times, Nucl. Phys. B 255 (1985) 201, http://dx.doi.org/10.1016/0550-3213(85)90134-8.
[38] S. Bhattacharyya, A. Grassi, M. Marino, A. Sen, A one-loop test of quantum supergravity, Class. Quantum Gravity 31 (2014) 015012, http://dx.doi.org/10.1088/0264-9381/31/1/015012, arXiv:1210.6057 [hep-th].

14

Machine-learning the string landscape

Yang-Hui He [a,b,c,*]

[a] Merton College, University of Oxford, UK
[b] Department of Mathematics, City, University of London, UK
[c] School of Physics, NanKai University, China

ARTICLE INFO

Editor: N. Lambert

ABSTRACT

We propose a paradigm to apply machine learning various databases which have emerged in the study of the string landscape. In particular, we establish neural networks as both classifiers and predictors and train them with a host of available data ranging from Calabi–Yau manifolds and vector bundles, to quiver representations for gauge theories, using a novel framework of recasting geometrical and physical data as pixelated images. We find that even a relatively simple neural network can learn many significant quantities to astounding accuracy in a matter of minutes and can also predict hithertofore unencountered results, whereby rendering the paradigm a valuable tool in physics as well as pure mathematics.

1. Introduction

Whereas theoretical physics now inevitably resides in an Age where new physics, new mathematics and new data coexist in a symbiosis transcending disciplines, string theory has spearheaded this vision. That it engenders the cross-fertilization between physics and pure mathematics is without dispute, that it also has been a testing ground for computational mathematics and "big data" is perhaps less known. With the advent of increasingly powerful computers, from this fruitful dialogue has also arisen a plethora of data, ripe for mathematical experimentation. This emergence of data began with the incipience of string phenomenology [1] where compactification of the heterotic string on Calabi–Yau threefolds (CY3) was widely believed to hold the ultimate geometric unification. A race, spanning the 1990s, to explicitly construct examples of Calabi–Yau (CY) manifolds ensued, beginning with the so-called complete intersection CY manifolds (CICYs) [2], proceeding to the hypersurfaces in weighted projective space [3], to elliptic fibrations [4] and ultimately culminating in the impressive (at least some 10^{10}) list of CY3s from reflexive polytopes [5].

With the realization that the landscape of stringy vacua might in fact exceed the number of inequivalent CY3s [6] by hundreds of orders of magnitude, there was a vering of direction toward a more multi-verse or anthropic philosophy. Nevertheless, hints have emerged that the vastness of the landscape might well be mostly infertile (cf. the swamp-land of [7]) and that we could live in a very special universe [8–10], a "des res" corner within a barren vista.

Thus, undaunted by the seeming over-abundance of possible vacua, fortified by the rapid growth of computing power and inspirited by the omnipresence of big data, the first two decades of the new millennium saw a return to the earlier principle of creating and mining geometrical data; the notable fruits of this combined effort between pure and computational algebraic geometers as well as formally and phenomenologically inclined physicists have included (q.v. [11] for a review of the various databases): (1) Continuing with Kreuzer–Skarke (KS) database [14–22]; (2) Generalizing the CICY construction [23–28]; (3) Finding elliptic and K3 fibred CY for F-theory and string dualities [13, 14,28–32]; (4) D-brane world-volume theories as supersymmetric quiver gauge theories [33–41].

All of the above cases are accompanied by typically accessible data of considerable size, representing a concrete glimpse onto the string landscape, to which we shall refer as **landscape data**. For instance, the heterotic line bundles on CICYs are on the order of 10^{10}, the spectral-cover bundles on the elliptically fibred CY3, 10^{6}, the brane-configurations in the CY volume studies, 10^{5}, type II intersecting brane models, 10^{9}, etc. Even by today's measure, these constitute a fertile playground of data, the likes of which Google and IBM are constantly analysing. A natural course of action, therefore, is to do unto this landscape data, what Google et al. do each second of our lives: to machine-learn.

Let us be precise about what we mean by *deep machine-learning* this landscape. Much of the aforementioned data have been the

* Correspondence to: Merton College, University of Oxford, UK.
E-mail address: hey@maths.ox.ac.uk.

brain-child of the marriage between physicists and mathematicians, especially incarnated by applications of computational algebraic geometry, numerical algebraic geometry and combinatorial geometry to problems which arise from the classification in the physics and recast into a finite, algorithmic problem in the mathematics (cf. [12]). Obviously, computing power is a crucial limitation. Unfortunately, in computational algebraic geometry – on which most of the data heavily rely, ranging from bundles stability in heterotic compactification to Hilbert series in brane gauge theories – a decisive step is finding a Groebner basis, which is notoriously known to be unparallelizable and double-exponential in running time. Thus, much of the challenge in establishing the landscape data had been to either circumvent the direct calculation of the Groebner bases by harnessing of the geometric configuration – e.g., using the combinatorics when dealing with toric varieties. Still, many of the combinatorial calculations, be they triangulation of polytopes or finding dual cones, are still exponentially expensive.

The good news for our present purpose is that, *much of the data have already been collected*. Oftentimes, as we shall find out in our forthcoming case-studies, tremendous effort is needed for deceptively simple questions. Hence, to draw inferences from *actual* theoretical data by deep-learning therefrom would not only help identify undiscovered patterns but also aid in predicting results which would otherwise cost formidable computations. Subsequently, we propose our

> **Paradigm:** To set-up neural networks (NN) to deep-learn the landscape data, to recognize unforeseeable patterns (as classifiers) and to extrapolate to new results (as predictors).

Of course, this paradigm is useful not only to physicists but to also to mathematicians; for instance, could our NN be trained well enough to approximate bundle cohomology calculations? This, and a host of other examples, we will now examine.

Methodology Neural networks are known for their complexity, involving usually a complicated directed graph each node of which is a "perceptron" (an activation function imitating a neuron) and amongst the multitude of which there are many arrows encoding input/output. Throughout this letter, we will use a rather simple multi-layer perceptron (MLP) consisting of 5 layers, three of which are hidden, with activation functions typically of the form of a logistic sigmoid or a hyperbolic tangent. The input layer is a linear layer of 100 to 1000 nodes, recognizing a tensor (as we will soon see, algebro-geometric objects such as Calabi–Yau manifolds or polytopes are generically configurations of integer tensors) and the output layer is a summation layer giving a number corresponding to a Hodge number, or to rank of a cohomology group, etc. Such an MLP can be implemented, for instance, on the latest versions of Wolfram Mathematica. With 500–1000 training rounds, the running time is merely about 5–20 minutes on an ordinary laptop. It is reassuring and pleasantly surprising that even such a relatively simple NN can achieve the level of accuracy shortly to be presented.

This letter is a companion summary of the longer paper [42] where the interested reader can find more details of the computations and the data.

2. Results

With simple NNs, we proceed to analyse our landscape data, a fertile ground constituting more than 2 decades of many international collaborations between physics and mathematicians. Using 4 concrete case studies, we first "learn" from the inherent structure and then "predict" unseen properties; considering how difficult some of the calculations involved had been in establishing the databases, the usefulness of our paradigm is evident.

2.1. *Case study 1: CY hypersurfaces in $W\mathbb{P}^4$*

One of the first datasets [3] to experimentally illustrate mirror symmetry was that of hypersurfaces in weighted projective space $W\mathbb{P}^4$. The ambient space $W\mathbb{P}^4_{[w_0:w_1:w_2:w_3:w_4]}$ with weights $w_{i=0,\dots,4} \in \mathbb{Z}_+$ is in general singular, but a generic enough homogeneous polynomial of degree $\sum_{i=0}^{4} w_i$ which misses the singularities defines a hypersurface therein which is a smooth CY3 X. There are 7555 inequivalent such configurations, each specified by a 5-vector $\vec{w}_{i=0,\dots,4}$. The Euler characteristic χ of X is easily given in terms of the vector. However, as is usually the case, the individual Hodge numbers $(h^{1,1}, h^{2,1})$ are less amenable to a simple combinatorial formula. The original computation resorted to Landau–Ginzberg techniques to obtain the list of Hodge numbers [3]. One could in principle use adjunction and Euler sequences, and singularity resolution, but this is not an easy task to automate.

Suppose we have a simple question: *how many such CY3s have a relatively large number of complex deformations*? We can, for instance, consider $h^{2,1} > 50$ to be "large" and let training data be of the form $w_i \to 1$ or 0 depending on whether $h^{2,1}(X) > 50$. Training the NN, with say 500 rounds, takes under a minute on an ordinary laptop. The result is an optimised continuous real output between 0 and 1, the rounding of which can then be compared with the actual data. An accuracy of 96.2 % is achieved almost effortlessly! To appreciate the *predictive* power of the network, suppose that we only had partial data. This is particularly relevant when for instance, due to computational limitations, a classification is not yet complete, or when a quantity in question has not been or could not be yet computed.

Therefore, let us pretend that we have only data available for the first 3000 out of the 7555 $(X, h^{2,1})$ pairs. We repeat the procedure on the 3000, and then test against the full 7555. We find that 6078 cases were actually correct. Thus, with rather incomplete training data, the NN has learnt, in under a minute, our question and predicted new results to 80% accuracy.

Emboldened, let us move onto another question, of importance to string phenomenology: *Given a configuration, can one tell whether χ is a multiple of 3?* In the early days of heterotic string compactification, this question was decisive on whether the model admitted 3 generation of particles in the low-energy effective gauge theory. Again, we can define a binary function taking the value of 1 if $\chi \bmod 3 \equiv 0$ and 0 otherwise. Training with the NN, we achieve 82% accuracy with 1000 training rounds, taking about 2 minutes; these figures are certainly expected to improve with increasing number of training rounds and with more layers or more nodes in the NN.

The astute reader might question at this stage why we have adhered to *binary queries*. Why not train the NN to answer a direct query, i.e., to try for instance to learn and predict the value of $h^{2,1}$ itself? This is a matter of spread in the present dataset: we have only some 10^4 inputs yet we can see that the values of $h^{1,1}$ ranges from 1 to almost 500. We do not have enough data here to make more accurate statements. This is precisely in line with our philosophy, the power of deep-learning the landscape lies in rapid *estimates*, in identifying patterns and drawing inferences and in avoiding intense computations.

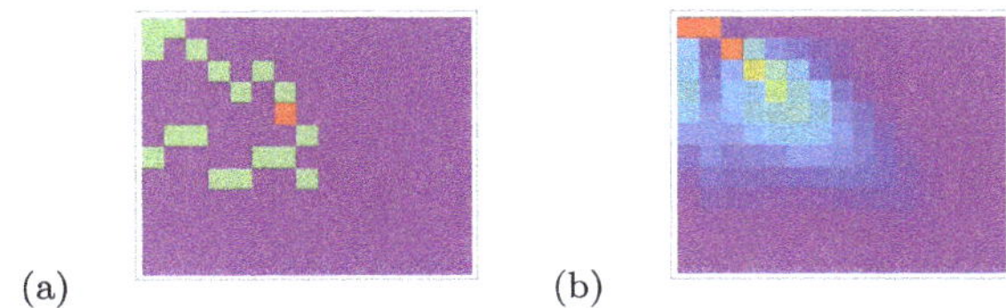

Fig. 1. We realize the set of 7890 CICYs (Calabi–Yau threefolds as complete intersections in products of projective spaces) as 12×15 matrices, padding with zeros where necessary. Then all CICY configurations are such matrices with entries in $\{0, 1, 2, 3, 4, 5\}$. We consider these as pixel colours and draw a typical CICY in (a), with 0 being purple. In (b), we average over all such matrices component-wise, and draw the "average" CICY as a pixelated image. (For interpretation of the references to colour in this figure legend, the reader is referred to the web version of this article.)

2.2. Case study 2: CICYs

Having warmed up, let us move onto complete intersection Calabi–Yau threefolds (CICYs) in products of projective spaces. This is both the first Calabi–Yau database (or, for that matter, the first database in algebraic geometry) [2] and the most heavily studied recently for string phenomenology [23–26,28]. It has the obvious advantage that the ambient space is smooth by choice.

Briefly, CICYs embed as K homogeneous polynomials in $\mathbb{P}^{n_1} \times \ldots \times \mathbb{P}^{n_m}$, of multi-degree q^r_j, with complete intersection meaning that $K = \sum\limits_{r=1}^{m} n_r - 3$ and CY condition implying $\sum\limits_{j=1}^{K} q^r_j = n_r + 1 \ \forall r = 1, \ldots, m$. The construction of CICYs is thus reduced to a combinatorial problem of classifying the integer matrices. The most famous CICY is, of course, $[4|5]$ or simply the matrix $[5]$, denoting the quintic hypersurface in $\mathbb{P}^4$. It was shown that such configurations are finite in number and the best available computer at the time (1990's), viz., the super-computer at CERN [2], was employed. A total of 7890 inequivalent manifolds were found, corresponding to matrices with entries $q^r_j \in [0, 5]$, of size ranging from 1×1 to maximum number of rows and columns being 12 and 15, respectively.

This representation is much in the standard way to represent an image: to pixelate it into blocks of $m \times n$, each of which carrying a colour info, for example, a 3-vector encapturing the RGB data. Therefore, we can represent all the 7890 CICYs into 12×15 matrices over $\mathbb{Z}/6\mathbb{Z}$, embedded starting from the upper-left corner, say, and padding with zeros everywhere else, as illustrated in Fig. 1. To view a CICY as a pixelated image, and indeed, to use image processing to address problems in geometry and mathematical physics, is an entirely new idea worthy of extensive exploration.

Can we deep-learn, say the full list of Hodge numbers? As usual, the Euler number is relatively easy to obtain and there is a combinatorial formula in terms of the integers q^r_j, whilst the individual Hodge numbers $(h^{1,1}, h^{2,1})$ involve some non-trivial adjunction and sequence-chasing, which luckily had been performed for us [2]. Again, we set up a list of training rules (padded configuration matrix $\to h^{1,1}$) and find that the NN can be trained to an accuracy of 99.91% in under 10 minutes! What about the NN as a predictor, which is obviously a more salient question? Suppose the NN were trained with the first 5000 of the data, then, checking against the full dataset comprising of configurations/images the NN has never before seen, we achieve 77% accuracy. Considering (1) that we have only trained the NN for a mere 6 minutes, (2) that it has seen only a little over half of the data, (3) that it is rather elementary with only 5 forward layers, and (4) that the variation of the output is integral ranging from 0 to 19, with no room for continuous tuning, such accuracy with so little effort is quite amazing.

2.3. Case study 3: bundle cohomology

The subject of vector bundle cohomology has, since the so-named "generalized embedding" [1] of heterotic compactification on smooth CY3 X endowed with a (poly-)stable holomorphic vector bundle V, become one of the most active dialogues between algebraic geometry and theoretical physics. The realization [9] that the theoretical possibility of [1] can be concretely achieved by a judicious choice of (X, V) to give the exact MSSM spectrum induced much activity in establishing relatively large datasets to see how often this might occur statistically [11,18,22,24,31], culminating in [25,26] which found some 200 out of a scan of 10^{10} bundles which have exact MSSM content.

Upon this vast landscape let us take an insightful glimpse by taking the dataset of [31], which are $SU(n)$ vector bundles V on elliptically fibred CY3. By virtue of a spectral-cover construction [4,30], these bundles are guaranteed to be stable and hence preserves $\mathcal{N} = 1$ supersymmetry in the low-effective action, together with GUT gauge groups E_6, $SO(10)$ and $SU(5)$ respectively for $n = 3, 4, 5$. We take the base of the elliptic fibration – of which there is a finite list [29] – as the r-th Hirzebruch surface ($r = 0, 1, \ldots, 12$ denoting the inequivalent ways which $\mathbb{P}^1$ can itself fibre over $\mathbb{P}^1$ to give a complex surface), in which case the stable $SU(n)$ bundle is described by 5 numbers (r, n, a, b, λ), with $(a, b) \in \mathbb{Z}_+$ and $\lambda \in \mathbb{Z}/2$ being coefficients which specify the bundle via the spectral cover. This ordered 5-vector will constitute our neural input. The database of viable models was set up in [31], viable meaning that the bundle-cohomology groups of V are such that $h^0(X, V) = h^3(X, V) = 0$ and $|h^1(X, V) - h^2(X, V)| \equiv 0 \pmod 3$, where the first is a necessary condition for stability and the second, that the GUT theory has the potential to allow for 3 net generations of particles upon breaking to MSSM by Wilson lines. Over all the Hirzebruch-based CY3, 14,264 models were found; a sizeable play-ground.

Suppose the output be a 2-vector, indicating (I) what the gauge group is, as denoted by n, and (II) whether there are more generations than anti-generations, as denoted by the sign of the difference $h^1(X, V) - h^2(X, V)$; this is clearly a phenomenologically interesting question. With 1000 training rounds on a NN with an output linear layer, and with the dataset consisting of entries in the form $(r, n, a, b, \lambda) \to (n, \mathrm{Sign}(h^1(X, V) - h^2(X, V)))$, in about 10 minutes, we achieve 100% accuracy (i.e., the neural network has completely learnt the data). Training with partial data, say 8000 points, a little over half, achieves 68.9% predicative accuracy over the entire set.

2.4. Case study 4: quiver gauge theories

As a final example let us tackle affine varieties in the context of quiver representations. Physically, these correspond to world-volume gauge theories coming from D-brane probes of geometric singularities in string theory, as well as the space of vacua for classes of supersymmetric gauge theories in various dimension; they have been data-mined since the early days of AdS/CFT (cf. [33, 34]). When the geometry concerned is an affine toric CY variety, the realization of brane-tiling [35] has become the correct way to understand the gauge theory and since then databases have begun to be compiled [36,37].

The input data consists of a quiver (directed graph) and a relation imposed by a polynomial super-potential (q.v. [11] for a rapid review). We can succinctly encode the above information into two matrices, which again can be turned into a pixelated image: (1) D-term matrix Q_D, which comes from the kernel of the incidence matrix d of the quiver, each column of which corresponds to an arrow with -1 as head and $+1$ as tail and 0 otherwise;

(2) F-term matrix Q_F each column of which documents where and with what exponent the field corresponding to the arrow appears in ∂W. Concatenating Q_D and Q_F gives the so-call total charge matrix Q_t of the moduli space as a toric variety (q.v. §2 of [34] for the precise procedure). The combinatorics and geometry of the above is a long story spanning a lustrum of research to uncover followed by a decade of still-ongoing investigations.

In the first database of [36], a host of examples were tabulated. A total of 375 quiver theories much like the above were catalogued (a catalogue which has recently been vastly expanded in [37]). Though not very large, this gives us a playground to test some of our ideas. The input data is the total charge matrix Q_t, the maximal of whose number of rows and columns are, respectively 33 and 36, and all taking values in $\{-3, -2, \ldots, 3, 4\}$. Now, suppose we wish to know the number of points of the toric diagram associated to the moduli space, which is clearly an important quantity. In principle, this can be computed (albeit computationally intensive): the integer kernel of Q_t should give a matrix whose columns are the coordinates of the toric diagram, with multiplicity (associated to the perfect matchings of the bipartite tiling). Training with our NN with the full list achieves, in under 5 minutes, 99.5% accuracy.

2.5. A sanity check

Lest the readers' optimism be elevated to unreasonable heights by the string of successes with the NNs, it is imperative that we be aware of deep-learning's limitations. We therefore finish with a sanity check that a NN is not some omnipotent oracle capable of predicting *any* pattern. An example which must be doomed to failure is the primes (or, for that matter, the zeros of the Riemann zeta function). Indeed, if an NN could learn some unexpected pattern in the primes, this would be a rather frightening prospect for mathematics. We test the sequence of primes (i.e., data of the form `i -> Prime[i]`) with our NN, and achieve no better than a 0.1% accuracy. Our NN is utterly useless against this formidable challenge; we are better off trying a simple regression against some $n\log(n)$ curve, as dictated by the prime number theorem. This is a sobering exercise as well as a further justification of the various case studied above, that it is indeed meaningful to deep-learn the landscape data and that our visual representation of geometrical configurations is an efficient methodology.

3. Discussion

There are many questions in theoretical physics, or even in pure mathematics, for which one would only desire a qualitative, approximate, or partial answer, and whose full solution would often either be beyond the current scope, conceptual or computational, or would have taken considerable effort to attain. Typical such questions could be "what is the likelihood of finding a universe with three generations of particles within the landscape of string vacua or inflationary scenarios", or "what percentage of known Calabi–Yau manifolds has Hodge numbers within a prescribed range"? Attempting to address these profound questions have, with the ever-increasing power of computers, engendered our community's version of "big data", which though perhaps humble compared to some other fields, do comprise, especially considering the abstract nature of the problems at hand, of significant information often resulting from intense dialogue between teams of physicists and mathematicians for many years.

On the still-ripening fruits of this labour the philosophy of the last decade or so, particularly for the string phenomenology and computational geometry community, has been to (I) create larger and larger datasets and (II) scan through them to test the likelihood of certain salient features. Now that the data is augmenting in size and availability, it is only natural to follow the standard procedures of the data-mining community. In this letter, we have proposed the paradigm of applying deep-learning, via neural networks, such data. The purpose is twofold, the neural network can act as

Classifiers: by association of input configuration with a requisite quantity, and pattern-match over a given dataset;

Predictors: by extrapolating to hithertofore unencountered configurations, having deep-learnt a given (partial) dataset.

This is, of course, the archetypal means by which Google deep-learns the internet and hand-writing recognition software adapts to the reader's esoteric script.

It is intriguing that by going through a wealth of concrete examples from what we have dubbed **landscape data**, some of whose creation the author had been a part, this philosophy remains enlightening. Specifically, we have taken test cases from a range of problems in mathematical physics, algebraic geometry and representation theory, such as CY datasets, classification of stable vector bundles, and catalogues of quiver varieties and brane tilings. We subsequently saw that even relatively simple NN can deep-learning to extraordinary accuracy.

In some sense, this is not surprising, there is underlying structure to any classification problem in our context, which may not be manifest. Indeed, what is novel is to look at the likes of a CICY or a quiver theory as a *pixelated image*, no different from a hand-written digit, for whose analysis machine-learning has become the de facto method and a blossoming industry. *The landscape data, be they work of human hands, elements of Nature or conceptions of Mathematics, have inherent structure, sometimes more efficiently uncovered by AI via deep-learning.* Thereby, one can rapidly obtain results, before embarking on finding a reductionist framework for a fundamental theory explaining the results or proceed to intensive computations from first principles. This paradigm is especially useful when classification problems become intractable, which is often the case, here a pragmatic approach would be to deep-learn partial classification results and predict future outcome.

Under this rubric, the possibilities are endless. Several immediate and pertinent directions spring to mind. First, the largest dataset in algebraic geometry/string theory is the Kreuzer–Skarke list [5,20,21] of reflexive polytopes in dimension 4 from each of which many CY manifolds (compact and non-compact) can be constructed. To discover hidden patterns is an ongoing enterprise [14, 17] and the help of deep-learning would be a most welcome one. Next, the issue of bundle stability and cohomology is a central problem in heterotic phenomenology as well as algebraic geometry. In many ways, this is a perfect problem for machine-learning: the input is usually encodable into an integer matrix or a list of matrices, representing the coefficients in an expansion into effective divisor classes, the output is simply a vector of integers (in the case of cohomology) or a binary answer (with respective to a given Kahler class, the bundle is either stable or not). The brute-force way involves the usual spectral sequences and determining all coboundary maps or finding the lattices of subsheafs, expensive by any standards. In the case of stability checking, this is an enormous effort to arrive at a yes/no query. With increasing number of explicitly known examples of stable bundles constructed from first principles, to deep-learn this and then estimate the probability of a given bundle being stable would be tremendous time-saver.

To give an idea the high non-triviality of our venture, suppose we wanted to know how many CY3 can be constructed from the famous Kreuzer–Skarke (KS) list of 473 million reflexive polytopes. Only recently [21] was a systematic triangulation carried out on a cluster, up to $h^{1,1} = 7$ (above which even the state-of-art computer is powerless), and $\simeq 100,000$ manifolds were found from $\simeq 25,000$ polytopes. The KS list has $h^{1,1}$ going up to 496, thus we have not even touched the tip of the iceberg in answering the simplest question of enumerating CY3s. Here, the NN would be extremely useful in predicting an estimate, having learnt the data from [21], which already took ~ 5000 core-hours with traditional methods on the cluster; this is currently under investigation.

We hope the reader has been persuaded by not only the scope but also the feasibility of our proposed paradigm, a paradigm of increasing importance in an Age where even the most abstruse of mathematics or the most theoretical of physics cannot avoid compilations of and investigations on perpetually growing datasets. The case studies of deep-learning such landscape of data here presented are but a few nuggets in an unfathomably vast gold-mine, rich with new science yet to be discovered.

Acknowledgements

We are indebted to the Science and Technology Facilities Council, UK, for grant ST/J00037X/1, the Chinese Ministry of Education, for a ChangJiang Chair Professorship at NanKai University, and the city of Tian-Jin for a Qian-Ren Award, as well as Merton College, University of Oxford for continued support.

References

[1] P. Candelas, G.T. Horowitz, A. Strominger, E. Witten, Vacuum configurations for superstrings, Nucl. Phys. B 258 (1985) 46.
[2] P. Candelas, A.M. Dale, C.A. Lutken, R. Schimmrigk, Complete intersection CY manifolds, Nucl. Phys. B 298 (1988) 493;
P. Candelas, C.A. Lutken, R. Schimmrigk, Complete intersection Calabi–Yau manifolds. 2. Three generation manifolds, Nucl. Phys. B 306 (1988) 113;
M. Gagnon, Q. Ho-Kim, An exhaustive list of complete intersection Calabi–Yau manifolds, Mod. Phys. Lett. A 9 (1994) 2235;
T. Hubsch, Calabi–Yau Manifolds: a Bestiary for Physicists, World Scientific, ISBN 981021927X, 1992.
[3] P. Candelas, M. Lynker, R. Schimmrigk, CY manifolds in weighted P(4), Nucl. Phys. B 341 (1990) 383.
[4] A. Grassi, D.R. Morrison, Group representations and the Euler characteristic of elliptically fibered Calabi–Yau threefolds, arXiv:math/0005196 [math-ag];
R.Y. Donagi, Principal bundles on elliptic fibrations, Asian J. Math. 1 (1997) 214, arXiv:alg-geom/9702002.
[5] A.C. Avram, M. Kreuzer, M. Mandelberg, H. Skarke, The web of Calabi–Yau hypersurfaces in toric varieties, Nucl. Phys. B 505 (1997) 625, arXiv:hep-th/9703003;
Victor V. Batyrev, Lev A. Borisov, On Calabi–Yau complete intersections in toric varieties, arXiv:alg-geom/9412017;
M. Kreuzer, H. Skarke, Reflexive polyhedra, weights and toric Calabi–Yau fibrations, Rev. Math. Phys. 14 (2002) 343, arXiv:math/0001106 [math-ag].
[6] S. Kachru, R. Kallosh, A.D. Linde, S.P. Trivedi, De Sitter vacua in string theory, Phys. Rev. D 68 (2003) 046005, arXiv:hep-th/0301240.
[7] C. Vafa, The string landscape and the swampland, arXiv:hep-th/0509212.
[8] F. Gmeiner, R. Blumenhagen, G. Honecker, D. Lust, T. Weigand, One in a billion: MSSM-like D-brane statistics, J. High Energy Phys. 0601 (2006) 004, arXiv:hep-th/0510170.
[9] V. Braun, Y.-H. He, B.A. Ovrut, T. Pantev, The exact MSSM spectrum from string theory, J. High Energy Phys. 0605 (2006) 043, arXiv:hep-th/0512177.
[10] P. Candelas, X. de la Ossa, Y.-H. He, B. Szendroi, Triadophilia: a special corner in the landscape, Adv. Theor. Math. Phys. 12 (2008) 429, arXiv:0706.3134.
[11] Y.H. He, Calabi–Yau geometries: algorithms, databases, and physics, Int. J. Mod. Phys. A 28 (2013) 1330032, arXiv:1308.0186 [hep-th].
[12] Y.-H. He, P. Candelas, A. Hanany, A. Lukas, B. Ovrut, Computational algebraic geometry in string, Gauge theory. Special issue, Adv. High Energy Phys. 2012 (2012) 431898, http://dx.doi.org/10.1155/2012/431898, Hindawi publishing.
[13] V. Braun, Toric elliptic fibrations and F-theory compactifications, J. High Energy Phys. 1301 (2013) 016, arXiv:1110.4883.
[14] W. Taylor, On the Hodge structure of elliptically fibered Calabi–Yau threefolds, J. High Energy Phys. 1208 (2012) 032, arXiv:1205.0952 [hep-th].
[15] A.P. Braun, J. Knapp, E. Scheidegger, H. Skarke, N.O. Walliser, PALP – a user manual, arXiv:1205.4147 [math.AG].
[16] P. Candelas, A. Constantin, H. Skarke, An abundance of K3 fibrations from polyhedra with interchangeable parts, arXiv:1207.4792 [hep-th].
[17] Y.H. He, V. Jejjala, L. Pontiggia, Patterns in Calabi–Yau distributions, arXiv:1512.01579 [hep-th].
[18] P. Candelas, R. Davies, New Calabi–Yau manifolds with small Hodge numbers, Fortschr. Phys. 58 (2010) 383, arXiv:0809.4681 [hep-th].
[19] W.A. Stein, et al., http://www.sagemath.org.
[20] Calabi Yau data, hep.itp.tuwien.ac.at/~kreuzer/CY/.
[21] R. Altman, J. Gray, Y.H. He, V. Jejjala, B.D. Nelson, A Calabi–Yau database: threefolds constructed from the Kreuzer–Skarke list, J. High Energy Phys. 1502 (2015) 158, arXiv:1411.1418, CY Database: www.rossealtman.com.
[22] Y.H. He, S.J. Lee, A. Lukas, Heterotic models from vector bundles on toric Calabi–Yau manifolds, J. High Energy Phys. 1005 (2010) 071, arXiv:0911.0865 [hep-th];
Y.H. He, M. Kreuzer, S.J. Lee, A. Lukas, Heterotic bundles on Calabi–Yau manifolds with small Picard number, J. High Energy Phys. 1112 (2011) 039, arXiv:1108.1031.
[23] L.B. Anderson, F. Apruzzi, X. Gao, J. Gray, S.J. Lee, A new construction of Calabi–Yau manifolds: generalized CICYs, Nucl. Phys. B 906 (2016) 441, arXiv:1507.03235 [hep-th].
[24] L.B. Anderson, Y.H. He, A. Lukas, Heterotic compactification, an algorithmic approach, J. High Energy Phys. 0707 (2007) 049, arXiv:hep-th/0702210 [hep-th].
[25] L.B. Anderson, J. Gray, A. Lukas, E. Palti, Heterotic line bundle standard models, J. High Energy Phys. 1206 (2012) 113, arXiv:1202.1757 [hep-th].
[26] L.B. Anderson, A. Constantin, J. Gray, A. Lukas, E. Palti, A comprehensive scan for heterotic SU(5) GUT models, J. High Energy Phys. 1401 (2014) 047, arXiv:1307.4787.
[27] J. Gray, A.S. Haupt, A. Lukas, All complete intersection Calabi–Yau four-folds, J. High Energy Phys. 1307 (2013) 070, arXiv:1303.1832 [hep-th].
[28] P. Gao, Y.H. He, S.T. Yau, Extremal bundles on Calabi–Yau threefolds, Commun. Math. Phys. 336 (3) (2015) 1167, arXiv:1403.1268 [hep-th].
[29] D.R. Morrison, C. Vafa, Compactifications of F theory on Calabi–Yau threefolds. 1 & 2, Nucl. Phys. B 473 (1996) 74, arXiv:hep-th/9602114, Nucl. Phys. B 476 (1996) 437, arXiv:hep-th/9603161.
[30] R. Friedman, J. Morgan, E. Witten, Vector bundles and F theory, Commun. Math. Phys. 187 (1997) 679, arXiv:hep-th/9701162.
[31] M. Gabella, Y.H. He, A. Lukas, An abundance of heterotic vacua, J. High Energy Phys. 0812 (2008) 027, arXiv:0808.2142 [hep-th].
[32] M. Cvetič, J. Halverson, D. Klevers, P. Song, On finiteness of type IIB compactifications: magnetized branes on elliptic Calabi–Yau threefolds, J. High Energy Phys. 1406 (2014) 138, arXiv:1403.4943 [hep-th].
[33] A. Hanany, Y.H. He, Nonabelian finite gauge theories, J. High Energy Phys. 9902 (1999) 013, arXiv:hep-th/9811183;
A. Hanany, Y.H. He, A monograph on the classification of the discrete subgroups of SU(4), J. High Energy Phys. 0102 (2001) 027, arXiv:hep-th/9905212.
[34] B. Feng, A. Hanany, Y.H. He, D-brane gauge theories from toric singularities and toric duality, Nucl. Phys. B 595 (2001) 165, arXiv:hep-th/0003085.
[35] S. Franco, A. Hanany, D. Martelli, J. Sparks, D. Vegh, B. Wecht, Gauge theories from toric geometry and brane tilings, J. High Energy Phys. 0601 (2006) 128, arXiv:hep-th/0505211.
[36] J. Davey, A. Hanany, J. Pasukonis, On the classification of brane tilings, J. High Energy Phys. 1001 (2010) 078, arXiv:0909.2868 [hep-th].
[37] S. Franco, Y.H. He, C. Sun, Y. Xiao, A comprehensive survey of brane tilings, arXiv:1702.03958 [hep-th].
[38] J. Davey, A. Hanany, N. Mekareeya, G. Torri, M2-branes and Fano 3-folds, J. Phys. A 44 (2011) 405401, arXiv:1103.0553 [hep-th].
[39] A. Hanany, R.K. Seong, Brane tilings and reflexive polygons, Fortschr. Phys. 60 (2012) 695, arXiv:1201.2614 [hep-th].
[40] Y.H. He, R.K. Seong, S.T. Yau, Calabi–Yau volumes and reflexive polytopes, arXiv:1704.03462.
[41] S. Franco, S. Lee, R.K. Seong, C. Vafa, Brane brick models in the mirror, J. High Energy Phys. 1702 (2017) 106, arXiv:1609.01723.
[42] Y.H. He, Deep-learning the landscape, arXiv:1706.02714.

15

Majorana neutrino as Bogoliubov quasiparticle

Kazuo Fujikawa [a,b], Anca Tureanu [a,*]

[a] *Department of Physics, University of Helsinki, P.O. Box 64, FIN-00014 Helsinki, Finland*
[b] *Quantum Hadron Physics Laboratory, RIKEN Nishina Center, Wako 351-0198, Japan*

ARTICLE INFO

Editor: J. Hisano

ABSTRACT

We suggest that the Majorana neutrino should be regarded as a Bogoliubov quasiparticle that is consistently understood only by use of a relativistic analogue of the Bogoliubov transformation. The unitary charge conjugation condition $\mathcal{C}\psi\mathcal{C}^{\dagger} = \psi$ is not maintained in the definition of a quantum Majorana fermion from a Weyl fermion. This is remedied by the Bogoliubov transformation accompanying a redefinition of the charge conjugation properties of vacuum, such that a C-noninvariant fermion number violating term (condensate) is converted to a Dirac mass. We also comment on the chiral symmetry of a Majorana fermion; a massless Majorana fermion is invariant under a global chiral transformation $\psi \to \exp[i\alpha\gamma_5]\psi$ and different Majorana fermions are distinguished by different chiral $U(1)$ charge assignments. The reversed process, namely, the definition of a Weyl fermion from a well-defined massless Majorana fermion is also briefly discussed.

1. Introduction

The Majorana fermions received much attention recently not only in particle physics in $d = 1 + 3$ [1–4] but also in condensed matter physics in $d = 1 + 3$ or less dimensions [5–7]. In the massless case, it is generally believed that the Majorana fermion and the Weyl fermion are identical in $d = 1 + 3$ as is seen by writing them in the two-component spinor notation. Nevertheless, it is not obvious at all how a self-conjugate object is identical to a complex chiral object. In this paper, we discuss some basic properties of Majorana and Weyl fermions using a relativistic analogue of Bogoliubov transformation in $d = 1 + 3$ space–time. Surprisingly, this analysis leads to the idea that the Majorana neutrino should be regarded as a Bogoliubov quasiparticle that is consistently understood, as it will be explained further, only by use of the Bogoliubov transformation. The Majorana neutrino could thus become the first Bogoliubov quasiparticle observed in particle physics.

2. Bogoliubov transformation

It is customary to define a Majorana fermion which satisfies

$$\psi_M(x) = C\overline{\psi_M}^T(x) \tag{1}$$

from a chiral Weyl fermion which satisfies

$$\gamma_5\psi_W(x)_R = \psi_W(x)_R, \tag{2}$$

in the manner [1–4]

$$\psi_M(x) = \psi_W(x)_R + C\overline{\psi_W(x)_R}^T, \tag{3}$$

such that $\psi_M(x) = C\overline{\psi_M}^T(x)$. Here, C is the charge conjugation matrix. Our notational conventions follow those in [8]. If one starts with a Dirac fermion $\psi_D(x)$, we have $\psi_W(x)_R = \psi_D(x)_R$ but we do not allow to use $\psi_W(x)_L = \psi_D(x)_L$ in addition to $\psi_W(x)_R$ for a moment.

In quantum field theory the simple matrix operation (1) has to correspond to the application of a unitary $\mathcal{C}$ operator to the quantum fields. In the quantum framework, the definition of a charge conjugated spinor as $\psi^c = C\overline{\psi}^T$ can be regarded as a classical operation, for which a quantum realization $\mathcal{C}$ has to exist.

To satisfy the operator relation $\mathcal{C}\psi_M(x)\mathcal{C}^{\dagger} = \psi_M(x)$, it is often assumed that the charge conjugation is given by [1–4] $(\psi_W(x)_R)^C = C\overline{\psi_W(x)_R}^T$ or, in the operator notation,

$$\mathcal{C}\psi_W(x)_R\mathcal{C}^{\dagger} = C\overline{\psi_W(x)_R}^T, \tag{4}$$

by presuming a suitable operator $\mathcal{C}$ defined on an unspecified vacuum. However, this leads to a puzzling result for the unitary charge conjugation operator using $\psi_W(x)_R = \frac{(1+\gamma_5)}{2}\psi_W(x)_R$ and [9]:

* Corresponding author.
E-mail address: anca.tureanu@helsinki.fi (A. Tureanu).

$$\mathcal{C}\psi_W(x)_R\mathcal{C}^\dagger = \frac{(1+\gamma_5)}{2}\mathcal{C}\psi_W(x)_R\mathcal{C}^\dagger = \frac{(1+\gamma_5)}{2}C\overline{\psi_W(x)_R}^T = 0. \quad (5)$$

Moreover, the well-known C- and P-violating weak interaction Lagrangian is written as

$$\begin{aligned}\mathcal{L}_{\rm Weak} &= (g/\sqrt{2})\bar{e}_L\gamma^\mu W_\mu^{(-)}(x)\nu_L + h.c.\\ &= (g/\sqrt{2})\bar{e}_L\gamma^\mu W_\mu^{(-)}(x)[(1-\gamma_5)/2]\nu_L + h.c.\end{aligned} \quad (6)$$

If one assumes again $\mathcal{C}\psi_W(x)_R\mathcal{C}^\dagger = C\overline{\psi_W(x)_L}^T$ as C-transformation law, one obtains ambiguous results, namely, the first expression implies that $\mathcal{L}_{\rm Weak}$ is invariant under C, while the second expression implies $\mathcal{L}_{\rm Weak} \to 0$ [9]. The quantity $(\psi_W(x)_R)^C = C\overline{\psi_W(x)_R}^T$ represents a convenient auxiliary object, but not a charge conjugation of $\psi_W(x)_R$. We are unable to maintain natural operator charge conjugation in (3) in this construction.

On the other hand, one may define the charge conjugation by

$$\mathcal{C}_\psi\psi_W(x)_R\mathcal{C}_\psi^\dagger = C\overline{\psi_W(x)_L}^T \quad (7)$$

by noting that $\psi_W(x)_R = \frac{1+\gamma_5}{2}\psi_D(x)$ and using the conventional charge conjugation $\mathcal{C}_\psi\psi_D(x)\mathcal{C}_\psi^\dagger = C\overline{\psi_D(x)}^T$ of a Dirac field $\psi_D(x)$. To make the statement precise, we denote the $\mathcal{C}$ for ψ_D as $\mathcal{C}_\psi$. In this case we do not encounter any obvious contradictions, but we have for the would-be Majorana field in (3)

$$\begin{aligned}\mathcal{C}_\psi\psi_M(x)\mathcal{C}_\psi^\dagger &= \mathcal{C}_\psi\psi_W(x)_R\mathcal{C}_\psi^\dagger + \mathcal{C}_\psi C\overline{\psi_W(x)_R}^T\mathcal{C}_\psi^\dagger\\ &= C\overline{\psi_W(x)_L}^T + \psi_W(x)_L\\ &\neq \psi_M(x).\end{aligned} \quad (8)$$

Namely, we can not satisfy the Majorana condition of $\psi_M(x)$. This implies that the subtle aspect of the definition of the operator charge conjugation is not solved by the mere change of the convention, but rooted at a more fundamental level. Practically, it is important that this difficulty persists independently of the masses of Weyl and Majorana fermions, for example, in the analysis of the seesaw mechanism.

We can however satisfy consistent operatorial CP,

$$\mathcal{CP}\psi_M(x)(\mathcal{CP})^\dagger = i\gamma^0\psi_M(t,-\vec{x}), \quad (9)$$

which is the fundamental symmetry in a parity violating theory. We adopt the parity operator with the action $\mathcal{P}\psi(x)\mathcal{P}^\dagger = i\gamma^0\psi(t,-\vec{x})$ as the natural choice for Majorana fields, since it preserves the reality of the field in the Majorana representation.

As a way to treat the charge conjugation in a transparent manner, we introduced a relativistic analogue of Bogoliubov transformation, $(\psi,\psi^c)\to(N,N^c)$, defined as

$$\begin{pmatrix} N(x)\\ N^c(x)\end{pmatrix} = \begin{pmatrix}\cos\theta\,\psi(x) - \gamma_5\sin\theta\,\psi^c(x)\\ \cos\theta\,\psi^c(x) + \gamma_5\sin\theta\,\psi(x)\end{pmatrix}, \quad (10)$$

with a suitable parameter θ [9]. Note that $N^c = C\bar{N}^T$ and $\psi^c = C\bar{\psi}^T$, and the transformation satisfies the (classical) consistency condition $N^c = C\bar{N}^T$ using the expressions given by the right-hand sides. This Bogoliubov transformation maps a linear combination of a Dirac fermion ψ and its charge conjugate ψ^c to another Dirac fermion N and its charge conjugate N^c, and thus the original Fock vacuum for ψ is mapped to a new vacuum for N, at $t=0$, for example. It is significant that we use Dirac fermions, either massless or massive, in the definition of the present Bogoliubov transformation. We can then show that the kinetic terms are invariant

$$\begin{aligned}\mathcal{L} &= \frac{1}{2}\{\bar{N}i\not\partial N + \bar{N^c}i\not\partial N^c\}\\ &= \frac{1}{2}\{\bar{\psi}i\not\partial\psi + \bar{\psi^c}i\not\partial\psi^c\}.\end{aligned} \quad (11)$$

Moreover, using (10), we can show that the anticommutators are preserved, i.e.,

$$\begin{aligned}&\{N(t,\vec{x}),N^c(t,\vec{y})\} = \{\psi(t,\vec{x}),\psi^c(t,\vec{y})\},\\ &\{N_\alpha(t,\vec{x}),N_\beta(t,\vec{y})\} = \{N_\alpha^c(t,\vec{x}),N_\beta^c(t,\vec{y})\} = 0.\end{aligned} \quad (12)$$

Thus, the canonicity condition of the Bogoliubov transformation is satisfied, irrespective of the masses of the fields ψ and N, more generally than that implied by the free fields in (11).

It is important that the Bogoliubov transformation (10) preserves the CP symmetry as a unitary operation on quantum fields, although it does not preserve the transformation properties under $i\gamma^0$-parity and C separately [9]. To be precise, the Bogoliubov transformation is a canonical transformation and thus we expect that the dynamical properties on the vacuum for ψ and the new vacuum for $N(x)$ are mostly equivalent, but the charge conjugation property is critically changed. A transformation analogous to (10) has been successfully used in the analysis of neutron–antineutron oscillations [10] and the analysis of the hitherto unrecognized sizable fine-tuning [9] in the see-saw mechanism [11–13].

We now suggest the use of the above Bogoliubov transformation as a means to evade the difficulty associated with the charge conjugation of Weyl and Majorana fermions in a more general context. The transformation (10) with $\theta=\pi/4$ gives

$$\begin{aligned}\frac{1}{\sqrt{2}}(N(x)+N^c(x)) &= \psi_R(x) + C\overline{\psi_R}^T(x),\\ \frac{1}{\sqrt{2}}(N(x)-N^c(x)) &= \psi_L(x) - C\overline{\psi_L}^T(x),\end{aligned} \quad (13)$$

namely, two Majorana fermions,

$$\begin{aligned}\psi_M^{(1)} &= \frac{1}{\sqrt{2}}(N(x)+N^c(x)),\\ \psi_M^{(2)} &= \frac{1}{\sqrt{2}}(N(x)-N^c(x)),\end{aligned} \quad (14)$$

are naturally defined in terms of the new field $N(x)$ introduced by the Bogoliubov transformation (Bogoliubov quasifermion), with the property:

$$\mathcal{C}_N N\mathcal{C}_N^\dagger = C\overline{N}^T. \quad (15)$$

Here we denoted the charge conjugation operator for N as $\mathcal{C}_N$.

Those Majorana fermions satisfy $\mathcal{C}_N\psi_M^{(1)}\mathcal{C}_N^\dagger = \psi_M^{(1)}$ and $\psi_M^{(1)} = C\overline{\psi_M^{(1)}}^T$, and $\mathcal{C}_N\psi_M^{(2)}\mathcal{C}_N^\dagger = -\psi_M^{(2)}$ and $\psi_M^{(2)} = -C\overline{\psi_M^{(2)}}^T$, the first being even and the second odd eigenfield of the charge conjugation operator $\mathcal{C}_N$. The fields $\psi_M^{(1)}$ and $\psi_M^{(2)}$ correspond to the conventional definitions of Majorana fermions in terms of Weyl fermions on the right-hand side of (13), that do not support the operator charge conjugation $\mathcal{C}_\psi$. Incidentally, the definition of Majorana fermions by itself implies a certain "condensation" of the fermion number in the vacuum (see Ref. [14]). The Bogoliubov transformation helps define the eigenstates of the charge conjugation operator $\mathcal{C}_N$ in a consistent manner.

In the case of massless Majorana and Weyl fermions, we do not encounter an explicit fermion number violation in the Lagrangian, for example,

$$\begin{aligned}\mathcal{L} &= \frac{1}{2}\overline{\psi_M^{(1)}}(x)i\not\partial\psi_M^{(1)}(x)\\ &= \overline{\psi_R(x)}i\not\partial\psi_R(x),\end{aligned} \quad (16)$$

unlike the Bardeen–Cooper–Schrieffer (BCS) theory [5,6] or the see-saw mechanism [11–13], where there is an energy or mass gap. Yet the chiral Weyl fermion $\psi_R(x)$, in its strict definition, is not the eigenstate of charge conjugation $\mathcal{C}$ and parity $\mathcal{P}$ transformations, although $\mathcal{CP}$ is well-defined. Thus the definition of the exact eigenstate of $\mathcal{C}_\psi$, namely the free Majorana fermion, has certain conflicts with the definition of the charge conjugation for $\psi_R(x)$.[1] These conflicts are resolved by the charge conjugation $\mathcal{C}_N$ after the Bogoliubov transformation, which is precisely what the first relation of (13) implies. We however implicitly assumed the existence of a Dirac fermion in defining the Bogoliubov transformation, which needs to be remembered when we consider applications.

We can also solve (10) with $\theta = \pi/4$ in terms of the Majorana fermions (14) as

$$\begin{pmatrix} \psi(x) \\ \psi^c(x) \end{pmatrix} = \begin{pmatrix} \left(\frac{1+\gamma_5}{2}\right)\psi_M^{(1)}(x) + \left(\frac{1-\gamma_5}{2}\right)\psi_M^{(2)}(x) \\ \left(\frac{1-\gamma_5}{2}\right)\psi_M^{(1)}(x) - \left(\frac{1+\gamma_5}{2}\right)\psi_M^{(2)}(x) \end{pmatrix}. \tag{17}$$

The Majorana fermions $\psi_M^{(1)}(x)$ and $\psi_M^{(2)}(x)$, if chosen as the primary dynamical degrees of freedom, belong to definite representations of the basic symmetries P and T and thus C, due to the CPT symmetry of field theory on the Minkowski space–time. With this choice of fundamental fields, the natural quantum realization of the charge conjugation in (17) is $\mathcal{C}_N$, under which $\mathcal{C}_N\psi_M^{(1)}(x)\mathcal{C}_N^\dagger \to \psi_M^{(1)}(x)$ and $\mathcal{C}_N\psi_M^{(2)}(x)\mathcal{C}_N^\dagger \to -\psi_M^{(2)}(x)$. However, on the left-hand side of (17) this operation does not send ψ to ψ^c, which would be expected if the operator charge conjugation is preserved. On the other hand, the classical consistency condition $\psi^c(x) = C\overline{\psi(x)}^T$ is satisfied. This inconsistency is precisely the difficulty we encountered in the construction of the Majorana fermions *via Weyl fermions* in (3). The Majorana fermions $\psi_M^{(1,2)}(x)$ are consistently defined in terms of the Bogoliubov $N(x)$ and $N^c(x)$, but not consistently in terms of the chiral projected components $\psi_{L,R}(x)$.

As for the physical implication of the above analysis, it may be natural to accept the description of the Majorana neutrino using the Bogoliubov transformation as a physical one in the Standard Model, since we start with the chiral Weyl fermions as the basic building blocks of gauge theory. As for the Dirac fermion $\psi(x)$, which plays an important role in (13), it is effectively produced in the see-saw mechanism by the addition of the right-handed neutrino. One may thus regard the possible Majorana neutrino as a first Bogoliubov quasiparticle, which is consistently understood only by the use of the Bogoliubov transformation.

3. Chiral symmetry and related issues

We now come back to the analysis of mostly massless fermions. From the point of view of conserved symmetries, the massless Dirac fermion has the $U(1)$ fermion number and chiral $U(1)$ symmetries. The Weyl fermion picks up the chiral choices $(1 \pm \gamma_5)/2$ as the conserved symmetry. The massless Majorana fermion retains only the chiral γ_5 symmetry of the Dirac fermion as a conserved symmetry, as is seen by the construction $\psi_M(x) = (1/2)[\psi_D(x) + C\overline{\psi_D}^T(x)]$, which implies $\psi_D \to e^{i\alpha\gamma_5}\psi_D \;\Rightarrow\; \psi_M \to e^{i\alpha\gamma_5}\psi_M$, or more directly

$$e^{i\alpha\gamma_5}\psi_M = C\overline{e^{i\alpha\gamma_5}\psi_M}^T \tag{18}$$

if one uses $\psi_M = C\overline{\psi_M}^T$. The quantity $i\gamma_5$ is real in the Majorana representation of γ matrices. If one insists on the eigenstates of the chiral symmetry, the chiral fields $\psi_M(x)_{L,R}$ with $\psi_M(x) = \psi_M(x)_L + \psi_M(x)_R$ are picked up.

The chiral symmetry implies the invariance under

$$\psi_M(x)_L \to e^{-i\alpha}\psi_M(x)_L, \quad \psi_M(x)_R \to e^{i\alpha}\psi_M(x)_R. \tag{19}$$

It is important that we have the same α in $e^{\pm i\alpha}\psi_M(x)_{L,R}$. In contrast, we have two parameters α and β in the case of a Dirac fermion:

$$\psi_D(x)_L \to e^{-i\alpha}\psi_D(x)_L, \quad \psi_D(x)_R \to e^{-i\beta}\psi_D(x)_R, \tag{20}$$

but for a Weyl fermion $\psi_W(x)_L = \psi_D(x)_L$, we have obviously only a single parameter. When one regards the quantities defined from Weyl fermions

$$\begin{aligned} \psi_M^{(1)}(x) &= \psi_W(x)_R + C\overline{\psi_W(x)_R}^T, \\ \psi_M^{(2)}(x) &= \psi_W(x)_L - C\overline{\psi_W(x)_L}^T, \end{aligned} \tag{21}$$

as Majorana fermions, the distinct chiral symmetries of these Majorana fermions arise from those two different chiral symmetries.

In general, each massless Majorana fermion has its own chiral symmetry generated by the generic chiral transformation $e^{i\alpha\gamma_5}$, and different Majorana fermions are distinguished by the different charge assignments to these $U(1)$ chiral transformations.

As for the neutrinoless double beta decay, one may consider

$$\mathcal{L} = (g/\sqrt{2})\bar{e}_L\gamma^\mu W_\mu \frac{(1-\gamma_5)}{2}\psi_M + h.c., \tag{22}$$

which projects the Majorana neutrino to the chiral state $\psi_M(x)_L$. It is well-known that the neutrinoless double beta decay takes place only with massive Majorana neutrinos [16]. We here briefly comment on this criterion from the point of view of the chiral symmetry breaking by the mass term. The chiral symmetry breaking may be characterized by

$$\langle 0|\overline{\psi(x)}\psi(x)|0\rangle = \lim_{x\to y}\langle 0|T^\star\overline{\psi(x)}\psi(y)|0\rangle \neq 0, \tag{23}$$

although we are dealing with explicit chiral symmetry breaking. The use of $\langle 0|\overline{\psi(x)}\psi(x)|0\rangle$ as an indicator of chiral symmetry is well-known [15,17]. For a Majorana fermion, $\psi(x) = \nu_L(x) + \nu_R(x)$, it is confirmed that this is equivalent to

$$\lim_{x\to y}\langle 0|T^\star\nu_L^T(x)C\nu_L(y)|0\rangle \neq 0 \tag{24}$$

by noting $\overline{\nu_R} = \nu_L^T C$, with a suitable regularization such as the dimensional regularization, where C is the charge conjugation matrix. Chiral symmetry for either a massless Majorana neutrino or a Weyl neutrino predicts the vanishing result of this correlation to be consistent with the conclusion in [16]. As for the Majorana propagator generated by the neutrinoless double beta decay amplitude discussed in [16], one may combine two ends of the neutrino line of the propagator at a point and then recognize the consistency with the criterion (24).

As for the extra CP violating phases in the case of massive Majorana neutrinos [18–20], they are eliminated by the chiral freedom in the case of massless Majorana neutrinos.

We finally mention a reversed process, namely, the definition of a Weyl fermion from a massless Majorana fermion that satisfies $\mathcal{C}\psi_M(x)\mathcal{C}^\dagger = \psi_M(x)$ as well as $\psi_M(x) = C\overline{\psi_M}^T(x)$. We emphasize that the Majorana fermion defined in (3) does not satisfy both these conditions. In this scheme, we have

[1] One may recall that the free Lagrangian $\mathcal{L} = \overline{\psi_R(x)}i\not\partial\psi_R(x) = \overline{\psi_R(x)}i\not\partial[(1+\gamma_5)/2]\psi_R(x)$ suffers from the ambiguity if one adopts the charge conjugation in (4), although we have already rejected $(\psi_R(x))^C = C\overline{\psi_R(x)}^T$ as a transformation rule of charge conjugation.

$$\begin{aligned}\psi_M(x) &= \psi_M(x)_R + \psi_M(x)_L \\ &= \psi_M(x)_R + C\overline{\psi_M(x)}_R^T \\ &= \psi_M(x)_L + C\overline{\psi_M(x)}_L^T,\end{aligned} \quad (25)$$

with the constraint $\psi_M(x)_L = C\overline{\psi_M(x)}_R^T$. A salient feature of the present scheme is that chiral components satisfy the operator charge conjugation properties

$$\mathcal{C}\psi_M(x)_R\mathcal{C}^\dagger = \psi_M(x)_R, \quad \mathcal{C}\psi_M(x)_L\mathcal{C}^\dagger = \psi_M(x)_L, \quad (26)$$

and one can avoid the inconsistency we encountered in (3), since in the present case the Majorana field is considered elementary. One can thus use one of $\psi_M(x)_L$, $\psi_M(x)_R$ and $\psi_M(x) = C\overline{\psi_M}^T(x)$ as a primary dynamical degree of freedom,

$$\begin{aligned}\mathcal{L}_M &= \frac{1}{2}\overline{\psi_M(x)}i\,\partial\!\!\!/\psi_M(x) \\ &= \overline{\psi_M(x)_R}i\,\partial\!\!\!/\psi_M(x)_R \\ &= \overline{\psi_M(x)_L}i\,\partial\!\!\!/\psi_M(x)_L.\end{aligned} \quad (27)$$

One may interpret the superposition $\psi_M(x) = \psi_M(x)_R + \psi_M(x)_L$ in a manner analogous to the case of the photon polarizations. The interaction chooses one of these possibilities. The field $\psi_M(x)$ which is transformed under parity as

$$\mathcal{P}\psi_M(x)\mathcal{P}^\dagger = i\gamma^0\psi_M(t, -\vec{x}), \quad (28)$$

is an analogue of linear polarization but only one linear polarization (Majorana fermion) appears in the present case. The variables $\psi_M(x)_L$ and $\psi_M(x)_R$ in $\psi_M(x) = \psi_M(x)_L + \psi_M(x)_R$ are interchanged by parity

$$\begin{aligned}\mathcal{P}\psi_M(x)_R\mathcal{P}^\dagger &= i\gamma^0\psi_M(t, -\vec{x})_L, \\ \mathcal{P}\psi_M(x)_L\mathcal{P}^\dagger &= i\gamma^0\psi_M(t, -\vec{x})_R,\end{aligned} \quad (29)$$

and they are the analogues of the circular polarization, but only one of them is allowed at a time, either left-handed or right-handed (Weyl fermions). The Majorana fermion has no *a priori* preference for a left- or right-handed state, and the measurement (interaction) will pick up one of $\psi_M(x)_R$, $\psi_M(x)_L$ or $\psi_M(x) = \psi_M(x)_R + \psi_M(x)_L$. This subject, besides an interest in the quantum information carried by ultra-relativistic particles [21], may become relevant in the future if one takes the Majorana fermion as a fundamental entity of Nature.

In conclusion, we have discussed the basic properties of Majorana and Weyl fermions and the relativistic analogue of Bogoliubov transformation in $d = 1 + 3$. The present study of the Majorana fermion as a Bogoliubov quasiparticle may be compared to condensed matter physics where the Bogoliubov quasiparticle is well-known but the notion of Majorana particles is new [5–7]. In the Standard Model it is natural to start with Weyl fermions, and we argued that a change of vacuum as is indicated by the Bogoliubov transformation is inevitable to understand the possible Majorana neutrino consistently. The Majorana neutrino is then regarded as a Bogoliubov quasiparticle for the first time in particle physics.

We thank Masud Chaichian for very helpful discussions. This work is supported in part by the Magnus Ehrnrooth Foundation. The support of the Academy of Finland under the Projects no. 136539 and 272919 is gratefully acknowledged.

Appendix A. Explicit example

We now illustrate the idea stated above, including the definitions of the charge conjugation operator and vacuum, using an explicit example of the single flavor seesaw model defined by [9]

$$\begin{aligned}\mathcal{L} &= (1/2)\{\overline{\nu}(x)[i\,\partial\!\!\!/ - m_D]\nu(x) + \overline{\nu^c}(x)[i\,\partial\!\!\!/ - m_D]\nu^c(x)\} \\ &\quad - (\epsilon_1/4)[\overline{\nu^c}(x)\nu(x) + \overline{\nu}(x)\nu^c(x)] \\ &\quad - (\epsilon_5/4)[\overline{\nu^c}(x)\gamma_5\nu(x) - \overline{\nu}(x)\gamma_5\nu^c(x)],\end{aligned} \quad (30)$$

where we used a Dirac-type variable

$$\nu(x) \equiv \nu_L(x) + n_R(x) \quad (31)$$

and $\epsilon_1 = m_R + m_L$ and $\epsilon_5 = m_R - m_L$, which are real if one assumes CP symmetry, for simplicity. The above Lagrangian (30) is CP conserving, although C, $\nu \to C\overline{\nu}^T$, and P ($i\gamma^0$-parity) are separately broken by the last term.

Let us first recapitulate the traditional approach to the single flavor seesaw and point out the issues with charge conjugation, and subsequently present on the same example our proposed approach using the Bogoliubov transformation and indicate how the issue is solved. Usually, one exactly diagonalizes the Lagrangian (30) (see, for example, Ref. [3]) as

$$\begin{aligned}\mathcal{L} &= \overline{\tilde{\nu}}(x)i\,\partial\!\!\!/\tilde{\nu}(x) \\ &\quad - (1/2)\left(\overline{\tilde{\nu}}_R M_1 \tilde{\nu}_L^c - \overline{\tilde{\nu}_R^c} M_2 \tilde{\nu}_L\right) + h.c.,\end{aligned} \quad (32)$$

where

$$\begin{aligned}\begin{pmatrix}\tilde{\nu}_L^c \\ \tilde{\nu}_L\end{pmatrix} &\equiv O\begin{pmatrix}\nu_L^c \\ \nu_L\end{pmatrix}, \\ \begin{pmatrix}\tilde{\nu}_R \\ \tilde{\nu}_R^c\end{pmatrix} &\equiv O\begin{pmatrix}\nu_R \\ \nu_R^c\end{pmatrix},\end{aligned} \quad (33)$$

with a suitable 2×2 orthogonal matrix O. The mass matrix in (30) is diagonalized as

$$O\begin{pmatrix}\frac{1}{2}(\epsilon_1 + \epsilon_5) & m_D \\ m_D & \frac{1}{2}(\epsilon_1 - \epsilon_5)\end{pmatrix}O^T = \begin{pmatrix}M_1 & 0 \\ 0 & -M_2\end{pmatrix}, \quad (34)$$

where

$$M_{1,2} = \sqrt{m^2 + (\epsilon_5/2)^2} \pm \epsilon_1/2 \quad (35)$$

are real eigenvalues. In our convention, $\nu_L^c = [(1 - \gamma_5)/2]\nu^c$ and $\nu_R^c = [(1 + \gamma_5)/2]\nu^c$ are left- and right-handed, respectively. Then one defines Majorana-type fields by

$$\begin{aligned}\tilde{\psi}_+(x) &= \tilde{\nu}_R(x) + \tilde{\nu}_L^c(x), \\ \tilde{\psi}_-(x) &= \tilde{\nu}_L(x) - \tilde{\nu}_R^c(x),\end{aligned} \quad (36)$$

and the Lagrangian becomes

$$\mathcal{L} = \frac{1}{2}\overline{\tilde{\psi}_+}(x)[i\,\partial\!\!\!/ - M_+]\tilde{\psi}_+(x) + \frac{1}{2}\overline{\tilde{\psi}_-}(x)[i\,\partial\!\!\!/ - M_-]\tilde{\psi}_-(x), \quad (37)$$

where we denoted $M_1 = M_+$ and $M_2 = M_-$, respectively. The mass M_- represents the tiny neutrino mass.

However, one cannot show that the fields $\tilde{\psi}_+$ and $\tilde{\psi}_-$ are truly Majorana in the quantum field theory sense, just as we failed to show that ψ_M in (3) is a Majorana field. This analysis shows that one can diagonalize the C-violating Lagrangian exactly in terms of Weyl fermions as in (32), but it is not invariant operatorially under C. Thus one cannot rewrite this C-violating Lagrangian in terms of Majorana fermions which are exact eigenstates of a quantum

C-operator. This is the origin of the puzzling aspects we encountered in (3). This feature is also shared with the simple case in (16).

To evade this problem, we recall that charge conjugation is well-defined for a Dirac field and that the Majorana field is supposed to be exactly invariant under charge conjugation. Therefore we propose to consider a different path, using the Bogoliubov transformation of Dirac fields as in (38). Thus the appearance of the Bogoliubov transformation is generic for the C-violating Lagrangian in the seesaw mechanism.[2]

Returning to the Lagrangian (30), we apply the Bogoliubov transformation $(\nu, \nu^c) \to (N, N^c)$, defined as

$$\begin{pmatrix} N(x) \\ N^c(x) \end{pmatrix} = \begin{pmatrix} \cos\theta\, \nu(x) - \gamma_5 \sin\theta\, \nu^c(x) \\ \cos\theta\, \nu^c(x) + \gamma_5 \sin\theta\, \nu(x) \end{pmatrix}, \tag{38}$$

with

$$\sin 2\theta = (\epsilon_5/2)/\sqrt{m^2 + (\epsilon_5/2)^2}. \tag{39}$$

We can then show that the anticommutators are preserved, i.e.,

$$\{N(t,\vec{x}), N^c(t,\vec{y})\} = \{\nu(t,\vec{x}), \nu^c(t,\vec{y})\},$$
$$\{N_\alpha(t,\vec{x}), N_\beta(t,\vec{y})\} = \{N^c_\alpha(t,\vec{x}), N^c_\beta(t,\vec{y})\} = 0, \tag{40}$$

hence the transformation is canonical.

After the Bogoliubov transformation, which diagonalizes the Lagrangian with $\epsilon_1 = 0$, $\mathcal{L}$ in (30) becomes

$$\mathcal{L} = \frac{1}{2}\left[\overline{N}(x)\left(i\,\partial\!\!\!/ - M\right)N(x) + \overline{N^c}(x)\left(i\,\partial\!\!\!/ - M\right)N^c(x)\right]$$
$$- \frac{\epsilon_1}{4}[\overline{N^c}(x)N(x) + \overline{N}(x)N^c(x)], \tag{41}$$

with the mass parameter

$$M \equiv \sqrt{m^2 + (\epsilon_5/2)^2}. \tag{42}$$

This implies that the Bogoliubov transformation maps the original theory to a theory characterized by the new (large) mass scale M, and $\epsilon_5/2$ corresponds to the energy gap, in analogy with the BCS theory. We emphasize that the Bogoliubov transformation (38) preserves the CP symmetry, although it does not preserve the transformation properties under $i\gamma^0$-parity and C separately. In the present single flavor model, this leads to the Lagrangian (41) of the Bogoliubov quasi-fermion $N(x)$ which is symmetric under the $i\gamma^0$-parity and C transformations, $N(t,\vec{x}) \to i\gamma^0 N(t,-\vec{x})$ and $N(x) \to C\overline{N}^T$, respectively.

The combinations

$$\psi_+(x) = \frac{1}{\sqrt{2}}(N(x) + N^c(x)), \quad \psi_-(x) = \frac{1}{\sqrt{2}}(N(x) - N^c(x)) \tag{43}$$

represent Majorana fields if one uses the charge conjugation operator defined by $N(x) \to N^c(x)$, and the Lagrangian (41) is exactly diagonalized in the form

$$\mathcal{L} = \frac{1}{2}\{\overline{\psi}_+[i\,\partial\!\!\!/ - M_+]\psi_+ + \overline{\psi}_-[i\,\partial\!\!\!/ - M_-]\psi_-\}, \tag{44}$$

with the masses

$$M_\pm = M \pm \epsilon_1/2 \equiv \sqrt{m^2 + (\epsilon_5/2)^2} \pm \epsilon_1/2, \tag{45}$$

such that $m_\nu = M_-$ corresponds to the small neutrino mass. As expected, the mass eigenvalues coincide with those obtained by the previously presented direct diagonalization of the mass matrix, eq. (35).

As for the explicit definitions of the charge conjugation operators and vacua, one may ideally define them directly. However, the direct definitions turn out to be very involved. We thus adopt the following consistency argument which can be formulated rigorously.

We recall that the Lagrangians (44) and (37) contain fields with the same masses and classically expected to be Majorana fermions, $\psi_+^c = \psi_+$ and $\psi_-^c = -\psi_-$, respectively. They satisfy the free equations for fermions:

$$[i\,\partial\!\!\!/ - M_+]\psi_+(x) = 0,$$
$$[i\,\partial\!\!\!/ - M_-]\psi_-(x) = 0. \tag{46}$$

These free Dirac equations are solved exactly, and the vacuum is defined by

$$\psi_+^{(+)}(x)|0\rangle_M = \psi_-^{(+)}(x)|0\rangle_M = 0, \tag{47}$$

where $\psi_\pm^{(+)}(x)$ stand for positive frequency components. One can also construct the operator charge conjugation $\mathcal{C}_M$ which satisfies

$$\mathcal{C}_M\psi_+(x)\mathcal{C}_M^\dagger = C\overline{\psi_+(x)}^T = \psi_+(x),$$
$$\mathcal{C}_M\psi_-(x)\mathcal{C}_M^\dagger = C\overline{\psi_-(x)}^T = -\psi_-(x), \tag{48}$$

with $\mathcal{C}_M|0\rangle_M = |0\rangle_M$ by following the procedure in the textbook [8].

We next invert (43) and (36) in the form

$$\begin{pmatrix} N(x) \\ N^c(x) \end{pmatrix} = \begin{pmatrix} \frac{1}{\sqrt{2}}(\psi_+(x) + \psi_-(x)) \\ \frac{1}{\sqrt{2}}(\psi_+(x) - \psi_-(x)) \end{pmatrix}, \tag{49}$$

and

$$\begin{pmatrix} \tilde{\nu}(x) \\ \tilde{\nu}^c(x) \end{pmatrix} = \begin{pmatrix} \left(\frac{1+\gamma_5}{2}\right)\psi_+(x) + \left(\frac{1-\gamma_5}{2}\right)\psi_-(x) \\ \left(\frac{1-\gamma_5}{2}\right)\psi_+(x) - \left(\frac{1+\gamma_5}{2}\right)\psi_-(x) \end{pmatrix}, \tag{50}$$

respectively. Both of these relations satisfy the classical charge conjugation conditions, $N^c = C\overline{N}^T$ and $\tilde{\nu}^c = C\overline{\tilde{\nu}}^T$, using the expressions on the right-hand sides with $C\overline{\psi_+(x)}^T = \psi_+(x)$ and $C\overline{\psi_-(x)}^T = -\psi_-(x)$.

We satisfy

$$\mathcal{C}_M N(x)\mathcal{C}_M^\dagger = \frac{1}{\sqrt{2}}\mathcal{C}_M\left(\psi_+(x) + \psi_-(x)\right)\mathcal{C}_M^\dagger = N^c(x), \tag{51}$$

but

$$\mathcal{C}_M\tilde{\nu}(x)\mathcal{C}_M^\dagger = \mathcal{C}_M\left(\left(\frac{1+\gamma_5}{2}\right)\psi_+(x) + \left(\frac{1-\gamma_5}{2}\right)\psi_-(x)\right)\mathcal{C}_M^\dagger$$
$$= \left(\frac{1+\gamma_5}{2}\right)\psi_+(x) - \left(\frac{1-\gamma_5}{2}\right)\psi_-(x)$$
$$\neq \tilde{\nu}^c(x). \tag{52}$$

One can thus consistently define $\mathcal{C}_N = \mathcal{C}_M$ and $|0\rangle_N = |0\rangle_M$ for the Bogoliubov quasiparticle $N(x)$, but the exact charge conjugation $\mathcal{C}_M$ for the exact solutions of the Majorana fermion does not induce the required charge conjugation of the variable $\tilde{\nu}(x)$, namely, $\mathcal{C}_{\tilde{\nu}} \neq \mathcal{C}_M$, which in turn implies that the vacuum defined by $\mathcal{C}_{\tilde{\nu}}|0\rangle_{\tilde{\nu}} = |0\rangle_{\tilde{\nu}}$ differs from $|0\rangle_M$, namely, $|0\rangle_{\tilde{\nu}} \neq |0\rangle_N$. The exact solutions of our Lagrangian (30) are given by two Majorana

[2] One the other hand, if one sets $\epsilon_5 = 0$ in the Lagrangian (30), the model becomes C-invariant although it still violates the fermion number. In such a case, one can directly rewrite the Lagrangian in terms of Majorana fermions in a consistent manner without the help of the Bogoliubov transformation. (This case corresponds to the model analyzed in [14].)

fermions with different masses, and the true vacuum $|0\rangle_M$ is defined by a direct product of the vacua for those two Majorana fermions. (The true vacuum $|0\rangle_M$ differs from the vacuum of $N(x)$ expected from the Dirac part of the Lagrangian (41) with mass M or the vacuum of $\tilde{\nu}(x)$ expected from the Dirac part of the Lagrangian (32) with a vanishing mass.) All the Dirac-type variables $N(x)$ and $\tilde{\nu}$ are expanded in terms of these two Majorana fermions, but the variable $\tilde{\nu}$, which contains chiral projection operators in the expansion, cannot be a representation of the charge conjugation operator $\mathcal{C}_M$.

Thus, the Bogoliubov transformation converts the "C-violating condensate" with the coefficient ϵ_5 in the fermion number violating condensates in (30) into a Dirac mass $M = \sqrt{m_D^2 + (\epsilon_5/2)^2}$. We regard this to be an analogue of the absorption of the chiral condensate by the massless Dirac fermion to become a massive Dirac fermion in the Nambu–Jona-Lasinio model [15], with the corresponding changes of vacuum states. For the very special case $\epsilon_5 = 0$ and thus the mixing angle $\theta = 0$ in (39), which is irrelevant for the seesaw mechanism but relevant for the neutron oscillations [10,14], the Lagrangian (30) becomes C-invariant and the Bogoliubov transformation is not required. In this case, effectively $N(x) = \nu(x)$, where $\nu(x)$ is the very original variable in (30); one can confirm that $\tilde{\psi}_+(x) = \tilde{\nu}_R(x) + \tilde{\nu}_L^c(x) = (\nu(x) + \nu^c(x))/\sqrt{2}$ and $\tilde{\psi}_-(x) = \tilde{\nu}_L(x) - \tilde{\nu}_R^c(x) = (\nu(x) - \nu^c(x))/\sqrt{2}$ in (33). In this special case with C conservation, we thus have $\mathcal{C}_\nu = \mathcal{C}_M$, with $\mathcal{C}_\nu \nu(x) \mathcal{C}_\nu^\dagger = C\overline{\nu(x)}^T = \nu^c(x)$, and the true vacuum is given by $|0\rangle_M$.

References

[1] M. Fukugita, T. Yanagida, Physics of Neutrinos and Application to Astrophysics, Springer, Berlin, Heidelberg, 2002.

[2] C. Giunti, C.W. Kim, Fundamentals of Neutrino Physics and Astrophysics, Oxford University Press, Oxford, 2007.

[3] S. Bilenky, Introduction to the Physics of Massive and Mixed Neutrinos, Lect. Notes Phys., vol. 817, Springer, Berlin, Heidelberg, 2010.

[4] Zhi-zhong Xing, Neutrino physics, in: Proceedings of the 1st Asia–Europe–Pacific School of High-Energy Physics, AEPSHEP, 2012, published in CERN Yellow Report CERN-2014-001, 177–217, arXiv:1406.7739.

[5] C. Chamon, R. Jackiw, Y. Nishida, S.-Y. Pi, L. Santos, Quantizing Majorana fermions in a superconductor, Phys. Rev. B 81 (2010) 224515, arXiv:1001.2760.

[6] C.W.J. Beenakker, Annihilation of colliding Bogoliubov quasiparticles reveals their Majorana nature, Phys. Rev. Lett. 112 (2014) 070604, arXiv:1312.2001.

[7] F. Wilczek, Majorana and condensed matter physics, in: The Physics of Ettore Majorana, Cambridge University Press, 2015, arXiv:1404.0637.

[8] J.D. Bjorken, S.D. Drell, Relativistic Quantum Fields, McGraw Hill, New York, 1965.

[9] K. Fujikawa, A. Tureanu, Naturalness in see-saw mechanism and Bogoliubov transformation, Phys. Lett. B 767 (2017) 199, arXiv:1609.03309.

[10] K. Fujikawa, A. Tureanu, Parity-doublet theorem for Majorana fermions and neutron oscillation, Phys. Rev. D 94 (2016) 115009, arXiv:1609.03203.

[11] P. Minkowski, $\mu \to e + \gamma$ at a rate of one out of 10^9 muon decays?, Phys. Lett. B 67 (1977) 421.

[12] T. Yanagida, in: O. Sawada, A. Sugamoto (Eds.), Proceedings of Workshop on Unified Theory and Baryon Number in the Universe, 1979, p. 95, KEK report 79-18;
M. Gell-Mann, P. Ramond, R. Slansky, in: P. van Nieuwenhuizen, D.Z. Freedman (Eds.), Supergravity, North-Holland, Amsterdam, 1979, p. 315.

[13] R.N. Mohapatra, G. Senjanovic, Neutrino mass and spontaneous parity nonconservation, Phys. Rev. Lett. 44 (1980) 912.

[14] L.N. Chang, N.P. Chang, Structure of the vacuum and neutron and neutrino oscillations, Phys. Rev. Lett. 45 (1980) 1540.

[15] Y. Nambu, G. Jona-Lasinio, Dynamical model of elementary particles based on an analogy with superconductivity. 1, Phys. Rev. 122 (1961) 345.

[16] J. Schechter, J.F. Valle, Neutrinoless double-β decay in $SU(2) \times U(1)$ theories, Phys. Rev. D 25 (1982) 2951.

[17] M. Gell-Mann, The symmetry group of vector and axial vector currents, Physics 1 (1964) 63.

[18] S.M. Bilenky, J. Hosek, S.T. Petcov, On the oscillations of neutrinos with Dirac and Majorana masses, Phys. Lett. B 94 (1980) 495.

[19] M. Doi, T. Kotani, H. Nishiura, K. Okuda, E. Takasugi, CP violation in Majorana neutrinos, Phys. Lett. B 102 (1981) 323.

[20] J. Schechter, J.W.F. Valle, Neutrino-oscillation thought experiment, Phys. Rev. D 23 (1981) 1666.

[21] K. Fujikawa, C.H. Oh, Chengjie Zhang, Spin operator and entanglement in quantum field theory, Phys. Rev. D 90 (2014) 025028, arXiv:1312.0326.

16

Note on soft theorems and memories in even dimensions

Pujian Mao [a], Hao Ouyang [a,b,*]

[a] Institute of High Energy Physics and Theoretical Physics Center for Science Facilities, Chinese Academy of Sciences, 19B Yuquan Road, Beijing 100049, PR China
[b] School of Physical Sciences, University of Chinese Academy of Sciences, 19A Yuquan Road, Beijing 100049, PR China

ARTICLE INFO

Editor: N. Lambert

ABSTRACT

Recently, it has been shown that the Weinberg's formula for soft graviton production is essentially a Fourier transformation of the formula for gravitational memory which provides an effective way to understand how the classical calculation arises as a limiting case of the quantum result. In this note, we propose a general framework that connects the soft theorems to the radiation fields obtained from classical computation for different theories in even dimensions. We show that the latter is nothing but Fourier transformation of the former. The memory formulas can be derived from radiation fields explicitly.

1. Introduction

Recently, a triangle of exact equivalence relation between three classes of physics has been established [1] (see, also, [2] for a comprehensive discussion). It was first observed by Strominger and collaborators in gravitational theory [3,4].[1] The precise ingredients of this triangle relation are Weinberg's soft graviton theorem [6], gravitational memory [7] and Bondi–Metzner–Sachs (BMS) Supertranslation [8]. Those physical phenomena are connected in the following way: The gravitational memory formula is nothing but a Fourier transformation of the soft graviton factor, which suggests that the classical result is a limiting case of quantum result; the soft graviton theorem is precisely equivalent to the Ward identity of the BMS supertranslation that is supposed to be a spontaneously broken symmetry of the quantum gravity S-matrix; gravitational memory can be understood as a transition between two inequivalent vacua of the gravitational field and the initial and final vacua are related by the specific BMS supertranslation. Shortly, this triangle relation has gained considerable attentions and has been extended to electromagnetism [9,10], Yang–Mills theory [11] and other theories [12,13].

Originally, the triangle relation was established in four dimension and only Weinberg's pole formula [6] was involved. In reality, soft theorems can be viewed as factorization properties that scattering amplitudes must obey in a low-energy expansion. For gauge and gravitational theory, the universal properties go through the subleading order and subsubleading order respectively in a low-energy expansion in arbitrary dimensions [14] (see, also, [15–21]). With the triangle equivalence in mind, one would expect that any new discovery at one corner will hint at new physics at other corners which are intensively discussed recently. In four dimension, new memory effects were reported [22,23] inspired by the subleading soft theorems. On the asymptotic symmetry side, the investigation is still ongoing and an agreement has not been reached. The subleading soft theorems can be understood as Ward identities of new symmetries [24]. But the nature of those new symmetries are not completely known namely it is not clear that the new reported symmetries are asymptotic symmetries of the theories of interest. Alternatively, a consistent picture was proposed [25] that the subleading soft theorems can be recovered as a Ward identity associated to the same symmetries that control the leading piece of the theorem. The situation in higher dimension is less clear even at the leading order. On the one hand, the asymptotic symmetry (*e.g.* the BMS supertranslation in gravitational theory) that is supposed to be responsible for the leading soft theorem seems to be debatable [26,27]. On the other hand, the memory effect in higher dimension was shown to be absent [28].

The aim of the present note is to clarify the connection between soft theorems and memories in even dimensions for different theories. Using the retarded and advanced Green's functions in even dimensions, we obtain the retarded and advanced solutions of the wave functions in Minkowski spacetime. The radiation fields [29] are defined by the difference of the retarded and advanced solutions. We find that the radiation fields are nothing but a Fourier transformation of the soft factors of the theory that we are solving for the radiation fields. In the asymptotic region, the memory for-

* Corresponding author.
E-mail addresses: maopj@ihep.ac.cn (P. Mao), ouyangh@ihep.ac.cn (H. Ouyang).

[1] The connection was already observed in Yang–Mills theory by Strominger [5], but the full triangle relation was first established in gravitational theory.

mulas can be reproduced explicitly by the radiation fields in four dimension. In higher dimensions, we show that in d dimension there is no memory effect at the order $r^{1-d/2}$ in the $1/r$ expansion which is in consistence with the results of [28]. However there are memories in dimensions higher than four if one is able to trace the subleading orders in the $1/r$ expansion. The nontrivial memory is of order r^{3-d}.[2] Consequently, we propose that the equivalence of the soft theorems and memories should be understood in the following way: The classical computation of the radiation fields arises as a limiting case of the quantum results namely soft theorems[3] and the memory effects are determined completely by the radiation fields in the asymptotic region.

The organization of this note is quite simple. In the next section we establish the general set-up. We compute the radiation fields from the classical field equations and show that the the radiation fields are nothing but Fourier transformation of the soft factors. In section 3, we discuss the relation of the radiation fields and the memories. The last section is devoted to discussions on open issues.

2. The set-up

The key observation of the equivalence of the soft theorems and memories is to consider the soft factors as the expectation value of fluctuation of fields [4,10,22,23]. This will be also applied in the present work. We start with the scalar field to establish our general set-up.

2.1. Scalar field

Soft scalar theorem has been recently derived in [13][4]:

$$M_{N+1}^{\text{scalar}} = \sum_{k=1}^{N} \frac{g_k}{p_k \cdot q} M_N^{\text{scalar}} + \mathcal{O}(\omega^0), \tag{1}$$

where g_k are the coupling constants, q is the momentum of the soft scalar and p_k are the momenta of the hard particles. We will consider massive hard particles since charged particle sources moving at the speed of light seem to be ill-defined [32] when solving the classical wave equations. The soft factor will be interpreted as the expectation value of the scalar field in the process of scattering in momentum space [4,10,22,23] which leads to[5]

$$\lim_{\omega\to 0} \omega\,\varphi_d(\omega,\vec{q}) = \lim_{\omega\to 0} \omega \frac{M_{N+1}^{\text{scalar}}}{M_N^{\text{scalar}}}, \tag{2}$$

where d denotes the dimension of the spacetime. Hence,

$$\varphi_d(\omega,\vec{q}) = \sum_{k=1}^{N} \frac{g_k}{p_k \cdot q}, \tag{3}$$

at the low energy limit. Performing a Fourier transformation, one obtains the scalar field in the position space as

$$\varphi_d(x) = \sum_{k=1}^{N} \int \frac{d^{d-1}q}{(2\pi)^{d-1}} \frac{1}{2\omega} \frac{\eta_k g_k}{q\cdot p_k} (e^{iq\cdot x} + \text{c.c.}) \tag{4}$$

where $\eta_k = 1$ or -1 for an outgoing or incoming particle.[6]

We now calculate $\varphi_d(x)$ in even dimension $d = 4+2n$. It is very useful to define the generating function

$$\Phi \equiv \sum_{n=0}^{\infty} \frac{1}{n!} (\pi s)^n \varphi_{4+2n}. \tag{5}$$

Then $\varphi_d(x)$ can be derived from the generating function easily by taking the limit

$$\varphi_{4+2n}(x) = \lim_{s\to 0} \frac{1}{\pi^n} \frac{d^n}{ds^n} \Phi \tag{6}$$

The generating function can be solved out explicitly

$$\begin{aligned}
\Phi &= -\sum_{k=1}^{N}\sum_{n=0}^{\infty} \eta_k g_k \int_{-\infty}^{\infty} d\omega_q \int_0^{\pi} d\theta \int_0^{\pi} d\phi \int d\Omega_{2n} \\
&\quad \times \frac{\omega_q^{2n} \sin^{2n+1}\theta \sin^{2n}\phi}{2(2\pi)^{3+2n} n!} (\pi s)^n e^{i\omega_q (r\cos\theta - t)} \\
&\quad \times (E_k - |\mathbf{p}_k| \sin\theta_k \sin\theta\cos\phi - |\mathbf{p}_k|\cos\theta_k\cos\theta)^{-1} \\
&= -\sum_{k=1}^{N}\sum_{n=0}^{\infty} \eta_k g_k \int_{-\infty}^{\infty} d\omega_q \int_0^{\pi} d\theta \int_0^{\pi} d\phi \frac{2n!(4\pi)^n}{(2n)!} \\
&\quad \times \frac{\omega_q^{2n} \sin^{2n+1}\theta \sin^{2n}\phi}{2(2\pi)^{3+2n} n!} (\pi s)^n e^{i\omega_q (r\cos\theta - t)} \\
&\quad \times (E_k - |\mathbf{p}_k| \sin\theta_k \sin\theta\cos\phi - |\mathbf{p}_k|\cos\theta_k\cos\theta)^{-1} \\
&= -\sum_{k=1}^{N} \int_{-\infty}^{\infty} d\omega_q \int_0^{\pi} d\theta \int_0^{\pi} d\phi \frac{\eta_k g_k}{(2\pi)^3} \\
&\quad \times \frac{\cos(\omega_q\sqrt{-s}\sin\theta\sin\phi)\sin\theta e^{i\omega_q(r\cos\theta - t)}}{E_k - |\mathbf{p}_k|\sin\theta_k\sin\theta\cos\phi - |\mathbf{p}_k|\cos\theta_k\cos\theta} \\
&= -\sum_{k=1}^{N} \frac{\eta_k g_k}{2} \int_{-\infty}^{\infty} d\omega_q \int_0^{\pi} d\theta \int_{-\pi}^{\pi} d\phi \frac{\sin\theta}{(2\pi)^3} \\
&\quad \times \exp(i\omega_q(r\cos\theta - t + \sqrt{-s}\sin\theta\sin\phi)) \\
&\quad \times (E_k - |\mathbf{p}_k| \sin\theta_k \sin\theta\cos\phi - |\mathbf{p}_k|\cos\theta_k\cos\theta)^{-1}.
\end{aligned} \tag{7}$$

Here we consider Φ as a function of (t, r, θ_k). It is useful to note that one can write this generating function as a three dimensional integral over the subspace spanned by $\mathbf{x}$, $\mathbf{p_k}$ and $\mathbf{n}$ with $\mathbf{n}\cdot\mathbf{x} = \mathbf{n}\cdot\mathbf{p} = 0$ and $|\mathbf{n}| = 1$:

$$\Phi = \sum_{k=1}^{N} \int \frac{d^3q}{(2\pi)^3} \frac{1}{2\omega} \frac{\eta_k g_k}{q\cdot p_k} (e^{iq\cdot(x+\sqrt{-s}n)} + \text{c.c.}), \tag{8}$$

where $n^\mu = (0, \mathbf{n})$. It follows that

$$\Phi = \sum_{k=1}^{N} g_k \Phi_k, \tag{9}$$

where[7]

[2] This was also noticed in [30].

[3] A well known example is from the soft bremsstrahlung process that was discussed in details in section 6.1 of [31].

[4] As already pointed in [13], the soft scalar theorem is only a tree-level result because scalar fields will not retain massless after including loop corrections.

[5] One may consider M_{N+1} as the expectation value of the scalar fluctuation produced in the process of $n \to N-n$ scattering while M_N may be regarded as the "vacuum" expectation value in this process.

[6] We take a different convention than [13] where all legs outgoing were assumed.

[7] The field (10) has the spacetime origin as a special point. From the point of view of the soft theorems, spacetime translation corresponds to a phase shift $\exp(iq\Delta x)$ in the soft factor.

$$\Phi_k = \frac{\eta_k}{4\pi\sqrt{(x\cdot p_k)^2 - p_k^2x^2 - p_k^2 s}} \times (\Theta(t-\sqrt{r^2-s}) - \Theta(t+\sqrt{r^2-s})). \quad (10)$$

As we will see shortly that this is nothing but the radiation field obtained from the solutions of the massless scalar wave equation.

In $d = 4+2n$ dimensional spacetime, the retarded and advanced Green's functions satisfying the equation

$$-\eta^{\mu\nu}\partial_\mu\partial_\nu G_d(x-x') = (\partial_t^2 - \sum_{i=1}^{3+2n}\partial_{x_i}^2)G_d(x-x') = \delta_d(x-x'), \quad (11)$$

are given by [33]

$$G_{4+2n}^{\rm ret} = \frac{1}{2\pi^{n+1}}\delta^{(n)}((t-t')^2 - |\mathbf{x}-\mathbf{x}'|^2)\Theta(t-t'), \quad (12)$$

$$G_{4+2n}^{\rm adv} = \frac{1}{2\pi^{n+1}}\delta^{(n)}((t-t')^2 - |\mathbf{x}-\mathbf{x}'|^2)\Theta(t'-t). \quad (13)$$

Considering the source corresponding to a particle created or destroyed at the origin

$$S_{dk} = \int_0^\infty d\tau\,\delta^d(x^\mu - \eta_k p_k^\mu\tau), \quad (14)$$

we obtain the retarded solution for the wave function

$$-\eta^{\mu\nu}\partial_\mu\partial_\nu\varphi_{dk}^{\rm ret} = S_{dk}, \quad (15)$$

as

$$\varphi_{4+2n\,k}^{\rm ret} = \int_0^\infty \frac{d\tau}{2\pi^{n+1}}\delta^{(n)}(-(x-v_k\tau)^2)\Theta(t-E_k\eta_k\tau). \quad (16)$$

Now we introduce the retarded generating function

$$\Phi_k^{\rm ret} \equiv \sum_{n=0}^\infty \frac{1}{n!}\varphi_{4+2n\,k}^{\rm ret}(\pi s)^n, \qquad = \int_0^\infty \frac{d\tau}{2\pi}\delta(-(x-p_k\tau)^2 + s)\Theta(t-E_k\eta_k\tau). \quad (17)$$

For massive particle source, we have

$$\Phi_k^{\rm ret} = \frac{\Theta(\eta_k(t-\sqrt{r^2-s}))}{4\pi\sqrt{(x\cdot p_k)^2 - p_k^2x^2 - p_k^2 s}}. \quad (18)$$

The generating function for the advanced solution with massive particle source can be derived in a similar way

$$\Phi_k^{\rm adv} = \frac{\Theta(\eta_k(t+\sqrt{r^2-s}))}{4\pi\sqrt{(x\cdot p_k)^2 - p_k^2x^2 - p_k^2 s}}. \quad (19)$$

A general source can be written as a linear superposition of such created and destroyed particles (14), so the solutions can be written as a superposition of individual ones. Using the relation $\Theta(\eta_k(t-\sqrt{r^2-s})) - \Theta(\eta_k(t+\sqrt{r^2-s})) = \eta_k[\Theta(t-\sqrt{r^2-s}) - \Theta(t+\sqrt{r^2-s})]$, the radiation field [29]

$$\Phi^{\rm rad} = \sum_{k=1}^N g_k\left(\Phi_k^{\rm ret} - \Phi_k^{\rm adv}\right) \quad (20)$$

is precisely the same as (9). However, (9) is derived as a low energy limit ($\omega \to 0$) of the quantum computation namely scattering amplitudes while (20) is a solution of classical wave equation. The equivalence of those two suggests that the classical calculation arises as a limiting case of the quantum result.

2.2. Electromagnetic field

Now we move on to the fields with spin. Let us first consider the electromagnetic theory. It has been realized long time ago that the scattering of a soft photon displayed universal properties through the subleading order in a low-energy expansion [14] (see, also, [15] for earlier, restricted to four dimension, versions)

$$M_{N+1}^{\rm photon} = \sum_{k=1}^N\left(e_k\frac{p_k^\mu}{p_k\cdot q} + e_k\frac{iJ_k^{\mu\nu}q_\nu}{p_k\cdot q}\right)\epsilon_\mu M_N^{\rm photon} + \mathcal{O}(\omega). \quad (21)$$

In analogy with the scalar case (3), one can consider the soft factors to be the classical fields in the momentum space at low energy limit

$$A^\mu(\omega,\vec{q}) = \sum_{k=1}^N\left(e_k\frac{p_k^\mu}{p_k\cdot q} + e_k\frac{iJ_k^{\mu\nu}q_\nu}{p_k\cdot q}\right). \quad (22)$$

Evaluating its Fourier transformation

$$A_d^\mu(x) = \sum_{k=1}^N\int\frac{d^{d-1}q}{(2\pi)^{d-1}}\frac{\eta_k}{2\omega}\left(e_k\frac{p_k^\mu}{p_k\cdot q} + e_k\frac{iJ_k^{\mu\nu}q_\nu}{p_k\cdot q}\right) \times (e^{iq\cdot x} + \text{c.c.}), \quad (23)$$

it is straightforward to get

$$A_d^\mu(x) = \sum_{k=1}^N e_k(p_k^\mu + J_k^{\mu\nu}\partial_\nu)\varphi_{dk}, \quad (24)$$

where φ_{dk} is defined by

$$\varphi_{dk}(x) = \lim_{s\to 0}\frac{1}{\pi^n}\frac{d^n}{ds^n}\Phi_k. \quad (25)$$

We will show that (24) is just the radiation fields obtained from Maxwell's equations in Lorentz gauge:

$$\partial^\nu\partial_\nu A_d^\mu = -j_d^\mu, \quad (26)$$

where j_d^μ is a conserved current.

A generic current for a collection of charged free point particles associated with the soft factor up to subleading order is given by

$$j_d^\mu(y) = \sum_{k=1}^N e_k\left[\int_0^\infty d\tau\,p_k^\mu\delta_d(y-\eta_k p_k\tau) + \int_0^\infty d\tau\,J_k^{\mu\nu}\partial_\nu\delta_d(y-\eta_k p_k\tau)\right], \quad (27)$$

where the second term of the right hand side is the dipole contribution. Conservation of the current implies conservation of the charge

$$\sum_{k=1}^N \eta_k e_k = 0. \quad (28)$$

In Lorentz gauge, the Maxwell's equations (26) take the form of a wave function (11) of each component of A^μ. Hence, we can immediately write down the retarded and advanced solutions with the help of the Green's functions (12) and (13). Finally, we get

$$A_d^{\mu\,\rm ret} = \sum_{k=1}^N e_k(p_k^\mu + J_k^{\mu\nu}\partial_\nu)\varphi_{dk}^{\rm ret},$$

$$A_d^{\mu\,\rm adv} = \sum_{k=1}^N e_k(p_k^\mu + J_k^{\mu\nu}\partial_\nu)\varphi_{dk}^{\rm adv}. \quad (29)$$

Then, the radiation field of electromagnetic theory

$$A_d^{\mu\,\mathrm{rad}} = A_d^{\mu\,\mathrm{ret}} - A_d^{\mu\,\mathrm{adv}}, \tag{30}$$

recovers the soft factor in position space (24) explicitly.

2.3. Linearized gravitational field

In this section, we demonstrate our last example which is the linearized gravitational theory. The soft graviton theorem goes through the subsubleading order [14,21] (see, also, [17,19] for earlier, restricted to four dimension, versions)[8]

$$M_{N+1}^{\mathrm{graviton}} = \sum_{k=1}^{N}\left(\frac{p_k^\mu p_k^\nu}{p_k\cdot q} + \frac{ip_k^\mu J_k^{\nu\alpha} q_\alpha}{p_k\cdot q} - \frac{1}{2}\frac{J_k^{\mu\alpha} q_\alpha J_k^{\nu\beta} q_\beta}{p_k\cdot q}\right) \times \epsilon_{\mu\nu} M_N^{\mathrm{graviton}} + \mathcal{O}(\omega^2). \tag{31}$$

We consider the soft factors as the classical field in the momentum space at low energy limit

$$\bar{h}^{\mu\nu}(\omega,\vec{q}) = \sum_{k=1}^{N}\left(\frac{p_k^\mu p_k^\nu}{p_k\cdot q} + \frac{ip_k^{(\mu} J_k^{\nu)\alpha} q_\alpha}{p_k\cdot q} - \frac{1}{2}\frac{J_k^{\mu\alpha} q_\alpha J_k^{\nu\beta} q_\beta}{p_k\cdot q}\right). \tag{32}$$

Similar to the scalar and electromagnetic case, the classical field in position space can be obtained simply by a Fourier transformation

$$\bar{h}_d^{\mu\nu}(x) = \sum_{k=1}^{N}(p_k^\mu p_k^\nu + p_k^{(\mu} J_k^{\nu)\alpha}\partial_\alpha + J_k^{\mu\alpha} J_k^{\nu\beta}\partial_\alpha\partial_\beta)\varphi_{dk}, \tag{33}$$

where Φ_k is again defined in (10).

Now we consider the linearized Einstein equations in harmonic gauge

$$\partial_\alpha\partial^\alpha \bar{h}_d^{\mu\nu} = -T_d^{\mu\nu}, \tag{34}$$

where $T_d^{\mu\nu}$ is a symmetric and conserved stress-energy tensor.

A generic stress-energy tensor for a collection of free point particles associated with the soft factor up to subsubleading order is given by

$$\begin{aligned} T_d^{\mu\nu}(y) = \sum_{k=1}^{N}[& \int_0^\infty d\tau\, p_k^\mu p_k^\nu \delta_d(y-\eta_k p_k \tau) \\ & + \int_0^\infty d\tau\, p_k^{(\mu} J_k^{\nu)\alpha}\partial_\alpha \delta_d(y-\eta_k p_k\tau) \\ & + \int_0^\infty d\tau\, J_k^{\mu\alpha} J_k^{\nu\beta}\partial_\alpha\partial_\beta\delta_d(y-\eta_k p_k\tau)], \end{aligned} \tag{35}$$

where the last terms of the right hand side are the dipole and quadrupole contributions [34]. The linearized Einstein equations also take the form of a wave function (11) of each component. Thus, the retarded and advanced solutions are obtained easily as

$$\begin{aligned} \bar{h}_d^{\mu\nu\,\mathrm{ret}} &= \sum_{k=1}^{N}(p_k^\mu p_k^\nu + p_k^{(\mu} J_k^{\nu)\alpha}\partial_\alpha + J_k^{\mu\alpha} J_k^{\nu\beta}\partial_\alpha\partial_\beta)\varphi_{dk}^{\mathrm{ret}}, \\ \bar{h}_d^{\mu\nu\,\mathrm{adv}} &= \sum_{k=1}^{N}(p_k^\mu p_k^\nu + p_k^{(\mu} J_k^{\nu)\alpha}\partial_\alpha + J_k^{\mu\alpha} J_k^{\nu\beta}\partial_\alpha\partial_\beta)\varphi_{dk}^{\mathrm{adv}}. \end{aligned} \tag{36}$$

Finally, the radiation field

$$\bar{h}_d^{\mu\nu\,\mathrm{rad}} = \bar{h}_d^{\mu\nu\,\mathrm{ret}} - \bar{h}_d^{\mu\nu\,\mathrm{adv}}, \tag{37}$$

is nothing but (33).

3. Radiation field and memory

We now turn our attention to the memory effect. We will recover the result of [28] from the radiation field. If one focuses on the future null infinity, the advanced field will not have memory effects on it. Thus our radiation field should be the same as the retarded field. Nevertheless, it is still meaningful to see this directly. Moreover, we will show that there are nontrivial memories in dimensions higher than four from certain subleading order in the $1/r$ expansion.

3.1. Scalar memory

The scalar field generated by the source (14) will produce a force on a distant test particle. The scalar force on a test particle is given by

$$f_d^\mu = \nabla^\mu \varphi_d. \tag{38}$$

In the limit of large r with $u = t - r$ finite, we expand scalar field in dimension $d = 4 + 2n$ as

$$\varphi_d = \sum_{m=0}^{\infty}\frac{\varphi_d^{(m)}}{r^{1+n+m}}, \tag{39}$$

where φ_d can be obtained from the generating function (10). The leading order behavior of φ_d is r^{-1-n} and the coefficient of leading order is

$$\varphi_d^{(0)} = \sum_k \frac{\eta_k g_k}{4\pi(2\pi)^n r^{n+1}\kappa_k}\frac{d^n}{du^n}\theta(u). \tag{40}$$

As proved in [27,28], the leading order does not give momentum memory effect for $d > 4$, which agrees with our result because the momentum memory effect is order r^{-1-2n} as we will show later. However, the momentum memory effect is related to the leading order of φ_d by equation of motion. Using the wave equation of the scalar field, we obtain

$$\partial_u \varphi_d^{(m+1)} = \frac{n^2+n-m^2-m-\nabla^2_{S^{2+2n}}}{2(m+1)}\varphi_d^{(m)}, \tag{41}$$

where $\nabla^2_{S^{2+2n}}$ is the Laplace–Beltrami operator on S^{2+2n}. The radiation field is zero at $u > 0$. Then it follows that

$$\int_{-\infty}^{\infty} du\,\partial_u\varphi_d^{(n)} = (\int du)^n \prod_{p=0}^{n}\frac{n^2+n-p^2-p-\nabla^2_{S^{2+2n}}}{2(p+1)}\varphi_d^{(0)}, \tag{42}$$

$$\int_{-\infty}^{\infty} du\,\partial_u\varphi_d^{(m)} = 0, \quad 0 < m < n, \tag{43}$$

$$\int_{-\infty}^{\infty} du\,\varphi_d^{(n-1)} = (\int du)^n \prod_{p=0}^{n-1}\frac{n^2+n-p^2-p-\nabla^2_{S^{2+2n}}}{2(p+1)}\varphi_d^{(0)}, \tag{44}$$

$$\int_{-\infty}^{\infty} du\,\varphi_d^{(m)} = 0, \quad 0 < m < n-1. \tag{45}$$

The momentum memory can be written as

[8] We are using natural units where $8\pi G_N = 1$.

$$\Delta P_d^\mu = \int du \nabla^\mu \varphi_d = -\Delta\varphi_d K^\mu + \int du \partial_r \varphi_d R^\mu + \sum_{i=1}^{2+2n} \int du \partial_{\phi_i} \varphi_d \nabla^\mu \phi_i, \quad (46)$$

$$\Delta\varphi_d = \frac{1}{r^{1+2n}} (\int du)^n \prod_{p=0}^{n} \frac{n^2+n-p^2-p-\nabla^2_{S^{2+2n}}}{2(p+1)} \varphi_d^{(0)}, \quad (47)$$

$$\int du \partial_r \varphi_d = \frac{-2n}{r^{1+2n}} (\int du)^n \prod_{p=0}^{n-1} \frac{n^2+n-p^2-p-\nabla^2_{S^{2+2n}}}{2(p+1)} \varphi_d^{(0)}, \quad (48)$$

$$\int du \partial_{\phi_i} \varphi_d = \frac{1}{r^{2n}} (\int du)^n \partial_{\phi_i} \prod_{p=0}^{n-1} \frac{n^2+n-p^2-p-\nabla^2_{S^{2+2n}}}{2(p+1)} \varphi_d^{(0)}, \quad (49)$$

where

$$T^\mu = (1,0), \quad R^\mu = (0, \frac{\mathbf{x}}{r}), \quad K^\mu = -\nabla u = T^\mu + R^\mu. \quad (50)$$

Therefore the momentum memory of the radiation field in dimension $4+2n$ is of order r^{-1-2n} and is determined by the leading order coefficient $\varphi_d^{(0)}$.

To see this more precisely, we calculate the memory using the generating function. The generating function of the scalar force on a test particle is given by

$$F^\mu \equiv \sum_{n=0}^{\infty} \frac{1}{n!} f^\mu_{4+2n} (\pi s)^n = \sum_{k=1}^{N} g_k (-\partial_u \Phi_k K^\mu + \partial_r \Phi_k R^\mu - r^{-1} \partial_{\kappa_k} \Phi_k L_k^\mu), \quad (51)$$

where

$$\kappa_k = E_k - \frac{\mathbf{p}_k \cdot \mathbf{x}}{r}, \quad L_k^\mu = -r\nabla\kappa_k = (0, \mathbf{p}_k - \frac{\mathbf{p}_k \cdot \mathbf{x}}{r^2}\mathbf{x}). \quad (52)$$

We consider the motion of a test particle near null infinity in the time interval $(-u_1, u_1)$ where $u_1 \ll r$. Then the generating function of the scalar memory which is the change in d-momentum due to this force is given by

$$\Delta P^\mu = \int_{-u_1}^{u_1} du f^\mu = \sum_{k=1}^{N} g_k (P_{ku} K^\mu + P_{kr} R^\mu + P_{k\kappa_k} L_k^\mu). \quad (53)$$

Let us define a new parameter $S = sr^{-2}$ and take the large r limit with $u_1 > 0$ fixed. Then, the leading term of ΔP^μ is of order r^{-1} and independent of u_1:

$$P_{ku} = -\frac{\eta_k}{4\pi r} \frac{1}{\sqrt{\kappa_k^2 + p_k^2 S}} (1 + O(\frac{u_1}{r})), \quad (54)$$

$$P_{kr} = \frac{\eta_k}{4\pi r} \frac{S}{\kappa_k^2 - p_k^2 - 2E_k\kappa_k + (E_k^2 + p_k^2) S} \times \left(\frac{E_k\kappa_k + p_k^2}{\sqrt{\kappa_k^2 + p_k^2 S}} + \frac{E_k - \kappa_k}{\sqrt{1-S}} \right) (1 + O(\frac{u_1}{r})), \quad (55)$$

$$P_{k\kappa_k} = -\frac{\eta_k}{4\pi r} \frac{1}{\kappa_k^2 - p_k^2 - 2E_k\kappa_k + (E_k^2 + p_k^2) S} \times \left(\frac{E_k S - \kappa_k}{\sqrt{\kappa_k^2 + p_k^2 S}} + \sqrt{1-S} \right) (1 + O(\frac{u_1}{r})). \quad (56)$$

For $d = 4, 6, 8$, the memories are as follows:

$$\Delta P_4^\mu = \sum_{k=1}^{N} -\eta_k \frac{g_k}{4\pi r \kappa_k} K^\mu, \quad (57)$$

$$\Delta P_6^\mu = \sum_{k=1}^{N} -\eta_k \frac{g_k}{8\pi^2 \kappa_k^3 r^3} (-p_k^2 K^\mu + 2\kappa_k^2 R^\mu - \kappa_k L_k^\mu), \quad (58)$$

$$\Delta P_8^\mu = \sum_{k=1}^{N} -\eta_k \frac{g_k}{16\pi^3 \kappa_k^5 r^5} \Big[-3(p_k^2)^2 K^\mu + \left(4E_k\kappa_k^3 + 4\kappa_k^4 - 4p_k^2\kappa_k^2\right) R^\mu - \left(2E_k\kappa_k^2 + \kappa_k^3 - 3p_k^2\kappa_k\right) L_k^\mu \Big]. \quad (59)$$

In four dimension, we obtain a momentum memory effect in the null direction K^μ as in [32]. In higher dimensions, there are momentum memory effects in radial and transverse directions at order r^{-1-2n} of the expansion.

3.2. Electromagnetic memory

We now turn our attention to electromagnetic memory. Electromagnetic field associated with the current (27) will produce velocity kick on a distant test particle. Expanding in powers of $1/r$ and using the Maxwell's equations, we can write the radiation electromagnetic field as

$$A_{d\mu} = \sum_{m=0}^{\infty} \frac{A_{d\mu}^{(m)}}{r^{1+n+m}}, \quad (60)$$

$$A_{d\mu}^{(m)} = (\int du)^m \prod_{p=0}^{m} \frac{n^2+n-p^2-p-\nabla^2_{S^{2+2n}}}{2(p+1)} A_{d\mu}^{(0)}, \quad (61)$$

$$A_{d\mu}^{(0)} = \sum_{k}^{N} \frac{\eta_k g_k v_{k\mu}}{4\pi (2\pi)^n r^{n+1} \kappa_k} \frac{d^n}{du^n} \theta(u). \quad (62)$$

The radiation field agrees with the results of the retarded field derived in [28] and the electromagnetic memory is determined completely in terms of $A_{d\mu}^{(0)}$. For simplicity, we consider the case that the current (27) only has the monopole contribution, which will be dominant in the memory.[9]

The generating function for the field strength is

$$\mathcal{F}^{\mu\nu} \equiv 2\nabla^{[\mu}\mathcal{A}^{\nu]} = 2\sum_{k=1}^{N} e_k \nabla^{[\mu}\Phi_k p_k^{\nu]}. \quad (63)$$

For a test particle with charge Q and d-velocity v^ν, the instantaneous kick in d-momentum is given by the generating function

[9] The dipole contribution will also lead to memory that corresponds to the subleading soft factor [23].

$$\Delta P^{\mu} \equiv Q \int_{-u_1}^{u_1} \mathcal{F}^{\mu\nu} v_{\nu},$$
$$= Q \sum_{k=1}^{N} e_k (E_k P_{kr} R^{\mu} + \kappa_k P_{ku} R^{\mu} + E_k P_{k\kappa_k} L_k^{\mu} - P_{ku} L_k^{\mu}), \tag{64}$$

at the rest frame of the test particle. Here P_{ku}, P_{kr} and $P_{k\kappa_k}$ are given in (54), (55) and (56). Similar to the scalar case, the memory in dimension $4+2n$ is of order r^{-1-2n}. The memory in arbitrary even dimension can be obtain by this generating function. Here we give the results in $d=4,6,8$ cases for our only-illustrative purposes. In the limit $s \to 0$, P_{kr} and $P_{k\kappa_k}$ vanish. Therefore in four dimension, the test particle obtains a change in its 4-momentum in the transverse direction

$$\Delta P_4^{\mu} = Q \sum_{k=1}^{N} \frac{e_k \eta_k}{4\pi \kappa_k r} L_k^{\mu}. \tag{65}$$

This agrees with the results obtained by Bieri and Garfinkle [35]. In six and eight dimension cases, the electromagnetic force also gives rise to a radial kick memory effect as

$$\Delta P_6^{\mu} = Q \sum_{k=1}^{N} \frac{e_k \eta_k}{8\pi^2 \kappa_k^3 r^3} \left[L_k^{\mu} (p_k^2 - 2\gamma_k \kappa_k) + R^{\mu} \kappa_k (p_k^2 - \gamma_k \kappa_k) \right], \tag{66}$$

$$\Delta P_8^{\mu} = Q \sum_{k=1}^{N} \frac{e_k \eta_k}{16\pi^3 \kappa_k^5 r^5} \Big[L_k^{\mu} (-4\gamma_k \kappa_k^3 - 4\gamma_k^2 \kappa_k^2 + 4\gamma_k \kappa_k p_k^2 - 3(p_k^2)^2) + R^{\mu} \kappa_k \left(\gamma_k \kappa_k^3 + 2\gamma_k^2 \kappa_k^2 - 3\gamma_k \kappa_k p_k^2 + 3(p_k^2)^2 \right) \Big]. \tag{67}$$

3.3. *Linearized gravitational field and memory*

The gravitational memory effect is a permanent relative displacement of nearby observers induced by gravitational waves. If two nearby observers are initially at rest and separated by spatial displacement l^{μ}, then gravitational wave changes their separation by

$$\frac{d^2}{dt^2} l^{\mu} = -R_{t\nu t}{}^{\mu} l^{\nu}. \tag{68}$$

Similar to the scalar and electromagnetic field cases, we can write $\bar{h}_{d\mu\nu}$ as

$$\bar{h}_{d\mu\nu} = \sum_{m=0}^{\infty} \frac{\bar{h}^{(m)}_{d\mu\nu}}{r^{1+n+m}}, \tag{69}$$

$$\bar{h}^{(m)}_{d\mu\nu} = (\int du)^m \prod_{p=0}^{m} \frac{n^2 + n - p^2 - p - \nabla^2_{S^{2+2n}}}{2(p+1)} \bar{h}^{(0)}_{d\mu\nu}, \tag{70}$$

$$\bar{h}^{(0)}_{d\mu\nu} = \sum_{k} \frac{\eta_k g_k v_{k\mu} v_{k\nu}}{4\pi (2\pi)^n r^{n+1} \kappa_k} \frac{d^n}{du^n} \theta(u), \tag{71}$$

which agrees with the results of [28] and the gravitational memory of the radiation field is determined explicitly in terms of $\bar{h}^{(0)}_{d\mu\nu}$.

We consider only the monopole contribution in the stress-energy tensor (35), which will be dominant in the memory.[10]

The gravitational memory is the double integral of nearby observers near null infinity

$$\frac{1}{2} \Delta h_{ij}^{TT} \equiv - \int_{-u_1}^{u_1} du_2 \int_{-u_1}^{u_2} du R_{titj}, \tag{72}$$

in the limit of $r \to \infty$. It is difficult to calculate the generating function for the gravitational memory. Nonetheless, we can check the memory for $d=4,6,8$ cases. In four dimension, we get

$$\frac{1}{2} \Delta h_{4ij}^{TT} = \sum_k \frac{\eta_k}{16 r \kappa_k} (R_i R_j p_k^2 + 2 L_{ki} L_{kj} - \delta_{ij} p_k^2), \tag{73}$$

which agrees with [36]. The results in six and eight dimensions are given by

$$\begin{aligned} \frac{1}{2} \Delta h_{6ij}^{TT} = \sum_k \frac{\eta_k}{64\pi^2 r^3 \kappa_k^3} \Big[& R_i R_j \left(4E_k^2 \kappa_k^2 + 16 E_k \kappa_k^3 + \kappa_k^2 p_k^2 - (p_k^2)^2 \right) + L_{ki} L_{kj} \left(8 E_k \kappa_k - 4 p_k^2 \right) \\ & + (R_j L_{ki} + L_{kj} R_i) \left(-12 E_k \kappa_k^2 + 3\kappa_k p_k^2 \right) \\ & - \delta_{ij} \left(4 E_k^2 \kappa_k^2 + \kappa_k^2 p_k^2 + (p_k^2)^2 \right) \Big]. \end{aligned} \tag{74}$$

$$\begin{aligned} \frac{1}{2} \Delta h_{8ij}^{TT} = \sum_k \frac{\eta_k}{192\pi^3 r^5 \kappa_k^5} \Big[& L_{ki} L_{kj} \left(36 E_k^2 \kappa_k^2 + 12 E_k \kappa_k^3 - 36 E_k \kappa_k p_k^2 + 2\kappa_k^2 p_k^2 + 18 (p_k^2)^2 \right) \\ & - (R_j L_{ki} + L_{kj} R_i) \left(60 E_k^2 \kappa_k^3 + 30 E_k \kappa_k^4 - 40 E_k \kappa_k^2 p_k^2 + 15 \kappa_k (p_k^2)^2 + 5 \kappa_k^3 p_k^2 \right) \\ & + R_i R_j (12 E_k^3 \kappa_k^3 + 138 E_k^2 \kappa_k^4 - 6 E_k^2 \kappa_k^2 p_k^2 - 38 E_k \kappa_k^3 p_k^2 + 9 \kappa_k^2 (p_k^2)^2 + 23 \kappa_k^4 p_k^2 + 3 (p_k^2)^3) \\ & - \delta_{ij} \left(12 E_k^3 \kappa_k^3 + 18 E_k^2 \kappa_k^4 - 6 E_k^2 \kappa_k^2 p_k^2 + 2 E_k \kappa_k^3 p_k^2 - \kappa_k^2 (p_k^2)^2 + 3 \kappa_k^4 p_k^2 + 3 (p_k^2)^3 \right) \Big]. \end{aligned} \tag{75}$$

4. Discussions

We have shown that one can re-interpret the equivalence of the soft theorems and memories as the equivalence between soft theorems and radiation fields where the soft theorems can be considered as a limiting case of quantum results that recovers the classical computations *i.e.* radiation fields. The memory formulas can be reproduced explicitly from the radiation fields near future null infinity. Now we would like to comment on some aspects that arise in previous sections.

It is amusing to find that the subleading orders of soft theorems are related to the multipole contributions of the source of the wave equations. The universal properties of soft emission stop at subleading and subsubleading order in electromagnetic theory and gravitational theory respectively. A very curious point is the

[10] The dipole contribution also creates memory that corresponds to the subleading soft factor [22], while the memory associated to the quadrupole contribution of the stress-energy tensor (*i.e.* the memory related to subsubleading soft factor) is not known.

radiation fields including multipoles contributions beyond dipole and quadrupole in the current (27) and the stress-energy tensor (35). There should be some reasonings that make the monopole and dipole (also quadrupole for gravitational theory) contributions special to be consistent with the soft theorems. We postpone relevant investigation elsewhere.

A puzzling issue to which we do not yet have an answer is that if there is observational effect associated to the memory in lower order in the $1/r$ expansion in higher dimensions. As we have shown in section 2, the formula of the equivalence is, in general, valid in any even dimension while the memory effect is quite sensitive to dimension of the spacetime. However, this may not be surprised. It is well known that the Newtonian limit of general relativity has different effects in different dimensions. For instance, general relativity in $2+1$ dimensions has a Newtonian limit without force between static point masses [37].

A more subtle one is the fact that the radiation fields that we are dealing with are solutions of source-free wave equations. This may be related to the argument that soft particles that we are concerning in the present work are free. There are indeed some other facts that hint such argument. On the one hand, it is well known that the effect of attaching several soft photon or soft graviton to an arbitrary scattering process is just to supply a product of factors of the Weinberg's pole formula, one for each soft particles [6].[11] Hence, adding multiple soft particles is like adding one by one and soft particles do not see each other namely they are free particles. On the other hand, the soft theorems are connected to asymptotic symmetries in the triangle equivalence. It is reasonable to consider that the soft particles arise from a proper treatment of representation theory of the asymptotic symmetry group *e.g.* BMS particles in gravitational theory. The BMS particles are well studied in three dimension (see [39] for a comprehensive discussion), and gain renewed attentions in four dimension recently [39,40]. It would be definitely meaningful to have more investigation elsewhere following this direction to explore the nature of soft particles.

Acknowledgements

The authors thank Eduardo Conde and Jun-Bao Wu for useful discussions and valuable comments on the draft. The authors would like to thank the anonymous referee for the suggestions and comments which are very helpful in improving the original manuscript. This work is supported in part by the National Natural Science Foundation of China (Grant No. 11575202).

References

[1] A. Strominger, Memory, Soft Theorems and Symmetries 2015, talk given at Strings 2015, Bengaluru, India.
[2] A. Strominger, Lectures on the infrared structure of gravity and gauge theory, arXiv:1703.05448 [hep-th].
[3] A. Strominger, On BMS invariance of gravitational scattering, J. High Energy Phys. 07 (2014) 152, arXiv:1312.2229 [hep-th];
T. He, V. Lysov, P. Mitra, A. Strominger, BMS supertranslations and Weinberg's soft graviton theorem, J. High Energy Phys. 05 (2015) 151, arXiv:1401.7026 [hep-th].
[4] A. Strominger, A. Zhiboedov, Gravitational memory, BMS supertranslations and soft theorems, J. High Energy Phys. 01 (2016) 086, arXiv:1411.5745 [hep-th].
[5] A. Strominger, Asymptotic symmetries of Yang–Mills theory, J. High Energy Phys. 07 (2014) 151, arXiv:1308.0589 [hep-th].
[6] S. Weinberg, Infrared photons and gravitons, Phys. Rev. 140 (1965) B516–B524.
[7] Y.B. Zel'dovich, A.G. Polnarev, Radiation of gravitational waves by a cluster of superdense stars, Sov. Astron. 18 (Aug. 1974) 17;
V.B. Braginsky, L.P. Grishchuk, Kinematic resonance and memory effect in free mass gravitational antennas, Sov. Phys. JETP 62 (1985) 427–430, Zh. Eksp. Teor. Fiz. 89 (1985) 744.
[8] H. Bondi, M.G.J. van der Burg, A.W.K. Metzner, Gravitational waves in general relativity. 7. Waves from axisymmetric isolated systems, Proc. R. Soc. Lond. A 269 (1962) 21–52;
R.K. Sachs, Gravitational waves in general relativity. 8. Waves in asymptotically flat space-times, Proc. R. Soc. Lond. A 270 (1962) 103–126;
R. Sachs, Asymptotic symmetries in gravitational theory, Phys. Rev. 128 (1962) 2851–2864.
[9] T. He, P. Mitra, A.P. Porfyriadis, A. Strominger, New symmetries of massless QED, J. High Energy Phys. 10 (2014) 112, arXiv:1407.3789 [hep-th];
M. Campiglia, A. Laddha, Asymptotic symmetries of QED and Weinberg's soft photon theorem, J. High Energy Phys. 07 (2015) 115, arXiv:1505.05346 [hep-th];
D. Kapec, M. Pate, A. Strominger, New symmetries of QED, arXiv:1506.02906 [hep-th].
[10] S. Pasterski, Asymptotic symmetries and electromagnetic memory, arXiv:1505.00716 [hep-th].
[11] T. He, P. Mitra, A. Strominger, 2D Kac–Moody symmetry of 4D Yang–Mills theory, J. High Energy Phys. 10 (2016) 137, arXiv:1503.02663 [hep-th];
P. Mao, J.-B. Wu, Note on asymptotic symmetries and soft gluon theorems, arXiv:1704.05740 [hep-th].
[12] M. Campiglia, A. Laddha, Asymptotic symmetries of gravity and soft theorems for massive particles, J. High Energy Phys. 12 (2015) 094, arXiv:1509.01406 [hep-th];
V. Lysov, Asymptotic fermionic symmetry from soft gravitino theorem, arXiv:1512.03015 [hep-th];
T.T. Dumitrescu, T. He, P. Mitra, A. Strominger, Infinite-dimensional fermionic symmetry in supersymmetric gauge theories, arXiv:1511.07429 [hep-th];
S.G. Avery, B.U.W. Schwab, Burg–Metzner–Sachs symmetry, string theory, and soft theorems, Phys. Rev. D 93 (2016) 026003, arXiv:1506.05789 [hep-th];
A. Campoleoni, D. Francia, C. Heissenberg, On higher-spin supertranslations and superrotations, J. High Energy Phys. 05 (2017) 120, arXiv:1703.01351 [hep-th].
[13] M. Campiglia, L. Coito, S. Mizera, Can scalars have asymptotic symmetries?, arXiv:1703.07885 [hep-th].
[14] Z. Bern, S. Davies, P. Di Vecchia, J. Nohle, Low-energy behavior of gluons and gravitons from gauge invariance, Phys. Rev. D 90 (8) (2014) 084035, arXiv:1406.6987 [hep-th].
[15] F.E. Low, Scattering of light of very low frequency by systems of spin 1/2, Phys. Rev. 96 (1954) 1428–1432;
M. Gell-Mann, M.L. Goldberger, Scattering of low-energy photons by particles of spin 1/2, Phys. Rev. 96 (1954) 1433–1438;
F.E. Low, Bremsstrahlung of very low-energy quanta in elementary particle collisions, Phys. Rev. 110 (1958) 974–977.
[16] T.H. Burnett, N.M. Kroll, Extension of the low soft photon theorem, Phys. Rev. Lett. 20 (1968) 86.
[17] D.J. Gross, R. Jackiw, Low-energy theorem for graviton scattering, Phys. Rev. 166 (1968) 1287–1292;
R. Jackiw, Low-energy theorems for massless bosons: photons and gravitons, Phys. Rev. 168 (1968) 1623–1633.
[18] J.S. Bell, R. Van Royen, On the Low–Burnett–Kroll theorem for soft-photon emission, Nuovo Cimento A 60 (1969) 62–68.
[19] C.D. White, Factorization properties of soft graviton amplitudes, J. High Energy Phys. 05 (2011) 060, arXiv:1103.2981 [hep-th];
F. Cachazo, A. Strominger, Evidence for a new soft graviton theorem, arXiv:1404.4091 [hep-th].
[20] E. Casali, Soft sub-leading divergences in Yang–Mills amplitudes, J. High Energy Phys. 08 (2014) 077, arXiv:1404.5551 [hep-th].
[21] B.U.W. Schwab, A. Volovich, Subleading soft theorem in arbitrary dimensions from scattering equations, Phys. Rev. Lett. 113 (10) (2014) 101601, arXiv:1404.7749 [hep-th];
C. Kalousios, F. Rojas, Next to subleading soft-graviton theorem in arbitrary dimensions, J. High Energy Phys. 01 (2015) 107, arXiv:1407.5982 [hep-th];
M. Zlotnikov, Sub-sub-leading soft-graviton theorem in arbitrary dimension, J. High Energy Phys. 10 (2014) 148, arXiv:1407.5936 [hep-th].
[22] S. Pasterski, A. Strominger, A. Zhiboedov, New gravitational memories, J. High Energy Phys. 12 (2016) 053, arXiv:1502.06120 [hep-th].
[23] P. Mao, H. Ouyang, J.-B. Wu, X. Wu, New electromagnetic memories and soft photon theorems, Phys. Rev. D 95 (2017) 125011, arXiv:1703.06588 [hep-th].
[24] D. Kapec, V. Lysov, S. Pasterski, A. Strominger, Semiclassical Virasoro symmetry of the quantum gravity $\mathcal{S}$-matrix, J. High Energy Phys. 08 (2014) 058, arXiv:1406.3312 [hep-th];
V. Lysov, S. Pasterski, A. Strominger, Low's subleading soft theorem as a symmetry of QED, Phys. Rev. Lett. 113 (11) (2014) 111601, arXiv:1407.3814 [hep-th];
M. Campiglia, A. Laddha, Asymptotic symmetries and subleading soft graviton theorem, Phys. Rev. D 90 (12) (2014) 124028, arXiv:1408.2228 [hep-th];
M. Campiglia, A. Laddha, New symmetries for the gravitational S-matrix, J. High Energy Phys. 04 (2015) 076, arXiv:1502.02318 [hep-th];
M. Campiglia, A. Laddha, Sub-subleading soft gravitons: new symmetries of quantum gravity?, Phys. Lett. B 764 (2017) 218–221, arXiv:1605.09094 [gr-qc];

[11] Yang–Mills theory has a very different behavior in multiple soft emissions [38].

M. Campiglia, A. Laddha, Subleading soft photons and large gauge transformations, J. High Energy Phys. 11 (2016) 012, arXiv:1605.09677 [hep-th];
M. Campiglia, A. Laddha, Sub-subleading soft gravitons and large diffeomorphisms, J. High Energy Phys. 01 (2017) 036, arXiv:1608.00685 [gr-qc];
D. Kapec, P. Mitra, A.-M. Raclariu, A. Strominger, A 2D stress tensor for 4D gravity, arXiv:1609.00282 [hep-th].

[25] E. Conde, P. Mao, Remarks on asymptotic symmetries and the subleading soft photon theorem, Phys. Rev. D 95 (2) (2017) 021701, arXiv:1605.09731 [hep-th];
E. Conde, P. Mao, BMS supertranslations and not so soft gravitons, J. High Energy Phys. 05 (2017) 060, arXiv:1612.08294 [hep-th].

[26] S. Hollands, A. Ishibashi, Asymptotic flatness and Bondi energy in higher dimensional gravity, J. Math. Phys. 46 (2005) 022503, arXiv:gr-qc/0304054;
K. Tanabe, N. Tanahashi, T. Shiromizu, On asymptotic structure at null infinity in five dimensions, J. Math. Phys. 51 (2010) 062502, arXiv:0909.0426 [gr-qc];
K. Tanabe, S. Kinoshita, T. Shiromizu, Asymptotic flatness at null infinity in arbitrary dimensions, Phys. Rev. D 84 (2011) 044055, arXiv:1104.0303 [gr-qc];
D. Kapec, V. Lysov, A. Strominger, Asymptotic symmetries of massless QED in even dimensions, arXiv:1412.2763 [hep-th];
D. Kapec, V. Lysov, S. Pasterski, A. Strominger, Higher-dimensional supertranslations and Weinberg's soft graviton theorem, arXiv:1502.07644 [gr-qc].

[27] S. Hollands, A. Ishibashi, R.M. Wald, BMS supertranslations and memory in four and higher dimensions, Class. Quantum Gravity 34 (15) (2017) 155005, arXiv:1612.03290 [gr-qc].

[28] D. Garfinkle, S. Hollands, A. Ishibashi, A. Tolish, R.M. Wald, The memory effect for particle scattering in even spacetime dimensions, Class. Quantum Gravity 34 (14) (2017) 145015, arXiv:1702.00095 [gr-qc].

[29] P.A.M. Dirac, Classical theory of radiating electrons, Proc. R. Soc. Lond. A 167 (1938) 148–169.

[30] Y.-Z. Chu, Gravitational wave memory in dS_{4+2n} and 4D cosmology, Class. Quantum Gravity 34 (3) (2017) 035009, arXiv:1603.00151 [gr-qc];
Y.-Z. Chu, More on cosmological gravitational waves and their memories, arXiv:1611.00018 [gr-qc].

[31] M.E. Peskin, D.V. Schroeder, An Introduction to Quantum Field Theory, Westview Press, Boulder, 1995.

[32] A. Tolish, R.M. Wald, Retarded fields of null particles and the memory effect, Phys. Rev. D 89 (6) (2014) 064008, arXiv:1401.5831 [gr-qc].

[33] S. Hassani, Mathematical Physics: A Modern Introduction to Its Foundations, Springer, 1999.

[34] J. Steinhoff, D. Puetzfeld, Multipolar equations of motion for extended test bodies in general relativity, Phys. Rev. D 81 (2010) 044019, arXiv:0909.3756 [gr-qc];
S. Marsat, Cubic order spin effects in the dynamics and gravitational wave energy flux of compact object binaries, Class. Quantum Gravity 32 (8) (2015) 085008, arXiv:1411.4118 [gr-qc].

[35] L. Bieri, D. Garfinkle, An electromagnetic analogue of gravitational wave memory, Class. Quantum Gravity 30 (2013) 195009, arXiv:1307.5098 [gr-qc].

[36] V.B. Braginsky, K.S. Thorne, Gravitational-wave bursts with memory and experimental prospects, Nature 327 (May 1987) 123–125.

[37] S. Carlip, Quantum Gravity in 2+1 Dimensions, Cambridge University Press, Cambridge, United Kingdom, 2003.

[38] F.A. Berends, W.T. Giele, Multiple soft gluon radiation in parton processes, Nucl. Phys. B 313 (1989) 595–633.

[39] B. Oblak, BMS particles in three dimensions, arXiv:1610.08526 [hep-th].

[40] B. Oblak, Berry phases on Virasoro orbits, arXiv:1703.06142 [hep-th].

17

On magnetic and vortical susceptibilities of the Cooper condensate

A. Gorsky [a,b], F. Popov [b,c,d,*]

[a] Institute of Information Transmission Problems of the Russian Academy of Sciences, Moscow, Russia
[b] Moscow Institute of Physics and Technology, Dolgoprudny 141700, Russia
[c] Institute of Theoretical and Experimental Physics, Moscow, Russia
[d] Department of Physics, Princeton University, Princeton, NJ 08544, USA

ARTICLE INFO

Editor: M. Cvetič

ABSTRACT

We discuss the susceptibility of the Cooper condensate in the s-wave $2+1$ superconductor in the external magnetic field and in the rotating frame. The extended holographic model involving the charged rank-two field is considered and it is argued that the susceptibility does not vanish. We interpret non-vanishing susceptibilities as the admixture of the p-wave triplet component in the Cooper condensate in the external field.

1. Introduction

The ground state of conventional superconductors (SC) involves the Cooper pairs in the different spin and orbital states forming the charged condensate. We shall be interested in the specific respond of the ground state of $2+1$ dimensional s-wave superconductor to the external magnetic field and rotation namely if the p-wave component of the condensate proportional to the external field is generated which is absent otherwise. The known examples of such phenomena one could have in mind are the generation of the triplet component via the Rashba term [1] or via spin-orbit interaction [2] in the singlet s-wave SC.

The interesting possibility of coexistence of the triplet and singlet order parameters in the SC occurs in the specific materials admitting the coexistence of the SC and antiferromagnetic (AF) orders. In this case at least in the model description the following relation takes place [3]

$$|\Delta_t| \propto |\Delta_s||M| \tag{1}$$

where Δ_t, Δ_s are the SC order parameters for triplet and singlet states while M is the AF magnetization order parameter. A bit loosely it can be claimed that the AF component induces the triplet Cooper pairing. In what follows our consideration has some similarities with this situation.

In the ground state of the conventional SC the magnetic field is screened by the supercurrent and is expelled from the bulk penetrating through the Abrikosov strings only. However one can imagine that the triplet spin component of the condensate can be generated by the external field in the bulk of the singlet SC. We shall look at the magnetic susceptibility of the Cooper condensate in the weak magnetic field defined as

$$< 0|\psi\sigma_{\mu\nu}\psi|0 >= g\chi_s < 0|\psi\psi|0 > F_{\mu\nu} \tag{2}$$

where g is dimensionful gauge coupling in $2+1$. We will be interested if $\chi_s \neq 0$ and discuss this issue from the holographic viewpoint.

There is the well-known analogy between the external magnetic field and the rotation which is encoded in the specific form of external metric, see for example, the recent discussion in [4]. Therefore it is natural to consider the responses of the ground state of superconductor at the external magnetic and gravimagnetic field in parallel. To this aim we shall also introduce and discuss the vortical susceptibility of the Cooper condensate in the rotating frame.

The partial motivation for this study is as follows. Consider the hadronic phase in QCD where the chiral symmetry is broken by the condensate $< \bar{\Psi}\Psi >$. The linear response of chiral condensate to the external magnetic field is parametrized as follows

$$< 0|\bar{\Psi}\sigma_{\mu\nu}\Psi|0 >= \chi_F < \bar{\Psi}\Psi > F_{\mu\nu}, \tag{3}$$

where χ_F is the magnetic susceptibility of the condensate introduced in [5]. The value of χ_F can be derived by the different means. In particular its value can be obtained from the anomalous CS terms in the conventional holographic model [6,8] and in the extended model with additional rank-two fields [7].

Recently the vortical susceptibility for the quark condensate was introduced and evaluated in the dense QCD via specific anomaly in the dense matter [9]

* Corresponding author.
E-mail address: fpopov@princeton.edu (F. Popov).

$$< 0|\bar{\Psi}\sigma_{\mu\nu}\Psi|0 >= \chi_G < \bar{\Psi}\Psi > G_{\mu\nu}, \quad (4)$$

where $G_{\mu\nu}$ is the curvature of the graviphoton field. The vortical susceptibility is the response of the chiral condensate to the external gravitational field corresponding to the rotation frame.

Of course there are some differences between QCD chiral condensate and superconducting condensates. The chiral condensate is neutral while the Cooper condensate is charged. In QCD the analogue of the Cooper condensate occurs only at high density in the color-flavor locking superconducting phase while the chiral condensate corresponds to the neutral exciton condensate in the condense matter context. Let us emphasize that in the superconductor case contrary to QCD we deal with the non-relativistic system.

In this Letter we consider the susceptibility of the SC χ_s combining the conventional and holographic means. To this aim we consider the $2+1$ SC described by the AdS_4 bulk geometry in the extended holographic model which involves complex scalar, U(1) gauge field and rank-two field. We argue that both magnetic and vortical susceptibilities do not vanish.

The paper is organized as follows. In Section 2 we remind the simplest holographic models of s-wave superconductor. In Section 3 we introduce the polarization of the Cooper condensate in the magnetic field. Section 4 involves the arguments showing that the magnetic susceptibility does not vanish for $2+1$ and $3+1$ cases. In Section 5 we make a few comments concerning the vortical susceptibility of the condensate. The results and open questions are summarized in the Conclusion.

2. The Holographic model

In this section we discuss the dual model for $2+1$ superconductor. In the case of s-wave superconductivity the relevant dual model reads as (see [10–12] for reviews)

$$S = \int d^4x\sqrt{-g}\Big[R + \frac{\sigma}{L^2} - F^2_{\mu\nu} - \left|\partial_\mu\Psi - igA_\mu\Psi\right|^2 - m^2\left|\Psi\right|^2\Big], \quad (5)$$

$$ds^2 = -f dt^2 + \frac{dr^2}{f} + r^2\left(dx^2+dy^2\right), f = \frac{r^2}{L^2}\left(1-\frac{r_0^2}{r^2}\right). \quad (6)$$

We work in the rigid background space–time – AdS with black hole, and do not consider the feedback on the gravity by scalar Ψ and electromagnetic field A_μ. The charged scalar Ψ is dual to the condensate $< \psi\psi >$ in an usual s-wave superconductor and A_μ is dual to the electric current. In the vicinity of the boundary we have the following asymptotic behavior of the fields.

$$\Psi = \frac{\Psi_1}{r} + \frac{\Psi_2}{r^2} + \dots \quad (7)$$

$$\Phi = A_0 = \mu - \frac{\rho}{r}, \qquad A_x = By + \frac{J_x}{r}, \quad J_x = 0, \quad (8)$$

where the zero-component of electromagnetic field provides the chemical potential μ and charge density of dual theory on the boundary. The dimension of $[< \psi\psi >] = 3, [\Psi] = 1$ and Ψ_2 corresponds to the value of condensate $< \psi\psi >$.

We study the behavior of $J_{\mu\nu} =< \psi\sigma_{\mu\nu}\psi >$ in the presence of external magnetic field $A_x = By$ and introduce an antisymmetric field $B_{\mu\nu}$ that will be a source for charged tensor current $J_{\mu\nu}$. The Lagrangian for antisymmetric field $B_{\mu\nu}$ has the following form

$$\Delta L = |dB - igA\wedge B|^2 - m^2\left|B_{\mu\nu}\right|^2 + \lambda\Psi^\dagger B_{\mu\nu}F^{\mu\nu} + \lambda\Psi B^\dagger_{\mu\nu}F^{\mu\nu}. \quad (9)$$

It is useful to compare the Lagrangian (9) with the Lagrangian for the antisymmetric tensor field considered in the extended holographic model for QCD [7,13]. Remind that the minimal holographic QCD model defined in 5d AdS-like space involves the gauge fields A_L, A_R in $U(N_F)\times U(N_F)$ supplemented by the Chern–Simons terms and massive self-dual rank-two field $B_{mn} = B_{+,mn} + iB_{-,mn}$ and massive complex scalar $X = X_+ + iX_-$ both in the bifundamental representation of the gauge group. The interaction terms in the Lagrangian involving the rank-two term looks as follows

$$L_{int} = \lambda_{QCD} X_\pm F_{V,mn} B^{mn}_\pm, \quad (10)$$

where F_V is the gauge curvature for the vector gauge field $A_V = A_L + A_R$. The non-vanishing coefficient λ_{QCD} in front of triple interaction term XBF implies the non-vanishing magnetic susceptibility χ_F of the quark condensate. It was shown in [6,14] that the non-vanishing susceptibility also follows from the anomaly in the axial current which corresponds to the nontrivial Chern–Simons term in the holographic action. In the PCAC approximation $\chi_F = -\frac{N_c}{4\pi^2 f_\pi^2}$ [14] and one could say that the external field induces the spin polarization of the condensate via the Goldstone modes.

3. Condensate polarization in magnetic field

To calculate the condensate of $< \psi\sigma_{\mu\nu}\psi >$ induced by the external magnetic field we will use the following procedure. First we consider the behavior of the field B_{12} in the pure AdS_4 space without presence of external electromagnetic field. After that we switch on electromagnetic field $F_{12} = B$ and calculate how the solution for B_{12} has changed. The susceptibility can be read off from the asymptotic behavior of this new solution. To make calculations easier we suppose that $r_0 = 0 = T$ and change to the variable $z = \frac{1}{r}$. From the dimensional analysis we choose $m^2 = -\frac{6}{L^2}$ and the equation of motion for B_{12} is

$$\partial_z\left[z^2\partial_z B_{12}\right] + m^2 B_{12} = \lambda\Psi F_{12}, \quad A_0 = \mu - \rho z. \quad (11)$$

This equation has the following solution

$$B_{12} = C_1 z^2 + \frac{C_2}{z^3}. \quad (12)$$

where $C_1 =< \psi\sigma_{12}\psi >$ and C_2 is a source for this operator in the boundary theory. After that we switch on an external magnetic field on the right-hand side of the equation (11)

$$\begin{aligned}
&\Psi = z^2 < \psi\psi > + \mathcal{O}(z^3), \quad z\to 0\\
&F_{12} = B + \mathcal{O}(z^2), \quad z\to 0,\\
&\Psi F_{12} = z^2 B < \psi\psi > + \mathcal{O}(z^3), \quad z\to 0.
\end{aligned} \quad (13)$$

This term modifies the solution in the following way

$$B_{12} = \left(C_1 - \frac{\lambda}{5}B\log z\right)z^2, \quad (14)$$

that we can attribute the new additional term to the non-vanishing magnetic susceptibility of the condensate. It gives us the following expression for susceptibility

$$< \psi\sigma_{\mu\nu}\psi > = -\frac{\lambda}{5}\log\frac{z_{UV}}{z_0} < \psi\psi > F_{\mu\nu}, \quad (15)$$

where z_{UV} is a UV cutoff and z_0 is some IR scale.

The IR scale z_0 entering the logarithm deserves some explanation. There is some natural scale in the holographic approach which yields the scale of the scalar condensate. It is related to the parameter of the gravity solution r_0 however to identify it more precisely it is necessary to perform more refined analysis. On the other hand it is possible to get some intuition concerning this scale if we consider the inhomogeneous external magnetic field.

Hence we look at plane wave for magnetic field

$$F_{12} = B\exp\left(-i\omega t + i\vec{k}\vec{x}\right).$$

The equation of motion for the field $B_{12}(z)$ reads as

$$\partial_z\left[z^2\partial_z B_{12}\right] + (m^2 + p^2z^2)B_{12} = \lambda\Psi B,$$
$$\text{where}\quad p^2 = \omega^2 - k_1^2 - k_2^2, \tag{16}$$

and has the following solution in the limit $z \to 0$

$$B_{12}(z) = -\frac{\lambda}{5}F_{12} < \psi\psi > \log(pa)\,z^2, \quad z \to 0. \tag{17}$$

This yields the following formula for the condensate

$$< \psi\sigma_\mu\psi > = -\frac{\lambda}{5}\log(pa)B_\mu < \psi\psi >, \tag{18}$$

where a is a microscopic scale for superconductor (e.g. an inter-atomic distance) and $\sigma_\mu = \frac{1}{2}\epsilon_{\mu\nu\rho}\sigma_{\nu\rho}$, $B_\mu = \frac{1}{2}\epsilon_{\mu\nu\rho}B_{\nu\rho}$.

4. Why is λ not equal to zero?

In this Section we shall present the semi-qualitative arguments in favor of $\lambda \neq 0$. First, comment on the derivation of the similar constant in QCD. It has been evaluated from the correlator of the tensor and vector currents $< VT >$ in the boundary theory [7,13] which yields at large Q^2 in the massless QCD

$$\int dx e^{iQx} < V_\mu(0)T_{\nu\rho}(x) >$$
$$\propto < \bar\Psi\Psi > (\eta_{\mu\nu}Q_\rho - \eta_{\mu\rho}Q_\nu)Q^{-2}. + O(Q^{-4}), \tag{19}$$

where the Euclidean OPE of two currents is taken into account. The correlator at low virtualities is saturated by the ρ-meson state which has non-vanishing residues both for vector and tensor currents.

On the other hand the same correlator can be evaluated holographically using the standard recipe that is varying the classical action in the bulk theory over the boundary values of the vector and tensor fields

$$< V(0)T(x) > = \frac{\delta^2 S_{cl}}{\delta A(0)\delta B(x)}. \tag{20}$$

The bulk term $\lambda_{QCD}XBF$ contributes and taking into account the boundary behavior of the neutral scalar $X(z) = < \bar\Psi\Psi > z^3 + \ldots$ one gets upon comparison of two expressions for $\frac{<\bar\Psi\Psi>}{Q^2}$ terms

$$\lambda_{QCD} = -\frac{3N_c}{4\pi^2}. \tag{21}$$

In our case similar calculation goes as follows. Consider the correlator $< VT_c >$ once again and hunt for the $\frac{<\psi\psi>}{Q^2}$ term assuming that Q^2 is large enough. The charged tensor current looks as $\psi\sigma\psi$ and two fermion legs can be send to the condensate yielding the non-vanishing contribution from the tree diagram

$$\int dx e^{iQx} < V_\mu(0)T_{\nu\rho}(x) >$$
$$\propto < \psi\psi > (\eta_{\mu\nu}Q_\rho - \eta_{\mu\rho}Q_\nu)Q^{-2} + O(Q^{-4}) \tag{22}$$

At the bulk side we focus at the ΨBF term again and take into account the boundary behavior of the charged scalar $\Psi(z) = z^2 < \psi\psi > + O(z^3)$. Evaluating the bulk action with this boundary condition we get the contribution proportional to the s-wave condensate as well. Equating the leading terms in correlators evaluated in the bulk and in the boundary superconductor we obtain that $\lambda \neq 0$. However at the holographic side we have the additional factor $\log Q^2$ which obstructs the estimate of the numerical value of λ.

Here we present also similar calculation for $3+1$ superconductor when the dual action reads as

$$S = \int d^5x\sqrt{-g}\Big[R + \frac{\sigma}{L^2} - \frac{1}{4}F_{\mu\nu}^2 + |D_\mu X|^2 - m_\Psi^2|X|^2$$
$$+ |dB - iA \wedge B|^2 - m_B^2|B|^2$$
$$+ \lambda X F_{\mu\nu}B^{\dagger,\mu\nu} + \lambda X^\dagger F_{\mu\nu}B^{\mu\nu}\Big],$$
$$ds^2 = \frac{1}{z^2}\left[-dt^2 + dx_i^2 + dz^2\right], \quad A_0 = \mu - \rho z^2, \tag{23}$$

where m_B and m_Ψ are chosen to satisfy tree-level dimensions for dual operators on the boundary

$$z^5\partial_z\left(z^{-3}\partial_z\right)X + m_\Psi^2 X = 0,$$
$$X \sim z^3 < \psi\psi > + zJ_\psi, \quad m_\Psi^2 = 3,$$
$$z\partial_z(z\partial_z)B_{12} + m_B^2 B_{12} = 0,$$
$$B_{12} \sim z^3 < \psi\sigma_{12}\psi > + \frac{J_B}{z^3}, \quad m_B^2 = -9, \tag{24}$$

and we have set the radius of AdS to be $L = 1$. We can calculate the correlator of tensor and electric current $< VT >$ imposing the following boundary conditions for the fields in AdS

$$B_{12} = \frac{J_B(x)}{z^3}, \quad A_1 = J_1(x)z^2, z \to 0. \tag{25}$$

If we assume that $J_1, J_B \sim e^{\pm ipx}$ we get the following equations for B_{12} and A_1 fields

$$z\partial_z\left(z^{-1}\partial_z\right)A_1 + p^2A_1 = 0,$$
$$z\partial_z(z\partial_z)B_{12} + \left(m_B^2 + p^2z^2\right)B_{12} = 0. \tag{26}$$

The solutions can be written as linear combinations of Bessel functions with proper boundary conditions

$$B_{12} = -\frac{\pi p^3}{16}J_B e^{ipx}Y_3(pz), \quad A_1 = \frac{2z}{p}J_B J_1(pz)e^{-ipx}. \tag{27}$$

That yields the following expression for the correlator

$$< V(-p)T(p) > = -\frac{\lambda\pi p_2}{8}p^2\int_0^\infty dz J_1(pz)Y_3(pz)X(z)$$
$$= -\frac{\lambda\pi p_2}{8}p^2\int_0^\infty dw J_1(w)Y_3(w)X(w/p) \approx_{p\gg 1}$$
$$\approx_{p\gg 1} -\frac{\lambda\pi p_2}{8p^2} < \psi\psi > \int_0^\infty dw w^3 J_1(w)Y_3(w). \tag{28}$$

Comparing this answer with (22) we get that $\lambda \neq 0$ for $3+1$ case as well.

5. On the vortical susceptibility of the Cooper condensate

Let us make a few comments concerning the similar response of the s-wave superconductor on the rotation. We introduce the corresponding susceptibility postponing the holographic study for the separate publication.

There is a well-known analogy between an external magnetic field and the rotating frame, manifested in the non-relativistic case by the substitution $e_f\vec{B} \leftrightarrow m\vec{\Omega}$ (see, for example [4]). It is thus natural to introduce a response of the Cooper condensate in superconductor to the rotation which can be parametrized as follows

$$< 0|\Psi\sigma_{\mu\nu}\Psi|0 >= \chi_{s,G} < \Psi\Psi > G_{\mu\nu}. \qquad (29)$$

We treat the rotation via the curvature of an external graviphoton field $G_{\mu\nu}$ and denote the corresponding vortical susceptibility of the s-wave Cooper condensate as $\chi_{s,G}$.

Let us recall that the graviphoton field is introduced as the specific form of the background metric

$$ds^2 = (1 + 2\phi_g)dt^2 - (1 - 2\phi_g)d^2\vec{x} + 2\vec{A}_g d\vec{x}dt. \qquad (30)$$

The gravimagnetic field corresponds to the angular velocity of rotation at small velocity

$$\vec{B}_g \propto \vec{\Omega}, \qquad (31)$$

however at large velocities the relation between the gravimagnetic field and the angular velocity is more complicated.

In the rotating superconductor the magnetic field in the bulk is generated

$$\vec{B} = -\frac{2m\vec{\Omega}}{e}. \qquad (32)$$

As we have shown in the previous Section this magnetic field induces the triplet component in the material with non-vanishing susceptibility. Hence this argument suggests that p-wave component is also generated in the s-wave condensate under rotation. This can be considered in the $2+1$ case where such component can be generated in the plane everywhere besides the droplets where the condensate vanishes.

We postpone the holographic analysis of the rotating case when the angular velocity is introduced via the rotating black hole. The analysis is expected to be parallel to the discussion in [9].

6. Conclusion

In this Letter we discuss in the holographic framework the effect of the external magnetic field on the s-wave superconductor at small temperature. We argue that apart from the conventional Meissner effect there is the possibility of generation of the homogeneous p-wave component in the volume of superconductor due to the polarization of the Cooper condensate. The magnetic and rotational susceptibilities of the Cooper condensate are introduced. We consider the linear approximation when the triplet component is proportional to the external field and argue that the susceptibilities do not vanish.

In the usual setting the external magnetic field influences the volume of superconductor via the Abrikosov vortices and the condensate vanishes at their cores. The effect of condensate polarization we consider seems to have nothing to do with the vortices since we do not assume the vanishing of the s-wave condensate anywhere. The appearance of the p-wave admixture in s-wave superconductor caused by the Rashba term or by antiferromagnetic component seems to be the most related phenomena. It would be also important to fit our observation with discussion in [15–17].

Certainly it is interesting to make the numerical estimates of the magnetic and vortical susceptibilities of s-wave SC and take into account the temperature dependence. It would be also interesting to investigate the susceptibilities of the exciton condensate in the external fields which is more close analogue to QCD case.

We are grateful to David Huse, Alexander Krikun, Phuang Ong and Nicolai Prokof'ev for the useful comments. The work by Fedor Popov has been supported by the RScF grant 16-12-10151. The work of A.G. is supported in part by the grant RFBR-15-02-02092 and Basis Foundation fellowship.

References

[1] L. Gor'kov, E. Rashba, Superconducting 2D system with lifted spin degeneracy: mixed singlet-triplet state, Phys. Rev. Lett. 87 (3) (2001) 037004.

[2] V. Cvetkovic, O. Vafek, Space group symmetry, spin-orbit coupling and the low energy effective Hamiltonian for iron based superconductors, Phys. Rev. B 88 (2013) 134510, arXiv:cond-mat/0103449.

[3] D. Eloy Almeida, R.M. Fernandes, E. Miranda, Induced spin-triplet pairing in the coexistence state of antiferromagnetism and singlet superconductivity: collective modes and microscopic properties, arXiv:1705.01955.

[4] Egor Babaev, Boris Svistunov, Rotational response of superconductors: magneto-rotational isomorphism and rotation-induced vortex lattice, Phys. Rev. B 89 (2014) 104501, arXiv:1311.5418.

[5] B.L. Ioffe, A.V. Smilga, Nucleon magnetic moments and magnetic properties of vacuum in QCD, Nucl. Phys. B 232 (1984) 109, http://dx.doi.org/10.1016/0550-3213(84)90364-X.

[6] A. Gorsky, A. Krikun, Magnetic susceptibility of the quark condensate via holography, Phys. Rev. D 79 (2009) 086015, http://dx.doi.org/10.1103/PhysRevD.79.086015, arXiv:0902.1832 [hep-ph].

[7] S.K. Domokos, J.A. Harvey, A.B. Royston, Successes and failures of a more comprehensive hard wall AdS/QCD, J. High Energy Phys. 1304 (2013) 104, http://dx.doi.org/10.1007/JHEP04(2013)104, arXiv:1210.6351 [hep-th].

[8] A. Gorsky, P.N. Kopnin, A. Krikun, A. Vainshtein, More on the tensor response of the QCD vacuum to an external magnetic field, Phys. Rev. D 85 (2012) 086006, http://dx.doi.org/10.1103/PhysRevD.85.086006, arXiv:1201.2039 [hep-ph].

[9] A. Aristova, D. Frenklakh, A. Gorsky, D. Kharzeev, Vortical susceptibility of finite-density QCD matter, J. High Energy Phys. 1610 (2016) 029, http://dx.doi.org/10.1007/JHEP10(2016)029, arXiv:1606.05882 [hep-ph].

[10] S.A. Hartnoll, A. Lucas, S. Sachdev, Holographic quantum matter, arXiv:1612.07324 [hep-th].

[11] R.G. Cai, L. Li, L.F. Li, R.Q. Yang, Introduction to holographic superconductor models, Sci. China, Phys. Mech. Astron. 58 (6) (2015) 060401, http://dx.doi.org/10.1007/s11433-015-5676-5, arXiv:1502.00437 [hep-th].

[12] J. Zaanen, Y.W. Sun, Y. Liu, K. Schalm, Holo Graphic Duality in Condensed Matter Physics, Cambridge Univ. Press, 2015.

[13] R. Alvares, C. Hoyos, A. Karch, An improved model of vector mesons in holographic QCD, Phys. Rev. D 84 (2011) 095020, http://dx.doi.org/10.1103/PhysRevD.84.095020, arXiv:1108.1191 [hep-ph].

[14] A. Vainshtein, Perturbative and nonperturbative renormalization of anomalous quark triangles, Phys. Lett. B 569 (2003) 187, http://dx.doi.org/10.1016/j.physletb.2003.07.038, arXiv:hep-ph/0212231.

[15] P. Basu, J. He, A. Mukherjee, M. Rozali, H.H. Shieh, Competing holographic orders, J. High Energy Phys. 1010 (2010) 092, arXiv:1007.3480;
Z.Y. Nie, R.G. Cai, X. Gao, H. Zeng, Competition between the s-wave and p-wave superconductivity phases in a holographic model, J. High Energy Phys. 1311 (2013) 087, http://dx.doi.org/10.1007/JHEP11(2013)087, arXiv:1309.2204 [hep-th];
Z.Y. Nie, R.G. Cai, X. Gao, L. Li, H. Zeng, Phase transitions in a holographic $s+p$ model with backreaction, Eur. Phys. J. C 75 (11) (2015) 559, arXiv:1501.00004.

[16] E. Kiritsis, L. Li, Holographic competition of phases and superconductivity, J. High Energy Phys. 1601 (2016) 147, http://dx.doi.org/10.1007/JHEP01(2016)147, arXiv:1510.00020 [cond-mat.str-el].

[17] R.G. Cai, R.Q. Yang, Holographic model for the paramagnetism/antiferromagnetism phase transition, Phys. Rev. D 91 (8) (2015) 086001, http://dx.doi.org/10.1103/PhysRevD.91.086001, arXiv:1404.7737 [hep-th];
R.G. Cai, R.Q. Yang, F.V. Kusmartsev, Holographic model for antiferromagnetic quantum phase transition induced by magnetic field, Phys. Rev. D 92 (2015) 086001, arXiv:1501.04481.

On the entropy associated with the interior of a black hole

Baocheng Zhang

School of Mathematics and Physics, China University of Geosciences, Wuhan 430074, China

ARTICLE INFO

Editor: M. Cvetič

ABSTRACT

The investigation about the volume of a black hole is closely related to the quantum nature of the black hole. The entropy is a significant concept for this. A recent work by Majhi and Samanta (2017) [9] after us presented a similar conclusion that the entropy associated with the volume is proportional to the surface area of the black hole, but the proportionality coefficient is different from our earlier result. In this paper, we clarify the difference and show that their calculation is unrelated to the interior of the black hole.

1. Introduction

Recently, Christodoulou and Rovelli (CR) suggested a definition of "volume" for a collapsed black hole [1], which bridges between the interior of black holes and thermodynamics [2]. The calculation of entropy associated with the volume is vital for building this connection. We firstly calculated such entropy statistically and found that the entropy is proportional to the surface area of the black hole [2]. In particular, the thermodynamics corresponding to the volume entropy is balanced by the contribution from the volume. Thus, the first law of black hole thermodynamics is not modified, unlike the situation in which the cosmological constant is considered [3–5]. We also found that the semi-classical result is not enough to interpret the black hole entropy statistically, which is the motivation for us to continue the calculation at Planck level with the aid of spatial noncommutativity. Thus, we found an example that a black hole has infinite volume but wrapped by finite surface [6], which supports the conjecture [7,8] that the black hole entropy can be independent of the black hole interior.

While Majhi and Samanta reinvestigated this volume entropy after us, they got a different result and pointed out that the difference was due to our "improper" treatments [9]. We have to say that all our "improper" treatments were also adopted in the calculation made by Majhi and Samanta. Firstly, their vital result is the expression of the energy, i.e. Eq. (39) in their paper [9], which derives from the relativistic dispersion relation or the primary constraint in their paper. In fact, this constraint condition is consistent with WKB approximation. In this paper, we will present this point and discuss the feasibility of this approximation although we had discussed this in our earlier paper [2]. The second point is that when we calculated the entropy from the free energy, we didn't make the derivative with respect to the inverse temperature for the CR volume. Actually, we once calculated this term and found that it gives a very small constant, so in our earlier paper, we wrote such words "Temporarily ignoring the exotic feature of the CR volume" when we calculated the entropy statistically from the free energy. In this paper, we will present this point in detail. The final point pointed out by Majhi and Samanta is that the Stefan–Boltzmann law is not valid in the final stage of black hole evaporation, which had been discussed in detail in our earlier paper. Actually, this is the reason that we introduce the spatial noncommutativity to continue to finish the calculation, i.e. see our paper [6]. Similarly, the work made by Majhi and Samanta had the same question and cannot extend to the final stage of evaporation.

So, since their treatment is consistent with ours, why does the expression of the entropy associated with the volume obtained by Majhi and Samanta is different from ours? Because their calculation is not related to the interior (volume) of the Schwarzschild black hole. This is the reason that we decide to write this paper, not only for clarifying our calculation, but also for clarifying the fact that the entropy obtained in Ref. [9] is not that corresponding to the interior of a Schwarzschild black hole as they claimed. Throughout this paper, we use units with $G = c = \hbar = k_B = 1$.

2. CR volume

We start with the CR definition [1] for a collapsed black hole, which is based on finding the maximal space-like hypersurface Σ bounded by a given surface S. For example, given a 2-dimensional sphere S in flat spacetime by $t = 0$ and $r^2 = x^2 + y^2 + z^2 = R^2$, the volume surrounded by the hypersurface $t = t(r)$ becomes

E-mail address: zhangbc.zhang@yahoo.com.

$$V = \int_0^R 4\pi r^2 \sqrt{1 - \left(\frac{dt}{dr}\right)^2}\, dr, \tag{2.1}$$

which gives $V = 4\pi R^3/3$ for a sphere by maximizing the hypersurface with $t =$ constant. A similar discussion can be applied to collapsed matter described by the Eddington–Finkelstein coordinates

$$ds^2 = -f(r)dv^2 + 2dvdr + r^2 d\varphi^2 + r^2 \sin^2\varphi d\phi^2, \tag{2.2}$$

where $f(r) = 1 - 2M/r$ and the advanced time $v = t + \int dr/f(r) = t + r + 2M \ln|r - 2M|$. The collapsed matter forms a Schwarzschild black hole in the end with its event horizon at $r = 2M$ serving as the required surface that bounds many space-like hypersurfaces. Any one hypersurface Σ can be coordinatized by λ, φ, ϕ [1], and the line element of the induced metric on it can be expressed as

$$ds_\Sigma^2 = \left[-f(r)\dot{v}^2 + 2\dot{v}\dot{r}\right] d\lambda^2 + r^2 d\varphi^2 + r^2 \sin^2\varphi\, d\phi^2 \tag{2.3}$$

where the dot represents a partial derivative with regard to the parameter λ, and $-f(r)\dot{v}^2 + 2\dot{v}\dot{r} > 0$ for a spacelike hypersurface. Its volume takes the usual form

$$\begin{aligned} V_\Sigma &= \int d\lambda d\varphi d\phi \sqrt{r^4\left[-f(r)\dot{v}^2 + 2\dot{v}\dot{r}\right]\sin^2\varphi} \\ &= 4\pi \int d\lambda \sqrt{r^4\left[-f(r)\dot{v}^2 + 2\dot{v}\dot{r}\right]}. \end{aligned} \tag{2.4}$$

The maximization is obtained at $r = 3M/2$ by the method of auxiliary manifold [1] or the method of maximal slicing of mathematical relativity [2]. Carrying out the integration with maximization condition [1,10–12], it is found the CR volume at late time

$$V_{\text{CR}} \sim 3\sqrt{3}\pi M^2 v. \tag{2.5}$$

This interesting result shows that volume is determined by advanced time, which depends on the future behavior, e.g. evaporation [13].

3. Entropy associated with CR volume

Based on the above statement about CR volume, we calculated the statistical entropy in this volume with the phase-space [14] labeled by position $\{\lambda, \varphi, \phi\}$ and momentum $\{p_\lambda, p_\varphi, p_\phi\}$. The total number of quantum states arises from integrating $d\lambda d\varphi d\phi dp_\lambda dp_\varphi dp_\phi/(2\pi)^3$ over the complete phase space. The integration is carried out by considering a massless scalar field Φ in the spacetime with the metric

$$ds^2 = -dT^2 + [-f(r)\dot{v}^2 + 2\dot{v}\dot{r}]d\lambda^2 + r^2 d\varphi^2 + r^2 \sin^2\varphi d\phi^2, \tag{3.6}$$

which comes from Eq. (2.2) by the transformation $dv = \frac{-1}{\sqrt{-f}} dT + d\lambda$ and $dr = \sqrt{-f}\, dT$.

But Majhi and Samanta took the phase space labeled by $\{r, v\}$ and momentum $\{p_r, p_v\}$ and considered the particle moving in such a background metric

$$ds^2_{ansatz} = -dt^2 + r^4\left(-f(r)dv^2 + 2dvdr\right). \tag{3.7}$$

They claimed that in such ansatz, the non-static property could be avoided.

The first and most important problem which requires to be clarified: is the metric given in Eq. (3.7) related to the interior of the Schwarzschild black hole? Obviously, it is unrelated. It is noted that the effective metric $ds^2_{eff} = r^4\left(-f(r)dv^2 + 2dvdr\right)$ is only an auxiliary manifold taken for the calculation of the maximization of the volume (2.4) [1]. So it is not the maximal hypersurface. The maximal hypersurface is expressed in Eq. (2.3) with $r = 3M/2$. As for the sphere in flat spacetime, its volume can be defined according to the maximization of Eq. (2.1). When the motion of particles needs to be discussed in this sphere, can it be taken in such background $ds^2_{eff} = r^4\left(-dt^2 + dr^2\right)$? Of course, it can not!

Then, is the metric (3.7) static as claimed by Majhi and Samanta? It is difficult to give a positive answer. In the statistical calculation for the modes of the scalar field, the interior spacetime has to be used. The metric (3.6) is obviously equivalent to the interior metric (2.2), but the metric (3.7) is not. If ones want to used the metric (3.7) to represent the interior of the black hole, a transformation has to be made to connect the metric (3.7) with Eddington–Finkelstein coordinates, which will be found that the coordinate t will not be independent on the coordinates r and v. Thus, the background given by the ansatz is not static. In general, the non-static characteristics is essential for the black hole interior, and cannot be taken out by the transformation of coordinates. So, either the metric (3.7) is not related to the interior of the black hole, which is the case made in Ref. [9], or it is not static, which will lead to the failure of the Hamiltonian analysis. Moreover, the metric (3.7) is strange, because both the coordinates t and v are both timelike.

The second problem is that the azimuthal coordinates and their conjugated momenta were not considered in phase space by Majhi and Samanta. This is improper, since the scalar field is moving in the spacetime including the azimuthal directions. If the azimuthal degrees of freedom is ignored in the construction of phase space, to say the least, how is the integral result 4π in the expression (2.4) of CR volume incorporated into the entropy associated with the volume? In particular, the spherical symmetry is represented by these azimuthal coordinates, which will constrain the possible modes for the scalar field and lead to a smaller statistical result than having no such consideration. This is the reason why our result is smaller than that obtained by Majhi and Samanta.

Now it is clear that the calculation made by Majhi and Samanta is not related to the interior of the black hole and is incomplete even in their ansatz. In what follows, we will clarify our calculation, and at first interpret the feasibility of WKB approximation. The main reason to make the approximation is that our calculation is to be carried out at $v >> M$ and $r = 3M/2$, which is essentially the hypersurface of $T =$ constant and is unaffected by the non-static nature of the metric. Actually, the WKB approximation is held only if the evolution of spacetime is slow enough near the maximal hypersurface. It is better to understand this point by maximal slicing [15–17] in which the evolution avoids the singularity, but the physical process should be equivalent since in essence the complete theory should slice-independent. As well-known, in that case, at the final stage of the hypersurfaces' evolution, a phenomenon called "collapse of the lapse" happens, which means that near the maximal hypersurface, the proper time between two neighboring hypersurfaces tends to zero and no evolution happens there. Thus, the hypersurface is nearly static and nearly no lapse into the next one. So slow evolution is just the requirement by the WKB approximation.

Under WKB approximation, the Klein–Gordon equation in curved spacetime gives the energy expression for the scalar field,

$$E^2 - \frac{1}{-f(r)\dot{v}^2 + 2\dot{v}\dot{r}} p_\lambda^2 - \frac{1}{r^2} p_\varphi^2 - \frac{1}{r^2 \sin^2\varphi} p_\phi^2 = 0, \tag{3.8}$$

which is just the primary constraint $p^2 = 0$ for massless field obtained using the Hamiltonian analysis of Ref. [9], but the Eq. (3.6) while not Eq. (3.7) is taken as the required spacetime metric. With the expression of energy, the number of quantum states and free energy can be calculated with the standard statistical method [14]. So, the entropy associated with the CR volume is expressed as

$$S_{\rm CR} = \beta^2 \frac{\partial F}{\partial \beta} = \frac{\pi^2 V_{\rm CR}}{45\beta^3} - \frac{\pi^2}{180\beta^4}\frac{\partial V_{\rm CR}}{\partial \beta}, \tag{3.9}$$

where the free energy $F = -\frac{\pi^2 V_{\rm CR}}{180\beta^4}$ [2]. With the CR volume in Eq. (2.5) and $\beta = T^{-1} = 8\pi M$, the second term is constant and the value is approximately 10^{-7} which is less and less than the first term. This is the reason why we didn't include the second term in our earlier paper. Here, $\beta = T^{-1} = 8\pi M$ is an important assumption that is guaranteed by the condition that the system is in equilibrium. It is consistent with our calculation with the Stefan–Boltzmann law. This assumption could also be confirmed and understood by the first law of black hole thermodynamics which was given in our earlier paper [2]. As presented above, the calculation of CR volume is made in the case of $v >> M$ in which the black hole has formed by the collapse of the matter and is static[1] for external observers. Thus, ones can construct the first law of thermodynamics, which defines an equilibrium state that means that the inside and outside of the black hole separated by the horizon is in equilibrium. In particular, it is noted that the interior of the black hole is static when our statistical calculation is made at the hypersurface of T = constant. The avoidance of non-static nature make the definition of the temperature in the interior feasible. Since the exterior temperature is the Hawking temperature for the observer at infinity, the interior temperature can also be taken as the Hawking temperature for the same observer by the requirement of equilibrium. In particular, such assumption remains the validity of the first law and provides an interpretation for the vacuum pressure at horizon from the perspective of thermodynamics which will be discussed in the next section.

So the volume entropy is found

$$S_{\rm CR} \sim \frac{3\sqrt{3}}{45 \times 8^3} M^2 = \frac{3\sqrt{3}}{90 \times 8^4 \pi} A, \tag{3.10}$$

where $A = 16\pi M^2$ is the event horizon area for a Schwarzschild black hole.

4. Thermodynamic significance

The entropy $S_{\rm CR}$ Eq. (3.10) associated with $V_{\rm CR}$ remains insufficient to interpret S_H because the prefactor in front of A is much smaller than 1/4, but the thermodynamics associated with the volume has to be considered in the first law [18,19]. In the original calculation of Hawking [13], the concept of particles is used in the asymptotically flat region far away from a black hole where particles can be unambiguously defined. While the particle flux carries away positive energy, an accompany flux of negative energy ones falls into the hole across the horizon, which can only be understood by the zero point fluctuations of the local energy density in a quantum theory. This phenomena is called vacuum polarization [20] which causes a quantum pressure $P = 1/(90 \times 8^4\pi^2 M^4)$ at the horizon [21–24]. It gives $PdV_{\rm CR} \sim 10^{-5}dM$ that nicely balances out $TdS_{\rm CR} \sim 10^{-5}dM$ on the level of 10^{-5} and supports the introduction of the volume $V_{\rm CR}$ thermodynamically interpretable together with the quantum pressure of vacuum polarization. It presents not only the connection of volume with quantum properties of the gravitational field, but also the robustness of the first law of black hole thermodynamics. At the same time, this in turn shows that little information can be leaked out from radiation emission except those about the vacuum polarization and thus indirectly confirms the thermal result of Hawking radiations. But the result of Majhi and Samanta indicated that the first law has to be broken, since the terms from the volume in their expression cannot be balanced out from the two sides of the equation for the first law. Thus, from this perspective of semi-classical calculation, their result is also not credible.

In conclusion, our result about the entropy associated with the CR volume is self-consistent in the semiclassical region, and the reason of the difference from the calculation made by Majhi and Samanta is that they choose an erroneous ansatz for the spacetime metric, in which they calculated an entropy. But its connection with the interior of the Schwarzschild black hole is unjustified.

Acknowledgements

We thank Dr. Wu for reminding us of the paper by Majhi and Samanta. This work is supported by the NSFC with Grant No. 11374330, No. 91636213 and No. 11654001.

References

[1] M. Christodoulou, C. Rovelli, Phys. Rev. D 91 (2015) 064046.
[2] B. Zhang, Phys. Rev. D 92 (2015), 081501(R).
[3] M. Cvetič, G.W. Gibbons, D. Kubizňák, C.N. Pope, Phys. Rev. D 84 (2011) 024037.
[4] B.P. Dolan, Class. Quantum Gravity 28 (2011) 235017.
[5] C. Liang, L. Gong, B. Zhang, Class. Quantum Gravity 34 (2017) 035017.
[6] B. Zhang, Li You, Phys. Lett. B 765 (2017) 226.
[7] T. Jacobson, D. Marolf, C. Rovelli, Int. J. Theor. Phys. 44 (2005) 1807.
[8] M.K. Parikh, Phys. Rev. D 73 (2006) 124021.
[9] B.R. Majhi, S. Samanta, Phys. Lett. B 770 (2017) 314.
[10] I. Bengtsson, E. Jakobsson, Mod. Phys. Lett. A 30 (2015) 1550103.
[11] Y.C. Ong, J. Cosmol. Astropart. Phys. 04 (2015) 003.
[12] Y.C. Ong, arXiv:1503.08245 [gr-qc].
[13] S.W. Hawking, Nature 248 (1974) 30;
S.W. Hawking, Commun. Math. Phys. 43 (1975) 199.
[14] B. Cowan, Topics in Statistical Mechanics, Royal Holloway, Imperial College Press, London, UK, 2005.
[15] F. Estabrook, H. Wahlquist, S. Christensen, B. DeWitt, L. Smarr, E. Tsiang, Phys. Rev. D 7 (1973) 2814.
[16] I. Cordero-Carrió, J.M. Ibáñez, J.A. Morales-Lladosa, J. Math. Phys. (N. Y.) 52 (2011) 112501.
[17] E. Gourgoulhon, arXiv:gr-qc/0703035.
[18] J.M. Bardeen, B. Carter, S.W. Hawking, Commun. Math. Phys. 31 (1973) 161.
[19] Raphael Bousso, Rev. Mod. Phys. 74 (2002) 825.
[20] V.P. Frolov, I.D. Novikov, Black Hole Physics: Basic Concepts and New Developments, Kluwer Academic Publishers, Dordrecht, Netherlands, 1998.
[21] W.G. Unruh, Phys. Rev. D 14 (1976) 870.
[22] P. Candelas, Phys. Rev. D 21 (1980) 2185.
[23] D.N. Page, Phys. Rev. D 25 (1982) 1499.
[24] T. Elster, Phys. Lett. A 94 (1983) 205.

On the flux vacua in F-theory compactifications

Yoshinori Honma [a], Hajime Otsuka [b,*]

[a] National Center for Theoretical Sciences, National Tsing-Hua University, Hsinchu 30013, Taiwan
[b] Department of Physics, Waseda University, Tokyo 169-8555, Japan

ARTICLE INFO

Editor: M. Cvetič

ABSTRACT

We study moduli stabilization of the F-theory compactified on an elliptically fibered Calabi–Yau fourfold. Our setup is based on the mirror symmetry framework including brane deformations. The complex structure moduli dependence of the resulting 4D $\mathcal{N}=1$ effective theory is determined by the associated fourfold period integrals. By turning on appropriate G-fluxes, we explicitly demonstrate that all the complex structure moduli fields can be stabilized around the large complex structure point of the F-theory fourfold.

1. Introduction

String theory compactifications to four dimensional spacetime provide a multitude of massless scalar fields. Unless these extra moduli fields are stabilized, one cannot predict anything for low energy physics including gravity. Moreover, the recent observational data for the acceleration of the universe motivated us to construct de Sitter vacua from a UV-complete quantum theory of gravity. Under these circumstances, the moduli stabilization and comprehensive study of flux vacua have become one of the major topics in string theory.

In the moduli stabilization, determination of the scalar potential of 4D $\mathcal{N}=1$ effective theories arising from spacetime compactifications is of particular interest. In the language of 4D $\mathcal{N}=1$ supersymmetry, there are two kinds of contributions to the scalar potential of moduli fields, namely the Kähler potential and the superpotential. The main problem of string compactifications is how to derive these quantities quantum mechanically from the geometry of internal compact spaces.

On the other hand, the mirror symmetry in string theory is known to be a useful tool to understand exact properties of moduli fields of geometries as first demonstrated to the quintic Calabi–Yau threefold in [1]. As has been explicitly performed in the literature, mirror symmetry can be applied to consider the closed string moduli stabilization. Inclusion of open string sector in the presence of the brane for the compact Calabi–Yau manifolds was initiated in [2] and has been subsequently applied in many contexts. In this framework, a brane is fixed on a specific submanifold and the system does not have a continuous open string moduli dependence. This means that the effective superpotential due to the wrapped branes cannot be evaluated from this kind of undeformed setup.

For the case of compact Calabi–Yau threefolds, the inclusion of brane deformations was first carried out in [3]. By using a Hodge theoretic approach, they computed the brane superpotential depending on both open and closed string moduli. Thereafter, alternative and more efficient methods to evaluate the brane superpotential has been constructed (see [4–7] for details). Remarkably, these generalizations have led to a duality between open string on a threefold with branes and closed string on a fourfold without branes, which can be naturally incorporated into the framework of the F-theory [8].

The F-theory conjecture implies that the physics of Type IIB string compactifications with branes on a complex three-dimensional Kähler manifold can be encoded in the geometry of an elliptically fibered Calabi–Yau fourfold. In contrast to various string compactifications on Calabi–Yau threefolds, the moduli stabilization of F-theory has not been fully established. The aim of this work is to fill this gap by utilizing the mirror symmetry techniques to study the F-theory vacua in the large complex structure limit, where the dynamics of moduli fields has not been investigated explicitly. For other earlier attempts in a similar spirit, see [9–11] initiated the M-theory and F-theory compactifications with G_4 fluxes, [12] investigated the orientifold limit [13,14] of F-theory and [15] based on the K3 × K3 backgrounds.

* Corresponding author.
E-mail addresses: yoshinori.honma255@cts.nthu.edu.tw (Y. Honma), h.otsuka@aoni.waseda.jp (H. Otsuka).

2. F-theory compactifications including fluxes

First we describe basic ingredients for spacetime compactification in the F-theory framework. For more details, we refer the reader to [16].

2.1. F-theory on Calabi–Yau fourfolds

Let us consider a class of 4D $\mathcal{N}=1$ effective theories arising from F-theory compactified on the elliptically fibered Calabi–Yau fourfold $X_4 \to B_3$. Here, B_3 is a complex three-dimensional Kähler base space with positive curvature. This setup can be also regarded as a Type IIB string theory compactified on B_3 with an axio-dilaton which varies over B_3 holomorphically.

In the F-theory perspective, the Kähler potential for complex structure moduli fields in 4D $\mathcal{N}=1$ effective theories can be represented by

$$K = -\ln \int_{X_4} \Omega \wedge \overline{\Omega}, \tag{1}$$

where Ω denotes a holomorphic $(4,0)$-form on X_4. Here and in what follows, we have adopted the reduced Planck unit $M_{\rm Pl} = 2.4 \times 10^{18}$ GeV $= 1$. It is also well-known that F-theory admits a superpotential of the form

$$W = \int_{X_4} G_4 \wedge \Omega, \tag{2}$$

in the presence of non-zero four-form fluxes G_4. This expression is inherited from a duality between F-theory and M-theory [9, 16–18]. To guarantee the compactness of a background, the above G_4 fluxes are required to satisfy the tadpole cancellation condition given by

$$\frac{\chi}{24} = n_{D3} + \frac{1}{2}\int_{X_4} G_4 \wedge G_4, \tag{3}$$

where χ is the Euler characteristic of X_4 and n_{D3} denotes the total charge of the space-time filling D3-branes.

Note that generically the variations of Ω in Calabi–Yau fourfolds do not completely span $H^4(X_4)$ and only its primary horizontal subspace given by

$$H^4_H(X_4, \mathbb{C}) = H^{4,0} \oplus H^{3,1} \oplus H^{2,2}_H \oplus H^{1,3} \oplus H^{0,4} \tag{4}$$

can contribute to the mirror symmetry calculations [19]. Here $H^{2,2}_H$ denotes the elements in $H^{2,2}$ which arise from the second derivatives of Ω with respect to the complex structure moduli of X_4. Correspondingly, the middle dimensional homology basis is also restricted to lie in the primary horizontal subspace of $H_4(X_4)$.

Concerning the dynamics of the Kähler moduli fields in effective theories, it is quite challenging to elicit exact interactions from internal geometry. One of the difficulties in this determination is due to the lack of understanding about the quantum moduli space of hypermultiplets (see [20] for recent developments) and the possibility of its consistent reduction to the 4D $\mathcal{N}=1$ supergravity formulation in general region of the moduli space. Here we simplify the situation and only add a classical term $-2\ln\mathcal{V}$ to the Kähler potential and assume that the radius of the background manifold is sufficiently large so that the classical Kähler moduli space has a no-scale structure [21]. This additional term corresponds to the volume of a background and is in general a nontrivial function of the Kähler moduli and the mobile D3-brane moduli [16]. We assume that $\mathcal{V}$ will be stabilized at a particular constant after the complex structure moduli stabilization, as first demonstrated in [22].

2.2. Moduli dependence

First we will describe general aspects of complex structure moduli dependence in F-theory compactifications. More concrete expressions based on a fixed background will be presented in the next subsection.

For a Calabi–Yau fourfold X_4 with $h^{3,1}(X_4)$ complex structure moduli, the period integrals of holomorphic $(4,0)$-form Ω defined by

$$\Pi_i = \int_{\gamma^i} \Omega \tag{5}$$

encode a closed string moduli dependence of the system. Here, γ^i with $i = 1, \ldots, h^4_H(X_4)$ denote a basis of primary horizontal subspace of $H_4(X_4)$. In terms of these fourfold periods, the Kähler potential for the complex structure moduli (1) can be written as

$$K = -\ln\Big(\sum_{i,j} \Pi_i \eta^{ij} \overline{\Pi}_j\Big), \tag{6}$$

where we have introduced a moduli independent intersection matrix η^{ij} and a dual basis $\hat{\gamma}^i$ in $H^4_H(X_4)$ as

$$\eta^{ij} = \int_{X_4} \hat{\gamma}^i \wedge \hat{\gamma}^j, \quad \int_{\gamma^i} \hat{\gamma}^j = \delta^{ij}. \tag{7}$$

Now we consider turning on a class of G_4 fluxes whose integer quantum numbers are given by

$$n_i = \int_{\gamma^i} G_4. \tag{8}$$

These fluxes generate a superpotential for the complex structure moduli of the form

$$W = \sum_{i,j} n_i \Pi_j \eta^{ij}. \tag{9}$$

Note that our choice of G_4 fluxes (8) only involved with $H^4_H(X_4)$. In general, there exists additional contributions to the system from other subspaces of $H^4(X_4)$ (see e.g. [23]). More rigorous treatment for the couplings arising from these remaining G_4 fluxes would be indispensable for studying the stabilization of Kähler moduli fields.

2.3. Topological data

As a simplest example of a fourfold X_4, we consider an elliptically fibered Calabi–Yau fourfold X_c^* which has been constructed in [4] from the quintic Calabi–Yau threefold with one toric brane (see also [6]). For details about F-theory fourfold construction, we refer the reader to [5] where a general analysis about the mirror pairs for the elliptic Calabi–Yau fourfolds has been clarified. Note that not every Calabi–Yau threefold can be uplifted to the consistent F-theory fourfold background in these prescriptions. As mentioned in [5], the existence of an elliptic fibration structure in the mirror of the underlying threefold is crucial for the fourfold uplifting.

The period integrals (5) for the fourfold X_c^* have been obtained in [4,6] by using toric geometry techniques and the result is

$$\begin{aligned}
&\Pi_1 = 1,\ \Pi_2 = z,\ \Pi_3 = -z_1,\ \Pi_4 = S,\\
&\Pi_5 = 5Sz,\ \Pi_6 = \frac{5}{2}z^2,\ \Pi_7 = 2z_1^2,\ \Pi_8 = -\frac{5}{2}Sz^2 - \frac{5}{3}z^3,\\
&\Pi_9 = -\frac{2}{3}z_1^3,\ \Pi_{10} = -\frac{5}{6}z^3,\ \Pi_{11} = \frac{5}{6}Sz^3 + \frac{5}{12}z^4 - \frac{1}{6}z_1^4,
\end{aligned} \tag{10}$$

where we have ignored further possible corrections to the leading interactions. The complex structure moduli of the fourfold $z, z - z_1, S$ are originated from a bulk quintic modulus, a brane modulus and the axio-dilaton in Type IIB description, respectively.

Note that our definition for the complex structure moduli fields $\{z\}$ deviates from the standard convention also used in [4], where the classical periods are expressed by logarithmic functions of the complex structure deformations. We redefined a logarithm of a standard complex structure modulus as a new single modulus just for later convenience.

The topological intersection matrix has been also clarified in [4] as

$$\eta = \begin{pmatrix} 0 & 0 & 0 & 0 & 1 \\ 0 & 0 & 0 & I_3 & 0 \\ 0 & 0 & \widetilde{\eta} & 0 & 0 \\ 0 & I_3 & 0 & 0 & 0 \\ 1 & 0 & 0 & 0 & 0 \end{pmatrix}, \quad \widetilde{\eta} = \begin{pmatrix} 0 & \frac{1}{5} & 0 \\ \frac{1}{5} & \frac{2}{5} & 0 \\ 0 & 0 & -\frac{1}{4} \end{pmatrix}, \tag{11}$$

and the Euler characteristic of the background is given by $\chi(X_c^*) = 1860$.

3. Illustrative example of moduli stabilization

3.1. Effective theory for moduli fields

Here we describe the explicit form of the 4D $\mathcal{N} = 1$ effective potentials for moduli fields arising from F-theory compactified on X_c^*. Substituting the fourfold data (10) and (11) into (6), one can easily check that the Kähler potential for moduli fields takes a form

$$K = -\ln\left[-i(S - \overline{S})\right] - \ln \widetilde{Y} - 2\ln \mathcal{V}, \tag{12}$$

where

$$\widetilde{Y} = \frac{5i}{6}(z - \bar{z})^3 + \frac{i}{S - \overline{S}}\left(\frac{5}{12}(z - \bar{z})^4 - \frac{1}{6}(z_1 - \bar{z}_1)^4\right), \tag{13}$$

and we have added the simplified Kähler moduli sector. Note that our simplification for Kähler moduli fields does not affect the later discussion about the vacuum structure of F-theory compactifications, as long as the masses of Kähler moduli fields are significantly smaller than the other moduli fields.

Similarly, the superpotential can be written as

$$\begin{aligned} W &= n_{11} + n_{10}S + n_8 z + n_6 S z + \frac{5}{2}\left(\frac{n_5}{5} + \frac{2n_6}{5}\right) z^2 \\ &\quad - \frac{5n_4}{6} z^3 - n_2\left(\frac{5}{2} S z^2 + \frac{5}{3} z^3\right) - n_9 z_1 - \frac{n_7}{2} z_1^2 \\ &\quad - \frac{2n_3}{3} z_1^3 + n_1\left(\frac{5}{6} S z^3 + \frac{5}{12} z^4 - \frac{1}{6} z_1^4\right), \end{aligned} \tag{14}$$

and the tadpole cancellation condition takes a form

$$\begin{aligned} \frac{1860}{24} &= n_{D3} + n_1 n_{11} + n_2 n_8 + n_3 n_9 + n_4 n_{10} \\ &\quad + \left(\frac{n_5 + n_6}{5}\right) n_6 - \frac{n_7^2}{8}. \end{aligned} \tag{15}$$

Obviously, n_7 must be $2 + 4k$ with $k \in \mathbb{Z}$ in order to satisfy the condition (15) while preserving the integrality of flux quanta. In a similar reason, $n_5 + n_6$ or n_6 is constrained to be $5k'$ with $k' \in \mathbb{Z}$.

3.2. F-theory flux vacua

Let us study the extremal conditions of moduli fields $\Phi^I = (z, z_1, S)$. The F-term scalar potential of our 4D $\mathcal{N} = 1$ effective theory for moduli fields has a form

$$V = e^K\left(K^{I\bar{J}} D_I W D_{\bar{J}} \overline{W}\right), \tag{16}$$

where $D_I = \partial_I + (\partial_I K)$ and $K^{I\bar{J}}$ is the inverse of the Kähler metric given by $K_{I\bar{J}} = \partial_I \partial_{\bar{J}} K$. Note that the no-scale structure of the Kähler moduli fields [21] is preserved at the classical level and a term proportional to $-3|W|^2$ in the standard 4D $\mathcal{N} = 1$ formula is canceled. Here we also define

$$F^I \equiv K^{I\bar{J}} D_{\bar{J}} \overline{W}, \tag{17}$$

for later convenience. In this notation, the extremal conditions for moduli fields become

$$\begin{aligned} e^{-K}\frac{\partial V}{\partial \overline{\Phi}^{\bar{I}}} &= \left[K_{\bar{I}} K_{J\bar{L}} - \partial_{\bar{I}} K_{J\bar{L}} + K_J K_{\bar{I}\bar{L}}\right] F^J \overline{F}^{\bar{L}} \\ &\quad + \overline{F}^{\bar{J}} \overline{W}_{\bar{J}\bar{I}} + (K_{\bar{J}\bar{I}} - K_{\bar{J}} K_{\bar{I}}) \overline{F}^{\bar{J}} \overline{W} + K_{J\bar{I}} F^J W = 0. \end{aligned} \tag{18}$$

Here we focus on the self-dual G_4 fluxes satisfying

$$G_4 = *_{X_4} G_4, \tag{19}$$

which correspond to the imaginary self-dual three-form fluxes in Type IIB compactifications. In our model, imposing the self-duality condition is equivalent to set $n_2 = n_3 = n_4 = n_8 = n_9 = n_{10} = 0$. In this setup, our $\mathcal{N} = 1$ effective theory has a solution to the F-term conditions $F^I = 0$, where the scalar potential becomes zero and the values of the moduli fields are fixed as

$$\begin{aligned} &\mathrm{Re}z = \mathrm{Re}z_1 = \mathrm{Re}S = 0, \\ &\mathrm{Im}z = \left(\frac{6n_{11}}{5n_1}\right)^{1/4} \frac{2\sqrt{n_6}}{(8n_6(n_5 + n_6) - 5n_7^2)^{1/4}}, \\ &\mathrm{Im}z_1 = \left(\frac{30n_{11}}{n_1}\right)^{1/4} \frac{\sqrt{n_7}}{(8n_6(n_5 + n_6) - 5n_7^2)^{1/4}}, \\ &\mathrm{Im}S = \left(\frac{6n_{11}}{5n_1}\right)^{1/4} \frac{n_5}{\sqrt{n_6}(8n_6(n_5 + n_6) - 5n_7^2)^{1/4}}. \end{aligned} \tag{20}$$

For example, there exists a Minkowski vacuum with $n_{D3} = 0$ in the following choice of non-zero G_4 fluxes:

$$n_1 = 1, n_5 = 15, n_6 = 10, n_7 = 2, n_{11} = 28. \tag{21}$$

The values of the moduli fields (20) in this vacuum are

$$\begin{aligned} &\mathrm{Re}z = \mathrm{Re}z_1 = \mathrm{Re}S = 0, \\ &\mathrm{Im}z \simeq 2.28, \ \ \mathrm{Im}z_1 \simeq 1.14, \ \ \mathrm{Im}S \simeq 1.71, \end{aligned} \tag{22}$$

and the vacuum expectation value of the superpotential becomes $W \simeq -72.97$. One can easily confirm that the mass eigenvalues of moduli fields are positive definite as

$$\mathcal{V}^{-2}(91.30,\ 35.05,\ 3.94,\ 2.96,\ 0.09,\ 0.07), \tag{23}$$

which means that all the complex structure moduli have been completely stabilized.

4. Conclusions

It has been known that the effective superpotential and the axio-dilaton dependence of Type IIB compactifications can be reformulated into a geometry and fluxes in F-theory. Meanwhile, exact calculations in such a situation has been studied in a framework of mirror symmetry with or without branes. In this work, we have shown that topological data extracted by mirror symmetry techniques can be directly applied to the F-theory compactifications. Especially we have demonstrated that all the complex structure moduli can be stabilized around the large complex structure point of F-theory fourfold.

Throughout this work, we have only focused on classical interactions of moduli fields. This means that we have not fully utilized the power of mirror symmetry and further quantum corrections to the effective couplings can be also easily calculated. It would be interesting to study the vacuum structure of F-theory including these corrections, which can be also computed as in [24].

Moreover, it would be fascinating to check whether the Kähler moduli can be stabilized as in the LARGE Volume Scenario [25] or the scenario of the Kachru, Kallosh, Linde and Trivedi [22], once our treatment of the Kähler moduli sector is extended.

Acknowledgements

We would like to thank K. Choi, C. S. Chu and H. Hayashi for useful discussions and comments. Y. H. was supported in part by the grant MOST-105-2119-M-007-018, 106-2119-M-007-019 from the Ministry of Science and Technology of Taiwan. H. O. was supported in part by Grant-in-Aid for Young Scientists (B) (No. 17K14303) from Japan Society for the Promotion of Science.

References

[1] P. Candelas, X.C. De La Ossa, P.S. Green, L. Parkes, A pair of Calabi–Yau manifolds as an exactly soluble superconformal theory, Nucl. Phys. B 359 (1991) 21.

[2] J. Walcher, Opening mirror symmetry on the quintic, Commun. Math. Phys. 276 (2007) 671.

[3] H. Jockers, M. Soroush, Effective superpotentials for compact D5-brane Calabi–Yau geometries, Commun. Math. Phys. 290 (2009) 249.

[4] M. Alim, M. Hecht, H. Jockers, P. Mayr, A. Mertens, M. Soroush, Hints for off-shell mirror symmetry in type II/F-theory compactifications, Nucl. Phys. B 841 (2010) 303.

[5] T.W. Grimm, T.W. Ha, A. Klemm, D. Klevers, Computing brane and flux superpotentials in F-theory compactifications, J. High Energy Phys. 1004 (2010) 015.

[6] H. Jockers, P. Mayr, J. Walcher, On N=1 4d effective couplings for F-theory and heterotic vacua, Adv. Theor. Math. Phys. 14 (5) (2010) 1433.

[7] T.W. Grimm, A. Klemm, D. Klevers, Five-brane superpotentials, blow-up geometries and $SU(3)$ structure manifolds, J. High Energy Phys. 1105 (2011) 113.

[8] C. Vafa, Evidence for F theory, Nucl. Phys. B 469 (1996) 403.

[9] K. Becker, M. Becker, M theory on eight manifolds, Nucl. Phys. B 477 (1996) 155.

[10] K. Dasgupta, G. Rajesh, S. Sethi, M theory, orientifolds and G-flux, J. High Energy Phys. 9908 (1999) 023.

[11] S. Gukov, C. Vafa, E. Witten, CFT's from Calabi–Yau four folds, Nucl. Phys. B 584 (2000) 69, Erratum: Nucl. Phys. B 608 (2001) 477.

[12] F. Denef, M.R. Douglas, B. Florea, A. Grassi, S. Kachru, Fixing all moduli in a simple f-theory compactification, Adv. Theor. Math. Phys. 9 (6) (2005) 861.

[13] A. Sen, F theory and orientifolds, Nucl. Phys. B 475 (1996) 562.

[14] A. Sen, Orientifold limit of F theory vacua, Phys. Rev. D 55 (1997) R7345.

[15] P. Berglund, P. Mayr, Non-perturbative superpotentials in F-theory and string duality, J. High Energy Phys. 1301 (2013) 114.

[16] F. Denef, Les Houches Lectures on constructing string vacua, arXiv:0803.1194 [hep-th].

[17] S. Sethi, C. Vafa, E. Witten, Constraints on low dimensional string compactifications, Nucl. Phys. B 480 (1996) 213.

[18] M. Haack, J. Louis, M theory compactified on Calabi–Yau fourfolds with background flux, Phys. Lett. B 507 (2001) 296.

[19] B.R. Greene, D.R. Morrison, M.R. Plesser, Mirror manifolds in higher dimension, Commun. Math. Phys. 173 (1995) 559.

[20] S. Alexandrov, J. Manschot, D. Persson, B. Pioline, Quantum hypermultiplet moduli spaces in N=2 string vacua: a review, Proc. Symp. Pure Math. 90 (2015) 181.

[21] S.B. Giddings, S. Kachru, J. Polchinski, Hierarchies from fluxes in string compactifications, Phys. Rev. D 66 (2002) 106006.

[22] S. Kachru, R. Kallosh, A.D. Linde, S.P. Trivedi, De Sitter vacua in string theory, Phys. Rev. D 68 (2003) 046005.

[23] A.P. Braun, T. Watari, The vertical, the horizontal and the rest: anatomy of the middle cohomology of Calabi–Yau fourfolds and F-theory applications, J. High Energy Phys. 1501 (2015) 047.

[24] Y. Honma, M. Manabe, Exact Kahler potential for Calabi–Yau fourfolds, J. High Energy Phys. 1305 (2013) 102.

[25] V. Balasubramanian, P. Berglund, J.P. Conlon, F. Quevedo, Systematics of moduli stabilisation in Calabi–Yau flux compactifications, J. High Energy Phys. 0503 (2005) 007.

Origin of the Drude peak and of zero sound in probe brane holography

Chi-Fang Chen *, Andrew Lucas *

Department of Physics, Stanford University, Stanford, CA 94305, USA

ARTICLE INFO

Editor: M. Cvetič

ABSTRACT

At zero temperature, the charge current operator appears to be conserved, within linear response, in certain holographic probe brane models of strange metals. At small but finite temperature, we analytically show that the weak non-conservation of this current leads to both a collective "zero sound" mode and a Drude peak in the electrical conductivity. This simultaneously resolves two outstanding puzzles about probe brane theories. The nonlinear dynamics of the current operator itself appears qualitatively different.

1. Introduction

One of the earliest applications of the AdS/CFT correspondence to condensed matter physics was to study holographic "probe branes" at finite density [1,2]. The holographic dual of such models is (in the simplest cases) widely believed to be $\mathcal{N}=2$ supersymmetric fundamental matter (analogous to quarks), localized on a defect within the $\mathcal{N}=4$ supersymmetric Yang–Mills plasma [3]. Because electrical transport in strongly interacting quantum systems remains a challenging problem in condensed matter, much of the work on these probe brane models focuses on the transport of the conserved U(1) baryon number. Unfortunately, the probe limit, where the background plasma is unaffected by the dynamics of the baryon matter, leads to certain simplifying features of transport that are absent in more "realistic" holographic models for strange metals [4,5].

It is still important to understand the transport properties of the probe brane models, however; they remain rare examples of solvable interacting quantum systems in higher dimensions. And at low temperature, the behavior of probe branes has remained rather mysterious for almost a decade. Firstly, at low temperatures one often finds a collective, propagating "sound mode" [6–10]. This cannot be ordinary sound, as the energy-momentum tensor is dominated by the 'decoupled' $\mathcal{N}=4$ plasma; it also cannot be (superfluid) second sound as the U(1) symmetry is not broken. From this line of thought, [6] subsequently concluded that this propagating mode was analogous to zero sound: the sloshing of the Fermi surface in a Fermi liquid at low temperature [11]. While there is no strong evidence for a well-defined baryonic Fermi surface, save for finite momentum spectral weight [12], we will follow the literature and call this propagating mode zero sound. Furthermore, when the zero sound waves are present, the low frequency electrical conductivity $\sigma(\omega)$ has an apparent Drude peak at low temperature [13]. Such a Drude peak would normally be associated with approximate conservation of momentum [14,5], but as we have already mentioned, that cannot be the case in probe brane models.

We demonstrate below that within linear response in probe brane holography, the charge current operator itself appears to be conserved at zero temperature when the dynamical critical exponent z of the background plasma obeys $z<2$. This emergent conservation law, and the resulting "hydrodynamics", is responsible for the Drude singularity in the electrical conductivity, as well as the propagation of the zero sound mode. At small but finite temperature, the charge current decays at a rate $\sim T^{2/z}$; this decay is responsible for both the breakdown of zero sound modes as well as the broadening of the Drude peak. We also emphasize that this low temperature hydrodynamics is distinct from the high temperature hydrodynamics of probe branes, which is conventional, and describes a single diffusive mode for charge.

This mechanism is analogous to the behavior of electrical conductivity [5,14,15] and ordinary sound waves [16] in a normal fluid with weak momentum relaxation; such similarity was qualitatively observed before [17]. Here, of course, the conserved momentum is replaced by the charge current operator itself. Unlike the hydrodynamics of ordinary fluids, we find that the nonlinear hydrodynamics of the current operator is often ill-posed: the gradient expansion always fails at low enough temperatures. Thus, as we resolve the mysteries of the zero sound modes and the Drude conductivity which arise within linear response, our work also calls

* Corresponding authors.
E-mail addresses: chifangc@stanford.edu (C.-F. Chen), ajlucas@stanford.edu (A. Lucas).

into question the ultimate fate of the zero sound mode, and of electrical transport more broadly, at the nonlinear level.

2. Probe branes at finite density

Consider a large N quantum field theory (QFT) in d spatial dimensions with a conserved U(1) current J^μ. If this QFT has a holographic dual, then J^μ is dual to a bulk gauge field A^a ($ab\cdots$ denote bulk indices; $\mu\nu\cdots$ denote boundary indices). For probe brane models, the generating functional for correlators of J^μ is simply the exponential of the Dirac–Born–Infeld (DBI) action for A^a:

$$S = \mathcal{K}\int \mathrm{d}^{d+2}x\sqrt{-\det\left(g_{ab}+2\pi\alpha' F_{ab}\right)} \tag{1}$$

where $F = \mathrm{d}A$, g_{ab} is the bulk spacetime metric and α' is the string tension, a dimensionful constant arising from the string theory interpretation of the holographic dual [5]. The constant $\mathcal{K}$ is related to the brane tension and the volumes of any compact wrapped spaces, and is unimportant for us. In this paper, we consider background metrics of "Lifshitz" form [18]:

$$\mathrm{d}s^2 = \frac{L^2}{r^2}\left[\frac{\mathrm{d}r^2}{f(r)} - \frac{f(r)}{r^{2z-2}}\mathrm{d}t^2 + \mathrm{d}x^i\mathrm{d}x_i\right] \tag{2}$$

The parameter z is called the dynamical critical exponent, and is related to the relative scaling of time and space in the dual critical theory; we restrict to theories with $z \geq 1$ [5]. The function $f(r)$ encodes a finite temperature T, and is given by

$$f(r) = 1 - \left(\frac{r}{r_\mathrm{h}}\right)^{d+z} \tag{3}$$

with

$$r_\mathrm{h}^z = \frac{d+z}{4\pi T}. \tag{4}$$

In the probe brane limit, g_{ab} is a fixed, non-dynamical background field.

If we are interested in studying matter at finite density, then we must look for saddle points of the action (1). These are solutions to

$$\partial_a\left(\sqrt{-\det(g+2\pi\alpha' F)}(g+2\pi\alpha' F)^{[ab]}\right) = 0. \tag{5}$$

In the above equation, $(g+2\pi\alpha' F)^{ab}$ refers to components of the matrix inverse of $g_{ab}+2\pi\alpha' F_{ab}$.

3. Hydrodynamics of the conserved current

Now, let us look for a solution to (5) in which ρ is a slowly varying function of the boundary theory coordinates x^μ. In other words, we perform a gradient expansion, keeping track of terms only to lowest order in the number of x^μ-derivatives (r-derivatives, denoted with $'$, will not be treated as perturbatively small). Working in radial gauge $A_r = 0$, we find that the μ-components of (5) give

$$\left(\frac{1}{\mathcal{L}}\left(\frac{L}{r}\right)^{2d} A_t'\right)' = \mathcal{O}\left(\partial_\mu^2\right), \tag{6a}$$

$$\left(\frac{1}{\mathcal{L}}\left(\frac{L}{r}\right)^{2d}\frac{f}{r^{2z-2}} A_i'\right)' = \mathcal{O}\left(\partial_\mu^2\right), \tag{6b}$$

where

$$\mathcal{L} = \frac{L^{d+2}}{r^{d+1+z}}\sqrt{1-\frac{(2\pi\alpha')^2}{L^4}\left[r^{2+2z}A_t'^2 - fr^4 A_x'^2\right] + \mathcal{O}\left(\partial_\mu^2\right)} \tag{7}$$

Note that either the presence of finite temperature T, or $z > 1$, breaks the symmetry between t and x^i. These equations can be exactly solved by

$$A_t' = -\rho(x^\mu)\frac{r^{2d}\mathcal{L}}{L^{d+2}}, \tag{8a}$$

$$A_i' = J_i(x^\mu)\frac{r^{2d}\mathcal{L}}{L^{d+2}}\frac{r^{2z-2}}{f}, \tag{8b}$$

where we can also write

$$\mathcal{L} = \frac{L^{d+2}}{r^{d+1+z}}\frac{1}{\sqrt{1+C^2\left(\rho^2 - \frac{r^{2z-2}}{f}(J^i)^2\right)r^{2d}}} \tag{9}$$

with

$$C^2 = \frac{(2\pi\alpha')^2}{L^4}. \tag{10}$$

Here $\rho(x^\mu)$ and $J^i(x^\mu)$ can be interpreted as the charge density and charge current in the dual theory through a rescaling of $\mathcal{K}$; we will assume this henceforth.

So far, this solution is "exact". However, it is clear that if $z > 1$ or $T > 0$, the nonlinear solutions with any finite J^i are *not well-posed*. This will lead to the breakdown of hydrodynamics at nonlinear order, but we defer this discussion to Section 3.4. At $T = 0$, and $z = 1$, we recover the "boost" solutions of [19], whose existence is demanded by Lorentz covariance.

3.1. Linear response

To proceed farther, let us assume that J^i is infinitesimally small, and only consider first order terms in J^i. This is a rather artificial limit to ensure that the solution (8) exists. However, this linear response regime is precisely where both zero sound and the Drude peak are observed, and so a careful understanding of this regime is sufficient to understand these phenomena. Thus, we proceed. We will see that within this linear response limit, we should treat J^i and $\partial_\mu\rho$ as infinitesimal quantities, and so they need only be kept to linear order.

Let us first perform the integral over r in A_t':

$$\begin{aligned} A_t(r) &= A_t^0 - \rho\int_0^r \mathrm{d}s\frac{s^{d-1-z}}{\sqrt{1+C^2\rho^2 s^{2d}}} \\ &= A_t^0 - C^{-1+z/d}\rho^{z/d}\mathcal{F}_1\left((C\rho)^{1/d}r\right) \end{aligned} \tag{11}$$

where A_t^0 is a constant of integration, physically dual to a background gauge field coupled to the current operator J^μ in the dual theory, and

$$\mathcal{F}_1(x) \equiv \int_0^x \mathrm{d}y\frac{y^{d-z-1}}{\sqrt{1+y^{2d}}}. \tag{12}$$

Implicitly, of course, ρ and J^i depend on x^μ, and we will drop the explicit dependence henceforth. For large x, we find the asymptotic expansion

$$\mathcal{F}_1(x) = c_1 - \frac{b_1}{x^z} + \cdots \tag{13}$$

with positive coefficient

$$c_1 = \frac{1}{2\sqrt{\pi}d}\Gamma\left(\frac{d-z}{2d}\right)\Gamma\left(\frac{z}{2d}\right) \tag{14}$$

and $b_1 > 0$; this will come in handy soon. Note that A_t^0 is not arbitrary, and should be chosen so that A_t vanishes on the horizon [5].

We now perform the r integral in A_i':

$$\begin{aligned} A_i(r) &= A_i^0 + J^i \int_0^r \mathrm{d}s \frac{s^{d+z-3}}{f(s)\sqrt{1+C^2\rho^2 s^{2d}}} \\ &= A_i^0 + J_i (C\rho)^{(2-z-d)/d} \mathcal{F}_2\left((C\rho)^{1/d} r\right) \\ &+ J_i \int_0^r \frac{\mathrm{d}s}{\sqrt{1+C^2\rho^2 s^{2d}}} s^{d+z-3}\left(\frac{s}{r_\mathrm{h}}\right)^{d+z}\left[1-\left(\frac{s}{r_\mathrm{h}}\right)^{d+z}\right]^{-1} \end{aligned} \tag{15}$$

where

$$\mathcal{F}_2(x) = \int_0^x \mathrm{d}y \frac{y^{d+z-3}}{\sqrt{1+y^{2d}}}. \tag{16}$$

Depending on the value of z, $\mathcal{F}_2$ has qualitatively different behavior. For $z < 2$, this integral is convergent and one finds

$$\mathcal{F}_2(x) = c_2 - \frac{b_2}{x^{2-z}} + \cdots \tag{17}$$

with

$$c_2 = \frac{1}{2\sqrt{\pi}d}\Gamma\left(\frac{d+z-2}{2d}\right)\Gamma\left(\frac{2-z}{2d}\right) \tag{18}$$

In the limit of low temperatures ($(C\rho)^{1/d} r_\mathrm{h} \gg 1$), the second term of (15) has a logarithmic divergence near the horizon, and so for $r \approx r_\mathrm{h}$ we cannot neglect the second term:

$$\begin{aligned} &\int_0^r \frac{\mathrm{d}s}{\sqrt{1+C^2\rho^2 s^{2d}}} s^{d+z-3}\left(\frac{s}{r_\mathrm{h}}\right)^{d+z}\left[1-\left(\frac{s}{r_\mathrm{h}}\right)^{d+z}\right]^{-1} \\ &\approx \frac{1}{C\rho}\int_0^r \mathrm{d}s \frac{s^{d+2z-3}}{r_\mathrm{h}^{d+z} - s^{d+z}} + \mathcal{O}\left(\frac{T^{2d/z}}{\rho^2}\right) \\ &\approx \frac{r_\mathrm{h}^{z-2}}{C\rho}\log\frac{r_*}{r_\mathrm{h}-r} + \text{constant}. \end{aligned} \tag{19}$$

The coefficient r_* is a constant which is not important. For $z \geq 2$, $\mathcal{F}_2$ diverges at large x. The coefficient of the logarithmic term in (19) also diverges in this limit. Thus when $z \geq 2$ our gradient expansion fails, and we will explain what happens in this limit briefly in Section 3.3.

Not every A_t given by (11) and A_x given by (15) is a solution of (5). We now must go to first order in the gradient expansion to find the physically allowed solutions. Using the r-component of (5), we obtain

$$0 = -\partial_t\left(\frac{1}{\mathcal{L}}\left(\frac{L}{r}\right)^{2d} A_t'\right) + \partial_i\left(\frac{f}{\mathcal{L}r^{2z-2}}\left(\frac{L}{r}\right)^{2d} A_i'\right). \tag{20}$$

Using (8) we immediately find

$$\partial_t \rho + \partial_i J^i = 0, \tag{21}$$

which implies that the expectation value of the current operator J^μ is conserved in the dual theory. This is of course a physical requirement, and not surprising.

A more subtle point is that the presence of a logarithmic divergence in A_i at any finite T would lead to a breakdown of our linear response theory. However, this is an artifact of the approximation (6). In fact, we will show that when $r \to r_\mathrm{h}$, the time derivatives in the x^i-components of (6) cannot be neglected. So we must treat the near horizon region more carefully: in this region we must replace (6b) with

$$\left(\frac{1}{\mathcal{L}}\left(\frac{L}{r}\right)^{2d}\frac{f}{r^{2z-2}}A_i'\right)' = \partial_t\left(\frac{1}{\mathcal{L}}\left(\frac{L}{r}\right)^{2d}\frac{\partial_t A_i - \partial_i A_t}{f}\right). \tag{22}$$

It is easiest to proceed by defining a gauge-invariant quantity

$$\mathcal{E}_i = \partial_i A_t - \partial_t A_i, \tag{23}$$

which obeys an approximate near-horizon equation of motion

$$\left(\frac{1}{\mathcal{L}}\left(\frac{L}{r}\right)^{2d}\frac{f}{r^{2z-2}}\mathcal{E}_i'\right)' \approx \partial_t\left(\frac{1}{\mathcal{L}}\left(\frac{L}{r}\right)^{2d}\frac{\partial_t \mathcal{E}_i}{f}\right). \tag{24}$$

We have dropped subleading contributions in $f \sim r_\mathrm{h} - r$. In particular, using (20), we observe that A_t' is generally small compared to A_x' near the horizon, which justifies taking a time derivative of (22) to obtain (24).

To linear order in $\mathcal{E}_i$ and $\partial_t \mathcal{E}_i$, we may simplify this equation in the near-horizon limit $r \approx r_\mathrm{h}$:

$$r_\mathrm{h}^{2-2z} f\left(f\mathcal{E}_i'\right)' \approx \partial_t^2 \mathcal{E}_i. \tag{25}$$

Near the horizon,

$$f(r) \approx \frac{d+z}{r_\mathrm{h}}(r_\mathrm{h} - r) + \cdots. \tag{26}$$

Defining

$$R \equiv \frac{1}{4\pi T}\log\frac{r_\mathrm{h}}{r_\mathrm{h}-r}, \tag{27}$$

and using (4) and (26), (25) becomes

$$\partial_R^2 \mathcal{E}_i = \partial_t^2 \mathcal{E}_i. \tag{28}$$

The solution obeying the physical boundary conditions (falling into the black hole) can be written as a Fourier transform

$$\mathcal{E}_i \approx \int \mathrm{d}\omega\, H_i(\omega) \mathrm{e}^{\mathrm{i}\omega(R-t)}. \tag{29}$$

Note that $\mathcal{E}_i$ is completely regular near the horizon.

Following [20], let us now consider – for fixed R – the limit where $H_i(\omega)$ only has support for vanishingly small frequencies. In this limit, we may Taylor expand (29) in the near-horizon limit:

$$\mathcal{E}_i(r,t,x^i) \approx H_i(t,r,x^i) - \frac{\partial_t H_i(t,r,x^i)}{4\pi T}\log\frac{r_\mathrm{h}}{r_\mathrm{h}-r} + \mathcal{O}\left(\partial_t^2\right). \tag{30}$$

This formula for $\mathcal{E}_i$ is valid for $r_\mathrm{h}\exp[-4\pi T/\omega] \lesssim r_\mathrm{h} - r \lesssim r_\mathrm{h}$. In the limit where $\omega \to 0$, the regime of validity of (11) and (15) is $r_\mathrm{h} - r \gtrsim \mathrm{O}(\omega)$.[1] Therefore, we can obtain a non-trivial constraint on the dynamics by demanding that the values of $\mathcal{E}_i$ obtained using (11) and (15) are consistent with (30). In particular, consider the approximation

$$\begin{aligned} \mathcal{E}_i \approx{}& \partial_i A_t^0 - \partial_t A_i^0 - \frac{c_1}{(C\rho)^{1-z/d}}\frac{z}{d}\partial_i\rho - \frac{c_2}{(C\rho)^{(d+z-2)/d}}\partial_t J_i \\ &- \frac{r_\mathrm{h}^{z-2}}{C\rho}\partial_t J_i \log\frac{r_*}{r_\mathrm{h}-r} \end{aligned} \tag{31}$$

[1] The precise prefactor can depend on the value of z.

We have dropped all terms which are subleading in powers of T in the above expression, for simplicity. Comparing (30) to (31) we conclude that

$$\partial_i A_t^0 - \partial_t A_i^0 - \frac{c_1}{(C\rho)^{1-z/d}} \frac{z}{d} \partial_i \rho - \frac{c_2}{(C\rho)^{(d+z-2)/d}} \partial_t J_i = 4\pi T \frac{r_h^{z-2}}{C\rho} J_i + q_i(x) \tag{32}$$

where $q(x)$ is a t-independent function. In physical circumstances, we must have $q_i = 0$ – consider sources which are switched off at $t = -\infty$, when the fluid is at rest. Then we clearly have $q_i = 0$, which will continue to hold for all times, even if we begin to turn on non-trivial A_μ^0.

Recognizing the source terms as simply the externally applied electric field E_i, we conclude that the ideal linearized hydrodynamics on probe branes is

$$\partial_t \rho + \partial_i J^i = 0, \tag{33a}$$

$$\partial_t J^i + v^2 \partial_i \rho = \chi E^i - \frac{J^i}{\tau}, \tag{33b}$$

where the decay rate of the weakly non-conserved J^i is

$$\frac{1}{\tau} = \frac{(4\pi T)^{2/z}}{(d+z)^{(2-z)/z} c_2 (C\rho)^{(2-z)/d}}, \tag{34}$$

the current–current susceptibility is

$$\chi = (C\rho)^{(d+z-2)/d}, \tag{35}$$

and the speed of sound is

$$v^2 = \frac{zc_1}{dc_2 (C\rho)^{(2-2z)/d}} \tag{36}$$

in agreement with [9]. Note that if $z = 1$, $c_1 = c_2$ and $v^2 = 1/d$ [6].

(33) is our main result. It demonstrates that the linearized low temperature hydrodynamics in probe brane models is mathematically equivalent to the hydrodynamics of an ordinary fluid with weak momentum relaxation [5,16]. This proves a "conjecture" of [5,17]. It will now be straightforward to see the emergence of the Drude peak and zero sound, and how both are linked to the same emergent conservation law.

3.2. The Drude peak

We begin with the Drude peak in the conductivity. To compute the electrical conductivity, we apply a spatially homogeneous time-dependent electric field $E_i(\omega)e^{-i\omega t}$. Using (33) together with Ohm's law:

$$J_i(\omega) = \sigma(\omega) E_i(\omega) \tag{37}$$

we obtain

$$\sigma(\omega) = \frac{\chi\tau}{1 - i\omega\tau}. \tag{38}$$

This functional form is called the Drude peak, and was numerically observed in [13]. The temperature dependence of τ is consistent with that found in [10,13]. (38) is consistent with the predictions of the memory matrix formalism [5,14,15], when the charge current itself is an almost conserved quantity.

3.3. Zero sound

Next, we turn to the zero sound modes. Looking for ρ and J^i proportional to $e^{ikx-i\omega t}$ which solve (33) when $E_i = 0$, we immediately find the dispersion relation

$$\omega\left(\omega + \frac{i}{\tau}\right) = v^2 k^2. \tag{39}$$

As already advertised, the speed of zero sound agrees with previous analytic results [6,9], and the finite temperature decay rate is consistent with the $z = 1$, $d = 3$ numerics of [10]. While our derivation above implicitly assumed that $\omega \ll T$, so that the Taylor expansion (30) was justified, we observe that the zero sound speed is not sensitive to whether $T = 0$ or $T > 0$, as numerically found in [10].

Let us now briefly turn to the fate of zero sound when $z > 2$. This was described in [9], and we repeat the result:

$$\omega = \eta k^{2z/(z+2)} + \cdots \tag{40}$$

with η a complex-valued coefficient with $\mathrm{Im}(\eta) < 0$. There are two important features of this result. Firstly, the fact that η is complex implies that the zero sound modes are no longer truly long-lived collective excitations; they are strongly damped. Secondly, for $z > 2$, this dispersion relation is not analytic in k, and this is associated directly with the breakdown of the gradient expansion that we observed in Section 3.1. The real-space equations governing these strongly damped modes will be non-local, in contrast to (33). When $z = 2$, the dispersion relation is of the form $k \sim \omega\sqrt{\log\omega}$, and so is also nonlocal; furthermore, the zero sound will be a purely dissipative mode as $\omega \to 0$.

3.4. Nonlinear dynamics

We now turn to the fate of this hydrodynamics at the nonlinear level. While we focus on the case $z = 1$ for simplicity, our comments are valid for $z > 1$ as well. We begin by discussing the $T = 0$ limit, and analyze the nonlinear corrections to the linearized equations of motion (6). (7) is replaced by

$$\mathcal{L} = \frac{L^{d+2}}{r^{d+2}} \sqrt{1 + C^2 r^4 A'_\mu A^{\mu\prime} + C^2 r^4 \partial_{[\mu} A_{\nu]} \partial^{[\mu} A^{\nu]}} + O(a_\mu^3). \tag{41}$$

Let us now consider a small but finite perturbation a_μ around the background

$$\bar{A}_i = 0, \quad \bar{A}_t = \int_r^\infty dr \frac{\rho r^{d-2}}{\sqrt{1 + C^2 \rho^2 r^{2d}}}; \tag{42}$$

thus $A_\mu = \bar{A}_\mu + a_\mu$. On the background solution, $1 - C^2 r^4 \bar{A}_t'^2 \sim r^{-2d}$ at large r. So following [2], analysis of (5) within linear response at $T = 0$ yields the following linear differential equation for a_μ at large r:

$$\left(r^2 a'_\mu\right)' + \partial^\nu \left(r^2 (\partial_\nu a_\mu - \partial_\mu a_\nu)\right) \approx 0, \tag{43}$$

which is approximately solved by

$$a_\mu \sim \mathrm{Re} \int d^{d+1} q \, \mathcal{A}_\mu(q) \frac{e^{iq\cdot x - \sqrt{q^2} r}}{r} \tag{44}$$

for $q_\mu \mathcal{A}^\mu = 0$, in the long wavelength limit. Unfortunately, a straightforward check reveals that for certain solutions $\mathcal{L}$ becomes imaginary on this ansatz. In particular,

$$C^2 r^4 a'_\mu a^{\mu\prime} + C^2 r^4 \partial_{[\mu} a_{\nu]} \partial^{[\mu} a^{\nu]} \propto r \tag{45}$$

as $r \to \infty$, with a coefficient of arbitrary sign. At $T = 0$, the geometry extends to $r = \infty$, and so we conclude that a non-perturbative correction to a_μ must be made in order for the dynamics to be well-posed: without such a correction, the argument of the square root in (41) is not always positive.[2] We stress that there is almost certainly a solution to the nonlinear equations of motion for A_μ which is real-valued; however, such a solution must necessarily differ non-perturbatively from the zero sound waves in the IR. As such, we do not expect (33) to be correct – even qualitatively – beyond linear response. We expect that the resulting equations of motion for J^μ in the boundary theory are nonlocal in space and time.

At finite temperature, we find a cure for the divergence observed in (45). In this limit, the geometry truncates at $r = r_\mathrm{h} < \infty$. Very close to the horizon, the dominant nonlinearities in $\mathcal{L}$ for small but finite amplitudes a_μ are (at $z = 1$)

$$\mathcal{L} = \frac{L^{d+2}}{r^{d+2}} \times \sqrt{1 + C^2 r^4 \left[-a_t'^2 - \frac{(\partial_t a_i - \partial_i a_t - f a_i')(\partial_t a_i - \partial_i a_t + f a_i')}{f} + \partial_{[i} a_{j]} \partial^{[i} a^{j]} \right]} + \mathrm{O}\left(a_\mu^3\right). \tag{46}$$

Upon first glance, there is a term above proportional to $1/f$ that diverges at the horizon. However, following our discussion near (29), we observe that the infalling boundary conditions are given by

$$f a_x' - \partial_t a_x + \partial_x a_t = 0, \tag{47}$$

and this removes the divergence at the horizon in (46). This cancellation generalizes to $z > 1$. Hence, we expect that the resulting theory of nonlinear zero sound waves is better behaved. Still, we do not know if the resulting nonlinear corrections to (33) admit a sensible hydrodynamic interpretation. In particular, following the same logic as (45), we observe that the amplitude of $\partial_\nu a_\mu$ at which the nonlinearities will qualitatively change the nature of the dynamics vanishes as $T \to 0$.

The breakdown of nonlinear hydrodynamics in the probe brane models at $T = 0$ appears analogous to the fact that the hydrodynamic gradient expansion in the Einstein–Maxwell holographic theory becomes singular as $T \to 0$, despite "appearances" that the mean free path is finite even at $T = 0$ [22]. In fact, in the Einstein–Maxwell system, one finds hydrodynamic sound and diffusion poles (for a conventional charged fluid) even at $T = 0$ [23, 24], but also observes non-analytic corrections to the gradient expansion at a finite subleading order. Such non-analytic corrections also arise in probe brane models, at least when $z > 1$ [9]. It is not clear whether the breakdown of the gradient expansion at subleading orders in derivatives is related to the failure of the perturbative expansion for small perturbations. More work to resolve these puzzles is warranted.

4. Conclusion

At low temperature, the total charge current is a long-lived quantity in holographic probe brane models with $z \le 2$. Solving the bulk equations of motion in a derivative expansion, we found an emergent hydrodynamics of the current operator, analogous to the response of weakly disordered fluids with almost conserved momentum. This hydrodynamics is responsible for both the holographic zero sound mode and the resulting Drude peak. In particular, this proves that the decay of zero sound at finite temperature is governed by the same decay rate as the Drude peak. This also leads to a curious example of a quantum field theory with two different "hydrodynamic" limits: one at high temperature, and a qualitatively different one at low temperature.

The nonlinear generalization of this novel low temperature hydrodynamics does not appear to be well-behaved. It is possible that the full, nonlinear equations of motion for the conserved current J^μ are non-local on the longest length scales. It will be interesting to determine the fate of zero sound at the nonlinear level, and the resulting nonlinear equations of motion for J^μ.

Acknowledgements

We thank Andreas Karch, Richard Davison and Sean Hartnoll for useful discussions. CFC was supported by the Physics/Applied Physics/SLAC Summer Research Program for undergraduates at Stanford University. AL was supported by the Gordon and Betty Moore Foundation's EPiQS Initiative through Grant GBMF4302.

References

[1] S. Kobayashi, D. Mateos, S. Matsuura, R.C. Myers, R.M. Thomson, Holographic phase transitions at finite baryon density, J. High Energy Phys. 02 (2007) 016, arXiv:hep-th/0611099.
[2] A. Karch, A. O'Bannon, Metallic AdS/CFT, J. High Energy Phys. 09 (2007) 024, arXiv:0705.3870.
[3] A. Karch, E. Katz, Adding flavor to AdS/CFT, J. High Energy Phys. 06 (2002) 043, arXiv:hep-th/0205236.
[4] J. Zaanen, Y. Liu, Y-W. Sun, K. Schalm, Holographic Duality in Condensed Matter Physics, Cambridge University Press, 2016.
[5] S.A. Hartnoll, A. Lucas, S. Sachdev, Holographic quantum matter, arXiv:1612.07324.
[6] A. Karch, D.T. Son, A.O. Starinets, Zero sound from holography, Phys. Rev. Lett. 102 (2009) 051602, arXiv:0806.3796.
[7] M. Kulaxizi, A. Parnachev, Comments on Fermi liquid from holography, Phys. Rev. D 78 (2008) 086004, arXiv:0808.3953.
[8] M. Kulaxizi, A. Parnachev, Holographic responses of fermion matter, Nucl. Phys. B 815 (2009) 125, arXiv:0811.2262.
[9] C. Hoyos-Badajoz, A. O'Bannon, J.M.S. Wu, Zero sound in strange metallic holography, J. High Energy Phys. 09 (2010) 086, arXiv:1007.0590.
[10] R.A. Davison, A.O. Starinets, Holographic zero sound at finite temperature, Phys. Rev. D 85 (2012) 026004, arXiv:1109.6343.
[11] D. Pines, P. Nozières, The Theory of Quantum Liquids, Volume I, W. A. Benjamin, 1966.
[12] R.J. Anantua, S.A. Hartnoll, V.L. Martin, D.M. Ramirez, The Pauli exclusion principle at strong coupling: holographic matter and momentum space, J. High Energy Phys. 03 (2013) 104, arXiv:1210.1590.
[13] S.A. Hartnoll, J. Polchinski, E. Silverstein, D. Tong, Towards strange metallic holography, J. High Energy Phys. 04 (2010) 120, arXiv:0912.1061.
[14] S.A. Hartnoll, D.M. Hofman, Locally critical umklapp scattering and holography, Phys. Rev. Lett. 108 (2012) 241601, arXiv:1201.3917.
[15] A. Lucas, S. Sachdev, Memory matrix theory of magnetotransport in strange metals, Phys. Rev. B 91 (2015) 195122, arXiv:1502.04704.
[16] A. Lucas, Sound waves and resonances in electron-hole plasma, Phys. Rev. B 93 (2016) 245153, arXiv:1604.03955.
[17] R.A. Davison, B. Goutéraux, Momentum dissipation and effective theories of coherent and incoherent transport, J. High Energy Phys. 01 (2015) 039, arXiv:1411.1062.
[18] S. Kachru, X. Liu, M. Mulligan, Gravity duals of Lifshitz-like fixed points, Phys. Rev. D 78 (2008) 106005, arXiv:0808.1725.
[19] A. Karch, A. O'Bannon, E. Thompson, The stress-energy tensor of flavor fields from AdS/CFT, J. High Energy Phys. 04 (2009) 021, arXiv:0812.3629.
[20] A. Lucas, Conductivity of a strange metal: from holography to memory functions, J. High Energy Phys. 03 (2015) 071, arXiv:1501.05656.
[21] S. Bhattacharyya, V.E. Hubeny, S. Minwalla, M. Rangamani, Nonlinear fluid dynamics from gravity, J. High Energy Phys. 02 (2008) 045, arXiv:0712.2456.

[2] One might ask whether this breakdown of the nonlinear dynamics could be cured by switching to infalling boundary coordinates, as in the conventional fluid-gravity correspondence [21]. Such a coordinate choice will not alleviate the problem here, however, because the metric is not dynamical.

[22] N. Banerjee, J. Bhattacharya, S. Bhattacharyya, S. Dutta, R. Loganayagam, P. Surówka, Hydrodynamics of charged black branes, J. High Energy Phys. 01 (2011) 094, arXiv:0809.2596.

[23] M. Edalati, J.I. Jottar, R.G. Leigh, Holography and the sound of criticality, J. High Energy Phys. 10 (2010) 058, arXiv:1005.4075.

[24] R.A. Davison, A. Parnachev, Hydrodynamics of cold holographic matter, J. High Energy Phys. 06 (2013) 100, arXiv:1303.6334.

Rigorous constraints on the matrix elements of the energy–momentum tensor

Peter Lowdon, Kelly Yu-Ju Chiu, Stanley J. Brodsky

SLAC National Accelerator Laboratory, Stanford University, 2575 Sand Hill Rd, CA 94025, USA

ARTICLE INFO

Editor: B. Grinstein

Keywords:
Energy–momentum tensor
Form factor
Anomalous gravitomagnetic moment

ABSTRACT

The structure of the matrix elements of the energy–momentum tensor play an important role in determining the properties of the form factors $A(q^2)$, $B(q^2)$ and $C(q^2)$ which appear in the Lorentz covariant decomposition of the matrix elements. In this paper we apply a rigorous frame-independent distributional-matching approach to the matrix elements of the Poincaré generators in order to derive constraints on these form factors as $q \to 0$. In contrast to the literature, we explicitly demonstrate that the vanishing of the anomalous gravitomagnetic moment $B(0)$ and the condition $A(0) = 1$ are independent of one another, and that these constraints are not related to the specific properties or conservation of the individual Poincaré generators themselves, but are in fact a consequence of the physical on-shell requirement of the states in the matrix elements and the manner in which these states transform under Poincaré transformations.

1. Introduction

The matrix elements of the energy–momentum tensor $T^{\mu\nu}$ are important measures of the non-perturbative structure of quantum field theories (QFTs). In particular, the form factors associated with these matrix elements have played a central role in the discussion of the spin structure of hadrons. By decomposing the angular momentum operator J^i between hadronic spin states, sum rules are derived which attempt to connect the total angular momentum of the hadron with the spin and orbital angular momentum of its constituents [1–9]. In doing so, the form factors at zero momentum become related to one another, and are interpreted as angular momentum observables. However, it is well known that the derivation of these sum rules are beset with technical difficulties, such as the construction of well-defined normalisable hadronic states, the handling of boundary terms, and the consistent definition of the Poincaré charges [6].

Axiomatic approaches to QFT provide an analytic framework from which one can analyse operator matrix elements and thus rigorously address the issues surrounding the derivation of these sum rules and their effect on the properties of the form factors. These approaches involve constructing a QFT via the definition of a series of physically motivated axioms [10–14] such as locality and relativistic covariance of the fields. Perhaps the most significant axiom is that quantised fields are operator-valued distributions, not functions. An operator-valued distribution φ is a continuous linear functional which maps (test) functions $f \in \mathcal{T}$ to operators $\varphi[f]$ that act on the space of states $\mathcal{H}$. In these formulations of QFT the space of test functions $\mathcal{T}$ is chosen to be $\mathcal{S}(\mathbb{R}^{1,3})$, the space of *Schwartz functions*[1] defined on Minkowski spacetime $\mathbb{R}^{1,3}$, and the fields φ are so-called *tempered distributions*. The operator $\varphi[f]$ has an integral representation: $\varphi[f] = \int d^4x\, \varphi(x) f(x)$, which gives meaning to the x-dependent field expression $\varphi(x)$. Although this representation is often convenient to use in calculations it should be treated carefully, since in general $\varphi(x)$ need not be continuous (e.g. the Dirac delta $\delta(x)$). From a physical perspective this corresponds to the fact that $\varphi(x)$ is not a well-defined operator, since $\varphi(x)$ would represent the performance of a measurement at a single spacetime point x, and this would require an infinite amount of energy [11].

Another important property of distributions, which will play a central role in the calculations in this paper, is the definition of differentiation. Given a distribution φ and test function f, one defines the derivative φ' of the distribution by $\varphi'[f] := -\varphi[f']$. Since the derivative of a test function is also a test function, this definition implies that the derivative of a distribution always exists, in contrast to functions. In this sense distributions represent

E-mail addresses: lowdon@slac.stanford.edu (P. Lowdon), yujuchiu@stanford.edu (K.Y.-J. Chiu), sjbth@slac.stanford.edu (S.J. Brodsky).

[1] Schwartz functions are functions of *rapid decrease* [15].

a generalisation of the space of functions.[2] In this paper we will demonstrate that by taking these distributional subtleties into account, one can avoid the potential inconsistencies that arise in the context of the spin sum rules and also shed new light on the behaviour of the form factors of the energy–momentum tensor.

The rest of the paper is structured as follows: first we outline the general analytic properties of the matrix elements and associated form factors of the energy–momentum tensor; using these results we then develop a novel method to constrain these form factors, and subsequently apply this method to the matrix elements of both the energy–momentum P^{μ} and angular momentum J^{i} operators. Finally, we conclude by summarising our key findings.

2. The form factors of the energy–momentum tensor

In order to understand the structure of the matrix elements involving the energy–momentum tensor, it is important to first outline how definite momentum eigenstates are defined in axiomatic formulations of QFT. Unlike ordinary states, $|p\rangle$ is not normalisable[3] and therefore cannot be an element of the space of states $\mathcal{H}$. Intuitively this makes sense, since quantum uncertainty would imply that $|p\rangle$ is completely delocalised. This feature is encoded by the requirement that quantised fields are operator-valued distributions, and hence it follows that definite momentum eigenstates must be distribution-valued states [13]. In order to create a normalisable state $|\Psi_{M}^{g}\rangle$ one must therefore smear $|p\rangle$ with a test function $g \in \mathcal{S}(\mathbb{R}^{1,3})$. Moreover, to guarantee that this state is physical, and hence on shell, one requires that the test function g is non-vanishing only on the upper hyperboloid $\Gamma_{M}^{+} = \{p^2 = M^2, p^0 > 0\}$, where M is the mass of the physical state. Defining $\delta_{M}^{(+)}(p) := 2\pi\theta(p^0)\delta(p^2 - M^2)$, this state can then be written in the form

$$|\Psi_{M}^{g}\rangle = \int \frac{d^4p}{(2\pi)^4}\,\delta_{M}^{(+)}(p)g(p)|p\rangle = \int \frac{d^3p}{(2\pi)^3 2p^0}\,g(p)\big|_{\Gamma_{M}^{+}}|p\rangle, \tag{1}$$

where $p^0 = \sqrt{\mathbf{p}^2 + M^2}$. From the second equality it is clear that $g(p)\big|_{\Gamma_{M}^{+}}$ is actually the QFT definition of a wavepacket. For the purpose of the calculations in this paper we will consider physical spin-$\frac{1}{2}$ hadronic momentum eigenstates $|p; m; M\rangle := \delta_{M}^{(+)}(p)|p; m\rangle$ with mass M and rest frame spin projection $m \in \left\{\frac{1}{2}, -\frac{1}{2}\right\}$ in the z-direction.[4] By explicitly including the factor $\delta_{M}^{(+)}(p)$ in the definition of $|p; m; M\rangle$, this ensures that these states only have support on the positive mass shell [13]. The inner product of these states is then defined in the following Lorentz-covariant manner:

$$\langle p'; m'; M|p; m; M\rangle = (2\pi)^4\delta^4(p' - p)\delta_{M}^{(+)}(p')\delta_{m'm} \tag{2}$$

Eq. (2) follows immediately from the standard definition of the norm[5] of $|p; m\rangle$.

Now that the momentum eigenstates have been defined, one can consistently characterise the form factors associated with the energy–momentum tensor $T^{\mu\nu}$. By using the various symmetries satisfied by $T^{\mu\nu}$, the matrix elements of spin-$\frac{1}{2}$ hadronic momentum eigenstates (with mass M) can be written[6] as follows [3]:

$$\begin{aligned}
&\langle p'; m'; M|T^{\mu\nu}(0)|p; m; M\rangle \\
&\quad = \bar{u}_{m'}(p')\bigg[\frac{1}{4}\gamma^{\{\mu}(p + p')^{\nu\}}A(q^2) \\
&\quad + \frac{1}{8M}(p + p')^{\{\mu}i\sigma^{\nu\}\rho}q_{\rho}B(q^2) \\
&\quad + \frac{1}{M}\left(q^{\mu}q^{\nu} - q^2 g^{\mu\nu}\right)C(q^2)\bigg]u_m(p)\,\delta_{M}^{(+)}(p)\delta_{M}^{(+)}(p'),
\end{aligned} \tag{3}$$

where $q = p' - p$, and u_m is the hadronic spinor. The $\delta_{M}^{(+)}$ factors reflect the fact that each state consistent with the normalisation in Eq. (2) explicitly involves this factor in their definition. One can rewrite Eq. (3) by applying the Gordon identity, and one obtains

$$\begin{aligned}
&\langle p'; m'; M|T^{\mu\nu}(0)|p; m; M\rangle \\
&\quad = \bar{u}_{m'}(p')\bigg[\frac{1}{8M}(p + p')^{\{\mu}(p + p')^{\nu\}}A(q^2) \\
&\quad + \frac{1}{8M}(p + p')^{\{\mu}i\sigma^{\nu\}\rho}q_{\rho}\left[A(q^2) + B(q^2)\right] \\
&\quad + \frac{1}{M}\left(q^{\mu}q^{\nu} - q^2 g^{\mu\nu}\right)C(q^2)\bigg]u_m(p)\,\delta_{M}^{(+)}(p)\delta_{M}^{(+)}(p').
\end{aligned} \tag{4}$$

An important consequence of the distributional nature of the hadronic states $|p; m; M\rangle$ is that the form factors $A(q^2)$, $B(q^2)$ and $C(q^2)$ are distributions, not functions. Although this feature can immediately be seen from Eq. (2), which is clearly a distribution in the two variables p' and p, this detail has largely been overlooked in the literature. The distributional nature of form factors implies that these objects are not in general point-wise defined [10]. Nevertheless, form factors $F(q^2)$ are seemingly measured at specific values of q^2. In order to reconcile these perspectives one must recognise that one cannot ever physically measure a form factor at a specific value of q^2, since this would require an experiment with infinite precision. In practice, a measurement of $F(q^2)$ at $q^2 = Q^2$ is really a measurement of an averaged-out quantity $\overline{F}(Q^2; \Delta)$ in some small (but non-vanishing) region $\left[Q^2 - \Delta, Q^2 + \Delta\right]$. Theoretically, this is described by the fact that one must integrate Eq. (3) with test functions in the variables p and p' in order to yield a finite result, since the definite momentum eigenstates are not physical states in $\mathcal{H}$. The smearing in p and p' subsequently implies a smearing in q, and this smooths out the form factors. One ends up with expressions $\overline{F}(Q^2; \Delta) := (F * f_{\Delta})(Q^2)$ which involve the convolution of the distribution F with some test function f_{Δ}, where Δ is related to the finite width associated with the wavepackets of the physical states. Since the convolution of a tempered distribution with a test function is always a smooth (infinitely differentiable) function [15], this explains why $\overline{F}(Q^2; \Delta)$ is always point-wise defined in Q^2.

Since the form factor decomposition explicitly involves the factor $\delta_{M}^{(+)}(p)\delta_{M}^{(+)}(p')$, it follows that the distribution $\langle p'; m'; M|T^{\mu\nu}(0)|p; m; M\rangle$ has support for $(p', p) \in \Gamma_{M}^{+} \times \Gamma_{M}^{+}$, and therefore $A(q^2)$, $B(q^2)$ and $C(q^2)$, as defined in Eq. (3), are restricted to have support for $q^2 \leq 0$. Another thing to note about Eq. (3) is that it involves the product of distributions, which by contrast to the product of functions, is generally ill defined. Nevertheless, under certain conditions it is possible to define the product of distributions in a consistent manner [13], and this is in fact the case for the product of $\delta^4(p' - p)$ and $\delta_{M}^{(+)}(p')$ in Eq. (2). We will assume here that the distributional products in Eq. (3) are similarly well defined, and are therefore also commutative, associative, and satisfy the Leibniz rule for derivatives.

Now that the structure and the distributional nature of the matrix elements of the energy–momentum tensor has been outlined, we will demonstrate in the following sections that one can obtain

[2] This explains why distributions are also often referred to as *generalised functions*.

[3] States of this form are often referred to as *improper states*.

[4] We assume here that $|p; m\rangle$ is a canonical spin state, as defined in Ref. [6].

[5] The norm of the improper states $|p; m\rangle$ is defined by $\langle p'; m'|p; m\rangle = (2\pi)^3 2p^0\delta^3(\mathbf{p}' - \mathbf{p})\delta_{m'm}$.

[6] Here we define $\sigma^{\mu\nu} = \frac{i}{2}[\gamma^{\mu}, \gamma^{\nu}]$, and $a^{\{\mu}b^{\nu\}} = a^{\mu}b^{\nu} + a^{\nu}b^{\mu}$.

both model and frame-independent constraints on these form factors by decomposing the matrix elements $\langle p'; m'; M|P^{\mu}|p; m; M\rangle$ and $\langle p'; m'; M|J^{i}|p; m; M\rangle$ in terms of these form factors, and then comparing these decompositions with the expressions obtained after the explicit action of the Poincaré generators P^{μ} and J^{i}.

3. The momentum matrix element

The calculation of the matrix element $\langle p'; m'; M|P^{\mu}|p; m; M\rangle$ requires one to first define the operator P^{μ} from $T^{\mu\nu}$. The standard definition of the energy–momentum operator is: $P^{\mu} = \int d^3x\, T^{0\mu}(x)$. However, this expression is ill-defined for several reasons.[7] First, since $T^{\mu\nu}$ is composed of quantised fields, it follows that it is also an operator-valued tempered distribution, and it therefore must necessarily be smeared with test functions in order to define a consistent operator P^{μ}. Moreover, since these test functions belong to the space $\mathcal{S}(\mathbb{R}^{1,3})$, the integral in the definition must be performed over both space and time. In order to solve this problem, one can define the energy–momentum operator as follows [16,17]:

$$P^{\mu} = \lim_{\substack{d\to 0\\ R\to\infty}} \int d^4x\, f_{d,R}(x) T^{0\mu}(x), \tag{5}$$

where $f_{d,R}(x) := \alpha_d(x_0) F_R(\mathbf{x}) \in \mathcal{S}(\mathbb{R}^{1,3})$, and the test functions α_d, F_R satisfy the conditions

$$\int dx_0\, \alpha_d(x_0) = 1, \qquad \alpha_d(x_0) \xrightarrow{d\to 0} \delta(x_0), \tag{6}$$

$$F_R(0) = 1, \qquad F_R(\mathbf{x}) \xrightarrow{R\to\infty} 1. \tag{7}$$

Not only does this definition guarantee that P^{μ} is convergent within matrix elements, but it also ensures [16] that P^{μ} is independent of the specific choice of test functions used in the limit. Using the definition in Eq. (5), one can write[8]

$$\begin{aligned}\langle p'; m'; M|P^{\mu}|p; m; M\rangle &= \lim_{\substack{d\to 0\\ R\to\infty}} \int d^4x\, f_{d,R}(x) e^{iq\cdot x} \langle p'; m'; M|T^{0\mu}(0)|p; m; M\rangle \\ &= \lim_{\substack{d\to 0\\ R\to\infty}} \widehat{f}_{d,R}(q) \langle p'; m'; M|T^{0\mu}(0)|p; m; M\rangle. \end{aligned}\tag{8}$$

$\widehat{f}_{d,R}(q)$ is the Fourier transform of $f_{d,R}(x)$, which due to the properties of $\mathcal{S}(\mathbb{R}^{1,3})$ is also a test function. Moreover, it follows from the conditions in Eqs. (6) and (7) that

$$\lim_{\substack{d\to 0\\ R\to\infty}} \widehat{f}_{d,R}(q) = (2\pi)^3 \delta^3(\mathbf{q}). \tag{9}$$

Using the definition of the norm in Eq. (2), and the fact that $|p; m; M\rangle$ is a momentum eigenstate, one has

$$\langle p'; m'; M|P^{\mu}|p; m; M\rangle = p^{\mu}(2\pi)^4\delta^4(p'-p)\delta_M^{(+)}(p')\delta_{m'm}. \tag{10}$$

Since Eqs. (8) and (10) are equivalent representations of $\langle p'; m'; M|P^{\mu}|p; m; M\rangle$, one can equate these expressions and use the form factor decomposition in Eq. (4) to derive distributional constraints on $A(q^2)$, $B(q^2)$ and $C(q^2)$. Doing so gives

$$\begin{aligned}\delta_M^{(+)}(p')\Bigg[&\lim_{\substack{d\to 0\\ R\to\infty}} \widehat{f}_{d,R}(q)\, \bar{u}_{m'}(p') \bigg\{ \frac{1}{8M}(p+p')^{\{0}(p+p')^{\mu\}} A(q^2) \\ &+ \frac{1}{8M}(p+p')^{\{0} i\sigma^{\mu\}\rho} q_{\rho} \left[A(q^2)+B(q^2)\right] \\ &+ \frac{1}{M}\left(q^0 q^{\mu} - q^2 g^{0\mu}\right) C(q^2) \bigg\} u_m(p) \delta_M^{(+)}(p) \\ &- p^{\mu}(2\pi)^4 \delta^4(p'-p)\delta_{m'm} \Bigg]_{p'\in\Gamma_M^+} = 0, \end{aligned}\tag{11}$$

which under the restriction that $p' \in \Gamma_M^+$ implies the equality

$$\begin{aligned}&\lim_{\substack{d\to 0\\ R\to\infty}} \widehat{f}_{d,R}(q)\, \bar{u}_{m'}(p') \bigg\{ \frac{1}{8M}(p+p')^{\{0}(p+p')^{\mu\}} A(q^2) \\ &\quad + \frac{1}{8M}(p+p')^{\{0} i\sigma^{\mu\}\rho} q_{\rho} \left[A(q^2)+B(q^2)\right] \\ &\quad + \frac{1}{M}\left(q^0 q^{\mu} - q^2 g^{0\mu}\right) C(q^2) \bigg\} u_m(p) \delta_M^{(+)}(p) \\ &= p^{\mu}(2\pi)^4 \delta^4(p'-p)\delta_{m'm}. \end{aligned}\tag{12}$$

Because the form factors depend only on the variable q, it is convenient to transform to the variables $\bar{p} = \frac{1}{2}(p'+p)$ and q. Under the restriction that $p' \in \Gamma_M^+$ one can write

$$\delta_M^{(+)}(p) = 2\pi\theta(\bar{p}^0)\frac{1}{2\bar{p}^0}\delta\left(q^0 - \frac{\bar{\mathbf{p}}\cdot\mathbf{q}}{\bar{p}^0}\right). \tag{13}$$

Substituting Eq. (13) into Eq. (12) then gives

$$\begin{aligned}&\lim_{\substack{d\to 0\\ R\to\infty}} 2\pi \widehat{f}_{d,R}(q)\delta\left(q^0 - \frac{\bar{\mathbf{p}}\cdot\mathbf{q}}{\bar{p}^0}\right) \bar{u}_{m'}\left(\bar{p}+\tfrac{1}{2}q\right) \bigg\{ \frac{1}{2M}\bar{p}^{\mu} A(q^2) \\ &\quad + \frac{1}{8M\bar{p}^0}\bar{p}^{\{0} i\sigma^{\mu\}\rho} q_{\rho}\left[A(q^2)+B(q^2)\right] \\ &\quad + \frac{1}{2M\bar{p}^0}\left(q^0 q^{\mu} - q^2 g^{0\mu}\right) C(q^2) \bigg\} u_m\left(\bar{p}-\tfrac{1}{2}q\right) \\ &= \left(\bar{p}^{\mu} - \tfrac{1}{2}q^{\mu}\right)(2\pi)^4\delta^4(q)\delta_{m'm}. \end{aligned}\tag{14}$$

Since $q^{\mu}\delta^4(q) = 0$, the right-hand side of Eq. (14) reduces to the two variable (tensor product) distribution $(2\pi)^4\delta_{m'm}\,\bar{p}^{\mu}\delta^4(q)$. Once the limit in d and R is taken, it follows from Eq. (9) that $\widehat{f}_{d,R}(q)\,\delta\left(q^0 - \frac{\bar{\mathbf{p}}\cdot\mathbf{q}}{\bar{p}^0}\right)$ tends towards the distribution $(2\pi)^3\delta^4(q)$. Using the distributional identity $h(q)\delta^4(q) = h(0)\delta^4(q)$, which holds for any infinitely differentiable function h, the various terms on the left-hand-side of Eq. (14) all end up with a different dependence on $\bar{p}$. In particular, the first term reduces to $(2\pi)^4\delta_{m'm}\,\bar{p}^{\mu} A(q^2)\delta^4(q)$. Since this is the only term which depends on $\bar{p}$ in the same manner as the right-hand side, one can equate these expressions to obtain the (distributional) constraint

$$A(q^2)\delta^4(q) = \delta^4(q). \tag{15}$$

The first thing to note with Eq. (15) is that the left-hand side involves a product of distributions, which due to the assumptions discussed in the previous section is well defined. By representing $\delta^4(q)$ as a limit of test functions $\delta_n^{\{0\}}(q)$, Eq. (15) can be written

$$\lim_{n\to\infty} \int d^4q\, \delta_n^{\{0\}}(q) A(q^2) = 1, \tag{16}$$

---

[7] This issue has been emphasised before in several different contexts [9,12,14,16,17].

[8] Here we have used the standard result: $T^{0\mu}(x) = e^{iP\cdot x}T^{0\mu}(0)e^{-iP\cdot x}$. However, in order for this relation to make rigorous sense one must define what one means by the operator $T^{0\mu}(0)$. $T^{0\mu}(0)$ cannot literally correspond to the $x \to 0$ limit of $T^{0\mu}(x)$ because $T^{0\mu}(x)$ is a (operator-valued) distribution, and it is therefore not point-wise defined. Instead, by $T^{0\mu}(x)$ one implicitly means the limit $n \to \infty$ of the convolution $\delta_n^{\{x\}} * T^{0\mu}$, where $\delta_n^{\{x\}}$ are a sequence of test functions whose support tend towards $\{x\}$ when $n \to \infty$. One then has the well-defined relation: $\delta_n^{\{x\}} * T^{0\mu} \equiv e^{iP\cdot x}T^{0\mu}[\delta_n^{\{0\}}]e^{-iP\cdot x}$ whose limit tends towards the intuitive result. Due to the continuity of $T^{0\mu}$ it then follows that the charge P_n^{μ} constructed using $\delta_n^{\{x\}} * T^{0\mu}$ converges to the definition of P^{μ} in Eq. (5) for $n \to \infty$.

which is a rigorous formulation of the well-known result $A(0) = 1$.

Since each of the remaining terms on the left-hand side of Eq. (14) depends on $\bar{p}$ in a different manner, and this dependence is not present in the distribution on the right-hand side, it follows that each of these distributions must individually vanish. Taking into account the fact that $\delta\left(q^0 - \frac{\bar{\mathbf{p}}\cdot\mathbf{q}}{\bar{p}^0}\right)$ sets $q^0 \to \frac{\bar{\mathbf{p}}\cdot\mathbf{q}}{\bar{p}^0}$, one obtains the constraints

$$q^j\left[A(q^2) + B(q^2)\right]\delta^4(q) = q^j B(q^2)\delta^4(q) = 0, \tag{17}$$

$$q^j q^l C(q^2)\delta^4(q) = 0, \tag{18}$$

where the last equality in Eq. (17) follows immediately from Eq. (15). Since both Eqs. (17) and (18) contain explicit factors of q and $\delta^4(q)$, without knowledge of the singular behaviour of $B(q^2)$ and $C(q^2)$ at $q = 0$ these constraints are not particularly informative.

4. The angular momentum matrix element

By continuing with the previous approach one can now calculate the constraints on the form factors $A(q^2)$, $B(q^2)$ and $C(q^2)$ from the matrix element $\langle p'; m'; M|J^i|p; m; M\rangle$. In order to do so one must define the operator J^i. For the same reasons as with the operator P^μ, the naive expression for the angular momentum operator $J^i = \frac{1}{2}\epsilon^{ijk}\int d^3x\left[x^j T^{0k}(x) - x^k T^{0j}(x)\right]$ is ill defined. Nevertheless, a consistent expression can be written in a similar manner

$$J^i = \frac{1}{2}\epsilon^{ijk}\lim_{\substack{d\to 0\\ R\to\infty}}\int d^4x\, f_{d,R}(x)\left[x^j T^{0k}(x) - x^k T^{0j}(x)\right], \tag{19}$$

and hence the angular momentum matrix element takes the form

$$\begin{aligned}
&\langle p'; m'; M|J^i|p; m; M\rangle \\
&= \epsilon^{ijk}\lim_{\substack{d\to 0\\ R\to\infty}}\int d^4x\, f_{d,R}(x)x^j e^{iq\cdot x}\langle p'; m'; M|T^{0k}(0)|p; m; M\rangle \\
&= -i\epsilon^{ijk}\lim_{\substack{d\to 0\\ R\to\infty}}\frac{\partial\widehat{f}_{d,R}(q)}{\partial q_j}\langle p'; m'; M|T^{0k}(0)|p; m; M\rangle.
\end{aligned} \tag{20}$$

However, since the states $|p; m; M\rangle$ transform non-trivially under rotations, the structure of the matrix element of J^i is more complicated than the corresponding expression for P^μ in Eq. (10). To derive this expression one can use the fact that one-particle states of spin s transform under (proper orthochronous) Lorentz transformations $\alpha \in \mathcal{L}_+^\uparrow$ as follows [11]:

$$U(\alpha)|p; k; M\rangle = \sum_l \mathcal{D}^s_{lk}(\alpha)|\Lambda(\alpha)p; l; M\rangle, \tag{21}$$

where $\mathcal{D}^s$ is the $(2s+1)$-dimensional Wigner rotation matrix, and $\Lambda(\alpha)$ is the four-vector representation of α.

For a general Lorentz transformation one has: $U(\alpha) = e^{i(\boldsymbol{\eta}\cdot\mathbf{K} - \boldsymbol{\beta}\cdot\mathbf{J})}$, where J^i and K^i are the angular momentum and boost operators, respectively. In particular, since we are interested in the matrix elements of J^i, one can consider the case where $\alpha = \mathcal{R}$ is a pure rotation, and hence: $U(\mathcal{R}) = e^{-i\boldsymbol{\beta}\cdot\mathbf{J}}$. Combining Eq. (21) together with the definition of the norm in Eq. (2), one obtains [6]

$$\begin{aligned}
&\langle p'; m'; M|J^i|p; m; M\rangle \\
&= i\frac{\partial}{\partial\beta_i}\langle p'; m'; M|U(\mathcal{R})|p; m; M\rangle_{\beta=0} \\
&= (2\pi)^4\delta_M^{(+)}(p')\left[S^i_{m'm} - i\delta_{m'm}\epsilon^{ijk}p^j\frac{\partial}{\partial p_k}\right]\delta^4(p'-p),
\end{aligned} \tag{22}$$

where $S^i_{m'm} := i\frac{\partial}{\partial\beta_i}\left[\mathcal{D}^s_{m'm}(\mathcal{R})\right]_{\beta=0}$ are the $(2s+1)$-dimensional spin matrices for spin-s [6]. In order to compare the form factor expansion of Eq. (20) with Eq. (22), one must consider the specific case of spin-$\frac{1}{2}$ states, in which case: $S^i_{m'm} = \frac{1}{2}\sigma^i_{m'm}$, where σ^i are the Pauli matrices. It should be noted that the remarkably simple expression in Eq. (22) is only true for canonical spin states [6], and is significantly more complicated for other spin states such as Wick helicity states[9] [18].

Now we are in a position to perform the same matching procedure as in Sec. 3 for the matrix element $\langle p'; m'; M|J^i|p; m; M\rangle$. Comparing the form factor expansion of Eq. (20) with Eq. (22), and substituting in Eq. (4) gives

$$\begin{aligned}
&(2\pi)^4\left[\frac{1}{2}\sigma^i_{m'm} + i\delta_{m'm}\epsilon^{ijk}\bar{p}^j\frac{\partial}{\partial q_k}\right]\delta^4(q) \\
&= \lim_{\substack{d\to 0\\ R\to\infty}} 2\pi i\epsilon^{ijk}\frac{\partial\widehat{f}_{d,R}}{\partial q_k}\delta\left(q^0 - \frac{\bar{\mathbf{p}}\cdot\mathbf{q}}{\bar{p}^0}\right)\bar{u}_{m'}\left(\bar{p} + \tfrac{1}{2}q\right) \\
&\quad\times\left\{\frac{1}{2M}\bar{p}^j A(q^2) + \frac{1}{8M\bar{p}^0}\bar{p}^{\{0}i\sigma^{j\}\rho}q_\rho\left[A(q^2) + B(q^2)\right]\right. \\
&\quad\left. + \frac{1}{2M\bar{p}^0}\left(q^0 q^j\right)C(q^2)\right\}u_m\left(\bar{p} - \tfrac{1}{2}q\right).
\end{aligned} \tag{23}$$

In order to simplify the right-hand side expression one can make use of the following set of relations [6]:

$$\begin{aligned}
&\frac{\partial}{\partial q_k}\left\{\left[\bar{u}_{m'}\left(\bar{p} + \tfrac{1}{2}q\right)u_m\left(\bar{p} - \tfrac{1}{2}q\right)\right]_{q^0 = \frac{\bar{\mathbf{p}}\cdot\mathbf{q}}{\bar{p}^0}}\right\}_{q=0} \\
&= \frac{i}{(\bar{p}^0 + M)}\epsilon^{kln}\,\bar{p}^l\,\sigma^n_{m'm},
\end{aligned} \tag{24}$$

$$\bar{u}_{m'}(\bar{p})\sigma^{jk}u_m(\bar{p}) = 2\epsilon^{jkl}\left[\bar{p}^0\sigma^l_{m'm} - \frac{\bar{p}^l(\bar{\mathbf{p}}\cdot\boldsymbol{\sigma}_{m'm})}{\bar{p}^0 + M}\right], \tag{25}$$

together with the distributional[10] identity [15]

$$h(q)\,\partial^k\delta^4(q) = h(0)\,\partial^k\delta^4(q) - (\partial^k h)(0)\,\delta^4(q). \tag{26}$$

Eq. (26) follows directly from the definition of the derivative of a distribution discussed in Sec. 1. Intuitively, one might have expected only the first term, as is the case with $h(q)\delta^4(q)$, but in fact this second term cannot be neglected. By taking this distributional subtlety into account this resolves the issues surrounding the treatment of boundary terms in the spin sum rules in the literature [2,6,7]. Applying Eqs. (24), (25) and (26) to Eq. (23), one finally obtains

$$\begin{aligned}
&(2\pi)^4\left[\frac{1}{2}\sigma^i_{m'm} + i\delta_{m'm}\epsilon^{ijk}\bar{p}^j\frac{\partial}{\partial q_k}\right]\delta^4(q) \\
&= (2\pi)\,\delta\left(q^0 - \frac{\bar{\mathbf{p}}\cdot\mathbf{q}}{\bar{p}^0}\right)\lim_{\substack{d\to 0\\ R\to\infty}}\left\{\frac{1}{2}\sigma^i_{m'm}\widehat{f}_{d,R}(q)A(q^2)\right. \\
&\quad + i\delta_{m'm}\epsilon^{ijk}\bar{p}^j\frac{\partial\widehat{f}_{d,R}}{\partial q_k}A(q^2) \\
&\quad + \left[\frac{\bar{p}^0}{2M}\sigma^i_{m'm} - \frac{\bar{p}^i(\bar{\mathbf{p}}\cdot\boldsymbol{\sigma}_{m'm})}{2M(\bar{p}^0 + M)}\right]\widehat{f}_{d,R}(q)B(q^2)
\end{aligned}$$

---

[9] In particular, for on-shell spin-$\frac{1}{2}$ Wick helicity states $|p; m; M\rangle_W$ one has the following relation: ${}_W\langle p'; m'; M|J^i|p; m; M\rangle_W = (2\pi)^4\delta_M^{(+)}(p')\left[m\,\delta_{m'm}\frac{(\delta^{i1}p^1 + \delta^{i2}p^2)|\mathbf{p}|}{(p^1)^2 + (p^2)^2} - i\delta_{m'm}\epsilon^{ijk}p^j\frac{\partial}{\partial p_k}\right]\delta^4(p'-p)$.

[10] In Eq. (26) and throughout the rest of the paper ∂^k signifies the distributional derivative with respect to q_k when acting on distributions, and the partial derivative when acting on infinitely differentiable functions h.

$$
\begin{aligned}
&- \epsilon^{ijk} \frac{\bar{p}^{\{0}\bar{u}_{m'}(\bar{p})\sigma^{j\}\rho}u_m(\bar{p})q_\rho}{8M\bar{p}^0} \frac{\partial \widehat{f}_{d,R}}{\partial q_k}\left[A(q^2)+B(q^2)\right] \\
&+ i\delta_{m'm}\epsilon^{ijk}\frac{q^0 q^j}{\bar{p}^0}\frac{\partial \widehat{f}_{d,R}}{\partial q_k}C(q^2) \\
&+ \epsilon^{ijk}\left[i\frac{q^j\bar{p}^k}{(\bar{p}^0)^2}\delta_{m'm} - \frac{\bar{p}_l q^l q^j(\bar{\mathbf{p}}\times\boldsymbol{\sigma}_{m'm})^k}{2M(\bar{p}^0)^2(\bar{p}^0+M)}\right]\widehat{f}_{d,R}(q)C(q^2)\Bigg\}.
\end{aligned} \tag{27}
$$

Just as with the P^μ matrix element, one can match the distributions in the variable $\bar{p}$ on both sides. Since only the first two terms depend in the same manner on $\bar{p}$ as the left-hand side, one obtains the following constraints:

$$A(q^2)\partial^k\delta^4(q) = \partial^k\delta^4(q), \tag{28}$$

$$A(q^2)\delta^4(q) = \delta^4(q), \tag{29}$$

$$B(q^2)\delta^4(q) = 0, \tag{30}$$

$$q^l\left[A(q^2)+B(q^2)\right]\partial^k\delta^4(q) = 0, \qquad (l\neq k) \tag{31}$$

$$q^j q^l C(q^2)\partial^k\delta^4(q) = 0, \qquad (l\neq k) \tag{32}$$

$$q^j C(q^2)\delta^4(q) = 0. \tag{33}$$

One can immediately see that the constraints derived in Sec. 3 are a subset of those above; Eq. (29) is identical to Eq. (15), and Eqs. (17) and (18) follow from Eqs (30) and (33) respectively. Thus the constraints imposed on the form factors by the Lorentz structure and support property of $\langle p';m';M|P^\mu|p;m;M\rangle$ are entirely encoded in $\langle p';m';M|J^i|p;m;M\rangle$. Combining Eqs. (28) and (29) gives

$$\partial^k\delta^4(q) = \partial^k\left[A(q^2)\delta^4(q)\right] = A(q^2)\partial^k\delta^4(q) + \delta^4(q)\partial^k A(q^2), \tag{34}$$

and hence A satisfies the constraint

$$\lim_{n\to\infty}\int d^4q\,\delta_n^{\{0\}}(q)\partial^k A(q^2) = 0, \tag{35}$$

which formally corresponds to the condition $\partial^k A(0) = 0$. In the case where $A(q^2)$ is non-singular at $q=0$, this condition follows immediately from the fact that $A(q^2)$ depends only on q^2, and hence derivatives with respect to q must vanish at $q=0$. In much the same way that Eq. (29) implies that $A(0)=1$, Eq. (30) implies the well-known result $B(0)=0$, i.e. the anomalous gravitomagnetic moment vanishes [19]. Eqs. (31)–(33) are not particularly informative, but it's interesting to note that the constraint imposed on $C(q^2)$ in Eq. (33) is identical to the constraint on $B(q^2)$ in Eq. (17). This demonstrates that the matrix element constraints are not sufficient to extract the behaviour of $C(q^2)$ near $q=0$.

In analyses of the form factors $A(q^2)$, $B(q^2)$ and $C(q^2)$ in the literature[11] usually only the matrix elements of the operator J^3 are considered. By then imposing that the states have fixed momentum along the z-axis, and introducing appropriate wavepacket functions to ensure that the states are normalisable, the sum rule: $\frac{1}{2} = \frac{1}{2}[A(0)+B(0)]$ is obtained. It then follows from the constraint $A(0)=1$ that $B(0)=0$. In contrast, in our approach we study the distributional properties of the matrix elements, and therefore no choice of frame, wavepacket, operator component, or spin component m is required. Since Eqs. (29) and (30) follow separately from the structure of the J^i matrix element, this implies that the vanishing of $B(0)$ is actually *independent* of the behaviour of $A(q^2)$, in contrast to the literature.

[11] See [7] and references within.

Interestingly, one can also analyse the matrix elements of the boost generator K^i in a completely analogous manner to J^i by using the transformation properties of the states $|p;m;M\rangle$ under pure boosts $U(\mathcal{B}) = e^{i\boldsymbol{\eta}\cdot\mathbf{K}}$. In this case for spin-$\frac{1}{2}$ states one has the relation [6]

$$
\begin{aligned}
&\langle p';m';M|K^i|p;m;M\rangle \\
&= (2\pi)^4\delta_M^{(+)}(p')\left[\frac{(\mathbf{p}\times\boldsymbol{\sigma}_{m'm})^i}{2(p^0+M)} + i\delta_{m'm}p^0\frac{\partial}{\partial p_i}\right]\delta^4(p'-p).
\end{aligned} \tag{36}
$$

After equating this general expression with the energy–momentum form factor decomposition in Eq. (4), and performing the same distributional matching procedure as for J^i, remarkably it turns out that one obtains precisely the same constraints as in Eqs. (28)–(33). Together, all of these findings demonstrate that the low-energy constraints imposed on the energy–momentum form factors are not related to the specific properties or conservation of the individual Poincaré generators themselves, as concluded in the literature [1,3–8,20], but are in fact a consequence of the physical on-shell requirement of the states in the matrix elements and the manner in which these states transform under Poincaré transformations.

Although the analysis in this paper has been performed in the instant form with canonical spin states, one could equally-well quantise the theory in the front form and using different types of spin states, and one would ultimately obtain the same form factor constraints.[12] Since the transformation properties and the positive mass-shell condition for physical states $|p;m;M\rangle$ are generic features of any QFT, it follows that the constraints in Eqs. (28)–(33) must hold for both free[13] *and* interacting theories. This is a similar situation as with the sum rules satisfied by the spectral densities of correlation functions [22].

In this paper we have focused solely on the form factors associated with the matrix elements of the energy–momentum tensor. However, because the distributional-matching approach employed is model independent, one could in principle also use this approach to investigate the structure of form factors related to any other currents, such as the electromagnetic or axial form factors. Moreover, since the decomposition in Eq. (22) is valid for arbitrary canonical spin, one could generalise this approach to analyse the form factors associated with matrix elements of non-spin-$\frac{1}{2}$ states.

5. Conclusions

Since the form factors associated with the matrix elements of the energy–momentum tensor, $A(q^2)$, $B(q^2)$ and $C(q^2)$, encode the dynamics of the states involved in these matrix elements, analysing the behaviour of these objects is key to understanding the non-perturbative characteristics of any QFT. A feature which has received considerable interest in the literature, especially in the context of hadronic spin, is the behaviour of these form factors as $q\to 0$. In this paper we apply a novel axiomatic QFT distributional-matching approach to the matrix elements of P^μ and J^i in order to derive $q\to 0$ constraints for these form factors. We find that these constraints imply the well-known results $B(0)=0$ and $A(0)=1$, but that in contrast with the consensus in the literature, these conditions are actually independent of one another. Furthermore, we also apply an identical procedure to the

[12] In particular, for light-front spin states in the front form one obtains the same expression as Eq. (22) for the J^3 matrix element [21].

[13] For free fields: $A(q^2)\equiv 1$, $B(q^2)\equiv 0$ and $C(q^2)\equiv 0$, which do indeed satisfy the constraints in Eqs. (28)–(33).

matrix elements of the boost generator K^i, and find that this leads to precisely the same constraints. These findings demonstrate that the constraints imposed on the energy–momentum form factors at zero momentum are not related to the specific properties or conservation of the individual Poincaré generators themselves, but are in fact a consequence of the physical on-shell requirement of the states in the matrix elements and the manner in which these states transform under Poincaré transformations.

Acknowledgements

We thank Elliot Leader for useful discussions. This work was supported by the U.S. Department of Energy under contract DE-AC02-76SF00515. P.L. is also supported by the Swiss National Science Foundation under contract P2ZHP2_168622.

References

[1] R.L. Jaffe, A. Manohar, The g_1 problem, Nucl. Phys. B 337 (1990) 509.
[2] X. Ji, J. Tang, P. Hoodbhoy, Spin structure of the nucleon in the asymptotic limit, Phys. Rev. Lett. 76 (1996) 740.
[3] X. Ji, Gauge-invariant decomposition of nucleon spin, Phys. Rev. Lett. 78 (1997) 610.
[4] X. Ji, Lorentz symmetry and the internal structure of the nucleon, Phys. Rev. D 58 (1998) 056003.
[5] G.M. Shore, B.E. White, The gauge-invariant angular momentum sum-rule for the proton, Nucl. Phys. B 581 (2000) 409.
[6] B.L.G. Bakker, E. Leader, T.L. Trueman, Critique of the angular momentum sum rules and a new angular momentum sum rule, Phys. Rev. D 70 (2004) 114001.
[7] E. Leader, C. Lorcé, The angular momentum controversy: what's it all about and does it matter?, Phys. Rep. 541 (2014) 163.
[8] M. Wakamatsu, Is gauge-invariant complete decomposition of the nucleon spin possible?, Int. J. Mod. Phys. A 29 (2014) 1430012.
[9] P. Lowdon, Boundary terms in quantum field theory and the spin structure of QCD, Nucl. Phys. B 889 (2014) 801.
[10] R.F. Streater, A.S. Wightman PCT, Spin and Statistics, and All That, W.A. Benjamin, Inc., 1964.
[11] R. Haag, Local Quantum Physics, Springer-Verlag, 1996.
[12] N. Nakanishi, I. Ojima, Covariant Operator Formalism of Gauge Theories and Quantum Gravity, World Scientific, 1990.
[13] N.N. Bogolubov, A.A. Logunov, A.I. Oksak, General Principles of Quantum Field Theory, Kluwer Academic Publishers, 1990.
[14] F. Strocchi, An Introduction to Non-Perturbative Foundations of Quantum Field Theory, Oxford University Press, 2013.
[15] R.S. Strichartz, A Guide to Distribution Theory and Fourier Transforms, CRC Press, Inc., 1994.
[16] D. Kastler, D.W. Robinson, J.A. Swieca, Conserved currents and associated symmetries; Goldstone's theorem, Commun. Math. Phys. 2 (1966) 108.
[17] F. Jegerlehner, On the existence of light-like charges in quantum field theory, Helv. Phys. Acta 46 (1974) 824.
[18] G.C. Wick, Angular momentum states for three relativistic particles, Ann. Phys. 18 (1962) 65.
[19] S.J. Brodsky, D.S. Hwang, B. Ma, I. Schmidt, Light-cone representation of the spin and orbital angular momentum of relativistic composite systems, Nucl. Phys. B 593 (2001) 311.
[20] J.F. Donoghue, B.R. Holstein, B. Garbrecht, T. Konstandin, Quantum corrections to the Reissner–Nordström and Kerr–Newman metrics, Phys. Lett. B 529 (2002) 132.
[21] K.Y. Chiu, S.J. Brodsky, Angular momentum conservation law in light-front quantum field theory, Phys. Rev. D 59 (2017) 065035.
[22] P. Lowdon, Spectral density constraints in quantum field theory, Phys. Rev. D 92 (2015) 045023.

Schwarzschild–de Sitter spacetime: The role of temperature in the emission of Hawking radiation

Thomas Pappas *, Panagiota Kanti

Division of Theoretical Physics, Department of Physics, University of Ioannina, Ioannina GR-45110, Greece

ARTICLE INFO

Editor: M. Cvetič

ABSTRACT

We consider a Schwarzschild–de Sitter (SdS) black hole, and focus on the emission of massless scalar fields either minimally or non-minimally coupled to gravity. We use six different temperatures, two black-hole and four effective ones for the SdS spacetime, as the question of the proper temperature for such a background is still debated in the literature. We study their profiles under the variation of the cosmological constant, and derive the corresponding Hawking radiation spectra. We demonstrate that only few of these temperatures may support significant emission of radiation. We finally compute the total emissivities for each temperature, and show that the non-minimal coupling constant of the scalar field to gravity also affects the relative magnitudes of the energy emission rates.

1. Introduction

The emission of Hawking radiation [1] from black holes has been a favourite topic in the literature since it combines the most fascinating objects of the General Theory of Relativity with the manifestation of a quantum effect in curved spacetime. One of the oldest black-hole solutions is the Schwarzschild–de Sitter (SdS) solution [2] describing a static, spherically-symmetric, uncharged black hole formed in the presence of a positive cosmological constant. The literature on the emission of Hawking radiation from such a black hole has been scarce: this topic was first discussed in the seminal work by Gibbons and Hawking [3], while an early attempt to calculate the particle production rate in SdS spacetime appeared in [4]. A subsequent work studied the interaction of the Hawking radiation emitted by a SdS black hole with a static source [5].

Remarkably, most of the works focusing on the emission of Hawking radiation from such a black hole have dealt with the higher-dimensional version of the SdS spacetime. The first such works studied the emission of either scalars [6,7] or fields with arbitrary spin [8] both on the brane and in the bulk, while another work [9] studied analytically the scalar greybody factor in an arbitrary number of dimensions. The emission of Hawking radiation from a purely 4-dimensional SdS black hole, in the form of scalar fields either minimally or non-minimally coupled to gravity, was studied only a few years ago [10], and soon afterwards the same study was extended in a higher-dimensional context [11,12]. In a few additional works [13–17], the greybody factors for fields propagating in variants of a Schwarzschild–de Sitter background was also studied.

However, the thermodynamics of the SdS spacetime has not been free of problems and open questions. The temperature of the black-hole horizon, or otherwise the *bare* temperature defined in terms of the corresponding surface gravity [3], fails to take into account the absence of an asymptotically-flat limit. A *normalised* temperature proposed in [18] resolves this problem. However, a new one soon emerged: the SdS spacetime possesses a second horizon, the cosmological horizon, that has its own temperature [19]. An observer located at a point between the two horizons will constantly interact with both of them, and thus will never be in a true thermodynamic equilibrium. This problem may be ignored in the limit of a small cosmological constant, when the two horizons are located far away, but it worsens when the cosmological constant takes a large value. In a new approach adopted in a series of works [20–24] (see also [25] for a review), the notion of the *effective temperature* for the SdS spacetime was proposed that implements both the black-hole and the cosmological horizon temperatures (for a number of additional works on SdS thermodynamics, see [26–45]).

For the study of the emission of Hawking radiation by a SdS black hole, it was either the normalised [6,8,12] or bare [7,10] temperatures that were used. Although the claim was made (see, for example, [21]) that an effective temperature should be used in-

* Corresponding author.
E-mail addresses: thpap@cc.uoi.gr (T. Pappas), pkanti@cc.uoi.gr (P. Kanti).

stead, until recently no such work existed in the literature. In [46], we undertook this task and performed a comprehensive study of the radiation spectra for a higher-dimensional SdS black hole, by using five different temperatures. We demonstrated that the energy emission rates depend strongly on the choice of the temperature as also does the corresponding bulk-over-brane energy ratio. Here, we perform a similar study for the case of a purely 4-dimensional SdS spacetime for the following two reasons: (i) the first effective temperature that appeared in the literature was formulated for a 3-dimensional space and thus may be implemented only in the context of a 4-dimensional analysis, (ii) our previous study [46] showed that some of the temperatures considered tend to acquire similar profiles as the number of spacelike dimensions increases; thus, we expect the largest differences to appear for the lowest dimensionality, i.e. for $D = 4$.

The outline of our paper is as follows: in Section 2, we present the gravitational background, and perform a detailed study of the different definitions of the temperature of a Schwarzschild–de Sitter spacetime. In Section 3, we consider a field theory of a scalar field that may be either minimally or non-minimally coupled to gravity, and solve its equation of motion in the aforementioned background in a exact numerical way. Its greybody factor is then used to derive the energy emission rates from the SdS black hole, for both a minimal and non-minimal coupling to gravity, and for the different temperatures. In Section 4, we calculate and compare the total emissivities of the black hole in each case and, in Section 5, we summarise our conclusions.

2. The gravitational background

We consider the Einstein–Hilbert action in four dimensions and assume also the presence of a positive cosmological constant Λ. Then, the gravitational action reads

$$S_G = \int d^4x \sqrt{-g}\left(\frac{R}{2\kappa^2} - \Lambda\right), \tag{1}$$

where R is the Ricci scalar, $\kappa^2 = 8\pi G$, and g the determinant of the metric tensor $g_{\mu\nu}$. By varying the above action with respect to $g_{\mu\nu}$, we obtain the Einstein's field equations that have the following form:

$$R_{\mu\nu} - \frac{1}{2} g_{\mu\nu} R = -\kappa^2 g_{\mu\nu}\Lambda\,. \tag{2}$$

It is well-known that the above equations admit a spherically-symmetric solution of the form [2]

$$ds^2 = -h(r)\,dt^2 + \frac{dr^2}{h(r)} + r^2\,(d\theta^2 + \sin^2\theta\,d\varphi^2)\,, \tag{3}$$

where the radial function $h(r)$ is given by the expression

$$h(r) = 1 - \frac{2M}{r} - \frac{\kappa^2\Lambda}{3} r^2\,. \tag{4}$$

The above solution describes a Schwarzschild–de Sitter (SdS) spacetime, with the parameter M being the black-hole mass. The horizons of the SdS spacetime follow from the equation $h(r) = 0$, which yields two real, positive roots for $0 < \Lambda M^2/9 < 1$ [3,47]. The smaller of the two roots stands for the black-hole horizon r_h, and the larger for the cosmological horizon r_c. In the critical limit $\Lambda M^2/9 = 1$, known also as the Nariai limit [48], the two horizons coincide and are given by $r_h = 1/\sqrt{\Lambda} = r_c$.

In principle, the temperature of a black hole is defined in terms of its surface gravity k_h at the location of the horizon [3,19] given by the covariant expression

$$k_h^2 = -\frac{1}{2}\lim_{r\to r_h}(D_\mu K_\nu)(D^\mu K^\nu)\,, \tag{5}$$

where D_μ is the covariant derivative and $K = \gamma_t \frac{\partial}{\partial t}$ the timelike Killing vector with γ_t a normalisation constant. For a spherically-symmetric gravitational background, k_h is simplified to [49]

$$k_h = \frac{1}{2}\frac{1}{\sqrt{-g_{tt}g_{rr}}}\,|g_{tt,r}|_{r=r_h}\,, \tag{6}$$

and the temperature of the Schwarzschild–de Sitter black hole (3) finally takes the form [3,49]

$$T_0 = \frac{k_h}{2\pi} = \frac{1-\Lambda r_h^2}{4\pi r_h}\,. \tag{7}$$

In the above, we have used the condition $h(r_h) = 0$ to replace M in terms of r_h and Λ, and set $\kappa^2 = 1$ for simplicity.

However, the SdS spacetime (3) does not have an asymptotic flat limit, where traditionally all parameters of a black hole are defined: the metric function $h(r)$ interpolates between two zeros, at r_h and r_c, reaching a maximum value at an intermediate point r_0 given by the expression $r_0^3 = 3M/\Lambda$; there, $h(r_0) = 1 - \Lambda r_0^2$, that indeed deviates from unity the larger Λ is. In order to fix this problem, a 'normalised' expression for the temperature of a SdS black hole was proposed in [18], given by

$$T_{BH} = \frac{1}{\sqrt{h(r_0)}}\frac{1-\Lambda r_h^2}{4\pi r_h}\,. \tag{8}$$

Mathematically, the inclusion of the factor $\sqrt{h(r_0)}$ is dictated by the non-trivial normalisation constant γ_t in the expression of the Killing vector K^μ when the latter is defined away from a flat spacetime. From the physical point of view, it is at the point r_0 that the effects of the black-hole and cosmological horizons cancel out and thus the point the closest to an asymptotically flat limit.

Nevertheless, the thermodynamics of a SdS black hole faces another problem: one may define the surface gravity k_c of the cosmological horizon in a similar way and, from that, the corresponding temperature [3,19]

$$T_c = -\frac{k_c}{2\pi} = -\frac{1-\Lambda r_c^2}{4\pi r_c}\,. \tag{9}$$

For small values of the cosmological constant, the two horizons are located far way from each other, and one may develop two independent thermodynamics [18,19,31]. But, as Λ increases while keeping M fixed, the two horizons approach each other finally becoming coincident at the critical limit; as the two temperatures, T_0 and T_c, are in principle different, an observer located at an arbitrary point of the causal region $r_h < r < r_c$ interacting with both horizons will never be in a true thermodynamical equilibrium.

As a result, the concept of the *effective temperature* of the Schwarzschild–de Sitter spacetime, that involves both temperatures T_0 and T_c, has emerged during the recent years. In the first approach that was taken in the literature [21], a thermodynamical first law for a Schwarzschild–de Sitter spacetime was written by applying the extended Iyer–Wald formalism [50]: in this, it was assumed that the black-hole mass plays the role of the internal energy of the system ($M = E$), the entropy is the sum of the entropies of the two horizons ($S = S_h + S_c$) and the volume is the one of the observable part of spacetime ($V = V_c - V_h$). Then, the coefficient of δS in the first law was identified with the effective temperature of the system and found to be:

$$T_{effEIW} = \frac{r_h^4\,T_c + r_c^4\,T_0}{(r_h + r_c)\,(r_c^3 - r_h^3)}\,. \tag{10}$$

In the second approach taken [20,22–24] (for a nice review on both approaches, see [25]), it was assumed instead that the black-hole

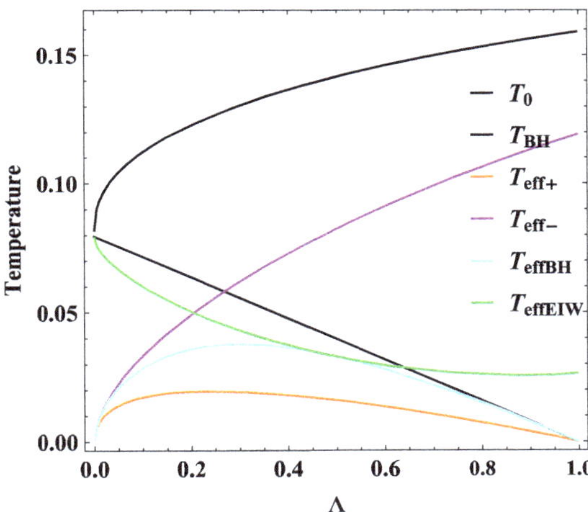

Fig. 1. Temperatures for a Schwarzschild–de Sitter black hole (from top to bottom in the low-Λ regime: T_{BH}, T_0, T_{effEIW}, T_{eff-}, T_{effBH}, and T_{eff+}) as a function of the cosmological constant Λ.

mass plays the role of the enthalpy of the system ($M = -H$), the cosmological constant that of the pressure ($P = \Lambda/8\pi$) while the entropy is still $S = S_h + S_c$. In that case, the effective temperature of the system was found to have the expression

$$T_{eff-} = \left(\frac{1}{T_c} - \frac{1}{T_0}\right)^{-1} = \frac{T_0 T_c}{T_0 - T_c} = -\frac{(1-\Lambda r_h^2)(1-\Lambda r_c^2)}{4\pi\,(r_h + r_c)(1-\Lambda r_h r_c)}\,. \tag{11}$$

However, the latter effective temperature (11) is not always positive-definite and may exhibit infinite jumps near the critical point in charged versions of the SdS spacetime [25]. An alternative expression for the effective temperature of the SdS spacetime was thus proposed in [25] (see also [20]) of the form

$$T_{eff+} = \left(\frac{1}{T_c} + \frac{1}{T_0}\right)^{-1} = \frac{T_0 T_c}{T_0 + T_c} = -\frac{(1-\Lambda r_h^2)(1-\Lambda r_c^2)}{4\pi\,(r_c - r_h)(1+\Lambda r_h r_c)}\,. \tag{12}$$

The above expression could follow from an analysis similar to that leading to T_{eff-} but assuming that the entropy of the system is the difference of the entropies of the two horizons, i.e. $S = S_c - S_h$; however, no physical reason exists for that. In a recent work of ours [46], we observed instead that, if one followed the same approach that led to T_{eff-} but merely replaced the 'bare' temperature T_0 with the 'normalised' one T_{BH}, one would obtain the following expression for the effective temperature of the SdS spacetime

$$T_{effBH} = \left(\frac{1}{T_c} - \frac{1}{T_{BH}}\right)^{-1} = \frac{T_{BH} T_c}{T_{BH} - T_c} = -\frac{(1-\Lambda r_h^2)(1-\Lambda r_c^2)}{4\pi\,(r_h\sqrt{h(r_0)} + r_c)(1-\Lambda r_h r_c)}\,. \tag{13}$$

As we will shortly see, the above expression shares several characteristics with the T_{eff+}; at the same time, it retains the usual assumption for the entropy of the system ($S = S_h + S_c$) and takes into account the absence of asymptotic flatness.

In Fig. 1, we depict the behaviour of all six temperatures (T_0, T_{BH}, T_{effEIW}, T_{eff-}, T_{eff+}, T_{effBH}) as functions of the cosmological constant Λ. Throughout the present analysis, we will allow Λ to vary in the complete regime $[0, \Lambda_c]$, where $\Lambda_c = 1/r_h^2$ is the maximum, critical value of the cosmological constant for which the two horizons coincide – for simplicity, we will keep the black-hole horizon fixed ($r_h = 1$), and thus allow Λ to take values in the range [0, 1]. Starting from the low-Λ regime, we observe that the six temperatures are split into two distinct groups that adopt two different asymptotic values as $\Lambda \to 0$. The first group is comprised by the two black-hole temperatures, T_0 and T_{BH}, and the effective temperature T_{effEIW}: the first two temperatures naturally reduce to the temperature of the Schwarzschild black hole $T_H = 1/4\pi r_h$ when the cosmological constant vanishes; as one may see from Eq. (10), T_{effEIW} also has the same smooth limit as $r_c \to \infty$. We may thus conclude that T_{effEIW} has been built on the assumption that the black-hole horizon should always be present while r_c may be located at either a finite or infinite distance. On the contrary, the three effective temperatures T_{eff-}, T_{eff+}, and T_{effBH} vanish when Λ goes to zero (but reduce to T_c when $r_h \to 0$); this is due to the fact that these effective temperatures were derived under the assumption that Λ is non-zero, standing for the pressure of the system – in this approach, it is the cosmological horizon that should always be present by construction whereas r_h may vanish or not.

As the cosmological constant increases, the normalised black-hole T_{BH} and the effective temperature T_{eff-} monotonically increase – as shown in Fig. 1 – while the bare black-hole one T_0 together with the effective temperature T_{effEIW} monotonically decrease. On the other hand, both effective temperatures T_{eff+} and T_{effBH} first increase with Λ and, after reaching a maximum value, start decreasing thus exhibiting a similar behaviour.

Towards the critical point ($\Lambda \to 1$), the six temperatures again split into two groups: the first one is now comprised of the temperatures T_{BH}, T_{eff-} and T_{effEIW} that retain an asymptotic, non-zero value at the critical limit. We may easily justify this behaviour by looking at Eqs. (8), (11) and (10): in the limit $r_h \to r_c \to 1/\sqrt{\Lambda}$, both the numerator and denominator in all three expressions go to zero in such a way that their ratio remains constant. On the other hand, the remaining three temperatures, T_0, T_{eff+}, and T_{effBH}, all vanish at the critical limit – indeed, it is now only the numerators in Eqs. (7), (12) and (13) that go to zero while all denominators are non-vanishing.

Comparing their overall behaviour, one may immediately see the dominance of the normalised temperature T_{BH} over the whole Λ-regime. For small cosmological constant, the values of the bare T_0 and effective temperature T_{effEIW} are also comparable. As the critical point is approached, the dominance of T_{BH} still holds, however, the effective T_{eff-} is also taking on a large value. Since the greybody factors for a specific type of field will be common, the aforementioned behaviour will also determine the corresponding spectra of Hawking radiation when each one of the aforementioned temperatures is used. In the next section, we will take up this task and consider a scalar field, minimally- or non-minimally coupled to gravity. We will study its emission in a 4-dimensional SdS spacetime, and determine the radiation spectra – and their profiles in terms of Λ – by using each one of the above temperatures.

3. The effect of the temperature on the EERs

In the recent work of ours [46], where the radiation spectra for a higher-dimensional SdS black hole were studied in detail using five different temperatures, it became clear that the temperature profiles depend strongly on the dimensionality of spacetime. In fact, the differences between the two black-hole temperatures, T_0 and T_{BH}, and between the three effective temperatures, T_{eff-}, T_{eff+} and T_{effBH}, were amplified as n decreased. Thus, here, we turn to the calculation of the energy emission rates (EERs) for the case of a purely 4-dimensional SdS black hole. We will also include a fourth effective temperature, T_{effEIW}, that was left out in the analysis of [46]: the reason for that was that its expression (10) explicitly involves the 3-dimensional spatial volume, and thus its generalisation in a higher-dimensional spacetime needs a careful consideration.

Let us consider the following field theory describing a massless, scalar field with a non-minimal coupling to gravity

$$S_\Phi = -\frac{1}{2}\int d^4x\sqrt{-g}\left[\xi\Phi^2 R + \partial_\mu\Phi\,\partial^\mu\Phi\right]. \tag{14}$$

In the above, $g_{\mu\nu}$ is the metric tensor defined in Eq. (3), and R the scalar curvature $R = 4\kappa^2\Lambda$. Also, ξ is a constant, with the value $\xi = 0$ corresponding to the minimal coupling and the value $\xi = 1/6$ to the conformal coupling. The equation of motion of the scalar field has the form

$$\frac{1}{\sqrt{-g}}\partial_\mu\left(\sqrt{-g}\,g^{\mu\nu}\partial_\nu\Phi\right) = \xi R\,\Phi. \tag{15}$$

If we assume a factorised ansatz for the field, i.e. $\Phi(t,r,\theta,\varphi) = e^{-i\omega t}\,P(r)\,Y(\theta,\varphi)$, where $Y(\theta,\varphi)$ are the scalar spherical harmonics, we obtain a radial equation for the function $P(r)$ of the form

$$\frac{1}{r^2}\frac{d}{dr}\left(hr^2\frac{dP}{dr}\right) + \left[\frac{\omega^2}{h} - \frac{l(l+1)}{r^2} - 4\xi\kappa^2\Lambda\right]P = 0. \tag{16}$$

We note that the non-minimal coupling term acts as an effective mass term for the scalar field [10–12]: any increase in its value increases the 'mass' of the field and thus suppresses both the greybody factors and radiation spectra especially in the low-energy regime [51–54]. The mass term also depends on the value of Λ: in the minimal coupling case, Λ enhances the EERs, however, for $\xi \neq 0$, the role of Λ is more subtle to infer and an exact computation of the radiation spectra is therefore necessary.

Equation (16) was first studied in [10] and later extended in a higher-dimensional context in [11,12]. Here, we follow the analysis of the last work, and focus on the 4-dimensional case with $n = 0$. The analytic study of Eq. (16) in the near-horizon regime leads to a general solution written in terms of a hypergeometric function. When expanded in the limit $r \to r_h$, the solution takes the form of an ingoing free wave, namely

$$R_{BH} \simeq A_1\,f^{\alpha_1} = A_1\,e^{-i(\omega r_h/A_h)\,\ln f}, \tag{17}$$

where $A_h = 1 - \Lambda r_h^2$, and f is a new radial variable defined by the relation: $r \to f(r) = h(r)/(1-\Lambda r^2/3)$. Upon setting the arbitrary constant A_1 to unity, the above asymptotic solution leads to the conditions [12]

$$R_{BH}(r_h) = 1, \qquad \left.\frac{dR_{BH}}{dr}\right|_{r_h} \simeq -\frac{i\omega}{h(r)}. \tag{18}$$

The above expressions serve as boundary conditions for the numerical integration of Eq. (16).

The solution of Eq. (16) near the cosmological horizon is again given in terms of hypergeometric functions. Taking the limit $r \to r_c$, we now find [11,12]

$$R_C \simeq B_1\,e^{-i(\omega r_c/A_c)\ln f} + B_2\,e^{i(\omega r_c/A_c)\ln f}, \tag{19}$$

where $A_c = 1 - \Lambda r_c^2$. The constant coefficients $B_{1,2}$ are easily identified with the amplitudes of the ingoing and outgoing free waves. As a result, the greybody factor, or transmission probability, for the scalar field may be expressed as

$$|A|^2 = 1 - \left|\frac{B_2}{B_1}\right|^2. \tag{20}$$

By numerically integrating Eq. (16), we may find the $B_{1,2}$ coefficients: we start the numerical integration close to the black-hole horizon, i.e. from $r = r_h + \epsilon$, where $\epsilon = 10^{-6}$–10^{-4}, and using the boundary conditions (18) we proceed towards the cosmological horizon. There, we isolate the constant amplitudes $B_{1,2}$ (for more information on this, see [12]) and determine the greybody factor $|A|^2$. According to the results of our numerical analysis, the greybody factor is suppressed with the non-minimal coupling constant ξ over the whole energy regime; for small values of ξ, the cosmological constant enhances the radiation spectra however, for large values of ξ, an increase in Λ may cause a suppression especially in the low-energy regime.

We may now proceed to derive the differential energy emission rate for scalar fields from a SdS black hole. The power emission spectrum is traditionally given by the expression [1,6]

$$\frac{d^2E}{dt\,d\omega} = \frac{1}{2\pi}\sum_l \frac{N_l\,|A|^2\,\omega}{\exp(\omega/T)-1}, \tag{21}$$

where ω is the energy of the emitted particle, and $N_l = 2l+1$ the multiplicity of states that have the same angular-momentum number. The above formula describes a thermal spectrum that takes into account the back-scattering of the emitted modes, via the presence of the greybody factor $|A|^2$. It has been used to describe the emission of Hawking radiation from a plethora of four- and higher-dimensional black holes (see [55–57] and references therein) as well as from a large number of stringy or D-brane backgrounds (see, for example [58,59]).

One should however be careful: although the authors of [3] anticipated the existence of a thermal spectrum of the form (21) for a SdS background, the presence of the second (i.e. cosmological) horizon prevented them from explicitly demonstrating that. The Schwarzschild and SdS spacetime have a number of similarities: they are both spherically-symmetric and static. These two features allow us to define positive-frequency basis modes – that are necessary for the study of Hawking radiation – in both backgrounds. In the Schwarzschild spacetime, the "up" modes [57,60] are defined at the past event horizon while the "in" modes are defined at asymptotic infinity. In the SdS spacetime, the asymptotically-flat regime is missing and replaced by the cosmological horizon, and this seems to create a problem. But the calculation of the particle production rate does not need a Minkowskian limit but only an inertial observer [60]. And such an observer is always present in a SdS spacetime residing at the point $r = r_0$: it is there that the effects of the black-hole and cosmological horizons exactly cancel, and the proper acceleration of the observer is zero. The only attempt in the literature to calculate the particle production rate in a SdS spacetime [4] defined the "in" modes close to the cosmological horizon, and found a non-thermal spectrum – that was a natural result, since a non-inertial observer fails to detect a thermal spectrum [60–62].

It is worth noting that the point r_0 is present for all values of the cosmological constant: when Λ goes to zero, r_0 becomes the asymptotic infinity; when $\Lambda \to \Lambda_c$, r_0 is approached on both sides by r_h and r_c until they all match at the critical limit. Therefore, we expect an inertial observer residing at the point $r = r_0$ to detect a Hawking radiation spectrum given indeed by Eq. (21) and for all values of the cosmological constant. In what follows, we will use Eq. (21) where the temperature T will be taken to be equal, in turn, to T_0, T_{BH}, T_{effEIW}, T_{eff-}, T_{eff+} and T_{effBH}, in order to derive the corresponding radiation spectra. The sum over the l-modes will be extended up to the $l = 7$ as all higher modes have negligible contributions to the total emission rate.

Let us start with the minimal-coupling case, with $\xi = 0$: the radiation spectra, for the six temperatures and for two indicative

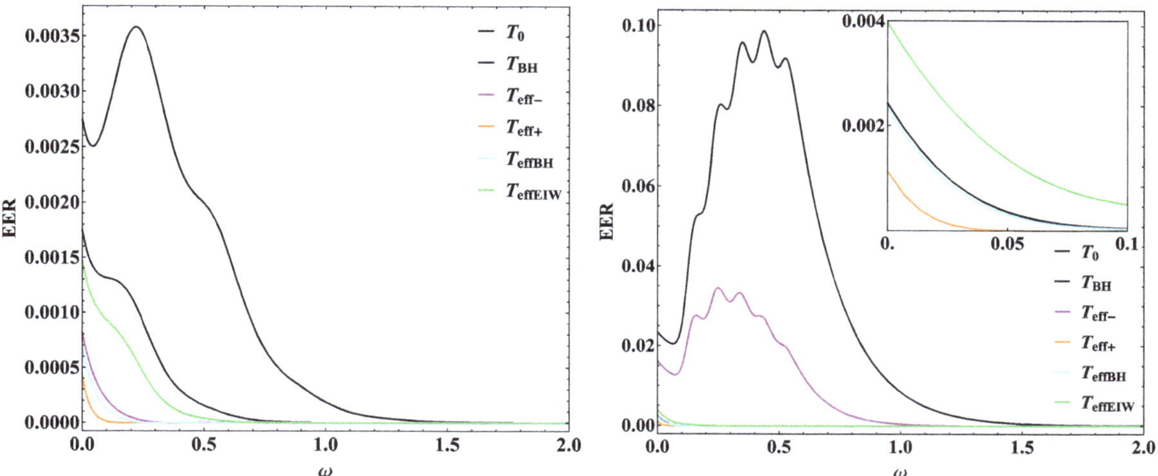

Fig. 2. Energy emission rates for minimally-coupled scalar fields from a Schwarzschild–de Sitter black hole for: **(Left plot)** $\Lambda = 0.1$ (in units of r_h^{-2}), and $T = T_{BH}$, T_0, T_{effEIW}, T_{eff-}, T_{effBH}, T_{eff+} (from top to bottom), and **(Right plot)** $\Lambda = 0.8$ and $T = T_{BH}$, T_{eff-} (from top to bottom, again).

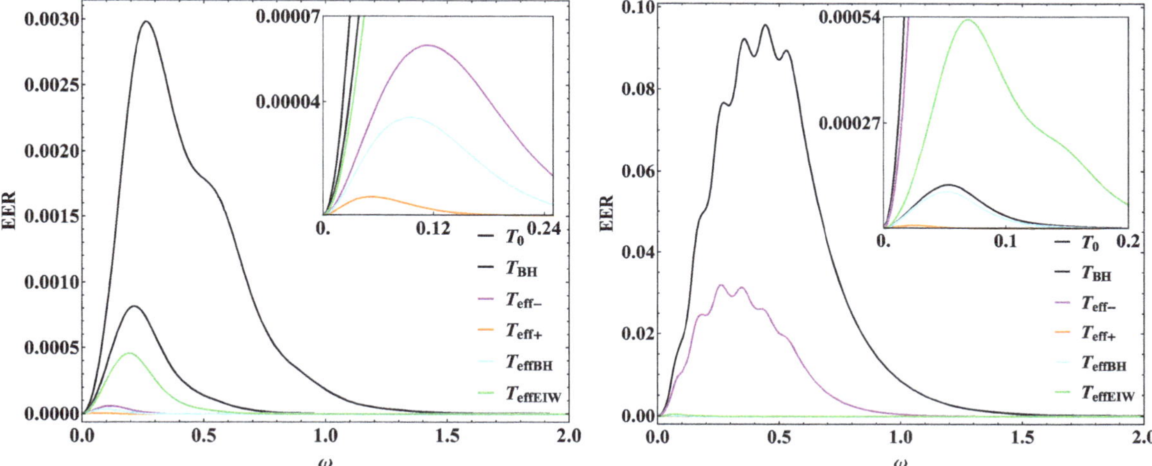

Fig. 3. Energy emission rates for non-minimally-coupled scalar fields with $\xi = 1/6$ from a Schwarzschild–de Sitter black hole for: **(Left plot)** $\Lambda = 0.1$ (in units of r_h^{-2}), and $T = T_{BH}$, T_0, T_{effEIW}, T_{eff-}, T_{effBH}, T_{eff+} (from top to bottom), and **(Right plot)** $\Lambda = 0.8$ and $T = T_{BH}$, T_{eff-} (from top to bottom, again).

values of the cosmological constant, are depicted in Fig. 2. Here, the effective mass term vanishes, and the emission curves exhibit the characteristic feature of the non-vanishing asymptotic limit as $\omega \to 0$; this is due to the following non-vanishing geometric limit

$$|A^2| = \frac{4r_h^2 r_c^2}{(r_c^2 + r_h^2)^2} + \mathcal{O}(\omega)\,, \tag{22}$$

adopted by the greybody factor for a massless, free scalar field propagating in a SdS black-hole background [6,9–12]. In fact, for the low value $\Lambda = 0.1$, the dominant emission channel lies in the low-energy regime. As Λ increases, the energy emission curves are significantly enhanced and reach their maximal points at intermediate values of energy as expected.

Focusing now on the radiation emission curves for the different temperatures, we observe that, for a low value of Λ, i.e. $\Lambda = 0.1$, and in accordance to the behaviour depicted in Fig. 1, the EER for the normalised temperature T_{BH} is clearly the dominant one [see the left plot of Fig. 2]; the ones for the bare T_0 and effective T_{effEIW} temperatures follow behind, while for the remaining effective ones – that all take a very small value in the low-Λ regime – the EERs are severely suppressed. For a value of Λ close to its critical one, namely for $\Lambda = 0.8$, the only significant EERs are now the ones for T_{BH} and T_{eff-}, as one may see in the right plot of Fig. 2: all the other temperatures adopt a much smaller value near the critical limit, and the corresponding spectra are thus suppressed.

Let us address now the case of a non-minimally coupled scalar field by assigning a non-vanishing value to the coupling constant ξ. As soon as we do that, the effective mass term re-appears and the non-zero low-energy asymptotic limit of the EERs disappears. This is obvious from the plots in Fig. 3, where the emission curves – drawn for the indicative value of $\xi = 1/6$ – have now assumed their traditional shape. The non-zero value of the coupling constant ξ causes a suppression in the energy emission rates over the whole energy regime: the emission of very low-energy particles has been severely suppressed, due to the disappearance of the low-energy asymptotic limit, but also the peaks of the curves are now lower. The same behaviour is observed if one increases further the value of the coupling constant. In Fig. 4, we depict the EERs for the value $\xi = 1/2$: all emission curves are further suppressed and the same pattern continues for even higher values of ξ. What one could note is that the suppression with ξ is much stronger when the cosmological constant takes a small value, while it becomes milder when Λ approaches its critical limit – we will return to

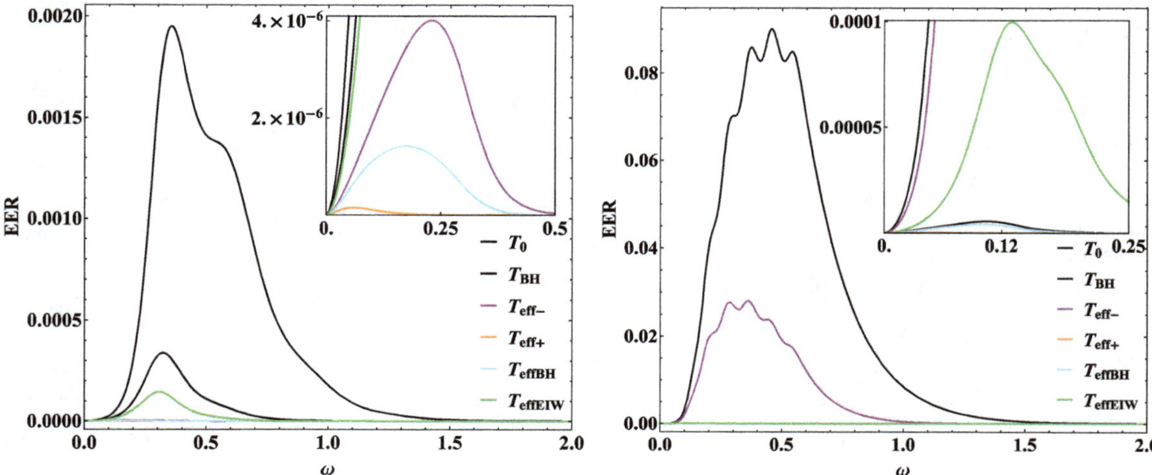

Fig. 4. Energy emission rates for non-minimally-coupled scalar fields with $\xi = 1/2$ from a Schwarzschild–de Sitter black hole for: **(Left plot)** $\Lambda = 0.1$ (in units of r_h^{-2}), and $T = T_{BH}$, T_0, T_{effEIW}, T_{eff-}, T_{effBH}, T_{eff+} (from top to bottom), and **(Right plot)** $\Lambda = 0.8$ and $T = T_{BH}$, T_{eff-} (from top to bottom, again).

this observation in the next section where the total emissivity of the black hole is computed.

The turning on of the non-minimal coupling constant ξ is also in a position to change the relative values of the EERs for the different temperatures, as is clear from both Figs. 3 and 4. However, the sequence and general behaviour of the emission curves remains the same: for small cosmological constant ($\Lambda = 0.1$), it is the group of temperatures (T_{BH}, T_0, T_{effEIW}) – i.e. the ones that adopt a non-vanishing value in the limit $\Lambda \to 0$ – that lead to significant EERs; close to the critical limit ($\Lambda = 0.8$), it is instead only T_{BH} and T_{eff-} that support a non-vanishing energy emission rate for the black hole. Another common feature is once again the dominance of the normalised temperature T_{BH} over the whole energy- and the whole-Λ regime.

4. Total emissivities

In this section, we will calculate the total emissivity of the SdS black-hole, i.e. the total energy emitted by the black hole per unit time over the whole energy regime, in the form of scalar fields. We will see that this important quantity depends strongly on the selected temperature with the results differing at times by orders of magnitude. Apart from choosing a different temperature each time, the total emissivities will be computed for three values of the non-minimal coupling constant, i.e. for $\xi = 0, 1/6, 1/2$, and for four values of the cosmological constant, namely $\Lambda = (0.1, 0.3, 0.5, 0.8)\, r_h^{-2}$. In this respect, our current analysis completes and extends the analysis of [10] where only small values of Λ were considered. Here, the values of Λ span the whole allowed regime up to its critical limit. This latter regime comprises in fact an important theoretical limit, that is usually ignored in the calculation of Hawking radiation. That was due to the fact that the traditional 'bare' temperature T_0 vanishes at the critical limit and the emission of Hawking radiation stops. However, as we showed, the normalised temperature T_{BH} as well as a number of the recently proposed effective temperatures assume there a non-vanishing value. This leads not only to emission of Hawking radiation but also to the maximisation of the total emissivities for the SdS black hole in that limit.

The total emissivities for the SdS black hole, for the aforementioned values of ξ and Λ, are presented in Tables 1–3. Let us first examine how the increase in the value of Λ affects our results in the minimal coupling case ($\xi = 0$). We observe that, in accordance to the results of the previous sections, the total emissivity

Table 1
Total emissivity for $\xi = 0$.

$\Lambda \to$	0.1	0.3	0.5	0.8
T_0	0.000444	0.000487	0.000335	0.000065
T_{BH}	0.001871	0.005432	0.011837	0.054554
T_{eff-}	0.000058	0.000636	0.002106	0.015937
T_{eff+}	0.000013	0.000047	0.000050	0.000014
T_{effBH}	0.000040	0.000200	0.000225	0.000060
T_{effEIW}	0.000266	0.000267	0.000222	0.000196

Table 2
Total emissivity for $\xi = 1/6$.

$\Lambda \to$	0.1	0.3	0.5	0.8
T_0	0.000228	0.000124	0.000057	$7.5796\ 10^{(-6)}$
T_{BH}	0.001358	0.003647	0.008889	0.050743
T_{eff-}	$9.8980\ 10^{(-6)}$	0.000191	0.001040	0.013964
T_{eff+}	$0.5696\ 10^{(-6)}$	$1.3972\ 10^{(-6)}$	$1.2237\ 10^{(-6)}$	$0.2977\ 10^{(-6)}$
T_{effBH}	$5.0160\ 10^{(-6)}$	0.000026	0.000027	$6.3134\ 10^{(-6)}$
T_{effEIW}	0.000112	0.000044	0.000026	0.000054

Table 3
Total emissivity for $\xi = 1/2$.

$\Lambda \to$	0.1	0.3	0.5	0.8
T_0	0.000087	0.000021	$6.1253\ 10^{(-6)}$	$0.5443\ 10^{(-6)}$
T_{BH}	0.000837	0.002126	0.006062	0.045459
T_{eff-}	$0.8973\ 10^{(-6)}$	0.000040	0.000433	0.011571
T_{eff+}	$0.0164\ 10^{(-6)}$	$0.0316\ 10^{(-6)}$	$0.0251\ 10^{(-6)}$	$0.0057\ 10^{(-6)}$
T_{effBH}	$0.3206\ 10^{(-6)}$	$1.7945\ 10^{(-6)}$	$1.8766\ 10^{(-6)}$	$0.4070\ 10^{(-6)}$
T_{effEIW}	0.000033	$4.1488\ 10^{(-6)}$	$1.8064\ 10^{(-6)}$	0.000012

for the bare temperature T_0 decreases, and drops to only 14% of its original value as Λ increases from 0.1 to 0.8. The total emissivity for the effective temperature T_{effEIW} also decreases, but now the decrease is milder, of the order of 73%. When the normalised temperature T_{BH} is used, the total emissivity of the black hole is actually enhanced, by a factor of 30 when Λ increases from 0.1 to 0.8. An increase is also noted for the total emissivity for T_{eff-} but now the enhancement factor is of the order of 300. Finally, the remaining two effective temperatures T_{eff+} and T_{effBH} see their emissivities first to increase and then to decrease as Λ increases.

As the non-minimal coupling constant increases from $\xi = 0$ to $\xi = 1/6$ and then to $\xi = 1/2$, all total emissivities decrease since the appearance of the effective mass term suppresses the EERs at all energy regimes. What is more significant is that the value of ξ strongly affects the suppression or enhancement factors for the total emissivities – computed for the different temperatures – as Λ varies. For example, the total emissivity for the bare T_0 drops to 3%, for $\xi = 1/6$, and to only 0.6%, for $\xi = 1/2$, of its original value as Λ goes from 0.1 to 0.8. Similarly, the total emissivity for the effective T_{effEIW} drops to 48% and 33%, respectively. On the other hand, the total emissivity for the dominant T_{BH} is now enhanced by a factor of 37, for $\xi = 1/6$, and by a factor of 54, for $\xi = 1/2$. The total emissivity for T_{eff-} is the one that is mostly affected: it increases by a factor of 1400, for $\xi = 1/6$, and by a factor of 12000, for $\xi = 1/2$, as Λ goes from 0.1 to 0.8. We expect the same pattern to continue as ξ increases further.

5. Conclusions

In this work, we have considered the Schwarzschild–de Sitter black hole and performed a study of the Hawking radiation spectra, emitted in the form of scalar fields either minimally or non-minimally coupled to gravity. The novel feature of our analysis is the use of six different temperatures for the SdS background, as the question of the proper temperature for such a spacetime is still debated in the literature. We have thus considered the *bare* temperature, defined in terms of the black-hole surface gravity, the *normalised* temperature, that takes into account the absence of an asymptotically-flat limit, and four *effective* temperatures defined in terms of both the black-hole and cosmological horizon temperatures.

We first studied the profiles of the above temperatures as a function of the cosmological constant Λ, from a zero value up to its maximum, critical limit. We have found that the temperatures are split in two groups depending on their behaviour in these two asymptotic Λ-regimes. In the limit of zero cosmological constant, the aforementioned temperatures either reduce to the temperature of the Schwarzschild black hole or vanish; near the critical limit, they either assume a non-vanishing asymptotic value or reduce again to zero.

Their different profiles inevitably affect the form of the energy emission rates for Hawking radiation. For small values of Λ, it is only the bare T_0, the normalised T_{BH} and the effective temperature T_{effEIW} that lead to significant radiation from the SdS black hole. In the opposite critical limit, it is only the spectra for T_{BH} and T_{eff-} that survive, with the one for the normalised T_{BH} being the dominant one over the whole energy regime. The computation of the total emissivities confirm the above behaviour in a quantitative way, and reveal that the value of the non-minimal coupling constant ξ determines the relative values of the EERs for the different temperatures as well as the enhancement or suppression factors for each one of them as the cosmological constant increases.

Acknowledgements

T.P. would like to thank the Alexander S. Onassis Public Benefit Foundation for financial support.

References

[1] S.W. Hawking, Commun. Math. Phys. 43 (1975) 199–220.
[2] F.R. Tangherlini, Nuovo Cimento 27 (1963) 636.
[3] G.W. Gibbons, S.W. Hawking, Phys. Rev. D 15 (1977) 2738.
[4] D. Kastor, J.H. Traschen, Class. Quantum Gravity 13 (1996) 2753.
[5] J. Castineiras, I.P. Costa e Silva, G.E.A. Matsas, Phys. Rev. D 68 (2003) 084022.
[6] P. Kanti, J. Grain, A. Barrau, Phys. Rev. D 71 (2005) 104002.
[7] J. Labbe, A. Barrau, J. Grain, PoS HEP 2005 (2006) 013, arXiv:hep-ph/0511211.
[8] S.F. Wu, S.y. Yin, G.H. Yang, P.M. Zhang, Phys. Rev. D 78 (2008) 084010.
[9] T. Harmark, J. Natario, R. Schiappa, Adv. Theor. Math. Phys. 14 (2010) 727.
[10] L.C.B. Crispino, A. Higuchi, E.S. Oliveira, J.V. Rocha, Phys. Rev. D 87 (10) (2013) 104034.
[11] P. Kanti, T. Pappas, N. Pappas, Phys. Rev. D 90 (12) (2014) 124077.
[12] T. Pappas, P. Kanti, N. Pappas, Phys. Rev. D 94 (2) (2016) 024035.
[13] P.R. Anderson, A. Fabbri, R. Balbinot, Phys. Rev. D 91 (6) (2015) 064061.
[14] C.A. Sporea, A. Borowiec, Int. J. Mod. Phys. D 25 (04) (2016) 1650043.
[15] J. Ahmed, K. Saifullah, arXiv:1610.06104 [gr-qc].
[16] S. Fernando, arXiv:1611.05337 [gr-qc].
[17] P. Boonserm, T. Ngampitipan, P. Wongjun, arXiv:1705.03278 [gr-qc].
[18] R. Bousso, S.W. Hawking, Phys. Rev. D 54 (1996) 6312.
[19] G.W. Gibbons, S.W. Hawking, Phys. Rev. D 15 (1977) 2752.
[20] S. Shankaranarayanan, Phys. Rev. D 67 (2003) 084026.
[21] M. Urano, A. Tomimatsu, H. Saida, Class. Quantum Gravity 26 (2009) 105010.
[22] S. Bhattacharya, A. Lahiri, Eur. Phys. J. C 73 (2013) 2673.
[23] S. Bhattacharya, Eur. Phys. J. C 76 (3) (2016) 112.
[24] H.F. Li, M.S. Ma, Y.Q. Ma, Mod. Phys. Lett. A 32 (02) (2016) 1750017, arXiv:1605.08225 [hep-th].
[25] D. Kubiznak, R.B. Mann, M. Teo, Class. Quantum Gravity 34 (6) (2017) 063001.
[26] L.J. Romans, Nucl. Phys. B 383 (1992) 395.
[27] D. Kastor, J.H. Traschen, Phys. Rev. D 47 (1993) 5370.
[28] R.G. Cai, Phys. Lett. B 525 (2002) 331; Nucl. Phys. B 628 (2002) 375.
[29] A.M. Ghezelbash, R.B. Mann, J. High Energy Phys. 0201 (2002) 005.
[30] X. Li, Y.G. Shen, Phys. Lett. A 324 (2004) 255.
[31] Y. Sekiwa, Phys. Rev. D 73 (2006) 084009.
[32] M. Cvetic, G.W. Gibbons, C.N. Pope, Phys. Rev. Lett. 106 (2011) 121301.
[33] B.P. Dolan, D. Kastor, D. Kubiznak, R.B. Mann, J. Traschen, Phys. Rev. D 87 (10) (2013) 104017.
[34] M.S. Ma, H.H. Zhao, L.C. Zhang, R. Zhao, Int. J. Mod. Phys. A 29 (2014) 1450050.
[35] H.H. Zhao, L.C. Zhang, M.S. Ma, R. Zhao, Phys. Rev. D 90 (6) (2014) 064018.
[36] L.C. Zhang, M.S. Ma, H.H. Zhao, R. Zhao, Eur. Phys. J. C 74 (9) (2014) 3052.
[37] M.S. Ma, L.C. Zhang, H.H. Zhao, R. Zhao, Adv. High Energy Phys. 2015 (2015) 134815, arXiv:1410.5950 [gr-qc].
[38] X. Guo, H. Li, L. Zhang, R. Zhao, Phys. Rev. D 91 (8) (2015) 084009.
[39] X. Guo, H. Li, L. Zhang, R. Zhao, Adv. High Energy Phys. 2016 (2016) 7831054.
[40] A. Araujo, J.G. Pereira, Int. J. Mod. Phys. D 24 (14) (2015) 1550099.
[41] D. Kubiznak, F. Simovic, Class. Quantum Gravity 33 (24) (2016) 245001.
[42] J. McInerney, G. Satishchandran, J. Traschen, Class. Quantum Gravity 33 (10) (2016) 105007.
[43] H.F. Li, M.S. Ma, L.C. Zhang, R. Zhao, Nucl. Phys. B 920 (2017) 211.
[44] H. Liu, X.h. Meng, arXiv:1611.03604 [gr-qc].
[45] B. Pourhassan, S. Upadhyay, H. Farahani, arXiv:1701.08650 [physics.gen-ph].
[46] P. Kanti, T. Pappas, Phys. Rev. D 96 (2) (2017) 024038, arXiv:1705.09108 [hep-th].
[47] C. Molina, Phys. Rev. D 68 (2003) 064007.
[48] H. Nariai, Sci. Rep. Tohoku Univ., I. 35 (1951) 62.
[49] J.W. York Jr., Phys. Rev. D 31 (1985) 775.
[50] R.M. Wald, Phys. Rev. D 48 (8) (1993) R3427;
V. Iyer, R.M. Wald, Phys. Rev. D 50 (1994) 846.
[51] D.N. Page, Phys. Rev. D 16 (1977) 2402.
[52] E.I. Jung, S.H. Kim, D.K. Park, Phys. Lett. B 586 (2004) 390; J. High Energy Phys. 0409 (2004) 005; Phys. Lett. B 602 (2004) 105.
[53] M.O.P. Sampaio, J. High Energy Phys. 0910 (2009) 008; J. High Energy Phys. 1002 (2010) 042.
[54] P. Kanti, N. Pappas, Phys. Rev. D 82 (2010) 024039.
[55] C.M. Harris, P. Kanti, J. High Energy Phys. 0310 (2003) 014.
[56] P. Kanti, Int. J. Mod. Phys. A 19 (2004) 4899–4951.
[57] P. Kanti, E. Winstanley, Fundam. Theor. Phys. 178 (2015) 229.
[58] S.R. Das, S.D. Mathur, Nucl. Phys. B 478 (1996) 561.
[59] J.M. Maldacena, A. Strominger, Phys. Rev. D 55 (1997) 861.
[60] N.D. Birrell, P.C.W. Davies, Quantum Fields in Curved Space, 1982.
[61] J.R. Ellis, P. Kanti, N.E. Mavromatos, D.V. Nanopoulos, E. Winstanley, Mod. Phys. Lett. A 13 (1998) 303.
[62] S. Singh, S. Chakraborty, Phys. Rev. D 90 (2) (2014) 024011;
S. Chakraborty, S. Singh, T. Padmanabhan, J. High Energy Phys. 1506 (2015) 192.

23

Special geometry and space–time signature

W.A. Sabra

Centre for Advanced Mathematical Sciences and Physics Department, American University of Beirut, Lebanon

ARTICLE INFO

Editor: M. Cvetič

ABSTRACT

We construct $\mathcal{N}=2$ four and five-dimensional supergravity theories coupled to vector multiplets in various space–time signatures (t,s), where t and s refer, respectively, to the number of time and spatial dimensions. The five-dimensional supergravity theories, $t+s=5$, are constructed by investigating the integrability conditions arising from Killing spinor equations. The five-dimensional supergravity theories can also be obtained by reducing Hull's eleven-dimensional supergravities on a Calabi–Yau threefold. The dimensional reductions of the five-dimensional supergravities on space and time-like circles produce $\mathcal{N}=2$ four-dimensional supergravity theories with signatures $(t-1,s)$ and $(t,s-1)$ exhibiting projective special (para)-Kähler geometry.

1. Introduction

The study of the geometry of the scalar manifolds of Euclidean $\mathcal{N}=2$ vector and hypermultiplets with or without coupling to supergravity has recently been considered in [1–4]. Our present work will only focus on theories with vector multiplets coupled to supergravity. In the standard supergravity theories with Lorentzian signature, it is well known that the scalar manifold is described by a projective special Kähler manifold in four dimensions and by a projective special real manifold in five dimensions. In [3], four-dimensional Euclidean supergravity theories were obtained by dimensionally reducing the five-dimensional Lorentzian theories of [5] over a time-like circle. It was established that the scalar geometries of the four-dimensional Euclidean vector multiplets can be obtained by replacing complex structures by para-complex structures. Four-dimensional Euclidean supergravity theories were also obtained as a dimensional reduction of the Euclidean ten-dimensional supergravity on a Calabi–Yau three-fold [6]. The Killing spinor equations of the Euclidean four-dimensional supergravities were also obtained in [7] and their gravitational solutions admitting Killing spinors were analysed in [8].

The theories of $\mathcal{N}=2$, five-dimensional Euclidean vector multiplets coupled to supergravity has recently been constructed in [9]. The Lagrangian of the Euclidean theory is the same as in the Lorentzian theory except that the gauge fields terms appear with the opposite sign. Multi-centred solutions of the gauged versions of these theories were recently studied in [10]. The dimensional reduction of the five-dimensional Euclidean theory on a circle produces the $\mathcal{N}=2$ four-dimensional Euclidean supergravity of [3] but with the non-conventional signs of the gauge terms.

In this work, our aim is to obtain $\mathcal{N}=2$ four and five-dimensional supergravity theories coupled to vector multiplets in various space–time signatures (t,s), where t and s refer, respectively, to the number of time and spatial dimensions. Space–times with various signatures are of mathematical and physical interest. For instance, spaces with $(2,2)$ signature have applications to string theory, M-theory, cosmology and twistor theory [11]. Moreover, the $(2,2)$ theory and its solutions without matter fields, have been considered in [12]. More recently, a classification of solutions with Killing spinors for the $(2,2)$ Einstein–Maxwell theory with a cosmological constant was given in [13].

We organise our work as follows. In Sec. 2, the five-dimensional supergravity theories are constructed through the analysis of the integrability conditions arising from generalised Killing spinor equations. The theories with various signatures are then obtained by reducing Hull's eleven-dimensional supergravity [14] on Calabi–Yau threefold. In Sec. 3, we obtain new $\mathcal{N}=2$ four-dimensional supergravity theories via the dimensional reductions of the five-dimensional supergravities on space and time-like circles. In particular, we obtain $\mathcal{N}=2$ supergravity with signature $(2,2)$ with scalar manifold described by a projective special para-Kähler manifold. We also obtain the Killing spinor equations for the reduced four-dimensional $\mathcal{N}=2$ supergravity theories. We end with a summary of our results.

E-mail address: ws00@aub.edu.lb.

2. (t, s) five-dimensional supergravity

We start our analysis with the original theory of $\mathcal{N}=2$, $D=5$ supergravity theory coupled to Abelian vector multiplets constructed in [5]. The theory contains the gravity multiplet and n vector multiplets and its bosonic Lagrangian is given by

$$\hat{\mathbf{e}}^{-1}\hat{\mathcal{L}}_5 = \frac{1}{2}\hat{R} - \frac{1}{2}G_{ij}\partial_{\hat{\mu}}h^i\partial^{\hat{\mu}}h^j - \frac{1}{4}G_{ij}F^i_{\hat{\mu}\hat{\nu}}F^{j\hat{\mu}\hat{\nu}} + \frac{\hat{\mathbf{e}}^{-1}}{48}C_{ijk}\epsilon^{\hat{\mu}\hat{\nu}\hat{\rho}\hat{\sigma}\hat{\tau}}F^i_{\hat{\mu}\hat{\nu}}F^j_{\hat{\rho}\hat{\sigma}}A^k_{\hat{\tau}}, \tag{2.1}$$

where C_{ijk} are real constants symmetric in i, j, k. Here $i, j, .. = 1, ..., n$, while the hatted indices refer to the space–time indices. The dynamics of (2.1) is encoded in the cubic potential

$$\mathcal{V} = \frac{1}{6}C_{ijk}h^ih^jh^k, \tag{2.2}$$

where the very special coordinates h^i are functions of the n real scalar fields belonging to the vector multiplets. The scalar manifold is described by the very special geometry constraint

$$\mathcal{V} = 1. \tag{2.3}$$

The gauge coupling metric takes the form

$$G_{ij} = -\frac{1}{2}\left(\partial_{h^i}\partial_{h^j}(\ln\mathcal{V})\right)_{\mathcal{V}=1} = \frac{1}{2}\left(9h_ih_j - C_{ijk}h^k\right), \tag{2.4}$$

where the dual fields h_i are given by

$$h_i = \frac{1}{6}C_{ijk}h^jh^k. \tag{2.5}$$

The Killing spinor equations arising from the vanishing of the fermionic fields and their supersymmetry transformations can be written in the form[1]

$$\begin{aligned}&\left[\hat{D}_{\hat{\mu}} + \frac{i}{8}h_i\left(\Gamma_{\hat{\mu}}{}^{\hat{\rho}\hat{\sigma}} - 4\delta^{\hat{\rho}}_{\hat{\mu}}\Gamma^{\hat{\sigma}}\right)F^i_{\hat{\rho}\hat{\sigma}}\right]\hat{\varepsilon} = 0,\\ &\left[i\left(F^i - h^ih_jF^j\right)_{\hat{\rho}\hat{\sigma}}\Gamma^{\hat{\rho}\hat{\sigma}} - 2\partial_{\hat{\mu}}h^i\Gamma^{\hat{\mu}}\right]\hat{\varepsilon} = 0.\end{aligned} \tag{2.6}$$

In what follows we obtain five-dimensional theories with various space–time signatures. One method to do so is through the analysis of the integrability of the Killing spinor equations. We start by allowing for a slight modification of the Killing spinor equations and write

$$\begin{aligned}&\left[\hat{D}_{\hat{\mu}} + \frac{\alpha}{8}h_i\left(\Gamma_{\hat{\mu}}{}^{\hat{\rho}\hat{\sigma}} - 4\delta^{\hat{\rho}}_{\hat{\mu}}\Gamma^{\hat{\sigma}}\right)F^i_{\hat{\rho}\hat{\sigma}}\right]\hat{\varepsilon} = 0,\\ &\left[\alpha\left(F^i - h^ih_jF^j\right)_{\hat{\rho}\hat{\sigma}}\Gamma^{\hat{\rho}\hat{\sigma}} - 2\partial_{\hat{\mu}}h^i\Gamma^{\hat{\mu}}\right]\hat{\varepsilon} = 0.\end{aligned} \tag{2.7}$$

After some calculation one can derive the following integrability condition

$$\begin{aligned}&2\alpha\left[\left(\hat{D}^{\hat{\nu}}\left(G_{ij}F^j_{\hat{\nu}\hat{\lambda}}\right) - h_ih^j\hat{D}^{\hat{\nu}}\left(G_{jl}F^l_{\hat{\nu}\hat{\lambda}}\right)\right)\Gamma^{\hat{\lambda}} + \frac{\alpha}{16}\left(h_iC_{jkl}h^l - C_{ijk}\right)F^j_{\hat{\lambda}_1\hat{\lambda}_2}F^k_{\hat{\lambda}_3\hat{\lambda}_4}\Gamma^{\hat{\lambda}_1\hat{\lambda}_2\hat{\lambda}_3\hat{\lambda}_4}\right]\hat{\varepsilon}\\ &+\alpha^2F^j_{\hat{\lambda}_1\hat{\lambda}_2}F^{k\hat{\lambda}_1\hat{\lambda}_2}\left(9h_ih_jh_k - \frac{1}{4}C_{jkl}h^lh_i + \frac{1}{4}C_{ijk} - \frac{3}{2}C_{ijm}h^mh_k\right)\hat{\varepsilon}\\ &+\left(3\hat{D}^{\hat{\nu}}\hat{D}_{\hat{\nu}}h_i + \frac{9}{2}h_i\hat{D}_{\hat{\nu}}h_j\hat{D}^{\hat{\nu}}h^j - \frac{1}{2}C_{ijk}\hat{D}_{\hat{\nu}}h^j\hat{D}^{\hat{\nu}}h^k\right)\hat{\varepsilon}\\ &= 0.\end{aligned} \tag{2.8}$$

The vanishing of the second and third terms in the above equation constitute the equations of motion for the scalar fields in a theory where the gauge kinetic terms coefficient is $\frac{\alpha^2}{4}$. Assuming that this is the case, then (2.8) reduces to

$$\left[\hat{D}^{\hat{\nu}}\left(G_{ij}F^j_{\hat{\nu}\hat{\lambda}}\right)\Gamma^{\hat{\lambda}} - \frac{\alpha}{16}C_{ijk}F^j_{\hat{\lambda}_1\hat{\lambda}_2}F^k_{\hat{\lambda}_3\hat{\lambda}_4}\Gamma^{\hat{\lambda}_1\hat{\lambda}_2\hat{\lambda}_3\hat{\lambda}_4}\right]\hat{\varepsilon} = 0. \tag{2.9}$$

If we modify the action (2.1) to take the form

$$\hat{\mathbf{e}}^{-1}\hat{\mathcal{L}}_5 = \frac{1}{2}\hat{R} - \frac{1}{2}G_{ij}\partial_{\hat{\mu}}h^i\partial^{\hat{\mu}}h^j + \frac{\alpha^2}{4}\left(G_{ij}F^i_{\hat{\mu}\hat{\nu}}F^{j\hat{\mu}\hat{\nu}} - \frac{\hat{\mathbf{e}}^{-1}}{12}C_{ijk}\epsilon^{\hat{\mu}\hat{\nu}\hat{\rho}\hat{\sigma}\hat{\tau}}F^i_{\hat{\mu}\hat{\nu}}F^j_{\hat{\rho}\hat{\sigma}}A^k_{\hat{\tau}}\right), \tag{2.10}$$

then the equations of motion for the gauge fields derived from (2.10) are given by

$$\hat{D}^{\hat{\nu}}\left(G_{ij}F^j_{\hat{\nu}\hat{\lambda}}\right) + \frac{1}{16}C_{ijk}\epsilon_{\hat{\lambda}}{}^{\hat{\lambda}_1\hat{\lambda}_2\hat{\lambda}_3\hat{\lambda}_4}F^j_{\hat{\lambda}_1\hat{\lambda}_2}F^k_{\hat{\lambda}_3\hat{\lambda}_4} = 0. \tag{2.11}$$

In theories with (1, 4), (3, 2) and (5, 0) signatures (odd numbers of time dimensions), (2.9) is consistent with (2.11) for the standard sign of the gauge terms, i.e., $\alpha = i$. For the mirror theories with signatures (4, 1), (2, 3) and (0, 5), consistency implies $\alpha = -1$ and thus the opposite sign of the gauge terms.

Five-dimensional $\mathcal{N}=2$ supergravity theories with (1, 4) signature can be obtained via the dimensional reduction of eleven-dimensional supergravity with the bosonic action [15]

$$S_{11} = \int_{M_{11}}\frac{1}{2}R\tilde{*}1 - \frac{1}{4}F_4\wedge\tilde{*}F_4 - \frac{1}{12}C_3\wedge F_4\wedge\tilde{*}F_4 \tag{2.12}$$

and signature (1, 10) on a Calabi–Yau three-fold, CY_3 [16]. Here $F_4 = dC_3$ and C_3 is a 3-form. The eleven-dimensional space-time manifold decomposes into $M_{11} = CY_3 \times M_5$, where M_5 is a Lorentzian five-dimensional manifold. Some useful details on the mathematics of CY_3 as well as the reduction on CY_3 can be found for example in [17–20].

We shall briefly present the basics of the reduction relevant to our discussion. One considers the deformations of the CY_3 metric that preserve the $SU(3)$ holonomy. These are the zero modes of the internal wave operator which correspond to deformations of the Kähler class and the complex structure. Ignoring the complex structure moduli, fluctuations of the CY_3 metric are then expanded as

$$\delta g_{A\bar{B}} = -iV^i_{A\bar{B}}\delta q^i \tag{2.13}$$

where q^i are the Kähler moduli taken to depend on the coordinates of M_5 and

$$V^i = V^i_{A\bar{B}}d\xi^A\wedge d\bar{\xi}^{\bar{B}}, \qquad i = 1, \dots, h_{(1,1)}, \tag{2.14}$$

are the basis of $h_{(1,1)}$ harmonic forms, ξ^A represent the three complex coordinates of CY_3. Note that the Kähler form and the volume of CY_3 are given by

$$\begin{aligned}J &= ig_{A\bar{B}}d\xi^A\wedge d\bar{\xi}^{\bar{B}} = q^iV^i,\\ \tilde{\mathcal{V}} &= \frac{1}{3!}\int_{CY_3}J\wedge J\wedge J = \frac{1}{6}C_{ijk}q^iq^jq^k.\end{aligned} \tag{2.15}$$

[1] Our conventions are as follows: The Clifford algebra is $\{\Gamma^a, \Gamma^b\} = 2\eta^{ab}$. The covariant derivative on spinors is $\hat{D}_{\hat{\mu}} = \partial_{\hat{\mu}} + \frac{1}{4}\omega_{\hat{\mu},ab}\Gamma^{ab}$ where $\omega_{\hat{\mu},ab}$ is the spin connection. Finally, antisymmetrization is with weight one, so $\Gamma^{a_1a_2\cdots a_n} = \frac{1}{n!}\Gamma^{[a_1}\Gamma^{a_2}\dots\Gamma^{a_n]}$.

The Kähler moduli space metric is given by

$$G_{ij}(q) = -\frac{3}{Cqqq}\left((Cq)_{ij} - \frac{3\,(Cqq)_i\,(Cqq)_j}{2\,(Cqqq)}\right), \tag{2.16}$$

where we have used the notation

$$Cqqq = C_{ijk}q^iq^jq^k, \qquad (Cqq)_i = C_{ijk}q^jq^k, \quad (Cq)_{ij} = C_{ijk}q^k.$$

Next one has to evaluate the eleven-dimensional Ricci curvature in terms of the Kähler moduli taking into consideration that for a CY_3, we have

$$g_{\bar{A}\bar{B}} = g_{AB} = R_{AB} = R_{\bar{A}\bar{B}} = R_{A\bar{B}} = 0. \tag{2.17}$$

In addition, we use the Kaluza–Klein ansatz for the three-form

$$C_3 = A^i \wedge V^i, \tag{2.18}$$

then the reduction of the action (2.12) after a rescaling of the five-dimensional metric and redefining scalars

$$g_{\hat{\mu}\hat{\nu}} \to \tilde{\mathcal{V}}^{-\frac{2}{3}} g_{\hat{\mu}\hat{\nu}}, \qquad h^i = \tilde{\mathcal{V}}^{-1/3}q^i, \tag{2.19}$$

gives[2]

$$S_5 = \int_{M_5} \frac{1}{2}R\tilde{*}1 - \frac{1}{2}G_{ij}(h)dh^i \wedge \hat{*}dh^j - \frac{1}{4}G_{ij}F_2^i \wedge \hat{*}F_2^j$$
$$- \frac{1}{12}C_{ijk}A^i \wedge F_2^j \wedge F_2^k, \tag{2.20}$$

where $G_{ij}(h)$ is given by (2.4) and $F_2^i = dA^i$. Note that $G_{ij}(h)$ is obtained from (2.16) by simply replacing the q^i with h^i.

The action of the eleven-dimensional supergravities constructed by Hull [14] can be written in the form

$$S_{11} = \frac{1}{2}\int_{M_{11}} R\tilde{*}1 + \frac{\alpha^2}{2}F_4 \wedge \tilde{*}F_4 - \frac{1}{6}C_3 \wedge F_4 \wedge \tilde{*}F_4, \tag{2.21}$$

where $\alpha^2 = -1$ for the theories with signatures $(1,10)$, $(5,6)$ and $(9,2)$, and $\alpha^2 = 1$ for the mirror theories with signatures $(10,1)$, $(6,5)$ and $(2,9)$. In the reduction of the theories with signatures $(1,10)$, $(5,6)$ and $(2,9)$, the CY_3 is of signature $(0,6)$ and thus M_5 is of signature $(1,4)$, $(5,0)$ and $(2,3)$. For the reduction of theories with signatures $(10,1)$, $(6,5)$ and $(9,2)$, the CY_3 is of signature $(6,0)$ and thus M_5 is of signature $(4,1)$, $(0,5)$ and $(3,2)$. All the five-dimensional supergravity theories obtained have the action

$$S_5 = \int_{M_5} \frac{1}{2}R\tilde{*}1 - \frac{1}{2}G_{ij}(h)dh^i \wedge \hat{*}dh^j + \frac{\alpha^2}{4}G_{ij}(h)F_2^i \wedge \hat{*}F_2^j$$
$$- \frac{1}{12}C_{ijk}A^i \wedge F_2^j \wedge F_2^k. \tag{2.22}$$

3. Four-dimensional supergravity

Starting with the action (2.22) of the $\mathcal{N}=2$ supergravity theory in five dimensions with (t,s) signature, we reduce the theory on a space-like and time-like circle. The Kaluza–Klein reduction ansatz is given by

$$\hat{\mathbf{e}}^a = e^{-\phi/2}\mathbf{e}^a, \qquad \hat{\mathbf{e}}^0 = e^{\phi}(dt - \sqrt{2}\mathcal{A}^0),$$
$$A^i = e^{-\phi}x^i\hat{\mathbf{e}}^0 + \sqrt{2}\mathcal{A}^i, \qquad h^i = e^{-\phi}y^i. \tag{3.1}$$

[2] We have ignored a kinetic term for the scalar field related to the volume of the Calabi–Yau that belongs to the hypermultiplet sector.

All the fields are taken to be independent of the compact dimension labelled by index 0, and the vector $\mathcal{A}^0$ has a vanishing component along the compact dimension. The non-vanishing components of the spin connection are given by

$$\hat{\omega}_{0,0\hat{a}} = -\epsilon e^{\frac{\phi}{2}}\partial_a\phi,$$
$$\hat{\omega}_{0,\hat{a}\hat{b}} = -\frac{\epsilon}{\sqrt{2}}e^{2\phi}\mathcal{F}^0_{ab},$$
$$\hat{\omega}_{\hat{a},0\hat{b}} = -\frac{\epsilon}{\sqrt{2}}e^{2\phi}\mathcal{F}^0_{ab},$$
$$\hat{\omega}_{\hat{a},\hat{b}\hat{c}} = e^{\frac{\phi}{2}}\left(\omega_{a,bc} + \frac{1}{2}\delta_{ac}\partial_b\phi - \frac{1}{2}\delta_{ab}\partial_c\phi\right), \tag{3.2}$$

where $\epsilon = 1$ corresponds to a reduction on a time-like circle, and $\epsilon = -1$ on a space-like circle. Note that all the indices on the right hand side of (3.2) are four dimensional, $\omega_{a,bc}$ are the spin connections of the four-dimensional theory with basis $\mathbf{e}^a$ and $\mathcal{F}^0 = d\mathcal{A}^0$.

The reduction of the action in (2.22) results in the following Lagrangian

$$\mathbf{e}^{-1}\mathcal{L}_4 = \frac{1}{2}R - g_{ij}\left(\partial_a x^i\partial^a x^j + \alpha^2\epsilon\partial_a y^i\partial^a y^j\right)$$
$$+\frac{\epsilon}{6}Cyyy(g_{ij}\mathcal{F}^i\cdot\mathcal{F}^j - 2\,(gx)_i\,\mathcal{F}^0\cdot\mathcal{F}^i$$
$$+ (gxx)\,\mathcal{F}^0\cdot\mathcal{F}^0 + \frac{1}{4}\mathcal{F}^0\cdot\mathcal{F}^0)$$
$$+\frac{\epsilon}{8}\epsilon^{abcd}\left((Cx)_{ij}\,\mathcal{F}^i_{ab}\mathcal{F}^j_{cd} - (Cxx)_i\,\mathcal{F}^0_{ab}\mathcal{F}^i_{cd}\right.$$
$$\left.+\frac{1}{3}(Cxxx)\,\mathcal{F}^0_{ab}\mathcal{F}^0_{cd}\right), \tag{3.3}$$

where $\mathcal{F}^i\cdot\mathcal{F}^j = \mathcal{F}^i_{ab}\mathcal{F}^{jab}$ and

$$g_{ij} = \frac{1}{2}\alpha^2\epsilon e^{-2\phi}G_{ij},$$
$$e^{3\phi} = \frac{1}{6}Cyyy. \tag{3.4}$$

Using (2.4) and (3.1) we get

$$g_{ij} = -\frac{3}{2}\epsilon\alpha^2\left(\frac{(Cy)_{ij}}{(Cyyy)} - \frac{3\,(Cyy)_i\,(Cyy)_j}{2\,(Cyyy)^2}\right). \tag{3.5}$$

In what follows we shall demonstrate that the action (3.3) describes a four-dimensional $\mathcal{N}=2$ supergravity theory with various signatures coupled to vector multiplets with the Lagrangian

$$\mathbf{e}^{-1}\mathcal{L} = \frac{1}{2}R - g_{ij}\partial_\mu z^i\partial^\mu\bar{z}^j$$
$$-\frac{\alpha^2}{4}\left(\mathrm{Im}\mathcal{N}_{IJ}\mathcal{F}^I\cdot\mathcal{F}^J + \mathrm{Re}\mathcal{N}_{IJ}\mathcal{F}^I\cdot\tilde{\mathcal{F}}^J\right), \tag{3.6}$$

with the prepotential

$$F = \frac{1}{6}C_{ijk}\frac{X^iX^jX^k}{X^0}. \tag{3.7}$$

The n complex scalar fields z^i of $\mathcal{N}=2$ vector multiplets are coordinates of a projective special (para)-Kähler manifold. In the symplectic formulation of the theory [21], one introduces the symplectic vectors

$$V = \begin{pmatrix} X^I \\ F_I \end{pmatrix}, \tag{3.8}$$

satisfying the symplectic constraint

$$i_\epsilon\left(\bar{X}^I F_I - X^I \bar{F}_I\right) = i_\epsilon\left(F_{IJ} - \bar{F}_{IJ}\right) X^I \bar{X}^J = -N_{IJ} X^I \bar{X}^J = 1. \tag{3.9}$$

Here $X^I = \mathrm{Re}X^I + i_\epsilon \mathrm{Im}X^I$, i_ϵ satisfies $\bar{i}_\epsilon = -i_\epsilon$ and $i_\epsilon^2 = \tau$, where $\tau = 1$ for the case when the scalar fields geometry is given by a projective special para-Kähler manifold and $\tau = -1$ when it is given by a projective special Kähler manifold, $F_I = \frac{\partial F}{\partial X^I}$ and $F_{IJ} = \frac{\partial^2 F}{\partial X^I \partial X^J}$. The constraint (3.9) can be solved by setting

$$X^I = e^{K(z,\bar{z})/2} Z^I(z) \tag{3.10}$$

where $K(z,\bar{z})$ is the Kähler potential. Then we have

$$e^{-K(z,\bar{z})} = -N_{IJ} Z^I(z) \bar{Z}^J(\bar{z}) . \tag{3.11}$$

The metric of the special (para)-Kähler manifold is given by

$$g_{ij} = \frac{\partial^2 K(z,\bar{z})}{\partial z^i \partial \bar{z}^j}, \tag{3.12}$$

and locally its $U(1)$ connection A is given

$$A = -\frac{i_\epsilon}{2}(\partial_i K dz^i - \partial_{\bar{i}} K d\bar{z}^i). \tag{3.13}$$

A convenient choice of inhomogeneous coordinates z^i are the special coordinates defined by

$$Z^0(z) = 1, \qquad Z^i(z) = z^i . \tag{3.14}$$

Defining

$$z^i = x^i - i_\epsilon y^i, \tag{3.15}$$

then for theories with cubic prepotentials given in (3.7), we obtain for the scalars kinetic term

$$g_{ij}\partial_\mu z^i \partial^\mu \bar{z}^j = \frac{3}{2Cyyy}\tau\left((Cy)_{ij} - \frac{3}{2}\frac{(Cyy)_i (Cyy)_j}{Cyyy}\right) \times \left(\partial_\mu x^i \partial^\mu x^i - \tau \partial_\mu y^i \partial^\mu y^i\right). \tag{3.16}$$

The gauge field coupling matrix is given by

$$\mathcal{N}_{IJ} = F_{IJ}(X) + i_\epsilon \tau \frac{(N\bar{X})_I (N\bar{X})_J}{\bar{X}N\bar{X}}, \tag{3.17}$$

which for theories with cubic prepotential gives

$$\begin{aligned}
\mathcal{N}_{00} &= \frac{1}{3}Cxxx + \tau i_\epsilon Cyyy\left(\frac{2}{3}gxx + \frac{1}{6}\right),\\
\mathcal{N}_{0i} &= -\frac{1}{2}(Cxx)_i - \frac{2}{3}\tau i_\epsilon Cyyy\,(gx)_i,\\
\mathcal{N}_{ij} &= (Cx)_{ij} + \frac{2}{3}\tau i_\epsilon g_{ij} Cyyy .
\end{aligned} \tag{3.18}$$

Using the above information we obtain

$$\begin{aligned}
&\left(\mathrm{Im}\mathcal{N}_{IJ}\mathcal{F}^I\cdot\mathcal{F}^J + \mathrm{Re}\mathcal{N}_{IJ}\mathcal{F}^I\cdot\tilde{\mathcal{F}}^J\right)\\
&= \tau\, Cyyy\left[-\frac{4}{3}(gx)_i\mathcal{F}^0\cdot\mathcal{F}^i + \frac{2}{3}g_{ij}\,\mathcal{F}^i\cdot\mathcal{F}^j + \left(\frac{2}{3}gxx + \frac{1}{6}\right)\mathcal{F}^0\cdot\mathcal{F}^0\right]\\
&\quad + \frac{1}{3}\left(Cxxx\mathcal{F}^0\cdot\tilde{\mathcal{F}}^0 - 3(Cxx)_i\mathcal{F}^0\cdot\tilde{\mathcal{F}}^i + 3(Cx)_{ij}\mathcal{F}^i\cdot\tilde{\mathcal{F}}^j\right).
\end{aligned} \tag{3.19}$$

After making the identification $\tau = -\alpha^2\epsilon$, and defining $\tilde{\mathcal{F}}^{iab} = \frac{\tau}{2}\epsilon^{abcd}\mathcal{F}^i_{cd}$, (3.6) is equivalent to (3.3).

Starting in five dimensions with the signatures $(1,4)$, $(3,2)$ and $(5,0)$ and $\alpha^2 = -1$, the reduction on a time-like circle results in four-dimensional $\mathcal{N} = 2$ supergravity theories with signatures $(0,4)$, $(2,2)$ and $(4,0)$. The Euclidean supergravity theory (signature $(0,4)$) is the one first obtained in [3]. The theory of $\mathcal{N} = 2$ supergravity with $(2,2)$ signature is new and shares some of the features of the Euclidean theory in the fact that the scalars are described by a projective special para-Kähler geometry. The reduction of the theories with signatures $(1,4)$ and $(3,2)$ on a space-like circle produces $\mathcal{N} = 2$ supergravity theories with signature $(1,3)$ and $(3,1)$. These are the well known original theories of $\mathcal{N} = 2$ supergravity [21] with projective special Kähler geometry. Similarly one obtains $\mathcal{N} = 2$ supergravity theories with signatures $(0,4)$, $(2,2)$ and $(4,0)$ via the reduction of the theories with $(4,1)$, $(2,3)$ and $(0,5)$ signatures on a space-like circle. We also obtain new $\mathcal{N} = 2$ supergravity theories with signatures $(3,1)$ and $(1,3)$ as reductions of the five-dimensional theories with signatures $(4,1)$ and $(2,3)$ on a time-like circle. These theories have the non-canonical sign of the gauge fields kinetic terms and have a projective special Kähler scalar manifold.

The Killing spinors equations of the five-dimensional supergravity theories with signatures $(3,2)$, $(1,4)$ and $(5,0)$ are given by (2.6). The reduction of these equations, using the results of [7], gives

$$\left[D_a - \frac{i}{2}\epsilon\Gamma_0 A_a + \frac{i}{4}e^{K/2}\Gamma.\mathcal{F}^I\left(\mathrm{Im}Z^J + i\epsilon\Gamma_0\mathrm{Re}Z^J\right)(\mathrm{Im}\mathcal{N})_{IJ}\Gamma_a\right]\varepsilon = 0, \tag{3.20}$$

and

$$\begin{aligned}
&\frac{i}{2}e^{K/2}(\mathrm{Im}\,\mathcal{N})_{IJ}\Gamma.\mathcal{F}^I\left[\mathrm{Im}(g^{ij}\bar{\mathcal{D}}_j\bar{Z}^I) + i\epsilon\Gamma_0\mathrm{Re}(g^{ij}\bar{\mathcal{D}}_j\bar{Z}^I)\right]\varepsilon\\
&+\Gamma^a\partial_a\left(\mathrm{Re}\,z^i - i\,\mathrm{Im}\,z^i\Gamma_0\right)\varepsilon\\
&=0,
\end{aligned} \tag{3.21}$$

where

$$\begin{aligned}
D_a &= \partial_a + \frac{1}{4}\omega_{a,bc}\Gamma^{bc},\\
A_a &= -\frac{i_\epsilon}{2}(\partial_i K\partial_a z^i - \partial_{\bar{i}} K\partial_a\bar{z}^i),\\
\bar{\mathcal{D}}_j\bar{Z}^I &= \partial_{\bar{j}}\bar{Z}^I + \partial_{\bar{j}}K\bar{Z}^I.
\end{aligned} \tag{3.22}$$

We also have $\hat{\varepsilon} = e^{-\phi/4}\varepsilon$, $(\Gamma_0)^2 = -\epsilon$ and $\Gamma^0 = -\epsilon\Gamma_0$. For $\epsilon = 1$, we obtain the Killing spinors for the four-dimensional $\mathcal{N} = 2$ supergravity theories with $(2,2)$, $(0,4)$ and $(4,0)$ while for $\epsilon = -1$ we obtain the Killing spinors for the $\mathcal{N} = 2$ supergravity theories with signatures $(3,1)$ and $(1,3)$.

The Killing spinors equations of the five-dimensional supergravity theories with signature $(2,3)$, $(4,1)$ and $(0,5)$ are given by

$$\begin{aligned}
&\left[\hat{D}_{\hat{\mu}} - \frac{1}{8}h_i\left(\Gamma_{\hat{\mu}}{}^{\hat{\rho}\hat{\sigma}} - 4\delta_{\hat{\mu}}^{\hat{\rho}}\Gamma^{\hat{\sigma}}\right)F^i_{\hat{\rho}\hat{\sigma}}\right]\hat{\varepsilon} = 0,\\
&\left(F^i - h^i h_j F^j\right)_{\hat{\rho}\hat{\sigma}}\Gamma^{\hat{\rho}\hat{\sigma}}\hat{\varepsilon} + 2\partial_{\hat{\mu}}h^i\Gamma^{\hat{\mu}}\hat{\varepsilon} = 0.
\end{aligned} \tag{3.23}$$

Those can be shown to reduce to

$$\left[D_a + \frac{1}{2}\epsilon\Gamma_0 A_a - \frac{1}{4}e^{K/2}\Gamma.\mathcal{F}^I\left(\mathrm{Im}Z^J - \epsilon\Gamma_0\mathrm{Re}Z^J\right)(\mathrm{Im}\mathcal{N})_{IJ}\Gamma_a\right]\varepsilon = 0, \tag{3.24}$$

and

$$-\frac{1}{2}e^{K/2}(\mathrm{Im}\,\mathcal{N})_{IJ}\Gamma.\mathcal{F}^{I}\left[\mathrm{Im}(g^{ij}\bar{\mathcal{D}}_{j}\bar{Z}^{I})-\epsilon\Gamma_{0}\mathrm{Re}(g^{ij}\bar{\mathcal{D}}_{j}\bar{Z}^{I})\right]\varepsilon$$
$$+\Gamma^{a}\partial_{a}\left(\mathrm{Re}\,z^{i}-\mathrm{Im}\,z^{i}\Gamma_{0}\right)\varepsilon$$
$$=0. \tag{3.25}$$

For $\epsilon=-1$, we obtain the Killing spinors for four-dimensional $\mathcal{N}=2$ superactivities with signatures $(2,2)$, $(4,0)$ and $(0,4)$. The Killing spinors for theories with signatures $(1,3)$, $(3,1)$ correspond to $\epsilon=1$.

4. Summary

In this work we have constructed $\mathcal{N}=2$ four and five-dimensional supergravity theories in various space–time signatures. The five-dimensional theories were constructed by employing the integrability conditions of the Killing spinor equations as well as by the reduction of the eleven-dimensional supergravities constructed by Hull [14] on a CY_3. Among the five-dimensional theories constructed, we obtained the Euclidean five-dimensional supergravity recently constructed in [9] and its mirror theory. The four-dimensional supergravity theories were then obtained as reductions of the five-dimensional theories on a time-like and space-like circles. One of the new four-dimensional supergravity theories obtained are the Lorentzian theories with signature $(1,3)$ with projective special Kähler geometry and with the wrong sign of the gauge coupling terms. Solutions of these $(1,3)$ theories with space-like Killing vectors were considered in [22]. There, these theories were labelled as fake theories. In the present work, however, they were shown to be genuine theories with higher dimensional origins. Also, in four dimensions a new theory with signature $(2,2)$ is obtained where the scalar manifold is described by a projective special para-Kähler manifold. A future direction is finding solutions to all these theories. The Killing spinor equations constructed should provide a starting point for a systematic analysis of their supersymmetric solutions. Also of interest is the reduction of the four-dimensional theories down to three dimensions and the investigations of the resulting c-maps along the lines of [4]. We hope to address these questions in forthcoming publications.

Acknowledgements

The author would like to thank J. Figueroa-O'Farrill for useful discussions. The author also thanks the School of Mathematics at the University of Edinburgh for hospitality when this work was completed. This work is supported in part by the National Science Foundation under grant number PHY-1620505.

References

[1] V. Cortes, C. Mayer, T. Mohaupt, F. Saueressig, Special geometry of Euclidean supersymmetry I: vector multiplets, J. High Energy Phys. 03 (2004) 028.

[2] V. Cortes, C. Mayer, T. Mohaupt, F. Saueressig, Special geometry of Euclidean supersymmetry II: hypermultiplets and the c-map, J. High Energy Phys. 06 (2005) 024.

[3] V. Cortes, T. Mohaupt, Special geometry of Euclidean supersymmetry III: the local r-map, instantons and black holes, J. High Energy Phys. 07 (2009) 066.

[4] V. Cortés, P. Dempster, T. Mohaupt, O. Vaughan, Special geometry of Euclidean supersymmetry IV: the local c-map, arXiv:1507.04620 [hep-th].

[5] M. Gunaydin, G. Sierra, P.K. Townsend, The geometry of N=2 Maxwell–Einstein supergravity and Jordan algebras, Nucl. Phys. B 242 (1984) 244.

[6] W.A. Sabra, O. Vaughan, 10D to 4D Euclidean supergravity over a Calabi–Yau three-fold, Class. Quantum Gravity 33 (2015) 1033.

[7] J.B. Gutowski, W.A. Sabra, Euclidean N=2 supergravity, Phys. Lett. B 718 (2012) 610.

[8] J.B. Gutowski, W.A. Sabra, Para-complex geometry and gravitational instantons, Class. Quantum Gravity 30 (2013) 195001.

[9] W.A. Sabra, O. Vaughan, Euclidean supergravity in five dimensions, Phys. Lett. B 760 (2016) 14.

[10] W.A. Sabra, Euclidean supergravity and multi-centered solutions, Phys. Lett. B 767 (2017) 253.

[11] I. Bars, Survey of two-time physics, Class. Quantum Gravity 18 (2001) 3113;
I. Bars, P. Steinhardt, N. Turok, Local conformal symmetry in physics and cosmology, Phys. Rev. D 89 (2014) 043515;
J.W. Barrett, G.W. Gibbons, M.J. Perry, C.N. Pope, P. Ruback, Kleinian geometry and the $N=2$ superstring, Int. J. Mod. Phys. A 9 (1994) 1457;
I. Bars, C. Deliduman, D. Minic, Lifting M-theory to two-time physics, Phys. Lett. B 457 (1999) 275;
H. Ooguri, C. Vafa, Selfduality and $N=2$ string magic, Mod. Phys. Lett. A 5 (1990) 1389;
R. Penrose, Twistor algebra, J. Math. Phys. 8 (1967) 345;
E. Witten, Perturbative gauge theory as a string theory in twistor space, Commun. Math. Phys. 252 (2004) 189.

[12] R.L. Bryant, Pseudo-Riemannian Metrics with Parallel Spinor Fields and Vanishing Ricci Tensor, Semin. Congr., vol. 4, Soc. Math. France, Paris, 2000, p. 53;
M. Dunajski, Anti-self-dual four manifolds with a parallel real spinor, Proc. R. Soc. Lond. A 458 (2002) 1205;
M. Dunajski, Einstein–Maxwell-dilaton metrics from three-dimensional Einstein–Weyl structures, Class. Quantum Gravity 23 (2006) 2833;
M. Dunajski, S. West, Anti-self-dual conformal structures in neutral signature;
S. Hervik, Pseudo-Riemannian VSI spaces II, Class. Quantum Gravity 29 (2012) 095011.

[13] D. Klemm, M. Nozawa, Geometry of Killing spinors in neutral signature, Class. Quantum Gravity 32 (2015) 185012.

[14] C.M. Hull, Duality and the signature of space–time, J. High Energy Phys. 11 (1998) 017.

[15] E. Cremmer, B. Julia, J. Scherk, Supergravity theory in eleven-dimensions, Phys. Lett. B 76 (1978) 409.

[16] A. Ceresole, R. D'Auria, S. Ferrara, 11-dimensional supergravity compactified on Calabi–Yau threefolds, Phys. Lett. B 357 (1995) 76.

[17] M. Bodner, A.C. Cadavid, S. Ferrara, (2, 2) vacuum configurations for type IIA superstrings: N=2 supergravity Lagrangians and algebraic geometry, Class. Quantum Gravity 8 (1991) 789.

[18] S. Ferrara, M. Bodner, A.C. Cadavid, Calabi–Yau supermoduli space, field strength duality and mirror manifolds, Phys. Lett. B 247 (1990) 25.

[19] P. Candelas, X. de la Ossa, Moduli space of Calabi–Yau manifolds, Nucl. Phys. B 355 (1991) 455.

[20] F. Bonetti, T.W. Grimm, Six-dimensional $(1,0)$ effective action of F-theory via M-theory on Calabi–Yau threefolds, J. High Energy Phys. 05 (2012) 019.

[21] A. Van Proeyen, $N=2$ supergravity in $d=4,5,6$ and its matter couplings, extended version of lectures given during the semester "Supergravity, superstrings and M-theory" at Institut Henri Poincaré, Paris, November 2000, http://itf.fys.kuleuven.ac.be/~toine/home.htm#B.1.

[22] W.A. Sabra, Phantom metrics with Killing spinors, Phys. Lett. B 750 (2015) 237;
M. Bu Taam, W.A. Sabra, Phantom space–times in fake supergravity, Phys. Lett. B 751 (2015) 297.

The black hole quantum atmosphere

Ramit Dey [a,b], Stefano Liberati [a,b], Daniele Pranzetti [a,b,*]

[a] SISSA, Via Bonomea 265, 34136 Trieste, Italy
[b] INFN, Sezione di Trieste, Italy

ARTICLE INFO

Editor: M. Cvetič

ABSTRACT

Ever since the discovery of black hole evaporation, the region of origin of the radiated quanta has been a topic of debate. Recently it was argued by Giddings that the Hawking quanta originate from a region well outside the black hole horizon by calculating the effective radius of a radiating body via the Stefan–Boltzmann law. In this paper we try to further explore this issue and end up corroborating this claim, using both a heuristic argument and a detailed study of the stress energy tensor. We show that the Hawking quanta originate from what might be called a quantum atmosphere around the black hole with energy density and fluxes of particles peaked at about $4MG$, running contrary to the popular belief that these originate from the ultra high energy excitations very close to the horizon. This long distance origin of Hawking radiation could have a profound impact on our understanding of the information and transplanckian problems.

1. Introduction

The discovery of Hawking radiation [1] changed our perspective towards black holes, giving us a deeper insight about the microscopic nature of gravity. At the same time, within the semiclassical framework, the current understanding of such process still leaves open several issues. Of course, a well known unresolved problem of black hole physics is the information loss paradox [2–4], i.e. the apparent incompatibility between the complete thermal evaporation of a black hole endowed with an event horizon and unitary evolution as prescribed by quantum mechanics.

For restoring unitarity of Hawking radiation and addressing the information loss problem correctly, it is important (among other things) to know from where the Hawking quanta originate. For example, if one assumes a near horizon origin of the Hawking radiation, then one way to restore unitarity is by conjecturing some sort of UV-dependent entanglement between partner Hawking quanta which would enable the late time Hawking flux to retrieve the information in the early stages of the evaporation process. Such scenario seems to lead to the so called "firewall" argument as the conjectured lack of maximal entanglement between the Hawking pairs makes the near horizon state singular and eventually demands some drastic modification of the near horizon geometry [5]. On the other hand, if one believes in a longer distance origin of the Hawking quanta, some effect must be operational at a larger scale for restoring unitarity rather than near the horizon, avoiding the "firewall".

A similar open issue is the transplanckian origin of Hawking quanta. Hawking's original calculation indicates that the quanta originate near the black hole horizon in a highly blue-shifted state requiring an assumption on the UV completion of the effective field theory used for the computation and on the lack of backreaction on the underlying geometry.[1] While it was debated for a while if Hawking quanta could originate initially, during the star collapse, and later released over a very long time, it was convincingly argued in [8] that this cannot be the case if an event horizon indeed forms. This leads to the conclusion that the Hawking quanta are generated in a region outside the horizon. A conclusion corroborated by studies of the Hawking modes correlation structure where it was shown that mode conversion happens over a long distance from the horizon [9]. A more recent claim in this direction, based on calculating the size of the radiating body via the Stefan–Boltzmann law, showed that the Hawking quanta originate in a near horizon quantum region, a sort of black hole "*atmosphere*" [10]. It is a well known fact that the typical wavelength of the radiated quanta is comparable to the size of the black hole,

* Corresponding author.
E-mail addresses: rdey@sissa.it (R. Dey), liberati@sissa.it (S. Liberati), dpranzetti@sissa.it (D. Pranzetti).

[1] See, for instance, Refs. [6,7] for a black hole evaporation analysis where these issues can be addressed in a quantum gravity context.

so one might think that the point particle description is not very accurate. However, as measured by a local observer near the horizon, the wavelength is highly blue-shifted when traced back from infinity to the horizon, thus validating the point particle description.

The Hawking process can be explained heuristically as-well, for example via a tunneling mechanism where the particle tunnels out of the horizon or the anti particle (propagating backwards in time) tunnels into the horizon and as a result of this we get the constant Hawking flux at infinity [11]. Alternatively, one popular picture is to imagine that the strong tidal force near the black hole horizon stops the annihilation of the particle and anti-particle pairs that are formed spontaneously from the vacuum. Once the antiparticle is "hidden" within the black hole horizon, having a negative energy effectively, the other particle can materialize and escape to infinity [12,13].

In this paper we shall explicitly make use of this latter heuristic picture as well as of a full calculation of the stress energy tensor in $1+1$ dimensions. We shall see that both methods seem to agree in suggesting that the Hawking quanta originate from the black hole *atmosphere* and not from a region very close to the horizon. In section 2, based on the heuristic picture of Hawking radiation described above and invoking the uncertainty principle and tidal forces, we show that most of the contribution to the radiation spectrum comes from a region far away from the horizon. In section 3 we further strengthen our claim by a detailed calculation of the renormalized stress energy tensor, which indicates a similar result.

2. A gravitational Schwinger effect argument

One ingredient of our heuristic argument to identify a quantum atmosphere outside the black hole horizon, where particle creation takes place, is the uncertainty principle. However, the use of the uncertainty principle alone, as originally suggested by Parker [14], does not contain any physically relevant information about the location of particle production and why smaller black holes should be hotter. Indeed, the uncertainty principle in this case provides a rough estimate of the region of particle production as inversely proportional to the energy of the Hawking quanta when they are produced, but it does not take into account any dynamical mechanism to estimate the probability of spontaneous emission.

Thus one can improve this argument by invoking a physical process of creation of the Hawking quanta and using the uncertainty principle as a complementary tool to estimate the region of origin of the quanta. In this section, we try to achieve this goal by relying on tidal forces.

Let us then consider a situation where a virtual pair, consisting of a particle and anti-particle, pops out of the vacuum spontaneously for a very short time interval and then annihilates itself. In the Schwinger effect [15] a static electric field is assumed to act on a virtual electron–positron pair until the two partners are torn apart once the threshold energy necessary to become a real electron–positron pair is provided by the field. Energy is conserved due to the fact that the electric potential energy has opposite sign for partners with opposite charge. However, in its gravitational counterpart a priori only vacuum polarization can be induced by a static field in the absence of an horizon.

In fact, only in the presence of the latter one has both the characteristic peeling structure of geodesics (diverging away from the horizon on both its sides) as well as the presence of an ergoregion behind it.[2] The presence of an ergoregion is crucial for energy conservation as it allows for negative energy states given that in it the norm of the timelike Killing vector, with respect to which we compute energy, changes sign.

Indeed, if a Schwinger-like process takes place near the black hole horizon, due to the tidal force of the black hole and the peeling of geodesics, the pair can get spatially separated and one partner can enter the black hole horizon following a timelike or null curve with negative energy while the other particle can escape to infinity and contribute to the Hawking flux. In this picture, we are implicitly assuming that virtual particles in the vicinity of a black hole horizon move along geodesics when they are just about to go on-shell.

Therefore, the physical scenario we want to envisage is that of a particle–antiparticle pair pulled apart by the black hole tidal force outside the horizon until they go on-shell as one of them reaches the horizon[3] located at $r_s = 2GM/c^2$ (actually an infinitesimal distance inside it so that the geodesic motion will drag it further inside) while the other particle is at a radial coordinate distance $r = r_*$. Once on-shell, the outgoing particle eventually reaches infinity and contributes to the Hawking spectrum. In order to do so though, it has to be created with an energy corresponding to the energy of the Hawking quanta at a distance $r_* > r_s$ from the center of the black hole as measured by a local static observer; this can be reconstructed by noticing that

$$\omega_r = \frac{\omega_\infty}{\sqrt{g_{00}}}, \tag{1}$$

where ω_∞ is the energy at infinity and we are using the $(+,-,-,-)$ signature. At infinity, the thermal spectrum of Hawking radiation gives

$$\omega_\infty = \gamma \frac{k_B T_H}{\hbar}, \tag{2}$$

where the Hawking temperature for a black hole of mass M reads $k_B T_H = \frac{\hbar c^3}{8\pi GM}$, and γ is a numerical factor spanning the energy range of the quanta giving rise to the radiation thermal spectrum. At the peak of the spectrum $\gamma \approx 2.82$.

Thus, we get

$$\omega_\infty = \gamma \frac{c^3}{8\pi GM} \tag{3}$$

and

$$\omega_r = \gamma \frac{c}{4\pi r_s} \frac{1}{\sqrt{1-\frac{r_s}{r}}}. \tag{4}$$

This energy is provided by the work done by the gravitational field to pull the two partners apart. We can compute this work in the static frame outside a black hole and compare it with $\omega(r_*)$. Using this relation, we can determine the region from which the Hawking quanta originate. This is the process we now want to implement. Although in the rest of this Section we present the detailed derivation of the relation between the outgoing particle energy and the radial distance at which it goes on-shell for the massive case, our result holds also for massless particles. We comment at the end of this Section on how the same Schwinger effect

[2] This is strictly true only for non-rotating black holes, for rotating ones the ergoregion lies outside of the horizon allowing for the classical phenomenon of superradiance. However, the quantum emission still requires the peculiar peeling structure of geodesics typical of the horizon.

[3] One could also consider the case where the ingoing particle tunnels through the horizon and goes on-shell well inside the horizon (as e.g. suggested by the results of [9]); however, since in our analysis below we are interested in the tidal force as computed in the outgoing particle rest frame, this should not affect the final expression for the force. Thus, from the point of view of an outside static observer, the work done by the gravitational field on the pair (in our heuristic derivation) is insensitive to the exact location where the ingoing particle becomes real.

argument can be implemented straightforwardly to the massless case.

Let us clarify that, in a general relativistic framework, the geodesic deviation equation does not describe the force acting on a particle moving along a geodesic. Rather, it expresses how the spacetime curvature influences two nearby geodesics, making them either diverge or converge, i.e. it effectively measures tidal effects. Therefore, we can interpret these effects as the pull of the gravitational force on particles and talk about the work done by the gravitational field only in an heuristic sense. Nevertheless, in the case considered here where the test particles have a mass much smaller than the black hole and we can neglect back-reaction effects, we expect this interpretation of the gravitational field effects to capture some relevant aspects of black hole physics. With these assumptions spelled out, let us proceed.

In the rest frame of the outgoing particle, one would see the antiparticle accelerating towards the horizon due to the tidal force. This radial acceleration in the rest frame of the particle can be computed using the geodesic deviation equation, namely

$$a^r\big|_{r_*} \equiv \frac{Dn^r}{D\tau^2}\bigg|_{r_*} = R^r{}_{\mu\nu\rho}u^\mu u^\nu n^\rho\big|_{r_*}\,, \tag{5}$$

where the r.h.s. is expressed in terms of the Riemann tensor components, n^r denotes the separation between the two radially infalling geodesics followed by the pair of particles and $u^\mu = [1,0,0,0]$ in the rest frame of the particle.

The separation between the particle and the anti-particle when the pair forms spontaneously (i.e. they go "on-shell") is given by their Compton wavelength, namely $n^\rho = [0, n^r, 0, 0]$ where $n^r \sim \lambda_C = \hbar/mc$, and $m \ll M$ is the particles rest mass (from now on we shall work in units where $\hbar = c = 1$). So in the end, Eq. (5) implies that the radial component of the tidal acceleration (as computed in the rest frame of the particle at coordinate r_*) is given by[4]

$$a^r|_{r_*} = \frac{2MG}{r_*^3}\lambda_C \tag{6}$$

Our aim is to determine the work done on the spontaneously created particle pair by the tidal force in the static frame outside the black hole. For this we need to compute the tidal force as measured by a static observer outside the black hole at the instant when the outgoing partner goes on shell. This can be achieved by considering the particle rest frame and the static observer frame as locally two inertial frames: The latter sees the particle as moving with outward velocity given by the radial component of the geodesic tangent vector $u^r = dr/d\tau$. Once this is known, we can derive the radial acceleration observed by the static observer by performing a boost with rapidity $\zeta = \tanh^{-1}(u^r)$.

We thus need to determine the instantaneous radial component of the free fall velocity of the outgoing particle when it goes on-shell. This can be computed from the geodesic equation and it is given by

$$u^r = \frac{dr}{d\tau} = \sqrt{\frac{2MG}{r}\left(1-\frac{r}{r_0}\right)}\,, \tag{7}$$

where r_0 comes as an integration constant corresponding to the coordinate distance at which the particle velocity goes to zero. Since we are interested in the value of the radial component of the geodesic tangent vector at the instant when the outgoing particle goes on-shell and becomes an Hawking quantum which eventually reaches infinity, we can take the integration constant $r_0 \to \infty$, i.e. Hawking quanta can be created with zero velocity only at infinity. Hence, we get

$$u^r\big|_{r_*} = \sqrt{\frac{2MG}{r_*}}\,. \tag{8}$$

We can now boost the acceleration vector $a^\mu = (0, a^r, 0, 0)$, where a^r given by (6), with a velocity parameter given by (8), in order to determine the tidal force in the static frame a^r_{st}. We get $a^r_{\text{st}} = a^r\cosh(\zeta) = a_r(1-2MG/r)^{-1}$ so that the radial component of the force under this transformation is given by

$$F^r_{\text{tidal-st}}\big|_{r_*} = \frac{ma^r_{\text{st}}}{(1-2MG/r)}\bigg|_{r_*} = \frac{m\lambda_C}{(1-2MG/r_*)^2}\frac{2MG}{r_*^3}\,, \tag{9}$$

where we have rescaled the mass in the rest frame by the appropriate Lorentz factor, $(1-2MG/r_*)^{-1}$. Finally, using the fact that $\lambda_C \sim 1/m$, the magnitude of the force is given by

$$||F^r_{\text{tidal-st}}|| = \frac{2MG}{r_*^3}\left(1-\frac{r_s}{r_*}\right)^{-\frac{3}{2}}\,. \tag{10}$$

In analogy with the Schwinger effect, we shall now assume that the work done by the tidal force to split the virtual pair can be approximated by the product of the force computed above with the distance over which it appears to have acted, i.e. the separation of the two Hawking quanta as they go on-shell as measured by a static observer at r_*. Given that we have assumed that the ingoing Hawking quantum goes on shell as soon as it can do so, i.e. at horizon crossing, this distance will coincide with the static observer's proper distance to the horizon $d(r_*)$.

Therefore, the work required by the tidal force to split the pair apart is given by[5]

$$W_{\text{tidal}} \sim ||F^r_{\text{tidal-st}}||\, d(r_*) = \frac{2MG}{r_*^3}\left(1-\frac{r_s}{r_*}\right)^{-\frac{3}{2}} d(r_*)\,, \tag{11}$$

where $d(r_*)$ is given by

$$d(r_*) = \int_{r_s}^{r_*}\sqrt{g_{rr}}dr' \tag{12}$$
$$= r_s\left(\sqrt{\alpha(\alpha-1)} + \frac{1}{2}\log\left[\alpha\left(1+\sqrt{1-\frac{1}{\alpha}}\right)^2\right]\right)\,,$$

and we have defined $\alpha \equiv r_*/r_s$.

We can then equate this work to the total energy of the two Hawking quanta being created, namely $W_{\text{tidal}} = 2\omega_r$. This gives us

$$\frac{2MG}{r_*^3}\left(1-\frac{2MG}{r_*}\right)^{-\frac{3}{2}} d(r_*) = \frac{\gamma}{2\pi r_s}\left(1-\frac{2MG}{r_*}\right)^{-\frac{1}{2}}\,. \tag{13}$$

Finally, from eq. (13) we get

$$\gamma = \frac{2\pi}{\alpha^2}\left(1-\frac{1}{\alpha}\right)^{-\frac{1}{2}} \tag{14}$$
$$\cdot\left(1 + \frac{1}{2\sqrt{\alpha^2-\alpha}}\log\left[\alpha\left(1+\sqrt{1-\frac{1}{\alpha}}\right)^2\right]\right)\,.$$

[4] For computation of the acceleration in the rest frame of the particle we need the Riemann tensor in the inertial frame of the particle. One can compute the Riemann tensor in the static Schwarzschild coordinates and then boost it using the free-fall velocity of the particle as measured in the static frame. A feature of the Schwarzschild geometry is that the components of the Riemann tensor remain invariant under such a boost [16]. Thus, in (5) we have $R_{rttr} = -2MG/r^3$.

[5] Alternatively, we could introduce a 4-vector $\ell^\mu = (0, \ell^r, 0, 0)$, with $||\ell|| = \sqrt{g_{\mu\nu}\ell^\mu\ell^\nu} = d(r_*)$, and compute the work as $W_{\text{tidal}} \sim g_{rr}F^r_{\text{tidal-st}}\ell^r\big|_{r_*}$. This would give the same result.

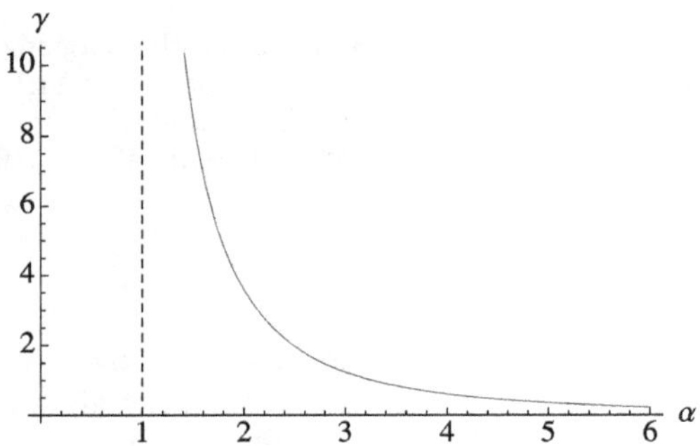

Fig. 1. This plot shows the variation of γ with respect to the radial distance from the center of the black hole. The red dashed line corresponds to the horizon location at $\alpha = 1$ where the expression for the tidal force work diverges, indicating that the quanta in the far UV tail of the Hawking spectrum originate from very near the horizon. (For interpretation of the references to color in this figure legend, the reader is referred to the web version of this article.)

The relation between γ and α, i.e. the radial distance scaled as r_*/r_s, is better illustrated in Fig. 1. It is clear from the plot that the part of the Hawking thermal spectrum around the peak ($\gamma \sim 2.82$), where most of the radiation is concentrated, corresponds to a region which extends far outside the horizon, up to around $2r_s$ (at the peak $r_* \approx 4.38\,MG$).

The plot above also shows how, in this tidal force derivation, the quanta with higher velocity (kinetic energy) are produced closer to the horizon. This is consistent with our analysis since the higher the initial radial velocity the stronger the Lorentz contraction of the outgoing particles distance from the horizon in their rest frame, given by λ_C, resulting in a shorter proper distance $d(r_*)$ at which they are detected.

Also, by using Eq. (12) and expressing the rest of Eq. (11) in terms of α, we can see that the work doable at fixed α by the tidal forces scales as the inverse of the mass of the black hole so making evident that smaller holes can produce hotter particles at the same relative distance from the horizon.

In the Schwinger effect argument we described in this Section we have considered the case of a massive test particle. However, in the physical context of a 4D Schwarzschild black hole, most of the radiation is emitted by massless particles. A generalization of our argument to the massless case can be achieved in a straightforward manner. In fact, despite the lack of a rest mass frame of one of the two partners, one can always study the Schwinger-like effect in a local inertial frame in the vicinity of the horizon and compute the radial acceleration (5) considering two radially infalling null geodesics with 4-velocity $u^\mu = [1, 1, 0, 0]$ in such given frame; due to the symmetries of the Riemann tensor, this leads to the same expression (6) but with the Compton wavelength λ_C replaced by the massless particle de Broglie wavelength λ_B. The acceleration as measured by a static observer outside the black hole then proceeds along the same lines as in the massive case, since the boost between the two frames, locally both inertial, is the same as in the massive case; we can thus compute the radial component of the tidal force as in (9), where on the r.h.s. we replace the combination $m\lambda_C$ with $E\lambda_B$, E being the massless particle energy, as measured by the static observer, which is related to the de Broglie wavelength by the standard relation $E = 1/\lambda_B$ (recall that we switched to units where $\hbar = c = 1$). In this way, we recover the expression (10), which is independent of the test particle mass. Therefore, the plot in Fig. 1 applies also to the case of radiation being emitted by massless particles.

Let us stress again the heuristic nature of our argument. We are considering the instantaneous value of the tidal force observed by the outgoing partner at a given coordinate distance r_* where it goes on-shell. However, we then use this instantaneous value to compute the work done by the gravitational field over a distance $d(r_*)$, as if the force was actually at work with the same constant value throughout the whole splitting process. A similar approach was also used in [17] to give an estimate of the wavelength of the Hawking quanta as produced by the gravitational tidal force.

So, although the analogy with the Schwinger effect for the electron–positron pair production by an electric field may be advocated to lend support to our description of Hawking quanta production from a quantum atmosphere that extends well beyond the horizon, we now want to present a more sound analysis based on the renormalized stress energy tensor in order to confirm this picture.

3. Stress-energy tensor

By analyzing the renormalized stress energy tensor (RSET) in the 2-dimensional case, one can understand Hawking radiation in a better way as this is a local object which can help to probe the physics in the vicinity of the black hole. The derivation of the RSET components has been considered in many places in the literature [18–22], here we build on these previous results and compute the energy density and flux as seen by an observer which has zero radial velocity (thus giving rise to no kinematical effects) and zero acceleration at the horizon.

3.1. Computation of RSET

Following [23], let us introduce a set of globally defined affine coordinates U, V on $\mathscr{I}^-_{\text{left}}$, $\mathscr{I}^-_{\text{right}}$ respectively. Restricting to the radial and time dimensions, the metric reads

$$ds^2 = C(U, V)dUdV\,. \tag{15}$$

In $(1+1)$ dimensions the renormalized stress energy tensor for any massless scalar field in terms of these affine null coordinates can be easily computed using the conformal anomaly [18–20,24,25]. The components of the RSET computed in some arbitrary vacuum state are given as:

$$\begin{aligned}\langle T_{UU}\rangle &= -\frac{1}{12\pi}C^{1/2}\partial_U^2 C^{-1/2}\\ &= \frac{1}{24\pi}\left[\frac{C_{,UU}}{C} - \frac{3}{2}\frac{(C_{,U})^2}{C^2}\right],\end{aligned} \tag{16}$$

$$\begin{aligned}\langle T_{VV}\rangle &= -\frac{1}{12\pi}C^{1/2}\partial_V^2 C^{-1/2}\\ &= \frac{1}{24\pi}\left[\frac{C_{,VV}}{C} - \frac{3}{2}\frac{(C_{,V})^2}{C^2}\right],\end{aligned} \tag{17}$$

$$\langle T_{UV}\rangle = \frac{RC}{96\pi} = \frac{1}{24\pi}\partial_U\partial_V \ln C\,, \tag{18}$$

where C is the conformal factor introduced in the above metric and R is the scalar curvature.

Now let us also introduce a null coordinate u affine on $\mathscr{I}^+_{\text{right}}$ such that

$$U = p(u)\,; \tag{19}$$

from this we get

$$\partial_U = \dot{p}^{-1}\partial_u\,. \tag{20}$$

In terms of the set (u, V), the metric reads

$$ds^2 = \bar{C}(u, V)dudV\,, \tag{21}$$

with

$$\bar{C}(u,V) = \dot{p}(u)C(U,V)\,. \tag{22}$$

Assuming that the observer is always outside the collapsing star, $\bar{C}(u,V)$ would be the metric component of a static spacetime. In terms of this newly defined null coordinate, a simple computation shows that T_{UU} is given as

$$\langle T_{UU}\rangle = -\frac{\dot{p}^{-2}}{12\pi}\left[\bar{C}^{1/2}\partial_u^2\bar{C}^{-1/2} - \dot{p}^{1/2}\partial_u^2\dot{p}^{-1/2}\right]. \tag{23}$$

Now T_{VV} will have only a static contribution if $V = v$ but if the affine null coordinate on $\mathscr{I}^+_{\text{left}}$ is defined as

$$V = q(v) \tag{24}$$

and we define $C'(U,v) = \dot{q}(v)C(U,V)$, T_{VV} is given as

$$\langle T_{VV}\rangle = -\frac{\dot{q}^{-2}}{12\pi}\left[C'^{1/2}\partial_v^2 C'^{-1/2} - \dot{q}^{1/2}\partial_v^2\dot{q}^{-1/2}\right]. \tag{25}$$

As mentioned earlier $\bar{C}(u,V)$ is the metric component of a static spacetime, so all the dynamics of the collapsing geometry is captured in the $\dot{p}$ term of (23). In the above analysis, by using another affine null coordinate, we can differentiate between the static contribution to the RSET and the one due to the dynamics associated with the collapse [23].

3.2. RSET for different vacuum states

Capturing the dependence at different radii of the RSET components would require a knowledge of the full $p(u)$ at any value of u, i.e. to specify a collapse history. However, this would lead to the inclusion of transient effects which are not relevant for the present discussion. For this reason, we shall here rely on the fact that, well after the collapse has settle down, the black hole geometry is formally indistinguishable from that of an eternal configuration [26,27] (where the form of $p(u)$ is simply fixed by the geometry, see (A.2)).

So, in order to extract physical information from the RSET, we shall compute the energy density and the flux experienced by an observer at constant Kruskal position long after the collapse has taken place in the two physically relevant states for Hawking radiation in the eternal black hole case, namely the Unruh and the Hartle–Hawking states. We shall start in this Section by explicitly evaluating the general expressions for the RSET components expectation values.

Using (A.2), we get the relations

$$\dot{p}(u) \equiv \partial_u p(u) = -\frac{p(u)}{2r_s}\,, \tag{26}$$

$$\ddot{p}(u) = \frac{p(u)}{4r_s^2} = -\frac{\dot{p}(u)}{2r_s}\,. \tag{27}$$

For computing the first term of (23) we can write

$$\bar{C}^{1/2}\partial_u^2\bar{C}^{-1/2} = \frac{3}{4}\bar{C}^{-2}\left(\partial_u\bar{C}\right)^2 - \frac{1}{2}\bar{C}^{-1}\partial_u^2\bar{C}\,. \tag{28}$$

Using the metric conformal factor C from (A.1) we get

$$\begin{aligned}\partial_u\bar{C} &= \partial_u[\dot{p}(u)C] = \ddot{p}C + \dot{p}\partial_u C\\ &= \dot{p}(u)\left(-\frac{1}{2r_s} + \frac{r^2 - r_s^2}{2r^2 r_s}\right)C\\ &= -\frac{r_s}{2r^2}\bar{C}\,,\end{aligned} \tag{29}$$

and

$$\partial_u^2\bar{C} = -\frac{1}{2}r_s\partial_u\left(\frac{\bar{C}}{r^2}\right) = \frac{r_s^2}{4r^4}\bar{C} - \frac{1}{2}\frac{r_s f(r)\bar{C}}{r^3}\,. \tag{30}$$

Using the above relation in (28) we have

$$\begin{aligned}&\bar{C}^{1/2}\partial_u^2\bar{C}^{-1/2} = \frac{3}{4}\bar{C}^{-2}\left[\frac{r_s^2}{4r^4}\bar{C}^2\right]\\ &-\frac{1}{2}\bar{C}^{-1}\left[\frac{r_s^2}{4r^4}\bar{C} - \frac{1}{2}\frac{r_s f(r)\bar{C}}{r^3}\right]\\ &= -\frac{3}{16}\frac{r_s^2}{r^4} + \frac{r_s}{4r^3} - \frac{3}{4}\frac{M^2G^2}{r^4} + \frac{MG}{2r^3}\,,\end{aligned} \tag{31}$$

where $f(r)$ is given in (A.8) and we used $r_s = 2MG$ in the last step. For the second term on the r.h.s. of (23), we have

$$\dot{p}^{1/2}\partial_u^2\dot{p}^{-1/2} = -\frac{\dot{p}^{1/2}}{2}\partial_u\left(\frac{\ddot{p}}{\dot{p}^{3/2}}\right) = \frac{1}{(8MG)^2}\,. \tag{32}$$

We are now ready to compute explicitly the expectation value of the different RSET components for the Hartle–Hawking ($|H\rangle$) and Unruh ($|U\rangle$) states.

We can start by observing that for the T_{UU} and T_{UV} components, the expectation values are the same in the two vacuum states [20]. Therefore, in the following we simply denote

$$\langle T_{UU}\rangle \equiv \langle H|T_{UU}|H\rangle = \langle U|T_{UU}|U\rangle\,, \tag{33}$$

$$\langle T_{UV}\rangle \equiv \langle H|T_{UV}|H\rangle = \langle U|T_{UV}|U\rangle\,. \tag{34}$$

By means of (31), (32), $\langle T_{UU}\rangle$ is given by

$$\begin{aligned}\langle T_{UU}\rangle &= \frac{\dot{p}^{-2}}{24\pi}\left[\frac{3}{2}\frac{M^2G^2}{r^4} - \frac{MG}{r^3} + \frac{1}{32M^2G^2}\right]\\ &= (768\pi M^2G^2)^{-1}\frac{V^2}{4r^2}e^{-r/MG}\\ &\cdot\left[1 + \frac{4MG}{r} + \frac{12M^2G^2}{r^2}\right].\end{aligned} \tag{35}$$

To compute $\langle T_{UV}\rangle$ we use (18), from which

$$\begin{aligned}\langle T_{UV}\rangle &= \frac{1}{24\pi}\partial_U\partial_V\ln C = \frac{1}{24\pi}(\dot{p}\dot{q})^{-1}\partial_u\partial_v\ln C\\ &= -\frac{1}{96\pi}(\dot{p}\dot{q})^{-1}C\partial_r^2 C.\end{aligned} \tag{36}$$

Using $C(t,r)$ from (A.1) and the exact values of $q(u)$ and $p(v)$, we get

$$\langle T_{UV}\rangle = -\frac{M^2G^2}{12\pi r^4}e^{-r/2MG}\,. \tag{37}$$

On the other hand, the dependence of $\langle T_{VV}\rangle$ on the state in which we are computing the expectation value is important. For the Hartle–Hawking state (eternal black hole scenario, non-singular vacuum state in both past and future horizons) in Kruskal coordinates the modes are given by $e^{-i\omega U}$, $e^{-i\omega V}$, where we defined V as

$$V \equiv q(v) = 2r_s e^{v/2r_s}\,. \tag{38}$$

Using this definition of V we can proceed in a similar way as for the computation of $\langle T_{UU}\rangle$. From (25), we obtain

$$\begin{aligned}\langle H|T_{VV}|H\rangle &= \frac{\dot{q}^{-2}}{24\pi}\left[\frac{3}{2}\frac{MG^2}{r^4} - \frac{MG}{r^3} + \frac{1}{32MG^2}\right]\\ &= (768\pi M^2G^2)^{-1}\frac{U^2}{4r^2}e^{-\frac{r}{MG}}\\ &\cdot\left[1 + \frac{4MG}{r} + \frac{12M^2G^2}{r^2}\right].\end{aligned} \tag{39}$$

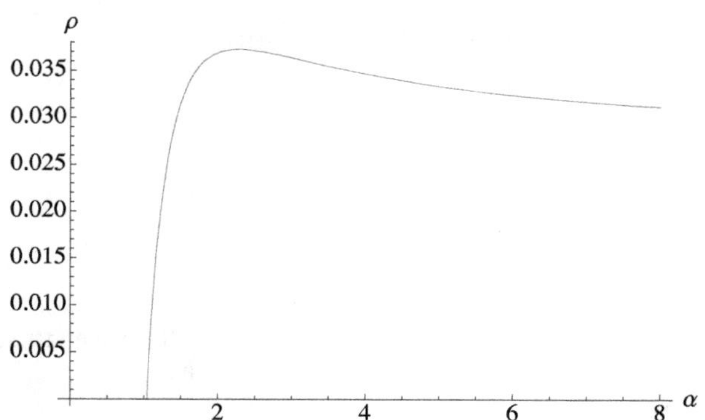

Fig. 2. Plot of the energy density at a given time as a function of the radial distance from the center of the black hole in Unruh state at a given instant of time.

For the Unruh state in Kruskal coordinates, the modes are given by $e^{-i\omega U}$, $e^{-i\omega v}$ and there is no regularization condition imposed in the past horizon. The expectation value of the T_{VV} component can be obtained from the relation

$$\langle U|T_{VV}|U\rangle = 16MG^2\dot{q}^{-2}\langle U|T_{vv}|U\rangle\,, \tag{40}$$

where $\langle U|T_{vv}|U\rangle$ can be computed from

$$\langle U|T_{vv}|U\rangle = -\frac{1}{12\pi}f(r)^{1/2}\partial_v^2 f(r)^{-1/2} \tag{41}$$

using $f(r) = \left(1-\frac{2MG}{r}\right)$, as follows from the metric of a black hole in static Schwarzschild coordinates. We have

$$\langle U|T_{vv}|U\rangle = \frac{1}{24\pi}\left[\frac{3M^2G^2}{2r^4} - \frac{MG}{r^3}\right], \tag{42}$$

and from (40) we get

$$\langle U|T_{VV}|U\rangle = \frac{1}{6\pi}\frac{M^2G^2}{V^2}\left[\frac{3M^2G^2}{2r^4} - \frac{MG}{r^3}\right]. \tag{43}$$

3.3. Energy density

We now have all the ingredients to extract physical information from the RSET. Let us first analyze the energy density as measured in the frame of an observer moving along fixed position in Kruskal coordinates.

Let us consider an observer at a given Kruskal position with 2-velocity $v^\mu = C^{-1/2}(1,0)$ (in $[T,X]$ coordinates).[6] The energy density, ρ, measured by this observer for the Unruh state is given by

$$\begin{aligned}\rho &= \langle U|T_{\mu\nu}|U\rangle v^\mu v^\nu = C^{-1}\langle U|T_{TT}|U\rangle \\ &= C^{-1}\langle U|T_{VV}+T_{UU}+2T_{UV}|U\rangle\,. \end{aligned}\tag{44}$$

Using (35), (37), (43) we can compute the energy density exactly and we plot it in Fig. 2 (where $\alpha \equiv r/r_s$).

The energy density (44) blows up at the horizon ($r=2M$) since we are computing the energy density as observed by a free falling (in Kruskal coordinates) observer in the Unruh state which is well known to be ill defined on the past horizon. Such divergence arises from the $1/V^2$ term in the component (43) when $V=0$, i.e. at the past horizon. The horizon location condition in Schwarzschild radial coordinate, $\alpha=1$, cannot distinguish between past and future horizons and thus the divergent contribution would enter in the plot above of the energy density expression (44) when evaluated at $\alpha=1$. However, a free falling observer at the future horizon would not see this divergence, which is just an artifact of Kruskal coordinates.[7] This is a well known fact already pointed in [19]. For this reason, we have removed the point $\alpha=1$ in the plot shown in Fig. 2.

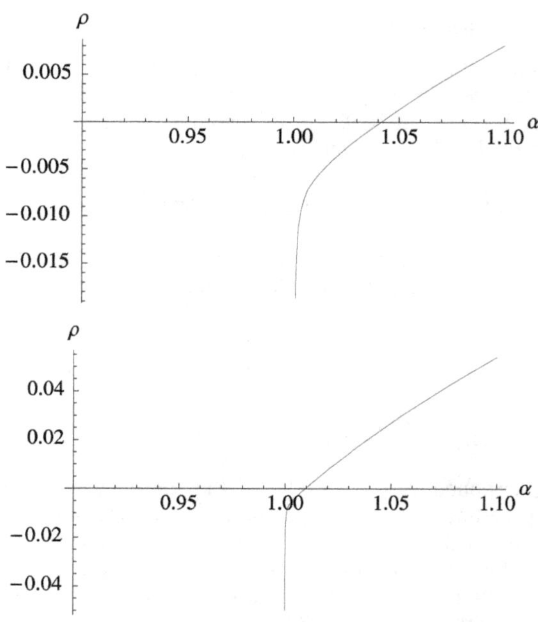

Fig. 3. Near horizon behavior of the energy density in the Unruh state at different times. The first plot corresponds to the same instant of time as the plot in Fig. 2; the second one to a close instant after.

Near the horizon the energy density becomes negative; these negative values are attained closer to the horizon as the energy density is measured at later times. We show this near horizon behavior in the two plots in Fig. 3, where the first is evaluated at the same time as the plot in Fig. 2 and the second one at a close instant after (a similar behavior was found also in [29]); the negative divergent behavior of the energy density at the horizon is clear from the plots.

However, let us remark again that this divergence is just fictitious for an observer crossing the future horizon $U=0$ at a given value of $V>0$ and it is an inevitable feature of plotting the energy density in the Unruh state as a function of r for a fixed instant of time t.

One way to avoid this misleading behavior of the energy density plot at the horizon could be to show it as a function of U for given $V = const > 0$; this would indeed remove the singularity from the plot since the point $\alpha=1$ would now correspond to $U=0$, i.e. to the future horizon where the Unruh state is regular. However, from such plot it would be very difficult to extrapolate the information about how the energy density is distributed in the

[6] This choice of trajectory is not geodesic; however the acceleration that the observer experiences is irrelevant compared to the Hawking temperature and one can show easily that the acceleration vanishes at the horizon. One might think that a free falling observer would have been a better choice. However, the problem with such choice would be the non-zero radial velocity of the free falling observer at the horizon, as well as near the horizon. In that case, it would then be difficult to separate out the Hawking radiation contribution from other kinematical effects [28].

[7] Let us stress that also the calculation in [23] of the RSET components in the collapse scenario shows that at the white hole horizon the Unruh state will necessarily be singular. This can be easily realized by applying time reversal to the subdominant terms in the dynamical contribution (32) derived in [23] (see Eq. (52) there), which then shows an exponentially growing flux at the white horizon which very rapidly would create a divergence in the T_{UU} component of the RSET soon after horizon formation.

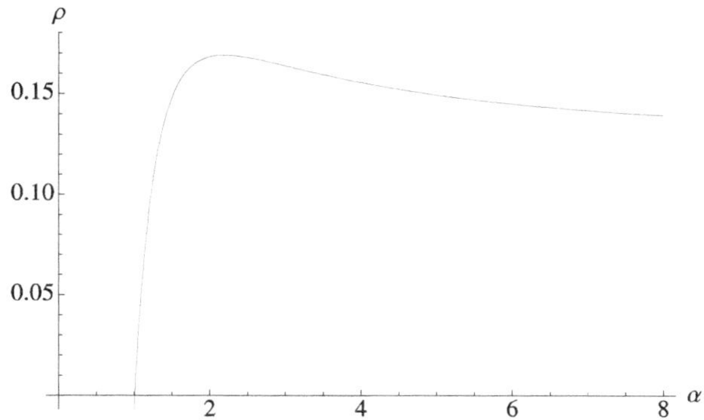

Fig. 4. Plot of the variation of energy density computed in Hartle–Hawking state with respect to the radial distance from the center of the black hole at fixed time measured in the static frame. Notice that close to the horizon the energy density is negative also in this case, but it remains finite at the horizon due to the non-divergent behavior of the T_{VV} component (39) in the Hartle–Hawking vacuum.

r coordinate for *fixed* time t, since fixing V and letting U run imply that different values of U correspond to different values of r *and* t.

The significant aspect of the plot in Fig. 2 for us is the peak in the distribution of ρ that is obtained outside the horizon which is at $r \approx 4.32MG$. Quite in agreement with our heuristic prediction based on the gravitational analogue of the Schwinger effect. Let us point out that, although we have shown the plot at a given instant of Killing time, the behavior of the energy density remains the same at any time, in particular the presence of the peak at the same location persists; the only difference is that the value of the energy density increases since it accumulates, given that we are not taking into account the effect of back-reaction.

To get a non-singular energy density plot for the free falling observer we should consider the Hartle–Hawking state. This is given by

$$\rho = \langle H|T_{\mu\nu}|H\rangle v^{\mu} v^{\nu} = C^{-1}\langle H|T_{TT}|U\rangle$$
$$= C^{-1}\langle H|T_{VV} + T_{UU} + 2T_{UV}|H\rangle\,. \qquad (45)$$

Using the expectation values given in (35), (37), (39), we can plot the energy density (45) with respect to radial distance parametrized by α. This is shown in Fig. 4, where we see a similar nature of the distribution with a peak outside the horizon; however, as expected, in this case the energy density is regular everywhere. Remarkably, the peak is located at $r \approx 4.37MG$, in close agreement with our heuristic findings.

These results strongly support our previous claim that the radiation density is maximized in a region outside the horizon. We now show that a similar behavior with a peak away from the horizon is exhibited also by the flux part of the RSET.

3.4. Flux

The flux of the Hawking radiation in the Unruh vacuum is given by [30][8]

$$F = -\langle U|T_{\mu\nu}|U\rangle v^{\mu} z^{\nu}\,, \qquad (46)$$

where v^{μ} is the velocity of the observer and z^{ν} is the contravariant component of the normal to the observer. Let us consider a static observer at fixed distance in a Kruskal frame with $v^{\mu} = C^{-1/2}[1,0]$ and indicate the normal vector as $z^{\nu} = [A, B]$. The latter has to satisfy the following conditions

[8] In the Hartle–Hawking vacuum the flux vanishes due to the thermal equilibrium of the state.

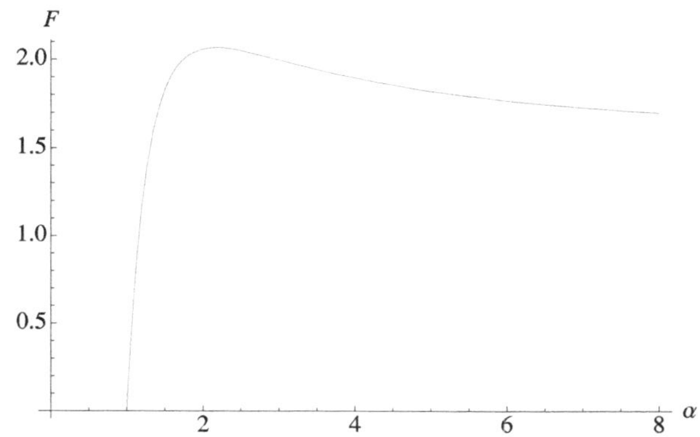

Fig. 5. This plot shows the variation of the flux of Hawking radiation with respect to the radial distance as measured by an observer in the Unruh state at a given instant of time.

$$g_{\mu\nu} z^{\mu} z^{\nu} = -1\,,\ z^{\mu} v_{\mu} = 0\,. \qquad (47)$$

Using the second relation we get $A = 0$ and from the first relation we get $B = C^{-1/2}$. Therefore, $z^{\nu} = C^{-1/2}[0,1]$.

Using these expressions for v^{μ}, z^{ν}, we get

$$F = -C^{-1}\langle U|T_{TX}|U\rangle = C^{-1}\langle U|[-T_{VV} + T_{UU}]|U\rangle\,. \qquad (48)$$

Plugging in the expectation values (35), (43) found above, we can plot the flux as a function of α. This is shown in Fig. 5. Also in this case the plot of the flux would receive a fictitious (for a free falling observer at the future horizon) divergent contribution from the component (43), and we have thus removed the point $\alpha = 1$ from the plot, thus avoiding the divergence at the past horizon $V = 0$. We see that the flux has a maximum at $r = 4.32MG$ and most of the contribution to the Hawking radiation comes from a region between the horizon and $r \approx 6MG$.

Let us remark that our findings are in line with the analysis of the 2-dimensional RSET done in [10] where it was shown that the ingoing and outgoing null components of the stress tensor would build up to their asymptotic values in a region outside the horizon. In our analysis we have been more precise in confirming this result by choosing an observer and explicitly computing the values of the energy density and the flux outside the horizon as measured by the observer.

4. Summary and discussion

It has been widely believed that Hawking radiation originates from the excitations close to the horizon and this eventually suggested some drastic modification of the states in the near horizon regime as a resolution to the information loss paradox [5,31–33]. One of the primary reasons for such an argument is based on the way Hawking did his original calculation, tracing back the modes all the way from future infinity to the past null infinity through the collapsing matter so that one has a vacuum state at the horizon for a free-falling observers.

The other disturbing feature about this argument is, when the modes are traced back they become highly blueshifted near the horizon and we are not well aware of the laws of physics in such high transplanckian domain. Some resolutions to the above problem have been proposed several times in the literature [34–36] but they all demand some challenging modification to our present knowledge of gravitation or quantum field theory.

Let us stress, however, that the UV departures from Lorentz invariance through the introduction of a fundamental cutoff postulated in [37,38] are relevant only very close to the horizon for large black holes (in units of the Lorentz breaking scale). Hence, even contemplating such scenario, our analysis in section 3 would be basically unchanged and unaffected away from the horizon, as also stressed in the similar analysis carried out in [39].

In this paper we have shown evidence that the Hawking quanta originate from a region which is far outside the horizon, which can be called a black hole *atmosphere*. More precisely, from the plots of the energy density and the flux in the Unruh state we get a maximum at $r \approx 4.32MG$, for the energy density in the Hartle–Hawking state the peak is at $r \approx 4.37MG$. This is strikingly close to our previous finding for an origin at about $r \approx 4.38MG$ for the peak of the thermal spectrum using the heuristic argument based on tidal forces. By large this is also in agreement with some previous claims using various other methods, such as calculating the effective radius of a radiating body using the Stefan–Boltzmann law or computing the effective Tolman temperature [10,40–42], as well as in close correspondence with the results of the study of the null component of the stress-energy tensor in the Unruh vacuum of [43].

Given the presence of a quantum atmosphere where the Hawking quanta are generated and which extends well beyond the black hole horizon, as originally suggested in [10], it would be interesting to investigate how its effective radius is affected by going to higher dimensions. Applying the Stefan–Boltzmann radiation law argument proposed in [10] for the $(3+1)$-dimensional case to $(D+1)$-dimensional Schwarzschild black holes, it was found in [40] that the effective radius gets squeezed towards the black hole horizon as the number of spatial dimensions increases.

Given that there is no derivation of the RSET components in dimensions higher than $(1+1)$, we cannot apply the argument presented in Section 3 to confirm this result. However, the heuristic derivation that we presented in Section 2 could be easily generalized for any arbitrary number of dimensions. Without presenting a complete derivation, we can understand in a qualitative way how the quantum atmosphere can be effected by going to arbitrary $(D+1)$ higher dimensions by considering the fact that the Hawking temperature scales as $T_{BH} = \frac{(D-2)\hbar}{8\pi MG}$, where D is the number of spatial dimensions. It can be shown that this dimensional scaling of the temperature, along with the modification of the Schwarzschild metric for an arbitrary D, would yield, for given $r = r_*$, a higher value of ω_r (1) as D increases. At the same time, it can be shown from dimensional arguments that the work done by the tidal force must decrease in value for the same given r as D increases. This implies that, for a fixed $D > 3$, the peak of the Hawking radiation spectrum corresponds to an higher value of energy than in the $D = 3$ case and, in order for the gravitational field to be able to provide enough work to reach such amount of energy, the outgoing partners comprising the bulk of the spectrum at infinity must go on-shell closer to the horizon. Our Schwinger effect argument thus confirms in a qualitative way the relation obtained in [40] for the decrease of the effective radius in the regime $D \gg 1$.

If the radiation has a long distance origin then we might not need to worry about the transplanckian issue at the horizon. Moreover, concerning the fundamental issue of unitarity of black hole evaporation, this result suggests to consider some effect operational at this new scale in order to eventually restore unitarity of Hawking radiation. A possible scenario is the one of non-violent nonlocality advocated in [44,45]; see also the proposal of [46,47]. We hope that the present contribution will stimulate further investigations in these directions.

Acknowledgments

We thank Renaud Parentani, Sebastiano Sonego and Matt Visser for illuminating discussions. We also acknowledge the John Templeton Foundation for the supporting grant #51876.

Appendix A. Kruskal frame

We want to examine the components of the RSET in a globally well defined coordinate system free of any pathological behavior (other than a true curvature singularity, like in the center of a black hole). For this purpose the Kruskal coordinate frame is an appropriate choice. The Kruskal metric is given as

$$ds^2 = \frac{r_s}{r} e^{-r/r_s} dU dV \,, \tag{A.1}$$

where r_s is the radius of the event horizon. For this coordinate system we have

$$U = p(u) = -2r_s e^{-u/2r_s} \,, \tag{A.2}$$

$$V = q(v) = 2r_s e^{v/2r_s} . \tag{A.3}$$

The affine null coordinate u, v in terms of radial distance from the center of the black hole, "r", and time, "t", as measured by a static observer is given as

$$u = t - r_* = t - \left[r + r_s \ln\left(\frac{r}{r_s} - 1\right)\right], \tag{A.4}$$

$$v = t + r_* = t + \left[r + r_s \ln\left(\frac{r}{r_s} - 1\right)\right], \tag{A.5}$$

also

$$\partial_u = \frac{\partial r_*}{\partial u}\partial_{r_*} = -\frac{1}{2}\partial_{r_*} = -\frac{1}{2} f(r)\partial_r \,, \tag{A.6}$$

$$\partial_v = \frac{\partial r_*}{\partial v}\partial_{r_*} = \frac{1}{2}\partial_{r_*} = \frac{1}{2} f(r)\partial_r , \tag{A.7}$$

where we used

$$\frac{dr_*}{dr} = [f(r)]^{-1} = \left(1 - \frac{r_s}{r}\right)^{-1} . \tag{A.8}$$

We can also define a set of time like and radial coordinates (T, X) as

$$T = \frac{1}{2}(V + U), \, X = \frac{1}{2}(V - U). \tag{A.9}$$

Using this metric (A.1) is given as

$$ds^2 = \frac{r_s}{r} e^{-r/r_s} (dT^2 - dX^2) . \tag{A.10}$$

References

[1] S.W. Hawking, Particle creation by black holes, Commun. Math. Phys. 43 (1975) 199.

[2] S.W. Hawking, The unpredictability of quantum gravity, Commun. Math. Phys. 87 (1982) 395.

[3] D.N. Page, Information in black hole radiation, Phys. Rev. Lett. 71 (1993) 3743, arXiv:hep-th/9306083.

[4] S.B. Giddings, Black holes and massive remnants, Phys. Rev. D 46 (1992) 1347, arXiv:hep-th/9203059.

[5] A. Almheiri, D. Marolf, J. Polchinski, J. Sully, Black holes: complementarity or firewalls?, J. High Energy Phys. 02 (2013) 062, arXiv:1207.3123.

[6] D. Pranzetti, Radiation from quantum weakly dynamical horizons in LQG, Phys. Rev. Lett. 109 (2012) 011301, arXiv:1204.0702.

[7] D. Pranzetti, Dynamical evaporation of quantum horizons, Class. Quantum Gravity 30 (2013) 165004, arXiv:1211.2702.

[8] W.G. Unruh, Origin of the particles in black-hole evaporation, Phys. Rev. D 15 (1977) 365.

[9] R. Parentani, From vacuum fluctuations across an event horizon to long distance correlations, Phys. Rev. D 82 (2010) 025008, arXiv:1003.3625.

[10] S.B. Giddings, Hawking radiation, the Stefan–Boltzmann law, and unitarization, Phys. Lett. B 754 (2016) 39, arXiv:1511.08221.

[11] M.K. Parikh, F. Wilczek, Hawking radiation as tunneling, Phys. Rev. Lett. 85 (2000) 5042, arXiv:hep-th/9907001.

[12] S. Hawking, W. Israel, General Relativity; an Einstein Centenary Survey, Cambridge University Press, ISBN 9780521222853, 1979.
[13] R.J. Adler, P. Chen, D.I. Santiago, The generalized uncertainty principle and black hole remnants, Gen. Relativ. Gravit. 33 (2101) (2001), arXiv:gr-qc/0106080.
[14] L. Parker, The Production of Elementary Particles by Strong Gravitational Fields, Springer US, Boston, MA, ISBN 978-1-4684-2343-3, 1977, pp. 107–226.
[15] J.S. Schwinger, On gauge invariance and vacuum polarization, Phys. Rev. 82 (1951) 664.
[16] C. Misner, K. Thorne, J. Wheeler, Gravitation, Gravitation, vol. 3, W. H. Freeman, ISBN 9780716703440, 1973.
[17] A. Grib, S. Mamayev, V. Mostepanenko, Vacuum Quantum Effects in Strong Fields, Friedmann Laboratory Pub., 1994, https://books.google.it/books?id=azBdcgAACAAJ.
[18] P.C.W. Davies, S.A. Fulling, W.G. Unruh, Energy–momentum tensor near an evaporating black hole, Phys. Rev. D 13 (1976) 2720.
[19] W.G. Unruh, Origin of the particles in black hole evaporation, Phys. Rev. D 15 (1977) 365.
[20] N. Birrell, P. Davies, Quantum Fields in Curved Space, Cambridge Monographs on Mathematical Physics, Cambridge University Press, ISBN 9780521278584, 1984.
[21] S. Singh, S. Chakraborty, Black hole kinematics: the "in"-vacuum energy density and flux for different observers, Phys. Rev. D 90 (2014) 024011, arXiv:1404.0684.
[22] S. Chakraborty, S. Singh, T. Padmanabhan, A quantum peek inside the black hole event horizon, J. High Energy Phys. 06 (2015) 192, arXiv:1503.01774.
[23] C. Barcelo, S. Liberati, S. Sonego, M. Visser, Fate of gravitational collapse in semiclassical gravity, Phys. Rev. D 77 (2008) 044032, arXiv:0712.1130.
[24] P. Candelas, Vacuum polarization in Schwarzschild space–time, Phys. Rev. D 21 (1980) 2185.
[25] T. Padmanabhan, Gravity and the thermodynamics of horizons, Phys. Rep. 406 (49) (2005), arXiv:gr-qc/0311036.
[26] I. Racz, R.M. Wald, Extension of space–times with Killing horizon, Class. Quantum Gravity 9 (1992) 2643.
[27] I. Racz, R.M. Wald, Global extensions of space–times describing asymptotic final states of black holes, Class. Quantum Gravity 13 (539) (1996), arXiv:gr-qc/9507055.
[28] L.C. Barbado, C. Barcelo, L.J. Garay, Hawking radiation as perceived by different observers, Class. Quantum Gravity 28 (2011) 125021, arXiv:1101.4382.
[29] M. Eune, Y. Gim, W. Kim, Something special at the event horizon, Mod. Phys. Lett. A 29 (2014) 1450215, arXiv:1401.3501.
[30] L.H. Ford, T.A. Roman, Motion of inertial observers through negative energy, Phys. Rev. D 48 (776) (1993), arXiv:gr-qc/9303038.
[31] K. Papadodimas, S. Raju, An infalling observer in AdS/CFT, J. High Energy Phys. 10 (2013) 212, arXiv:1211.6767.
[32] S.L. Braunstein, S. Pirandola, K. Życzkowski, Better late than never: information retrieval from black holes, Phys. Rev. Lett. 110 (2013) 101301, arXiv:0907.1190.
[33] A. Almheiri, D. Marolf, J. Polchinski, D. Stanford, J. Sully, An apologia for firewalls, J. High Energy Phys. 09 (2013) 018, arXiv:1304.6483.
[34] W.G. Unruh, Sonic analogue of black holes and the effects of high frequencies on black hole evaporation, Phys. Rev. D 51 (1995) 2827.
[35] S. Corley, T. Jacobson, Hawking spectrum and high frequency dispersion, Phys. Rev. D 54 (1996) 1568, arXiv:hep-th/9601073.
[36] T. Jacobson, Black-hole evaporation and ultrashort distances, Phys. Rev. D 44 (1991) 1731.
[37] W.G. Unruh, Sonic analog of black holes and the effects of high frequencies on black hole evaporation, Phys. Rev. D 51 (1995) 2827.
[38] T. Jacobson, Black hole evaporation and ultrashort distances, Phys. Rev. D 44 (1991) 1731.
[39] R. Brout, S. Massar, R. Parentani, P. Spindel, Hawking radiation without trans-Planckian frequencies, Phys. Rev. D 52 (1995) 4559, arXiv:hep-th/9506121.
[40] S. Hod, Hawking radiation and the Stefan–Boltzmann law: the effective radius of the black-hole quantum atmosphere, Phys. Lett. B 757 (2016) 121, arXiv:1607.02510.
[41] M. Eune, Y. Gim, W. Kim, Effective Tolman temperature induced by trace anomaly, arXiv:1511.09135, 2015.
[42] W. Kim, Origin of Hawking radiation: firewall or atmosphere?, Gen. Relativ. Gravit. 49 (2017) 15, arXiv:1604.00465.
[43] R. Parentani, R. Brout, Physical interpretation of black hole evaporation as a vacuum instability, Int. J. Mod. Phys. D 1 (1992) 169.
[44] S.B. Giddings, Black holes, quantum information, and unitary evolution, Phys. Rev. D 85 (2012) 124063, arXiv:1201.1037.
[45] S.B. Giddings, Nonviolent nonlocality, Phys. Rev. D 88 (2013) 064023, arXiv:1211.7070.
[46] Y. Nomura, F. Sanches, S.J. Weinberg, Black hole interior in quantum gravity, Phys. Rev. Lett. 114 (2015) 201301, arXiv:1412.7539.
[47] Y. Nomura, F. Sanches, S.J. Weinberg, Relativeness in quantum gravity: limitations and frame dependence of semiclassical descriptions, J. High Energy Phys. 04 (2015) 158, arXiv:1412.7538.

Thermodynamics of novel charged dilatonic BTZ black holes

M. Dehghani

Department of Physics, Ilam University, Ilam, Iran

ARTICLE INFO

Editor: N. Lambert

Keywords:
Charged BTZ black hole
Charged black hole with scalar hair
Maxwell's theory of electrodynamics
Three-dimensional dilatonic black holes

ABSTRACT

In this paper, the three-dimensional Einstein–Maxwell theory in the presence of a dilatonic scalar field has been studied. It has been shown that the dilatonic potential must be considered as the linear combination of two Liouville-type potentials. Two new classes of charged dilatonic BTZ black holes, as the exact solutions to the coupled scalar, vector and tensor field equations, have been obtained and their properties have been studied. The conserved charge and mass of the new black holes have been calculated, making use of the Gauss's law and Abbott–Deser proposal, respectively. Through comparison of the thermodynamical extensive quantities (i.e. temperature and entropy) obtained from both, the geometrical and the thermodynamical methods, the validity of the first law of black hole thermodynamics has been confirmed for both of the new black holes we just obtained. A black hole thermal stability or phase transition analysis has been performed, making use of the canonical ensemble method. Regarding the black hole heat capacity, it has been found that for either of the new black hole solutions there are some specific ranges in such a way that the black holes with the horizon radius in these ranges are locally stable. The points of type one and type two phase transitions have been determined. The black holes, with the horizon radius equal to the transition points are unstable. They undergo type one or type two phase transitions to be stabilized.

1. Introduction

Although the Einstein's tensorial theory of gravitation is in agreement with a large amount of observational tests, but it fails regarding some important issues [1–5]. Modification of the Einstein's theory of gravity is one of the main approaches to overcome the related failures. Among the various proposed modifications [6–13], the so-called scalar-tensor theories [14], as the modification arisen from string theory, have provided interesting results [15]. The Einstein's action is naturally modified by the scalar-tensor superstring terms at the high energy regime. In the low energy limit of the string theory, the Einstein's theory of gravity is recovered which is coupled to a dilatonic scalar field [16].

Black holes with scalar hair are interesting solutions of Einstein's theory of gravity and also of certain types of modified gravity theories. These solutions have been investigated by theoretical physicists in four and higher dimensional space times for a long time (see [17] and references therein). The first studies on the three-dimensional black holes, as the interesting predictions of Einstein's theory of relativity in lower dimensional space times, have been done by Banados, Teitelboim, and Zanelli (BTZ) [18]. Investigation of the three-dimensional black holes is one of the interesting subjects for recent gravitational studies [19]. Chan and Mann [20], are the first authors who investigated the charged three-dimensional dilatonic black holes in the presence of a minimally coupled logarithmic dilaton field.

It is a commonly believed that study of three-dimensional solutions help us to find a deeper insight into the fundamental ideas in comparison to higher dimensional black holes. Also, according to (A)dS/CFT correspondence, there is a dual between quantum gravity on A(dS) space and Euclidean conformal field theory on the lower dimensional space times [21,22]. From this point of view, study of physics in $(2+1)$-dimensional space times can be useful for understanding of quantum field theory on A(dS) spacetimes. Although this subject area has been considered extensively [23], it still has many unknown and interesting parts to be studied [24].

On the other hand, after the discoveries of Bekenstein, Bardeen, Carter and Hawking, it is well-known that black holes can be considered as the thermodynamical systems with a temperature proportional to the surface gravity and having pure geometrical entropy equal to one-fourth of the horizon area [25–27]. When a dilatonic scalar field is coupled to the three-dimensional Einstein–Maxwell theory, it is expected to produce new and interesting consequences for the black hole solutions. Thus, it is worth to find exact solutions of Einstein–Maxwell theory in the presence of a

E-mail address: m.dehghani@ilam.ac.ir.

dilatonic scalar field with an arbitrary coupling constant, and investigate how the thermodynamical properties of black holes are modified. Also it is interesting to investigate the black holes remnant and find out the impacts of dilatonic field on the thermal stability of the black hole solutions.

The main object of this paper is to introduce new charged dilatonic BTZ black holes as the exact solutions to the coupled scalar, vector and tensor field equations, and provide a detailed analysis of the thermodynamical properties as well as the thermal stability of the new three-dimensional electrically charged black holes in the presence of a dilatonic scalar field.

The paper is outlined in the following order. In Sec. 2, by varying the proper Einstein–Maxwell action coupled to a dilatonic scalar field, the related scalar, vector and tensor field equations have been obtained. By introducing a static spherically symmetric geometry two new classes of charged dilatonic BTZ black holes, as the exact solution to the field equations, have been obtained. The asymptotic behavior of the new black holes are neither flat nor like anti-de Sitter (AdS) black holes. Sec. 3 is dedicated to study of thermodynamical properties of the new charged dilatonic black hole solutions obtained in the previous section. The conserved masses and charges of the black holes have been calculated based on the Gauss's law and Abbott–Deser proposal, respectively. Also, the black holes temperature, entropy and electric potential have been calculated from both, the geometrical an thermodynamical approaches. The compatibility of the results of these two alternative approaches confirms the validity of the first law of black hole thermodynamics for both of the new black holes obtained here. Sec. 4 is devoted to study of thermal stability or phase transition of the new charged dilatonic BTZ black holes introduced here. A black hole stability analysis has been performed, making use of the canonical ensemble method and regarding the black hole heat capacity with the black hole charge as a constant. It has been found that the black holes under consideration are stable or may undergo phase transition if some simple conditions are satisfied. Some concluding remarks and discussions have been presented in Sec. 5.

2. Basic equations and black hole solutions

The action for three-dimensional charged hairy black holes can be written in the following general form [20,28]

$$I=-\frac{1}{16\pi}\int\sqrt{-g}d^3x\Big[\mathcal{R}-U(\phi)-2g^{\mu\nu}\nabla_\mu\phi\nabla_\nu\phi-\mathcal{F}e^{-2\alpha\phi}\Big]. \tag{2.1}$$

Here, $\mathcal{R}$ is the Recci scalar. ϕ is the scalar field coupled to itself via the functional form $U(\phi)$. The parameter α is the scalar-electromagnetic coupling constant and $\mathcal{F}=F^{\mu\nu}F_{\mu\nu}$ being the Maxwell invariant. $F_{\mu\nu}=\partial_\mu A_\nu-\partial_\nu A_\mu$ and A_μ is the electromagnetic potential. By varying the action (2.1) with respect to the gravitational, electromagnetic and scalar fields, we get the related field equations as

$$\mathcal{R}_{\mu\nu}-\frac{1}{2}\mathcal{R}g_{\mu\nu}+\frac{1}{2}g_{\mu\nu}U(\phi)=T^{(s)}_{\mu\nu}+T^{(em)}_{\mu\nu}, \tag{2.2}$$

$$T^{(s)}_{\mu\nu}=2\nabla_\mu\phi\nabla_\nu\phi-g_{\mu\nu}(\nabla\phi)^2,$$

$$T^{(em)}_{\mu\nu}=-\frac{1}{2}\mathcal{F}e^{-2\alpha\phi}g_{\mu\nu}+2e^{-2\alpha\phi}F_{\mu\alpha}F_\nu{}^\alpha,$$

$$\nabla_\mu\left[e^{-2\alpha\phi}F^{\mu\nu}\right]=0, \tag{2.3}$$

$$4\Box\phi=\frac{dU(\phi)}{d\phi}-2\alpha\mathcal{F}e^{-2\alpha\phi},\qquad \phi=\phi(r). \tag{2.4}$$

Assuming as a function of r, the only non-vanishing component of the electromagnetic field is $F_{tr}=-E(r)=h'(r)$, and we have

$$\mathcal{F}=-2E^2(r)=-2(h'(r))^2. \tag{2.5}$$

In overall the paper, prime means derivative with respect to the argument. The gravitational field equations (2.2) can be rewritten as

$$\mathcal{R}_{\mu\nu}=U(\phi)g_{\mu\nu}+2\nabla_\mu\phi\nabla_\nu\phi-\left(\mathcal{F}g_{\mu\nu}-2F_{\mu\alpha}F_\nu^{\ \alpha}\right)e^{-2\alpha\phi}. \tag{2.6}$$

We consider the following ansatz as the three-dimensional spherically symmetric solution to the gravitational field equations (2.6)

$$ds^2=-\Psi(r)dt^2+\frac{1}{\Psi(r)}dr^2+r^2R(r)^2d\theta^2. \tag{2.7}$$

It leads to the following independent differential equations

$$E_{00}\equiv\Psi''+\left(\frac{1}{r}+\frac{R'}{R}\right)\Psi'+2U=0, \tag{2.8}$$

$$E_{11}\equiv E_{00}+2\Psi\left(\frac{R''}{R}+\frac{2R'}{rR}+2\phi'^2\right)=0, \tag{2.9}$$

$$E_{22}\equiv\left(\frac{1}{r}+\frac{R'}{R}\right)\Psi'+\left(\frac{R''}{R}+\frac{2R'}{rR}\right)\Psi+U+2F_{tr}^2e^{-2\alpha\phi}=0. \tag{2.10}$$

Noting Eqs. (2.8) and (2.9) we obtain

$$\frac{R''}{R}+\frac{2}{r}\frac{R'}{R}+2\phi'^2=0. \tag{2.11}$$

The differential equation (2.11) can be written in the following form

$$\frac{2}{r}\frac{d}{dr}\ln R(r)+\frac{d^2}{dr^2}\ln R(r)+\left(\frac{d}{dr}\ln R(r)\right)^2+2\phi'^2=0. \tag{2.12}$$

From Eq. (2.12), one can argue that $R(r)$ must be an exponential function of $\phi(r)$. Therefore, we can write $R(r)=e^{2\beta\phi}$, in Eq. (2.12), and show that $\phi=\phi(r)$ satisfies the following differential equation

$$\beta\phi''+(1+2\beta^2)\phi'^2+\frac{2\beta}{r}\phi'=0. \tag{2.13}$$

The case of $\beta=\alpha$ has been considered in a previous work [29]. Here, we are interested on the case $\beta\neq\alpha$.

It is easy to write the solution of Eq. (2.13) in terms of a positive constant b as $\phi(r)=\gamma\ln\left(\frac{b}{r}\right)$, with $\gamma=\beta(1+2\beta^2)^{-1}$. Similar solutions have been used by Hendi et al. [28]. The authors of ref. [20], have started with a power law of the form $R(r)\propto r^n$ and $\phi(r)\propto\ln r$, and showed that black hole solutions can exist if n is restricted in some ranges. In the following subsection, we proceed to obtain the solution of the field equations with the condition $\beta\neq\alpha$.

2.1. Solutions with $\beta\neq\alpha$

We start with the scalar field

$$\phi=\gamma\ln\left(\frac{b}{r}\right),\quad \text{and}\quad \gamma=\frac{\beta}{1+2\beta^2}. \tag{2.14}$$

Making use of these solutions together with Eqs. (2.3) and (2.7), we have

$$\begin{cases} h(r)=-\frac{q}{A}r^{-A}, & \text{and}\quad A=2\gamma(\alpha-\beta),\\ F_{tr}=q\,r^{-(1+A)}, \end{cases} \tag{2.15}$$

where, q is an integration constant related to the total electric charge on black hole. It will be calculated in the following section. Note that in the case of $\beta = \alpha$, $h(r)$ is a logarithmic function of r [29]. In order to the potential function $h(r)$ be physically reasonable (i.e. zero at infinity), the statement $A = 2\gamma(\alpha - \beta)$ must be positive. Thus we suppose that $\alpha > \beta$.

Now, Eq. (2.10) can be rewritten as

$$\Psi' - \frac{2\beta\gamma}{r}\Psi + \frac{r}{1-2\beta\gamma}\left[U(\phi) + 2F_{tr}^2 e^{-2\alpha\phi}\right] = 0. \tag{2.16}$$

For solving this equation for the metric function $\Psi(r)$, we need to calculate the functional form of $U(\phi(r))$ as the function of radial coordinate. For this purpose we proceed to solve the scalar field equation (2.4). It can be written as

$$\frac{dU(\phi)}{d\phi} - 4\beta U(\phi) - 4(2\beta - \alpha)F_{tr}^2 e^{-2\alpha\phi} = 0. \tag{2.17}$$

Noting Eq. (2.15), the first order differential (2.17) can be solved as

$$U(\phi) = 2\Lambda e^{4\beta\phi} + 2\Lambda_0 e^{4\beta_0\phi}, \tag{2.18}$$

where

$$\Lambda_0 = \frac{q^2(\Upsilon - 1)}{b^{2(A+1)}} \quad \text{and} \quad \Upsilon = (1 + \alpha\beta - 2\beta^2)^{-1} \quad \text{and}$$
$$\beta_0 = \frac{1+\alpha\beta}{2\beta}. \tag{2.19}$$

It is notable that the solution given by Eq. (2.18) can be considered as the generalized form of the Liouville scalar potential. Also, it must be noted that in the absence of dilatonic field ϕ, we have $U(\phi = 0) = 2\Lambda = -2\ell^{-2}$ and the action (2.1) reduces to that of Einstein-Λ-Maxwell theory.

Now, making use of Eqs. (2.15), (2.16) and (2.18) the metric function $\Psi(r)$ can be obtained as

$$\Psi(r) = \begin{cases} -m\,r^{2/3} - 3\left(\frac{r}{b}\right)^{2/3}\left[2b^2\Lambda\ln\left(\frac{r}{\ell}\right) - \frac{3q^2}{(\alpha-1)^2}(br)^{\frac{2}{3}(1-\alpha)}\right], \\ \quad \text{for } \beta = 1,\ \alpha > 1 \\ -m\,r^{2\beta\gamma} - (1+2\beta^2)^2\left[\frac{\Lambda r^2}{1-\beta^2}\left(\frac{b}{r}\right)^{4\beta\gamma}\right. \\ \quad \left. + \frac{q^2\Upsilon b^{-2A}}{\beta(\beta-\alpha)}\left(\frac{b}{r}\right)^{2\gamma(\alpha-2\beta)}\right], \\ \quad \text{for } \beta \neq 1. \end{cases} \tag{2.20}$$

The plots of metric functions $\Psi(r)$, presented in Eq. (2.20), for $\beta = 1$ and $\beta \neq 1$ cases have been shown in Figs. 1 and 2, respectively. The effects of α, Q and b on the metric function $\Psi(r)$ have been shown in Fig. 1 for the case $\beta = 1$. Plots of Fig. 2 show the effects of parameters α, β and b on the metric function $\Psi(r)$ for the case $\beta \neq 1$ by considering the condition $\alpha > \beta$. From the curves of Figs. 1 and 2, it is understood that the metric function $\Psi(r)$ can produce two horizon, extreme and naked singularity black holes for both of $\beta = 1$ and $\beta \neq 1$ cases.

Now, we investigate the curvature singularities. As a matter of calculation, one can show that the Ricci and Kretschmann scalars can be written in the following forms

$$R = \begin{cases} \frac{2}{9r^2} + 6\Lambda\left(\frac{b}{r}\right)^{4/3} + \frac{2q^2}{b^{2(1+2\alpha)/3}}\left(\frac{2\alpha-5}{\alpha-1}\right)\left(\frac{b}{r}\right)^{2(1+\alpha)/3}, \\ \quad \text{for } \beta = 1, \\ 6\Lambda\left(\frac{b}{r}\right)^{\frac{4\beta^2}{1+2\beta^2}} + \frac{2\beta^2}{b^2(1+2\beta^2)^2}\left(\frac{b}{r}\right)^2 + \frac{2q^2(3\Upsilon-2)}{b^{2(1+A)}}\left(\frac{b}{r}\right)^{\frac{2(1+\alpha\beta)}{1+2\beta^2}}, \\ \quad \text{for } \beta \neq 1, \end{cases} \tag{2.21}$$

$$R^{\mu\nu\rho\lambda}R_{\mu\nu\rho\lambda} = \begin{cases} r^{-2\delta_1}\left(\zeta_0 + \zeta_1 r^{\delta} + \zeta_2 r^{2\delta}\right) + r^{-\delta_1}\left[\zeta_3 + r^{\delta}(\zeta_4 + \zeta_5\ln r)\right]\ln r, \\ \quad \text{for } \beta = 1, \\ r^{-4}\left[B_1 r^{\frac{4\gamma}{\beta}} + B_2 r^{4\gamma(\alpha-2\beta)} + B_3 r^{4\gamma(\alpha-\beta)} + B_4 r^{4\gamma\beta}\right. \\ \quad + B_5 r^{\frac{2\gamma}{\beta}(1-\alpha\beta+2\beta^2)} + B_6 r^{\frac{2\gamma}{\beta}(1+\alpha\beta-\beta^2)} + B_7 r^{\frac{2\gamma}{\beta}(1+\beta^2)} \\ \quad + B_8 r^{2\gamma\beta} + B_9 r^{2\gamma(3\beta-\alpha)} + B_{10} r^{2\gamma(4\beta-\alpha)} \\ \quad \left. + B_{11} r^{2\gamma(\beta+\alpha)} + B_{12} r^{2\gamma\alpha}\right], \\ \quad \text{for } \beta \neq 1, \end{cases} \tag{2.22}$$

where $\delta = \frac{2}{3}(\alpha - 1)$, $\delta_1 = \frac{2}{3}(\alpha + 1)$, ζ_i's and B_i's are functions of Λ, q, m, β, α and b. From Eqs. (2.21) and (2.22), one can argue that there is an essential singularity located at $r = 0$. Also, the black holes asymptotic behavior are neither flat nor AdS. Recently, three and higher dimensional asymptotically Lifshitz black holes have been studied by many authors [30].

3. Thermodynamics

In this section, we would like to check the validity of the first law of black hole thermodynamics for the new dilatonic black holes we just introduced. At first it must be noted that the conserved charge of the black hole can be obtained by calculating the total electric flux measured by an observer located at infinity with respect to the horizon (i.e. $r \to \infty$) [31–33]. Making use of Eq. (2.15) together with the help of Gauss's law, after some simple calculations we arrived at

$$q = \begin{cases} 2Q\,b^{\frac{2}{3}(\alpha-1)}, & \text{for } \beta = 1, \\ 2Q\,b^{A}, & \text{for } \beta \neq 1, \end{cases} \tag{3.1}$$

which reduces to that of charged BTZ black holes in the absence of dilatonic field.

The other conserved quantity to be calculated is the black hole mass. As mentioned before, it can be obtained in terms of the mass parameter m. The Abbott–Deser total mass of the charged dilatonic BTZ black holes introduced here can be obtained as [28,35]

$$m = \begin{cases} 24M\,b^{-2/3}, & \text{for } \beta = 1, \\ 8M(1+2\beta^2)b^{-2\beta\gamma}, & \text{for } \beta \neq 1, \end{cases} \tag{3.2}$$

which is compatible with the mass of charged BTZ black hole when the dilatonic potential disappears.

We can obtain the Hawking temperature associated with the black hole horizon $r = r_+$, which is the root(s) of $\Psi(r_+) = 0$, in terms of the surface gravity κ as

$$T = \frac{\kappa}{2\pi} = \frac{1}{4\pi}\frac{d}{dr}\Psi(r)|_{r=r_+}$$
$$= \begin{cases} -\frac{3}{2\pi}\left(\frac{b}{r_+}\right)^{1/3}\left[b\Lambda + \frac{q^2 b^{(1-4\alpha)/3}}{\alpha-1}\left(\frac{b}{r_+}\right)^{\frac{2}{3}(\alpha-1)}\right], & \text{for } \beta = 1, \\ -\frac{1+2\beta^2}{2\pi r_+}\left[\Lambda r_+^2\left(\frac{b}{r_+}\right)^{4\beta\gamma} + \frac{q^2\Upsilon}{b^{2A}}\left(\frac{b}{r_+}\right)^{2\gamma(\alpha-2\beta)}\right], & \text{for } \beta \neq 1. \end{cases} \tag{3.3}$$

Since, the terms in the brackets have opposite sign ($\Lambda < 0$), from thermodynamical point of view, the physical and un-physical black holes can appear. Also, it must be noted that extreme black holes occur if q and r_+ be chosen such that $T = 0$. Now, making use of Eq. (3.3) we can obtain the horizon radius of the extreme black holes as

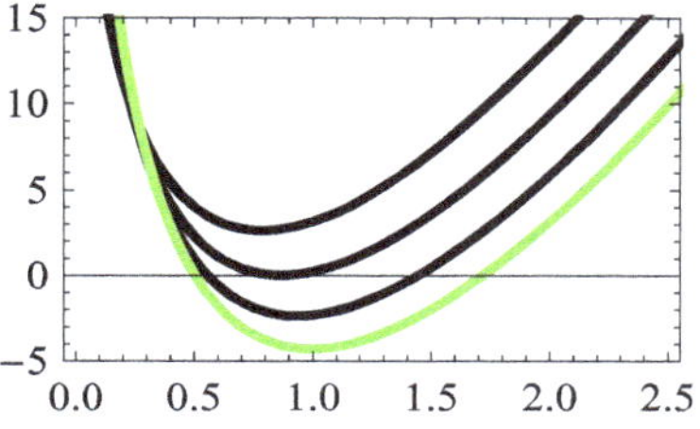
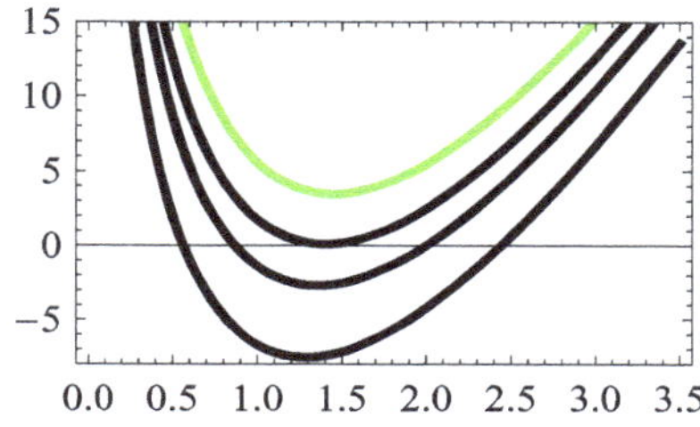
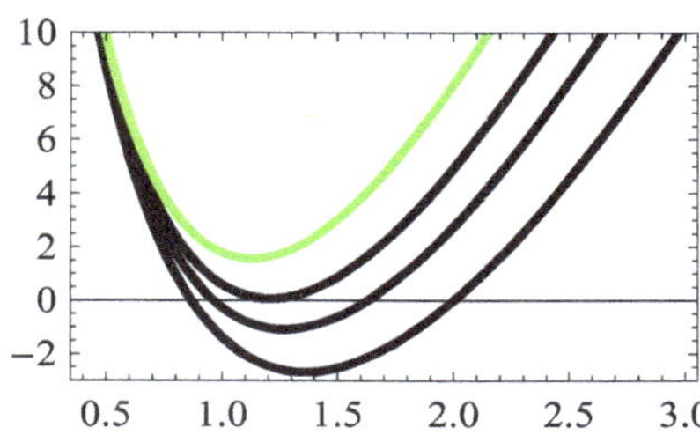

Fig. 1. $\Psi(r)$ versus r for $\beta=1$ and $\Lambda=-1$. Left: $Q=1$, $b=3$, $M=2.5$ and $\alpha=2.1, 2.195, 2.3, 2.4$ for black, red, blue and green curves, respectively. Middle: $M=3$, $b=3$, $\alpha=2.5$ and $Q=1.1, 1.2, 1.255, 1.32$ for black, red, blue and green curves, respectively. Right: $Q=1.2$, $M=3$, $\alpha=2.5$ and $b=3.0, 3.1, 3.18, 3.3$ for black, red, blue and green curves, respectively. (For interpretation of the references to color in this figure legend, the reader is referred to the web version of this article.)

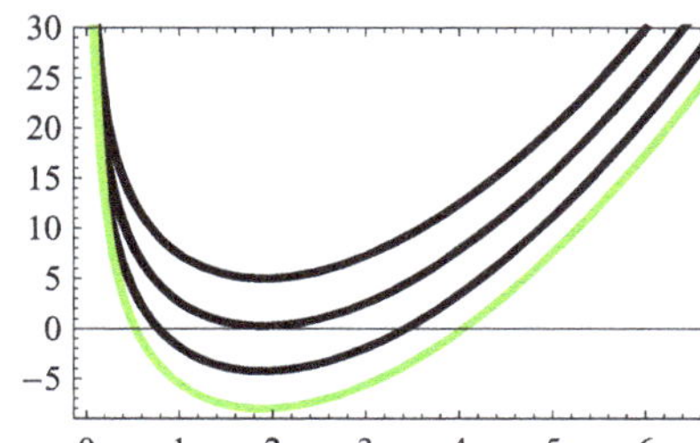
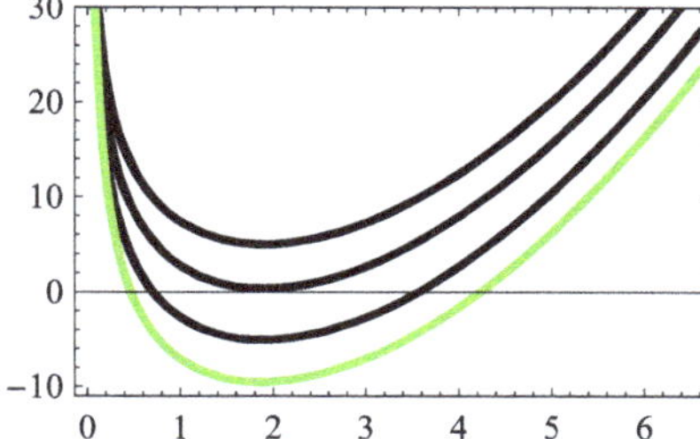
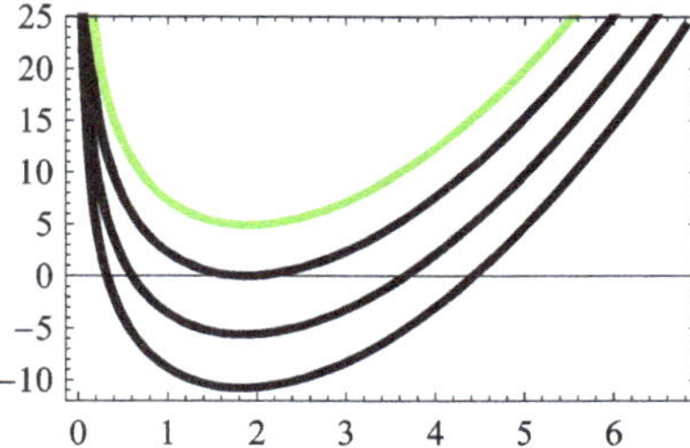

Fig. 2. $\Psi(r)$ versus r for $M=2$, $Q=1$ and $\Lambda=-1$. Left: $b=2$, $\beta=0.06$ and $\alpha=2.0, 2.3, 2.8, 3.4$ for black, blue, red and green curves, respectively. Middle: $b=2$, $\alpha=2$, and $\beta=0.06, 0.0702, 0.084, 0.1$ for black, blue, red and green curves, respectively. Right: $\alpha=2$, $\beta=0.06$, and $b=1.2, 1.4, 1.68, 2$ for black, blue, red and green curves, respectively. (For interpretation of the references to color in this figure legend, the reader is referred to the web version of this article.)

$$r_{ext}=\begin{cases}\left(\frac{q^2\ell^2}{\alpha-1}\right)^{\frac{3}{2(\alpha-1)}} b^{\frac{\alpha+2}{1-\alpha}}, & \text{for } \beta=1,\ \alpha>1,\\ b\left(\frac{q^2\ell^2\Upsilon}{b^{2(1+A)}}\right)^{\frac{\Upsilon}{2}(1+2\beta^2)}, & \text{for } \beta\neq 1.\end{cases} \tag{3.4}$$

It is evident that, in the case $\beta\neq 1$, extreme black holes exist if $\Upsilon>0$. In order to investigate the effects of scalar hair on the horizon temperature of the black holes correspond to $\beta=1$ the plot of black hole temperature versus horizon radius, for different values of α, has been shown in Fig. 3 (left). The physical black holes with positive temperature are those for which $r_+>r_{ext}$ and un-physical black holes, having negative temperature, occur if $r_+<r_{ext}$. The plot of T versus r_+, with $\beta\neq 1$, for different values of α and β has been shown in Fig. 4. They show that both, the physical (having positive temperature) and un-physical black holes (having negative temperature), will occur if the parameters are fixed, properly.

Next, we calculate the entropy of the black holes. It can be obtained from Hawking–Bekenstein entropy-area law, that is

$$S=\frac{A}{4}=\begin{cases}\frac{\pi b}{2}\left(\frac{r_+}{b}\right)^{1/3}, & \text{for } \beta=1,\\ \frac{\pi b}{2}\left(\frac{r_+}{b}\right)^{1-2\beta\gamma}, & \text{for } \beta\neq 1.\end{cases} \tag{3.5}$$

The black hole's electric potential Φ, measured by an observer located at infinity with respect to the horizon, can be obtained by using the following standard relation [17,31–33]

$$\Phi=A_\mu\chi^\mu|_{\text{reference}}-A_\mu\chi^\mu|_{r=r_+}, \tag{3.6}$$

where, $\chi=C\partial_t$ is the null generator of the horizon and C is an arbitrary constant [34]. Noting Eqs. (2.15) and (3.6) we can obtain the black hole's electric potential on the horizon. That is

$$\Phi=\begin{cases}\frac{3Cq}{2(\alpha-1)}r_+^{-\frac{2}{3}(\alpha-1)}, & \text{for } \beta=1,\\ \frac{Cq}{A}r_+^{-A}, & \text{for } \beta\neq 1,\end{cases} \tag{3.7}$$

in terms of the constant coefficient C, which will be determined in the following.

In order to investigate the consistency of these quantities with the thermodynamical first law, from Eqs. (2.20), (3.1) and (3.5), we can obtain the black hole mass as the function of extensive parameters S and Q. For this purpose we use the relation $\Psi(r_+)=0$. The corresponding Smarr-type mass formula is obtained as

$$M(S,Q)=\begin{cases}\frac{1}{4}\left[\frac{b^2}{\ell^2}\ln\left(\frac{\ell}{r_+(S)}\right)+\frac{6Q^2}{(\alpha-1)^2}\left(\frac{b}{r_+(S)}\right)^{\frac{2}{3}(\alpha-1)}\right], & \text{for } \beta=1,\\ -\frac{1+2\beta^2}{8}\left[\frac{\Lambda b^2}{1-\beta^2}\left(\frac{b}{r_+(S)}\right)^{2(3\beta\gamma-1)}+\frac{4\Upsilon Q^2}{\beta(\beta-\alpha)}\left(\frac{b}{r_+(S)}\right)^{A}\right], & \text{for } \beta\neq 1.\end{cases} \tag{3.8}$$

We can calculate the intensive parameters T and Φ, conjugate to the black hole entropy and charge, respectively. It is a matter of calculation to show that

$$\left(\frac{\partial M}{\partial S}\right)_Q=T \quad \text{for both} \quad \beta=1 \quad \text{and} \quad \beta\neq 1, \tag{3.9}$$

and

$$\left(\frac{\partial M}{\partial Q}\right)_S=\Phi, \tag{3.10}$$

provided that [34]

$$C=\begin{cases}(\alpha-1)^{-1}, & \text{for } \beta=1,\\ (1+\alpha\beta-2\beta^2)^{-1}, & \text{for } \beta\neq 1.\end{cases} \tag{3.11}$$

Therefore, we proved that the first law of black hole thermodynamics is valid, for both classes of the charged dilatonic BTZ black holes, in the following form

$$dM(S,Q)=TdS+\Phi dQ. \tag{3.12}$$

4. Thermal stability analysis in the canonical ensemble method

In this section, we would like to analyze the stability or phase transition of the either of the black hole solutions, regarding

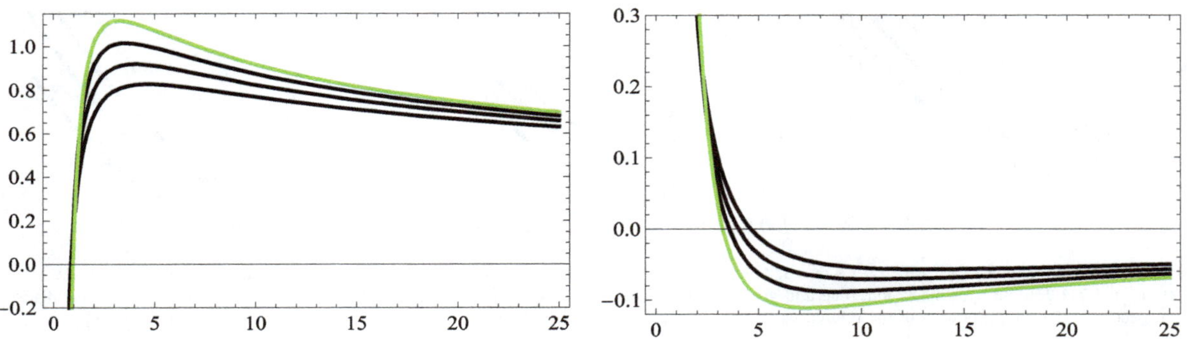

Fig. 3. Assuming $\beta = 1$, $\Lambda = -1$, $Q = 1$, $b = 3$ and $\alpha = 2, 2.2, 2.5, 3$, for black, red, blue, and green curves, respectively. Left: T versus r_+. Right: $(\partial^2 M/\partial S^2)_Q$ versus r_+. (For interpretation of the references to color in this figure legend, the reader is referred to the web version of this article.)

the black hole heat capacity with the fixed black hole charge, separately. It is well known that the positivity of heat capacity $C_Q = T\,(\partial S/\partial T)_Q = T/\left(\partial^2 M/\partial S^2\right)_Q$ or equivalently the positivity of $(\partial S/\partial T)_Q$ or $\left(\partial^2 M/\partial S^2\right)_Q$ with $T > 0$ are sufficient to ensure the local stability of the black hole. The unstable black holes undergo phase transitions to be stabilized. Type one phase transition takes place at the points where the black hole heat capacity vanishes. On the other hand, an unstable black hole undergoes type two phase transition at the divergent points of black hole heat capacity [17,31–33]. With these issues in mind, we proceed to analyze the thermal stability or phase transition of both of the new black hole solutions we just obtained here.

4.1. Black holes with $\beta = 1$

Making use of Eq. (3.7), the denominator of the black hole heat capacity can be calculated as

$$\left(\frac{\partial^2 M}{\partial S^2}\right)_Q = \frac{3}{\pi^2}\left(\frac{b}{r_+}\right)^{2/3}\left[\Lambda + q^2\left(\frac{2\alpha-1}{\alpha-1}\right)b^{-\frac{2}{3}(2\alpha+1)}\left(\frac{b}{r_+}\right)^{\frac{2}{3}(\alpha-1)}\right]. \tag{4.1}$$

Since $\alpha > 1$ and $\Lambda = -\ell^{-2}$, the terms in the brackets have opposite sign and it is understood from Eq. (4.1) that if

$$r_+ \equiv r_0 = \left[q^2\ell^2\left(\frac{2\alpha-1}{\alpha-1}\right)\right]^{\frac{3}{2(\alpha-1)}} b^{\frac{\alpha+2}{1-\alpha}}, \tag{4.2}$$

the denominator of the black hole heat capacity vanishes and black holes with the size satisfying this condition undergo type two phase transition. It must be noted that $r_0 > r_{ext}$ (see Eq. (3.4)). At the $r_+ = r_{ext}$ the black hole heat capacity vanishes and the type one phase transition takes place. In addition, for $r_+ > r_0$ the denominator of the heat capacity is negative and the physical black holes (black holes with positive temperature) will be thermodynamically unstable. On the other hand, for $r_{ext} < r_+ < r_0$ the denominator of the black hole heat capacity as well as the black hole temperature are positive and as a result the heat capacity of the black holes with the horizon in this range are positive and they will be locally stable. The plot of $\left(\partial^2 M/\partial S^2\right)_Q$ versus r_+ is shown in Fig. 3 (right). The plots show that the physical black holes with the horizon radius in the range $r_{ext} < r_+ < r_0$ are locally stable. Otherwise they are thermally unstable and can undergo phase transition to be stabilized.

4.2. Black holes with $\beta \neq 1$

It is a matter of calculation to show that

$$\left(\frac{\partial^2 M}{\partial S^2}\right)_Q = \frac{1+2\beta^2}{\pi^2}\left[\Lambda(2\beta^2-1)\left(\frac{b}{r_+}\right)^{2\beta\gamma} + \frac{q^2(1+\alpha\beta\Upsilon)}{b^{2(1+A)}}\left(\frac{b}{r_+}\right)^{\frac{2\Upsilon}{\beta}(1+\alpha\beta-\beta^2)}\right]. \tag{4.3}$$

The statement given in Eq. (4.3) denotes the denominator of the black hole heat capacity. It is consisted of two terms which, apart from scalar hair, show the contributions from Λ and from black hole charge, separately. One can argue that charged dilatonic BTZ black holes are unstable and undergo type two phase transition at the real roots of Eq. (4.3), which are located at

$$r_+ \equiv r_1 = b\left[\frac{q^2\ell^2(1+\alpha\beta\Upsilon)}{(2\beta^2-1)b^{2(1+A)}}\right]^{\frac{\Upsilon}{2}(1+2\beta^2)}. \tag{4.4}$$

Also, the type one phase transition takes place at the point $r_+ = r_{ext}$, given by Eq. (3.4), where the black hole heat capacity vanishes. The plots of $\left(\partial^2 M/\partial S^2\right)_Q$ versus r_+ are shown in Fig. 4 for various α and β values. They show that for the properly fixed parameters the three following cases are distinguishable

- r_{ext} exist but r_1 does not exist (Fig. 4 (left)). In this case $\left(\partial^2 M/\partial S^2\right)_Q$ does not vanish and no type two phase transition takes place. The type one phase transition takes place at $r_+ = r_{ext}$, where the temperature and hence the black hole heat capacity vanishes. Both T and $\left(\partial^2 M/\partial S^2\right)_Q$ are positive for $r_+ > r_{ext}$ and the charged dilatonic BTZ black holes with $\beta \neq 1$ are stable if their horizon radius, r_+, is greater than r_{ext}.
- r_1 exist but r_{ext} does not exist (Fig. 4 (middle)). The black holes have positive temperature and therefore are reasonable, thermodynamically. The black hole heat capacity does not vanish and no type one phase transition takes place. It diverges at $r_+ = r_1$ and black holes with the horizon radius equal to r_1 are unstable. They will undergo type two phase transition to be stabilized. The black hole heat capacity is positive for the charged dilatonic BTZ black holes with the horizon radius, r_+, greater than r_1 and they are thermodynamically stable.
- Both r_{ext} and r_1 are exist (Fig. 4 (right)). The black hole temperature vanishes at $r_+ = r_{ext}$. Therefore, the charged dilatonic BTZ black holes with $\beta \neq 1$ undergo type one phase transition at this point. The black holes with $r_+ < r_{ext}$, having negative temperature, are not physically reasonable. The denominator of the black hole heat capacity vanishes at $r_+ = r_1$. Therefore, $r_+ = r_1$ is the point of type two phase transition. For the

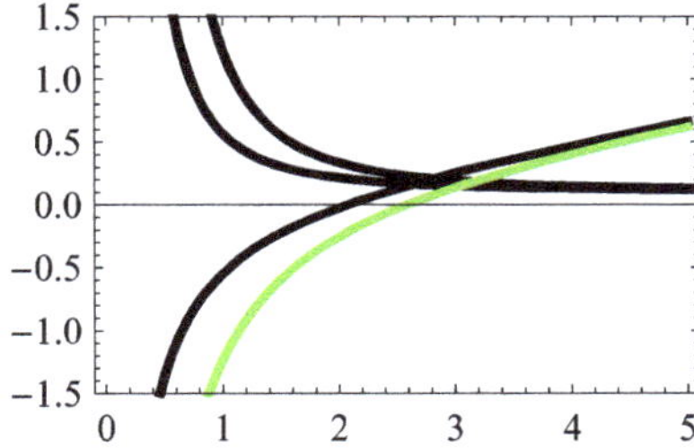

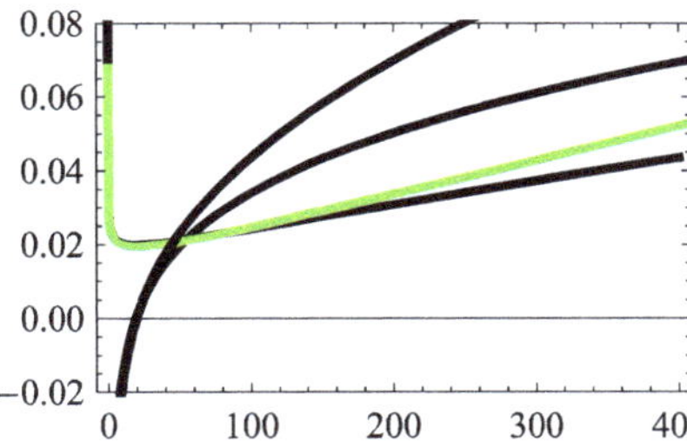

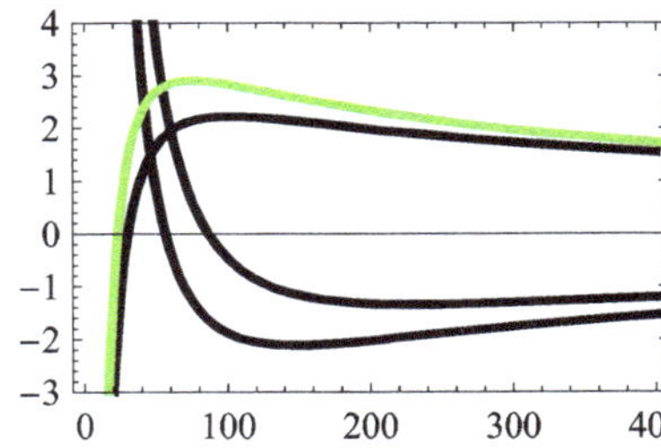

Fig. 4. Left: T (green $\alpha = 3.5$ and black $\alpha = 0.5$) and $\left(\partial^2 M/\partial S^2\right)_Q$ (blue $\alpha = 3.5$ and red $\alpha = 0.5$) versus r_+ for $\Lambda = -1$, $\beta = 0.04$, $b = 3$ and $Q = 1$. r_{ext} exist but r_1 does not exist. Middle: $0.02T$ (green $\alpha = 1.2$ and black $\alpha = 1$) and $\left(\partial^2 M/\partial S^2\right)_Q$ (blue $\alpha = 1.2$ and red $\alpha = 1$) versus r_+ for $\Lambda = -1$, $\beta = -0.8$, $b = 3$ and $Q = 1$. r_1 exist but r_{ext} does not exist. Right: $20T$ (green $\alpha = 3.2$ and black $\alpha = 3.1$) and $\left(\partial^2 M/\partial S^2\right)_Q$ (blue $\alpha = 3.2$ and red $\alpha = 3.1$) versus r_+ for $\Lambda = -1$, $\beta = 1.5$, $b = 3$ and $Q = 1$. Both r_{ext} and r_1 are exist. (For interpretation of the references to color in this figure legend, the reader is referred to the web version of this article.)

charged dilatonic BTZ black holes with the horizon radius in the range $r_{ext} < r_+ < r_1$, both T and $\left(\partial^2 M/\partial S^2\right)_Q$, are positive and they are locally stable.

5. Conclusion

Here, we studied the charged dilatonic BTZ black holes as the exact solutions to the Einstein–Maxwell theory coupled to a dilatonic scalar field. We solved the coupled scalar, electromagnetic and gravitational field equations in a static spherically symmetric space time and obtained two new classes of charged dilatonic BTZ black hole solutions. The solutions contain an essential (not coordinate) singularity located at the origin. Also, they do not behave asymptotically like the flat or AdS black holes. Furthermore, we showed that both of the new black hole solutions (correspond to $\beta = 1$ and $\beta \neq 1$) present naked singularity, extreme and two horizon black holes if the parameter α and β are chosen suitably (see Figs. 1 and 2).

We calculated the electric charges and masses of the black holes, as conserved quantities, making use of the Gauss's law and Abbott–Deser proposal, respectively. Also, we calculated the entropy, temperature and electric potential using the geometrical methods. On the other hand, through a Smarr-type mass formula, we constructed out the black holes masses as functions of both charge and entropy, as the thermodynamical extensive quantities. Making use of the Smarr-type mass formula we calculated the electric potential and temperature, as the thermodynamical intensive quantities, for both classes of the new BTZ black holes. We found that the thermodynamical quantities obtained from geometrical and thermodynamical approaches are identical for either of the black hole classes. It confirms the validity of the first law of black hole thermodynamics for both of the new black hole solutions in the form of Eq. (3.11).

Finally, making use of the canonical ensemble method, we performed the thermal stability analysis of both classes of the new BTZ black holes. Regarding the black hole heat capacity with constant black hole charge, we found that the black holes correspond to the case of $\beta = 1$ are stable if their horizon radius is in the range $r_{ext} < r_+ < r_0$ (see Eqs. (3.4) and (4.2)). This class of BTZ black holes undergo type one phase transition at $r_+ = r_{ext}$ and type two phase transition at $r_+ = r_0$ (see Fig. 3). In order to discuss the thermal stability of the second class of the black hole solutions according to the plots of Fig. 4 three possibilities are considered, separately. (*i*) If the parameters are chosen such that r_{ext} is exist but r_1 does not exist, there is type one phase transition located at r_{ext}. No type two phase transition can occur and black holes with the horizon radius greater than r_{ext} are stable. (*ii*) when the parameters are fixed in such a way that r_1 exist but r_{ext} does not exist, no type one phase transition takes place. There is type two phase transition point located at $r_+ = r_{ext}$. The black holes with the horizon radius greater than r_1 are locally stable. (*iii*) It is possible to fix the parameters such that both r_{ext} and r_1 are exist. In this case $r_+ = r_{ext}$ and $r_+ = r_1$ are points of type one and type two phase transitions, respectively. The black holes are locally stable if their horizon radiuses are in the range $r_{ext} < r_+ < r_0$.

The dynamical stability and quasi-normal modes of these novel dilatonic black hole solutions will be investigated in the forthcoming papers.

Acknowledgements

The author appreciates the Ilam University Research Council for official supports of this work.

References

[1] S. Perlmutter, et al., Astrophys. J. 517 (1999) 565;
S. Perlmutter, M.S. Turner, M. White, Phys. Rev. Lett. 83 (1999) 670;
A.G. Riess, et al., Astrophys. J. 607 (2004) 665.
[2] D.N. Spergel, et al., Astrophys. J. Suppl. Ser. 148 (2003) 175;
D.N. Spergel, et al., Astrophys. J. Suppl. Ser. 170 (2007) 377.
[3] M. Tegmark, et al., Phys. Rev. D 69 (2004) 103501;
U. Seljak, et al., Phys. Rev. D 71 (2005) 103515.
[4] D.J. Eisenstein, et al., Astrophys. J. 633 (2005) 560.
[5] B. Jain, A. Taylor, Phys. Rev. Lett. 91 (2003) 141302.
[6] D. Lovelock, J. Math. Phys. 12 (1971) 498;
D. Lovelock, J. Math. Phys. 13 (1972) 874;
N. Deruelle, L. Farina-Busto, Phys. Rev. D 41 (1990) 3696.
[7] K. Bamba, S.D. Odintsov, J. Cosmol. Astropart. Phys. 04 (2008) 024;
G. Cognola, E. Elizalde, S. Nojiri, S.D. Odintsov, L. Sebastiani, S. Zerbini, Phys. Rev. D 77 (2008) 046009;
C. Corda, Europhys. Lett. 86 (2009) 20004;
T.P. Sotiriou, V. Faraoni, Rev. Mod. Phys. 82 (2010) 451;
S. Nojiri, S.D. Odintsov, Phys. Rep. 505 (2011) 59.
[8] L.A. Gergely, Phys. Rev. D 74 (2006) 024002;
M. Demetrian, Gen. Relativ. Gravit. 38 (2006) 953;
L. Amarilla, H. Vucetich, Int. J. Mod. Phys. A 25 (2010) 3835.
[9] S. Capozziello, V.F. Cardone, S. Carloni, A. Troisi, Int. J. Mod. Phys. D 12 (2003) 1969.
[10] A.A. Starobinsky, Phys. Lett. B 91 (1980) 99;
M. Dehghani, J. High Energy Phys. 03 (2016) 203.
[11] S. Capozziello, V.F. Cardone, A. Troisi, J. Cosmol. Astropart. Phys. 0608 (2006) 001.
[12] K-i. Maeda, N. Ohta, Phys. Lett. B 597 (2004) 400;
K-i. Maeda, N. Ohta, Phys. Rev. D 71 (2005) 063520;
N. Ohta, Int. J. Mod. Phys. A 20 (2005) 1;
K. Akune, K-i. Maeda, N. Ohta, Phys. Rev. D 73 (2006) 103506.
[13] S. Nojiri, S.D. Odintsov, Phys. Rev. D 74 (2006) 086005;
S. Nojiri, S.D. Odintsov, Phys. Lett. B 631 (2005) 1;
S. Capozziello, S. Nojiri, S.D. Odintsov, Phys. Lett. B 632 (2006) 597;
S. Nojiri, S.D. Odintsov, Phys. Lett. B 576 (2003) 5;
S. Nojiri, S.D. Odintsov, Phys. Rev. D 68 (2003) 123512;
S. Nojiri, S.D. Odintsov, M. Sasaki, Phys. Rev. D 71 (2005) 123509.
[14] T.P. Sotiriou, Class. Quantum Gravity 23 (2006) 5117;
P. Klepac, J. Horsky, Gen. Relativ. Gravit. 34 (2002) 1979;
R.G. Cai, S.P. Kim, B. Wang, Phys. Rev. D 76 (2007) 024011;
Y. Ling, C. Niu, J.P. Wu, Z.Y. Xian, J. High Energy Phys. 11 (2013) 006;
M. Ghodrati, Phys. Rev. D 90 (2014) 044055.

[15] G.W. Gibbons, K. Maeda, Ann. Phys. (N.Y.) 167 (1986) 201;
V. Frolov, A. Zelnikov, U. Bleyer, Ann. Phys. (Berlin) 44 (1987) 371;
J.H. Horne, G.T. Horowitz, Phys. Rev. D 46 (1992) 1340.
[16] M.B. Green, J.H. Schwarz, E. Witten, Superstring Theory, Cambridge University Press, Cambridge, England, 1987.
[17] A. Sheykhi, M.H. Dehghani, N. Riazi, J. Pakravan, Phys. Rev. D 74 (2006) 084016;
A. Sheykhi, N. Riazi, Phys. Rev. D 75 (2007) 024021;
M.H. Dehghani, N. Farhangkhah, Phys. Rev. D 71 (2005) 044008;
A. Sheykhi, M.H. Dehghani, S.H. Hendi, Phys. Rev. D 81 (2010) 084040;
A. Sheykhi, S. Hajkhalili, Phys. Rev. D 89 (2014) 104019;
S.H. Hendi, M. Faizal, B. Eslampanah, S. Panahiyan, Eur. Phys. J. C 76 (2016) 296.
[18] M. Banados, C. Teitelboim, J. Zanelli, Phys. Rev. Lett. 69 (1992) 1849;
M. Banados, M. Henneaux, C. Teitelboim, J. Zanelli, Phys. Rev. D 48 (1993) 1506.
[19] W. Xu, L. Zhao, Phys. Rev. D 87 (2013) 124008;
H. Zhang, D.J. Liu, X.Z. Li, Phys. Rev. D 90 (2014) 124051;
T.P. sotiriou, I. Vega, D. Vernieri, Phys. Rev. D 90 (2014) 044046;
M. Dehghani, Phys. Rev. D 94 (2016) 104071;
M. Park, Phys. Rev. D 80 (2009) 084026;
A.S.D. Goncalo, P.S.L. Jose, Phys. Rev. D 78 (2008) 084020;
B. Sahoo, A. Sen, J. High Energy Phys. 0607 (2006) 008;
M. Park, Phys. Rev. D 77 (2008) 026011;
M. Park, Phys. Rev. D 77 (2008) 126012;
J. Parsons, S.F. Ross, J. High Energy Phys. 0904 (2009) 134.
[20] K.C.K. Chan, R.B. Mann, Phys. Rev. D 50 (1994) 6385;
K.C.K. Chan, Phys. Rev. D 55 (1997) 3564.
[21] R.G. Cai, Nucl. Phys. B 628 (2002) 375;
P.O. Mazur, E. Mottola, Phys. Rev. D 64 (2001) 104022.
[22] M.R. Setare, Mod. Phys. Lett. A 17 (2002) 2089.
[23] D. Park, J.K. Kim, J. Math. Phys. 38 (1997) 2616;
P.M. Sa', A. Kleber, J.P.S. Lemos, Class. Quantum Gravity 13 (1996) 125;
R. Yamazaki, D. Ida, Phys. Rev. D 64 (2001) 024009;
S. Fernando, Int. J. Theor. Phys. 51 (2012) 418;
A. Sheykhi, S.H. Hendi, S. Salarpour, Phys. Scr. 89 (2014) 105003.
[24] G. Cl'ement, Class. Quantum Gravity 10 (1993) 49;
G. Cl'ement, Phys. Lett. B 367 (1996) 70;
C. Martinez, C. Teitelboim, J. Zanelli, Phys. Rev. D 61 (2000) 104013;
C. Martinez, R. Troncoso, J. Zanelli, Phys. Rev. D 70 (2004) 084035.
[25] J.D. Bekenstein, Phys. Rev. D 7 (1973) 2333.
[26] J.M. Bardeen, B. Carter, S.W. Hawking, Commun. Math. Phys. 31 (1973) 161.
[27] S.W. Hawking, Commun. Math. Phys. 43 (1975) 199.
[28] S.H. Hendi, B. Eslam Panah, S. Panahiyan, A. Sheykhi, Phys. Lett. B 767 (2017) 214;
S.H. Hendi, B. Eslam Panah, S. Panahiyan, Prog. Theor. Exp. Phys. 2016 (10) (2016) 103A02.
[29] M. Dehghani, Phys. Rev. D 96 (2017) 044014.
[30] Y. SooMyung, Y-W. Kim, Y-J. Park, Eur. Phys. J. C 70 (2010) 335;
M. Kord Zangeneh, A. Dehyadegari, A. Sheykhi, M.H. Dehghani, J. High Energy Phys. 1603 (2016) 037;
A. Dehyadegari, A. Sheykhi, M. Kord Zangeneh, Phys. Lett. B 758 (2016) 226;
M. Kord Zangeneh, A. Dehyadegari, M.R. Mehdizadeh, B. Wang, A. Sheykhi, Eur. Phys. J. C 77 (2017) 423.
[31] A. Sheykhi, Phys. Rev. D 86 (2012) 024013;
M. Kord Zangeneh, M.H. Dehghani, A. Sheykhi, Phys. Rev. D 92 (2015) 104035.
[32] S.H. Hendi, S. Panahiyan, R. Mamasani, Gen. Relativ. Gravit. 47 (2015) 91;
S.H. Hendi, S. Panahiyan, M. Momennia, Int. J. Mod. Phys. D 25 (2016) 1650063;
S.H. Hendi, S. Panahiyan, B. Eslampanah, Eur. Phys. J. C 75 (2015) 296.
[33] M.H. Dehghani, H.R. Rastegar-Sedehi, Phys. Rev. D 74 (2006) 124018;
M.H. Dehghani, S.H. Hendi, Int. J. Mod. Phys. D 16 (2007) 1829;
M.H. Dehghani, S.H. Hendi, A. Sheykhi, H.R. Rastegar-Sedehi, J. Cosmol. Astropart. Phys. 02 (2007) 020;
M.H. Dehghani, N. Alinejadi, S.H. Hendi, Phys. Rev. D 77 (2008) 104025;
S.H. Hendi, J. Math. Phys. 49 (2008) 082501.
[34] M. Kord Zangeneh, A. Sheykhi, M.H. Dehghani, Phys. Rev. D 91 (2015) 044035;
M. Kord Zangeneh, A. Sheykhi, M.H. Dehghani, Eur. Phys. J. C 75 (2015) 497;
M. Kord Zangeneh, A. Sheykhi, M.H. Dehghani, Phys. Rev. D 92 (2015) 024050.
[35] L.F. Abbott, S. Deser, Nucl. Phys. B 195 (1982) 76;
R. Olea, J. High Energy Phys. 06 (2005) 023;
G. Kofinas, R. Olea, Phys. Rev. D 74 (2006) 084035.

The zeroth law in quasi-homogeneous thermodynamics and black holes

Alessandro Bravetti [a], Christine Gruber [b], Cesar S. Lopez-Monsalvo [c], Francisco Nettel [b,*]

[a] Instituto de Investigaciones en Matemáticas Aplicadas y en Sistemas, Universidad Nacional Autónoma de México, Ciudad Universitaria, Ciudad de México 04510, Mexico A.P. 70-543, 04510 Ciudad de México, Mexico
[b] Instituto de Ciencias Nucleares, Universidad Nacional Autónoma de México, A.P. 70-543, 04510 Ciudad de México, Mexico
[c] Conacyt-Universidad Autónoma Metropolitana Azcapotzalco Avenida San Pablo Xalpa 180, Azcapotzalco, Reynosa Tamaulipas, 02200 Ciudad de México, Mexico

ARTICLE INFO

Editor: M. Cvetič

Keywords:
Quasi-homogeneous thermodynamics
Zeroth Law
Black hole thermodynamics
Gibbs-Duhem

ABSTRACT

Motivated by black holes thermodynamics, we consider the zeroth law of thermodynamics for systems whose entropy is a quasi-homogeneous function of the extensive variables. We show that the generalized Gibbs–Duhem identity and the Maxwell construction for phase coexistence based on the standard zeroth law are incompatible in this case. We argue that the generalized Gibbs–Duhem identity suggests a revision of the zeroth law which in turns permits to reconsider Maxwell's construction in analogy with the standard case. The physical feasibility of our proposal is considered in the particular case of black holes.

1. Introduction

The thermodynamics of black holes contains several peculiarities in contrast to standard thermodynamics. One example is the different scaling behaviour when rescaling the thermodynamic variables. This can be directly verified noting that the entropy of a black hole is – in general – a quasi-homogeneous function of the extensive thermodynamic quantities describing the system [1], and its scaling behavior is dictated by the Smarr relation. Such systems are generically called *quasi-homogeneous*. As a consequence, it is usually recognized that using the formalism of homogeneous thermodynamics in the case of black holes is not fully justified and that a modification of the thermodynamic laws for systems with quasi-homogeneous entropy is called for [1].

It has been established that in systems where entropy and energy are not additive the standard way to define equilibrium has to be adjusted and, in such case, the thermodynamic temperature may not be the correct parameter to be equated at equilibrium [2–9]. In spite of this, it has been repeatedly argued in favor of the existence of first order phase transitions – i.e., coexistence processes – within the framework of black hole thermodynamics. Such arguments are based on the analogy with the van der Waals (vdW) phase diagram and use the Maxwell equal area law to find the coexistence curve *as if* the system was homogeneous (see e.g. [1,10–19] and the references therein).

In this work we consider systems whose entropy is a quasi-homogeneous function of the extensive variables and show that Maxwell's equal area law – based on the definition of thermodynamic equilibrium for homogeneous systems (cf. [20] and the discussion in Section 4.3 in [21]) – is inconsistent with the generalized Gibbs–Duhem (GGD) identity that must hold in such cases [22,23]. We show that this situation can be remedied introducing a new set of the variables defining equilibrium. Based on these generalized variables, we propose a definition of thermodynamic equilibrium originating from the GGD identity and we demonstrate that such revision is essential in Maxwell's construction for phase coexistence. It is worth mentioning that our *generalized zeroth law* reduces to the standard definition for homogeneous systems of degree one.

To illustrate our proposal we discuss two relevant cases: on the one hand, we show that for the Schwarzschild black hole the *new* temperature characterizing equilibrium is constant, i.e. it does not depend on its mass M. This coincides (up to a constant factor) with the result in [24], where such parameter is obtained using a gener-

* Corresponding author.
E-mail addresses: alessandro.bravetti@iimas.unam.mx (A. Bravetti), christine.gruber@correo.nucleares.unam.mx (C. Gruber), cslopezmo@conacyt.mx (C.S. Lopez-Monsalvo), fnettel@nucleares.unam.mx (F. Nettel).

alized zeroth law for non-extensive statistical mechanics developed in [6]. This proves that, at least in the Schwarzschild case, there is a consistency between different approaches. On the other hand, we consider the first order phase transition in the Kerr–Anti de Sitter (Kerr–AdS) family of black holes and show that the Maxwell construction as applied in the literature leads to a violation of the GGD. Using the new generalized intensive parameters and according to our definition of thermodynamic equilibrium, such transition seems to disappear. Given the importance of this example in the context of the AdS/CFT correspondence, we believe that this can be relevant for future investigations.

This paper is structured as follows. In Section 2 we review the thermodynamics of quasi-homogeneous systems as developed in [22,23]. In Section 3 we point out the aforementioned mathematical inconsistency between Maxwell's construction based on the standard zeroth law of thermodynamics and the Gibbs–Duhem relation in the case of quasi-homogeneous entropy, and continue by proposing a generalized form of the zeroth law, which is consistent with the corresponding GGD relation. To illustrate the new form of the zeroth law, we consider the examples of Schwarzschild and Kerr-AdS black holes in Section 4, before we conclude in Section 5. Throughout this work we use Planck units, in which $c = G = \hbar = k_B = 1$.

2. Quasi-homogeneous thermodynamics

In this section we briefly review some results of the thermodynamics of quasi-homogeneous systems obtained in [22,23]. Let us start by recalling some definitions. Unless otherwise stated, we will not use Einstein's sum convention.

Definition 2.1 *(Quasi-homogeneous function).* Let $r, \lambda \in \mathbb{R}$, $\lambda \neq 0$ and $\beta = (\beta_1, ..., \beta_n) \in \mathbb{R}^n$. A function w of a set of variables $\{q^i\}_{i=1}^n$ is said to be *quasi-homogeneous of degree r and type β* if

$$w(\lambda^{\beta_1} q^1, ..., \lambda^{\beta_n} q^n) = \lambda^r w(q^1, ..., q^n). \tag{1}$$

The particular case where $\beta_i = 1$ for every value of i yields the standard scaling relation of homogeneous functions of degree r, i.e.

$$w(\lambda q^1, ..., \lambda q^n) = \lambda^r w(q^1, ..., q^n). \tag{2}$$

In the following we will use S instead of w, because the function of interest in thermodynamics is the entropy. The variables $\{q^i\}_{i=1}^n$ are the extensive variables of the system, such as internal energy U, volume V or number of particles N. In standard thermodynamics of extensive systems the entropy is a homogeneous function of degree one of the extensive variables, i.e.,

$$S(\lambda U, \lambda V, \lambda N) = \lambda S(U, V, N), \tag{3}$$

while in black holes thermodynamics the entropy is a quasi-homogeneous function as in Definition 2.1.

Proposition 2.1. *Let $S = S(q^1, ..., q^n)$ be a quasi-homogeneous function of degree r and type β. Then, the conjugate variables to the q^i, defined by*

$$p_i\left(q^j\right) \equiv \frac{\partial}{\partial q^i} S\left(q^j\right), \tag{4}$$

are quasi-homogeneous functions of degree $r - \beta_i$ for every value of i.

Proof.

$$p_i\left(\lambda^{\beta_j} q^j\right) = \frac{\partial}{\partial(\lambda^{\beta_i} q^i)} S\left(\lambda^{\beta_j} q^j\right) = \frac{1}{\lambda^{\beta_i}} \frac{\partial}{\partial q^i}\left[\lambda^r S\left(q^j\right)\right] = \lambda^{r-\beta_i} \frac{\partial}{\partial q^i} S(q^j). \tag{5}$$

Therefore

$$p_i\left(\lambda^{\beta_j} q^j\right) = \lambda^{r-\beta_i} p_i(q^j). \quad \square \tag{6}$$

Note that if S is homogeneous of degree $r = 1$ [cf. equation (3) above], then the conjugate variables p_i are homogeneous functions of degree 0, i.e. $p_i(\lambda q^j) = p_i(q^j)$, i.e., they do not change when the system is re-scaled. Only in this case, the conjugate variables are *intensive* and we recover the usual thermodynamic quantities, e.g. $1/T$, p/T, μ/T. In all other cases we shall refer to the conjugate variables p_i as the *would-be intensive* quantities, as in [22,23].

Proposition 2.2 *(Euler's Theorem). Let $S = S(q^1, ..., q^n)$ be a quasi-homogeneous function of degree r and type β. Then*

$$rS(q^j) = \sum_{i=1}^n \beta_i \left[q^i p_i(q^j)\right]. \tag{7}$$

Proof. Consider the derivative of $S(\lambda^{\beta_j} q^j)$ with respect to the scaling parameter λ. On the one hand, since S is a quasi-homogeneous function of degree r and type β, we have

$$\frac{\partial}{\partial\lambda} S(\lambda^{\beta_j} q^j) = \frac{\partial}{\partial\lambda}\left[\lambda^r S(q^j)\right] = r\lambda^{r-1} S(q^j). \tag{8}$$

On the other hand, a direct calculation yields

$$\begin{aligned}\frac{\partial}{\partial\lambda} S(\lambda^{\beta_j} q^j) &= \sum_{i=1}^n \frac{\partial S(\lambda^{\beta_j} q^j)}{\partial(\lambda^{\beta_i} q^i)} \frac{\partial(\lambda^{\beta_i} q^i)}{\partial\lambda} \\ &= \sum_{i=1}^n \frac{\partial S(\lambda^{\beta_j} q^j)}{\partial(\lambda^{\beta_i} q^i)} \left(\beta_i \lambda^{\beta_i - 1} q^i\right) \\ &= \sum_{i=1}^n \left(\beta_i \lambda^{r-1} q^i\right) p_i(q^j),\end{aligned} \tag{9}$$

where the last equality follows from Definition (4) and Eqs. (5) and (6). Thus, combining the results of (8) and (9), Eq. (7) is obtained. $\square$

In standard thermodynamics the above result reduces to the well-known identity for the entropy,

$$S = \frac{1}{T} U - \frac{p}{T} V + \frac{\mu}{T} N. \tag{10}$$

With Proposition 2.2, we can write a GGD relation for the case of quasi-homogeneous thermodynamic systems.

Proposition 2.3 *(Generalized Gibbs–Duhem identity). Let $S(q^1, \ldots, q^n)$ be a quasi-homogeneous function of degree r and type β and let $\{p_i\}_{i=1}^n$ be the set of conjugate variables [cf. equation (4)]. Then,*

$$\sum_{i=1}^n \left[(\beta_i - r)\, p_i(q^j) dq^i + \beta_i q^i dp_i(q^j)\right] = 0. \tag{11}$$

Proof. Since S satisfies the hypothesis of Proposition 2.2, let us consider the differential of (7), namely

$$r\mathrm{d}S(q^j) = \sum_{i=1}^{n} \beta_i \mathrm{d}\left[q^i p_i(q^j)\right] . \tag{12}$$

The left hand side is simply

$$r\mathrm{d}S = r\sum_{i=1}^{n} \frac{\partial}{\partial q^i} S(q^j)\mathrm{d}q^i = r\sum_{i=1}^{n} p_i(q^j)\mathrm{d}q^i , \tag{13}$$

whereas the right hand side yields

$$\sum_{i=1}^{n} \beta_i \mathrm{d}\left[q^i p_i(q^j)\right] = \sum_{i=1}^{n} \beta_i \left[q^i \mathrm{d}p_i(q^j) + p_i(q^j)\mathrm{d}q^i\right] . \tag{14}$$

Subtracting (13) from (14) and collecting the β_i produces the desired result. □

In the case where S is homogeneous of degree r, equation (11) reduces to

$$(1-r)\sum_{i=1}^{n} p_i(q^j)\mathrm{d}q^i + \sum_{i=1}^{n} q^i \mathrm{d}p_i(q^j) = 0 . \tag{15}$$

From this result it follows that in standard thermodynamics (with $r = 1$), using the appropriate identifications of the variables, one obtains the Gibbs–Duhem relation

$$U\mathrm{d}\left(\frac{1}{T}\right) - V\mathrm{d}\left(\frac{p}{T}\right) + N\mathrm{d}\left(\frac{\mu}{T}\right) = 0 , \tag{16}$$

which is a mathematical identity stating that the intensive quantities are not all independent in equilibrium [20].

3. A mathematical inconsistency and its possible resolution

In this section we prove the mathematical inconsistency between the usual zeroth law of thermodynamics, the standard Maxwell construction for coexistence between different phases and the GGD identity, and provide a possible resolution through a re-definition of the equilibrium parameters.

We start from the crucial fact that in ordinary thermodynamics the Gibbs-Duhem identity (16) is mathematically consistent with Maxwell's law for phase coexistence. Here, one considers a single system splitting into two different phases remaining at equilibrium, i.e. sharing the same values of their intensive quantities, while the entropy and volume of the system change, causing a discontinuity in the extensive quantities and thus giving rise to a *first order phase transition*. Clearly in this case the definition of equilibrium between the phases in terms of equal values of the conjugate (intensive) quantities is consistent with (16).

From the above discussion on the role of the intensive variables in Maxwell's construction and its consistency with the Gibbs–Duhem relation (16), it is evident why such consistency is lost in the case of quasi-homogeneous systems, where equation (11) holds. Indeed for the two phases to be at equilibrium, the zeroth law would predict that no change in any of the would-be intensive variables p_i would happen, i.e., $\mathrm{d}p_i = 0$ for all i. This implies that the second term in (11) vanishes identically. However, in general the first term in (11) is different from zero, thus leading to an inconsistency. For instance, in the case of a homogeneous entropy of degree r, it follows from the first law $\mathrm{d}S = \sum_{i=1}^{n} p_i \mathrm{d}q^i$ that the first term of (15) is proportional to the change in the entropy during the transition, and hence to the latent heat, which cannot be zero in a first order phase transition.

This inconsistency leads to the two following possibilities: either one gives up the standard formulation of phase coexistence expressed by the Maxwell construction (at least in its usual form), or one has to re-define the conditions for equilibrium, i.e., the zeroth law. Due to the many indications arising from different perspectives pointing to the fact that the zeroth law needs to be revisited for systems with non-additive entropy and energy relations (see e.g. [2–9]), we opt for the latter route.

From the analysis of the homogeneity of the first derivatives of S – see (6) – let us propose the following

Definition 3.1 *(Generalized intensive variables).* Let $S(q^1,\ldots,q^n)$ be a quasi-homogeneous function of degree r and type β and let $\{p_i\}_1^n$ be the set of conjugate variables. Assume that $\beta_i \neq 0$ for every i. The quantities

$$\tilde{p}_i(q^j) \equiv \left[\left(q^i\right)^{\beta_i - r}\right]^{1/\beta_i} p_i(q^j) \tag{17}$$

are called the *generalized intensive variables.*

Indeed, these variables reduce to (4) when S is homogeneous of degree 1. Moreover, one can easily prove the following

Proposition 3.1. *The generalized intensive variables* (17) *are quasi-homogeneous functions of degree 0.*

Proof.

$$\begin{aligned}\tilde{p}_i(\lambda^{\beta_j} q^j) &= \left[\left(\lambda^{\beta_i} q^i\right)^{\beta_i - r}\right]^{1/\beta_i} p_i(\lambda^{\beta_j} q^j) \\ &= \lambda^{\beta_i - r}\left[\left(q^i\right)^{\beta_i - r}\right]^{1/\beta_i} \left[\lambda^{r-\beta_i} p_i(q^j)\right] \\ &= \left[\left(q^i\right)^{\beta_i - r}\right]^{1/\beta_i} p_i(q^j) = \tilde{p}_i(q^j) . \quad \square\end{aligned} \tag{18}$$

This is a desirable property for quantities defining a notion of equilibrium as they remain invariant under a scaling of the system. Note that these generalized variables could have been inferred from Eq. (75) in [22]. However, in that work they were not singled out nor were advocated as the correct ones to describe equilibrium.

Using the generalized intensive variables (17), we can re-write the GGD identity (11) as in [22]:

Proposition 3.2. *Let $S(q^1,\ldots,q^n)$ be a quasi-homogeneous function of degree r and type β and let $\left\{\tilde{p}_i\right\}_{i=1}^n$ be the set of generalized intensive variables. Then,*

$$\sum_{i=1}^{n} \beta_i \left(q^i\right)^{r/\beta_i} \mathrm{d}\tilde{p}_i(q^j) = 0 . \tag{19}$$

Proof. From Eq. (17) we have

$$p_i(q^j) = \tilde{p}_i(q^j)\left(q^i\right)^{r/\beta_i - 1} , \tag{20}$$

and we can thus rewrite the identity (11) in terms of the $\tilde{p}_i(q^j)$ as

$$\sum_{i=1}^{n} \beta_i \left[\left(1 - \frac{r}{\beta_i}\right)\tilde{p}_i \left(q^i\right)^{r/\beta_i - 1} \mathrm{d}q^i + q^i \mathrm{d}\left(\tilde{p}_i \left(q^i\right)^{r/\beta_i - 1}\right)\right] = 0 . \tag{21}$$

By explicit calculation of the second term, we can rewrite the above identity as

$$0=\sum_{i=1}^{n}\beta_i\Bigg[\left(1-\frac{r}{\beta_i}\right)\tilde{p}_i\left(q^i\right)^{r/\beta_i-1}dq^i$$
$$+q^i\Big[\left(q^i\right)^{r/\beta_i-1}d\tilde{p}_i+\left(\frac{r}{\beta_i}-1\right)\left(q^i\right)^{r/\beta_i-2}\tilde{p}_i\,dq^i\Big]\Bigg]$$
$$=\sum_{i=1}^{n}\beta_i\left(q^i\right)^{r/\beta_i}d\tilde{p}_i\,. \quad \square \tag{22}$$

Note that the GGD identity (11) only establishes the existence of a relation between the would-be intensive and the would-be extensive variables, without fixing the values of the generalized intensive variables uniquely. In this sense our choice of the generalized intensive variables (17) is not the only one possible. However, it is motivated by the following considerations. Firstly, (17) reduce to (4) when the entropy is homogeneous of degree 1. Moreover, these quantities are quasi-homogeneous functions of degree 0, thus being true intensive variables (under the appropriate rescalings of the extensive ones). Finally, as stated in Proposition 3.2, using these variables the GGD identity takes the same form as the standard one (cf. [22]). Indeed, Propositions 3.1 and 3.2 suggest the following modification of the notion of thermodynamic equilibrium:

Definition 3.2 *(Thermodynamic equilibrium).* Two systems whose entropy is a quasi-homogeneous function of the same degree and type are in thermodynamic equilibrium with each other if and only if they have the same values of the $\tilde{p}_i(q^j)$.

This is the *generalized zeroth law of thermodynamics* that we propose for any quasi-homogeneous system. Note that Definition 3.2 is mathematically consistent with the identity (11) – cf. (19) – even when considering processes of coexistence as in the case of the usual Maxwell equal area law.

Let us remark that with our prescription one can consider the example of a process of coexistence among different phases at equilibrium without any incongruence, as long as the definition of equilibrium is given by equating the quantities in (17). Note also that our simple redefinition gives a general prediction about the quantities that have to be constant at equilibrium.

In the next section we consider examples from black holes thermodynamics and show that for the Schwarzschild black hole our redefinition of the equilibrium condition yields a constant generalized temperature. This result coincides with a different instance of the generalized zeroth law of thermodynamics resulting from non-extensive statistical mechanics [24]. As a more relevant consequence we will also show that for the Kerr–AdS black hole our construction suggests that a reconsideration of the first order phase transition might be in order.

4. Quasi-homogeneous black hole thermodynamics

In this section we investigate some examples for the above ideas in the context of black hole thermodynamics. In principle, our generalization of the zeroth law can be applied to any black hole system, given that one can easily determine the degrees of homogeneity from the Smarr relation,

$$(D-3)M=(D-2)TS+(D-2)\Omega J-2PV+(D-3)\Phi Q \tag{23}$$

where D is the number of spacetime dimensions, M is the mass of the black hole, T is the Hawking temperature, S is the entropy and the other terms are work terms depending on the black hole family in question [1]. Here, we consider two in particular, namely the Schwarzschild and the Kerr–AdS black holes, to compare our results with previous proposals and to illustrate new features.

4.1. Schwarzschild

The Schwarzschild black hole is the most straightforward example, since its thermodynamics is described by only one extensive variable, i.e., its mass M. The entropy as a function of M is

$$S(M)=4\pi M^2\,, \tag{24}$$

which is a homogeneous function of degree $r=2$. From this the standard temperature is derived as

$$\frac{1}{T}=\frac{\partial S}{\partial M}=8\pi M\,. \tag{25}$$

It is immediate to see that this a homogeneous function of degree 1 with respect to M, and therefore not a real intensive quantity. With (25) and using (17), we can obtain the generalized temperature as

$$\tilde{T}=TM=\frac{1}{8\pi}\,, \tag{26}$$

i.e., a constant. Note that, a constant is – trivially – a real intensive quantity, as it does not change with any scaling of M. Note also that by (19) the generalized intensive quantities cannot be independent. This means that in this case, since there is only one such generalized intensive quantity, it must be a constant. This fact outlines that the Schwarzschild black hole is not a proper thermodynamic system. However, it is interesting to see that even in this case our formalism coincides with previous approaches. Indeed, a similar result, i.e., a constant generalized temperature, has been obtained previously for the Schwarzschild black hole [24] by using the generalized zeroth law derived from non-extensive statistical mechanics proposed in [6]. In this work, the most general conditions for thermal equilibrium of systems with non-additive energy and entropy are established by using a method based on the definition of the so-called formal logarithms of these quantities. However, the same method was also applied in [25] in the analysis of the Kerr black hole, resulting in a constant generalized temperature, regardless of the angular momentum, identical to the Schwarzschild case – an indication that the result may be unphysical, as the authors point out themselves. Moreover, from our formalism a dependence of the generalized temperature on the angular momentum is to be expected. Finally, in [26,27] the Rényi entropy was used as the formal logarithm of the Bekenstein–Hawking entropy. In this case the temperature for the Schwarzschild case depends on the mass M and is not intensive. The connection of our proposal to these approaches and the general question of the underlying behaviour of the energy and entropy is thus not quite clear and might be addressed in future works.

4.2. Kerr–AdS

Kerr black holes in asymptotically Anti–de Sitter space are thermodynamically determined by three extensive variables, namely their mass M, angular momentum J and pressure P, which is defined via the cosmological constant Λ of the spacetime as

$$P=-\frac{\Lambda}{8\pi}\,. \tag{27}$$

The cosmological constant is usually included as a pressure into the thermodynamic description of black holes [28,10,1], and thus it turns out that the internal energy of the black hole is

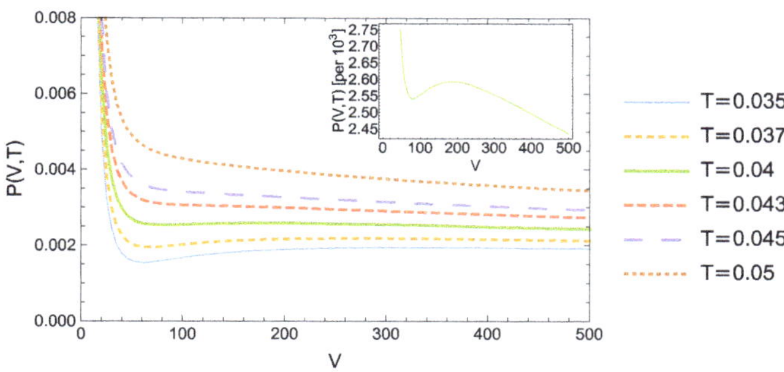

Fig. 1. Equation of state $P(V,T)$ for different values of T and with $J=1$. (For interpretation of the colors in this figure, the reader is referred to the web version of this article.)

$$U = M - PV\,, \tag{28}$$

and therefore the mass of the black hole is identified with the enthalpy

$$M \equiv H = U + PV\,. \tag{29}$$

For the Kerr–AdS black hole one obtains [10]

$$H(S,P,J) = \frac{1}{2}\sqrt{\frac{4\pi^2 J^2\left(\frac{8PS}{3}+1\right)+\left(\frac{8PS^2}{3}+S\right)^2}{\pi S}}\,, \tag{30}$$

and from this, provided $J \neq 0$, it is possible to calculate the expression for the internal energy as

$$U(S,V,J) = \left(\frac{\pi}{S}\right)^3\left[\left(\frac{3V}{4\pi}\right)\left\{\frac{S^2}{2\pi^2}+J^2\right\} - J^2\left\{\left(\frac{3V}{4\pi}\right)^2 - \left(\frac{S}{\pi}\right)^3\right\}^{1/2}\right]. \tag{31}$$

For simplicity and without loss of generality, we will limit further analyses to positive angular momenta, i.e., $J > 0$. The temperature and pressure can be easily obtained as

$$T = \frac{1}{8S^4}\left[\frac{6\pi^{3/2}J^2\left(9\pi V^2 - 8S^3\right)}{\sqrt{9\pi V^2 - 16S^3}} - 18\pi^2 J^2 V - 3S^2 V\right], \tag{32}$$

and

$$P = \frac{3}{8S^3}\left[2\pi^2 J\left(J - \frac{3\sqrt{\pi}JV}{\sqrt{9\pi V^2 - 16S^3}}\right) + S^2\right], \tag{33}$$

respectively.

The case of Kerr–AdS is particularly interesting for our purposes because its equation of state, i.e., the relation $P(V,T)$ at fixed J, qualitatively shows the same oscillatory behaviour as a vdW fluid, which is generally taken as an indication of the presence of a first order phase transition, sometimes referred to as the CCK phase transition [14,29,30]. To see this let us fix $J=1$ from now on and first look at Fig. 1, where we plot $P(V,T)$ as a function of V for various choices of T, with the inlet zooming in on one of the curves to show the characteristic vdW bump. The region of the bump is the area where one would apply the Maxwell equal area law in analogy to ordinary thermodynamics [10]. A different (equivalent) way to look at such transition is by considering the graph of the Gibbs free energy,

$$G(T,P,J) = U - TS + PV\,. \tag{34}$$

To illustrate the multi-valued behavior of the Gibbs free energy we plot in Figs. 2 and 3 the cuts along the lines of constant T and P, respectively, featuring the characteristic swallowtails.

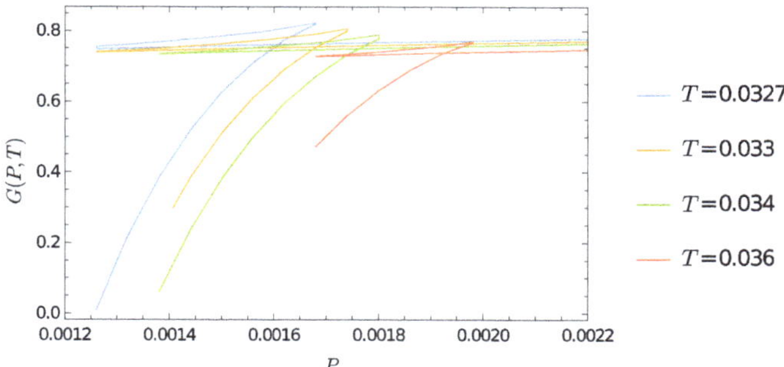

Fig. 2. Cuts of the Gibbs free energy at constant T. (For interpretation of the colors in this figure, the reader is referred to the web version of this article.)

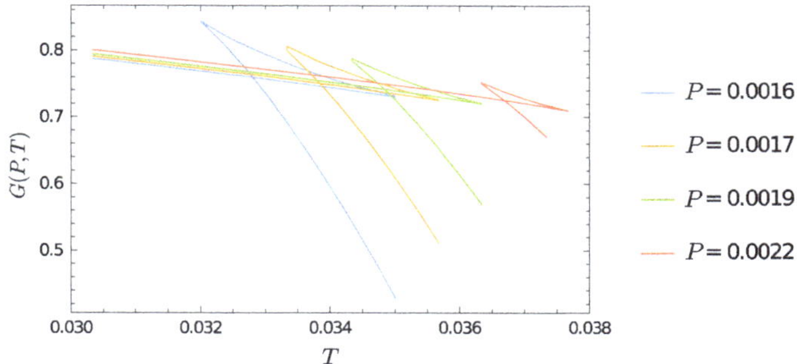

Fig. 3. Cuts of the Gibbs free energy at constant P. (For interpretation of the colors in this figure, the reader is referred to the web version of this article.)

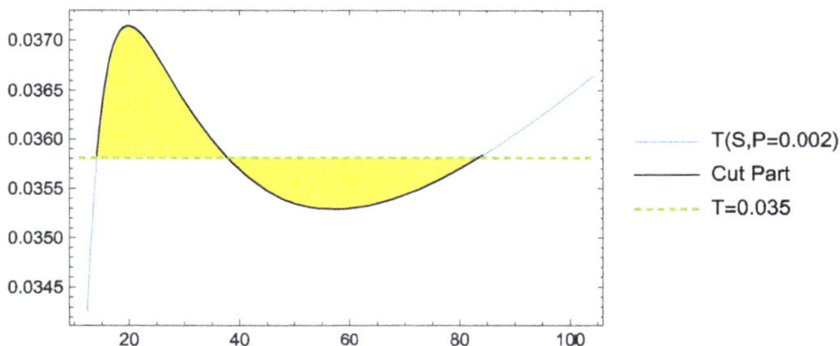

Fig. 4. Equation of state $T(S,P)$ for $J=1$ and $P=0.002$, together with the standard Maxwell construction. The two colored areas are equal. (For interpretation of the colors in this figure, the reader is referred to the web version of this article.)

Based on the above analogy with the vdW phase diagram, it has been argued that there is a first order phase transition between small and large Kerr–AdS black holes, for appropriate values of the temperature and pressure. Indeed, the standard Maxwell construction can be performed, and the form of the coexistence curve can be calculated [14,29–31]. In the following we use this example – which is considered to be well understood in the literature – to claim that a revision due to the GGD identity is called for.

We start by showing that the standard Maxwell construction in this case is inconsistent with the GGD identity. To do so, it is more convenient to use the equation of state $T(S,P)$, as plotted in Fig. 4. For appropriate values of T and P, this equation of state exhibits an oscillatory behavior, as for the case of $P(T,V)$ above (cf. Fig. 1). The corresponding value for the transition temperature is also calculated using Maxwell's equal area law.

Now we proceed to verify the GGD identity for this case. Using the Smarr relation (23) applied to four spacetime dimensions (and $Q=0$), we have

$$S = \frac{1}{2T}U + \frac{3}{2}\frac{P}{T}V - \frac{\Omega}{T}J\,. \tag{35}$$

Table 1
Values of T_{tr}, $\text{Maxwell}_{\text{Dev}}$ and GGD_{Dev} obtained numerically from Maxwell's equal area law for different choices of P_{tr}. More details in the text.

P_{tr}	T_{tr}	$\text{Maxwell}_{\text{Dev}}$	GGD_{Dev}
1.0×10^{-3}	2.6×10^{-2}	1.8×10^{-2}	6.5×10^{-1}
1.6×10^{-3}	3.2×10^{-2}	2.2×10^{-2}	4.6×10^{-1}
2.0×10^{-3}	3.5×10^{-2}	2.2×10^{-3}	3.6×10^{-1}
2.6×10^{-3}	3.9×10^{-2}	4.5×10^{-3}	2.0×10^{-1}

From this, we can determine the degrees of homogeneity of the variables T, P and Ω as $\beta_T = 1/2$, $\beta_P = 3/2$ and $\beta_\Omega = 1$ (cf. Eq. (7)). The overall degree of homogeneity of the entropy is $r = 1$. Therefore, the GGD (11) in the Kerr–AdS case reads

$$-\frac{1}{2T}\mathrm{d}U + \frac{1}{2}\frac{P}{T}\mathrm{d}V + \frac{1}{2}U\mathrm{d}\left(\frac{1}{T}\right) + \frac{3}{2}V\mathrm{d}\left(\frac{P}{T}\right) + J\mathrm{d}\left(\frac{\Omega}{T}\right) = 0. \tag{36}$$

Note that the last three terms have an analogous form with the standard Gibbs–Duhem relation (16) (although with different coefficients). Now let us use the standard Maxwell equal area law to prove an inconsistency with (36). By the usual argument, the two coexisting phases are in equilibrium and therefore all the last three terms vanish along the coexistence process. Thus we are left with the following expression

$$-\frac{1}{2T_{tr}}\Delta U + \frac{1}{2}\frac{P_{tr}}{T_{tr}}\Delta V = 0, \tag{37}$$

where ΔU and ΔV represent the jumps in these quantities along the coexistence line and T_{tr} and P_{tr} are the constant values of the temperature and pressure along the transition. Eq. (37) can be further simplified to

$$\Delta U - P_{tr}\Delta V = T_{tr}\Delta S - 2P_{tr}\Delta V = 0, \tag{38}$$

where in the last equality we made use of the first law (with J constant), that is, $\Delta U = T_{tr}\Delta S - P_{tr}\Delta V$. Now it is an easy exercise to use the values of T_{tr}, P_{tr}, ΔS and ΔV calculated using the standard Maxwell construction to show that Eq. (38) is not satisfied, i.e., that there is an inconsistency with the GGD identity. Table 1 shows the results of these calculations for different values of the transition pressure P_{tr}. In the first column we report the chosen values for the transition pressure P_{tr}. In the second column we provide the corresponding transition temperature T_{tr}, calculated using the standard Maxwell construction. In the third column we show that the area law is satisfied, by checking that the deviation from zero of the difference between the two areas in yellow in Fig. 4 is negligible. In the last column we demonstrate that the GGD identity (38) is not satisfied, by showing that the deviation from zero is large compared to that of the area law, and thus not negligible.

Since the analysis of the phase transition in terms of the standard definition of thermodynamic equilibrium leads to an inconsistency with the GGD identity, we now reconsider the phase transition in terms of the generalized intensive quantities defined in (17). From Eq. (17) and (35), we can infer the generalized intensive variables responsible for equilibrium as

$$\frac{1}{\tilde{T}} = \frac{1}{TU} \quad \text{and} \quad \frac{\tilde{P}}{\tilde{T}} = \frac{P}{T}V^{1/3}. \tag{39}$$

Combining the two expressions, we end up with the generalized thermodynamic equilibrium parameters

$$\tilde{T} = TU \quad \text{and} \quad \tilde{P} = PUV^{1/3}. \tag{40}$$

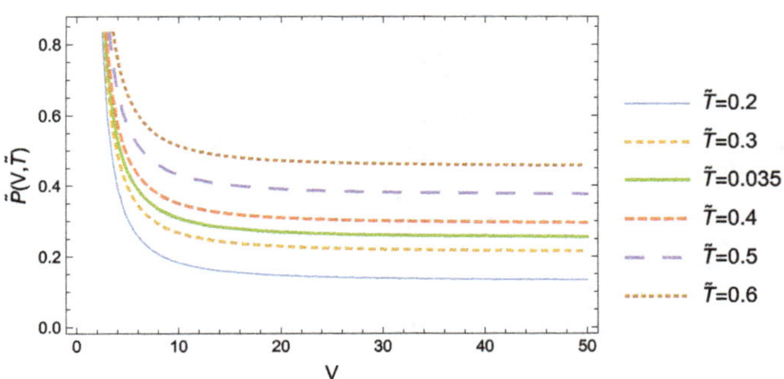

Fig. 5. Equation of state $\tilde{P}(V, \tilde{T})$ at constant $\tilde{T}$. (For interpretation of the colors in this figure, the reader is referred to the web version of this article.)

By construction, these functions are quasi-homogeneous of degree 0 and type $\beta = (1, 3/2, 1)$ with respect to the correspondingly rescaled extensive variables S, V and J, i.e.,

$$\tilde{T}(\lambda^1 S, \lambda^{3/2} V, \lambda^1 J) = \lambda^0 \tilde{T}(S, V, J), \tag{41}$$

$$\tilde{P}(\lambda^1 S, \lambda^{3/2} V, \lambda^1 J) = \lambda^0 \tilde{P}(S, V, J). \tag{42}$$

In terms of S and V (for $J = 1$) these read

$$\begin{aligned}\tilde{T}(S, V) = {} & \frac{3V\pi^{13/2}}{8S^7}\left[-36\left(\frac{S}{\pi}\right)^3 - 10\left(\frac{S}{\pi}\right)^5 + 27\left(\frac{V}{\pi}\right)^2 \right.\\ & \left. + 9\left(\frac{S}{\pi}\right)^2\left(\frac{V}{\pi}\right)^2\right] + \frac{3\pi^6}{64S^7}\sqrt{9\pi V^2 - 16S^3}\left[32\left(\frac{S}{\pi}\right)^3 \right.\\ & \left. - 72\left(\frac{V}{\pi}\right)^2 - 24\left(\frac{S}{\pi}\right)^2\left(\frac{V}{\pi}\right)^2 - 3\left(\frac{S}{\pi}\right)^4\left(\frac{V}{\pi}\right)^2\right]\end{aligned} \tag{43}$$

and

$$\begin{aligned}\tilde{P}(S, V) = {} & \frac{9\pi^{5/2}V^{4/3}}{32S^6}\sqrt{9\pi V^2 - 16S^3}\left(2\pi^2 + S^2\right)\\ & \times\left(6\pi^2 V + 3S^2 V - 2\pi^{3/2}\sqrt{9\pi V^2 - 16S^3}\right).\end{aligned} \tag{44}$$

Note that in order to show the quasi-homogeneity of these functions by rescaling the extensive variables, it is necessary to recover the terms containing J, including it as an extensive variable. Using these expressions, we can return to the plot of the equation of state, but now in terms of the new variables, plotting $\tilde{P}(V, \tilde{T})$ as a function of V for different choices of $\tilde{T}$. As can be seen in Fig. 5, the curves are monotonously decreasing, therefore the system appears to be stable and there is no necessity for the Maxwell construction. The same effect can be observed using the Gibbs free energy. We can re-express definition (34) in terms of the new intensive variables $\tilde{T}$ and $\tilde{P}$ and calculate the function $G(\tilde{T}, \tilde{P}, J)$, inverting Eqs. (43) and (44) numerically. The result can be seen in figures Figs. 6 and 7, where cuts at constant $\tilde{T}$ and $\tilde{P}$ show that the Gibbs free energy in terms of the generalized intensive variables is a single-valued smooth function.

We conclude that for the Kerr–AdS black hole the standard Maxwell equal area law is inconsistent with the GGD identity. Besides, the use of the generalized intensive variables proposed here as the parameters defining thermodynamic equilibrium seems to indicate that there is no first order phase transition between large and small black holes, as previously argued in the literature. However, our results deserve more investigation. Perhaps a comparison with explicit models directly constructed from statistical mechanics could shed more light on the validity of such statements. Alternatively, an analysis involving thermodynamic response functions

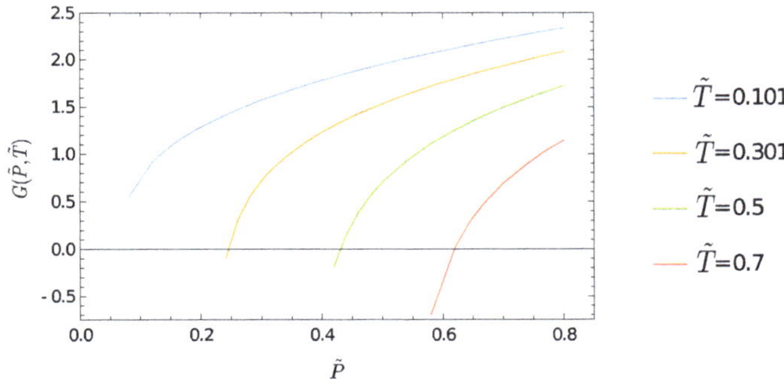

Fig. 6. Cuts of the Gibbs free energy at constant $\tilde{T}$. (For interpretation of the colors in this figure, the reader is referred to the web version of this article.)

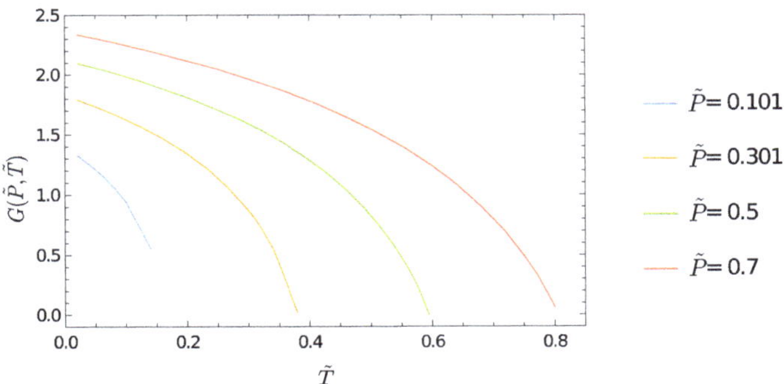

Fig. 7. Cuts of the Gibbs free energy at constant $\tilde{P}$. (For interpretation of the colors in this figure, the reader is referred to the web version of this article.)

could be interesting, although the significance of these response functions in the context of a generalized zeroth law should be re-evaluated.

5. Conclusions and future directions

In this work we consider a generalization of the zeroth law of thermodynamics for systems whose thermodynamic entropy is a quasi-homogeneous function of the (would-be) extensive variables (Definition 3.2). Originating from the generalized version of the Gibbs-Duhem identity, we show how to define the generalized intensive variables that can be used to define thermodynamic equilibrium in such general cases (Definition 3.1). Moreover, we prove that this new definition resolves an inconsistency between the use of the standard Maxwell equal area law and the GGD identity that is usually overlooked, especially in the literature regarding the thermodynamics of black holes. Within this context, we consider two examples where the application of our generalized zeroth law should be relevant, namely the Schwarzschild and the Kerr–AdS black holes. The former is important because with our approach we recover a previous result found in [24], derived from a different perspective. The latter example is of interest because in the usual treatment the Kerr–AdS family of black holes shows a behavior which is very similar to that of a van der Waals fluid, including a first order phase transition. However, we argue that the use of the standard Maxwell equal area law in such case is not fully consistent and that using the generalized intensive variables that we have introduced here in order to define thermodynamic equilibrium, such phase transition disappears. This statement however should be further investigated in other contexts in order to corroborate such a conclusion.

Our results are intended to be a step forward towards a deeper formal understanding of the thermodynamic properties of systems with quasi-homogeneous entropy. However, it also calls for more detailed investigations. One can use the arguments given here to understand whether other reported first order phase transitions in black holes are consistent with their respective GGD identities or not (cf. e.g. [11–19]). It would also be interesting to study the implications of the present analysis for the conditions of equilibrium between black holes and heat reservoirs, e.g. a Schwarzschild black hole in a hot flat space. Moreover, we would like to extend the comparison between our approach and the one presented in [6, 24–27] to other cases to see whether the agreement we found for the Schwarzschild black hole holds in more general contexts. Besides, it would be worth using explicit calculations as in [4,5] to check whether our prediction of the new thermodynamic parameters defining equilibrium can be tested by numerical experiments, and to compare our results with the formalism proposed in [32,33] presenting a different instance of a GGD relation for systems with long-range interactions. These directions will be the subject of future work.

Acknowledgements

A.B. was supported by a DGAPA–UNAM postdoctoral fellowship. C.G. was supported by an UNAM postdoctoral fellowship program. F.N. received support from PAPIIT-UNAM Grant IN-111617.

References

[1] B.P. Dolan, Black holes and Boyle's law–the thermodynamics of the cosmological constant, Mod. Phys. Lett. A 30 (03n04) (2015) 1540002.

[2] J. Oppenheim, Thermodynamics with long-range interactions: from Ising models to black holes, Phys. Rev. E 68 (1) (2003) 016108.

[3] S. Abe, General pseudoadditivity of composable entropy prescribed by the existence of equilibrium, Phys. Rev. E 63 (6) (2001) 061105.

[4] A. Ramírez-Hernández, H. Larralde, F. Leyvraz, Violation of the zeroth law of thermodynamics in systems with negative specific heat, Phys. Rev. Lett. 100 (12) (2008) 120601.

[5] A. Ramírez-Hernández, H. Larralde, F. Leyvraz, Systems with negative specific heat in thermal contact: violation of the zeroth law, Phys. Rev. E 78 (6) (2008) 061133.

[6] T. Biró, P. Ván, Zeroth law compatibility of nonadditive thermodynamics, Phys. Rev. E 83 (6) (2011) 061147.

[7] E. Lenzi, A. Scarfone, Extensive-like and intensive-like thermodynamical variables in generalized thermostatistics, Phys. A, Stat. Mech. Appl. 391 (8) (2012) 2543–2555.

[8] H.M. Haggard, C. Rovelli, Death and resurrection of the zeroth principle of thermodynamics, Phys. Rev. D 87 (8) (2013) 084001.

[9] L. Velazquez, Remarks about the thermodynamics of astrophysical systems in mutual interaction and related notions, J. Stat. Mech. Theory Exp. 2016 (3) (2016) 033105.

[10] B.P. Dolan, Where is the PdV term in the fist law of black hole thermodynamics?, preprint, arXiv:1209.1272, 2012.

[11] D. Kubizňák, R.B. Mann, P–V criticality of charged AdS black holes, J. High Energy Astrophys. 2012 (7) (2012) 1.

[12] E. Spallucci, A. Smailagic, Maxwell's equal area law and the Hawking-page phase transition, J. Gravity 2013 (2013).

[13] E. Spallucci, A. Smailagic, Maxwell's equal-area law for charged Anti-de Sitter black holes, Phys. Lett. B 723 (4) (2013) 436–441.

[14] B.P. Dolan, Vacuum energy and the latent heat of AdS–Kerr black holes, Phys. Rev. D 90 (8) (2014) 084002.

[15] S.-Q. Lan, J.-X. Mo, W.-B. Liu, A note on Maxwell's equal area law for black hole phase transition, Eur. Phys. J. C 75 (9) (2015) 419.

[16] S.-W. Wei, Y.-X. Liu, Clapeyron equations and fitting formula of the coexistence curve in the extended phase space of charged ads black holes, Phys. Rev. D 91 (4) (2015) 044018.

[17] S.-W. Wei, Y.-X. Liu, et al., Insight into the microscopic structure of an ads black hole from a thermodynamical phase transition, Phys. Rev. Lett. 115 (11) (2015) 111302.

[18] J.-X. Mo, G.-Q. Li, Coexistence curves and molecule number densities of ads black holes in the reduced parameter space, Phys. Rev. D 92 (2) (2015) 024055.

[19] D. Kubizňák, R.B. Mann, M. Teo, Black hole chemistry: thermodynamics with lambda, Class. Quantum Gravity 34 (6) (2017) 063001.

[20] H.B. Callen, Thermodynamics & an Intro. to Thermostatistics, John Wiley & Sons, 2006.

[21] A. Bravetti, C. Lopez-Monsalvo, F. Nettel, Contact symmetries and Hamiltonian thermodynamics, Ann. Phys. 361 (2015) 377–400.

[22] F. Belgiorno, Quasi-homogeneous thermodynamics and black holes, J. Math. Phys. 44 (3) (2003) 1089–1128.
[23] F. Belgiorno, S. Cacciatori, General symmetries: from homogeneous thermodynamics to black holes, Eur. Phys. J. Plus 126 (9) (2011) 1–19.
[24] V.G. Czinner, Black hole entropy and the zeroth law of thermodynamics, Int. J. Mod. Phys. D 24 (09) (2015) 1542015.
[25] V. Czinner, H. Iguchi, A zeroth law compatible model to Kerr black hole thermodynamics, Universe 3 (2017) 14, http://dx.doi.org/10.20944/preprints201612.0096.v1.
[26] T.S. Biró, V.G. Czinner, A q-parameter bound for particle spectra based on black hole thermodynamics with Rényi entropy, Phys. Lett. B 726 (4) (2013) 861–865.
[27] V. Czinner, H. Iguchi, Rényi entropy and the thermodynamic stability of black holes, Phys. Lett. B 752 (2016) 306.
[28] D. Kastor, S. Ray, J. Traschen, Enthalpy and the mechanics of AdS black holes, Class. Quantum Gravity 26 (19) (2009) 195011.
[29] M.M. Caldarelli, G. Cognola, D. Klemm, Thermodynamics of Kerr–Newman–AdS black holes and conformal field theories, Class. Quantum Gravity 17 (2) (2000) 399.
[30] N. Altamirano, D. Kubizňák, R.B. Mann, Z. Sherkatghanad, Thermodynamics of rotating black holes and black rings: phase transitions and thermodynamic volume, Galaxies 2 (1) (2014) 89–159.
[31] S.-W. Wei, P. Cheng, Y.-X. Liu, Analytical and exact critical phenomena of d-dimensional singly spinning Kerr–AdS black holes, Phys. Rev. D 93 (8) (2016) 084015.
[32] I. Latella, A. Pérez-Madrid, Local thermodynamics and the generalized Gibbs–Duhem equation in systems with long-range interactions, Phys. Rev. E 88 (Oct. 2013) 042135.
[33] I. Latella, A. Pérez-Madrid, A. Campa, L. Casetti, S. Ruffo, Thermodynamics of nonadditive systems, Phys. Rev. Lett. 114 (Jun 2015) 230601.

Three dimensional magnetic solutions in massive gravity with (non)linear field

S.H. Hendi [a,b,*], B. Eslam Panah [a,b,c], S. Panahiyan [a,d,e], M. Momennia [a]

[a] Physics Department and Biruni Observatory, College of Sciences, Shiraz University, Shiraz 71454, Iran
[b] Research Institute for Astrophysics and Astronomy of Maragha (RIAAM), P.O. Box 55134-441, Maragha, Iran
[c] ICRANet, Piazza della Repubblica 10, I-65122 Pescara, Italy
[d] Helmholtz-Institut Jena, Fröbelstieg 3, Jena D-07743, Germany
[e] Physics Department, Shahid Beheshti University, Tehran 19839, Iran

ARTICLE INFO

Editor: N. Lambert

ABSTRACT

The Noble Prize in physics 2016 motivates one to study different aspects of topological properties and topological defects as their related objects. Considering the significant role of the topological defects (especially magnetic strings) in cosmology, here, we will investigate three dimensional horizonless magnetic solutions in the presence of two generalizations: massive gravity and nonlinear electromagnetic field. The effects of these two generalizations on properties of the solutions and their geometrical structure are investigated. The differences between de Sitter and anti de Sitter solutions are highlighted and conditions regarding the existence of phase transition in geometrical structure of the solutions are studied.

1. Introduction

The Nobel Prize in physics 2016 has been assigned to the interesting consequences of topological invariant and topological phase transitions. Although general relativity is based on the local transformation, it is found that topological properties have significant impacts on our understanding of the universe. In this regard, topological defects have been observed in the various branches of physics. The crucial role of topological defects was observed in a new type of phase transition in two-dimensional systems [1]. Kosterlitz and Thouless applied the mentioned points to superconducting and superfluid films and they found their important roles in quantum nature of one-dimensional systems at very low temperatures. In addition, it was shown that phenomenological properties of different phases of physical systems could be explained by these topological defects. For example, in studying liquid crystal, it was shown that structural properties and phase transitions are affected by topological effects [2]. The applications of these defects in condensed matter with ordered media [3], magnetism and nanomagnetism [4], vortices in superfluid [5] and Bose–Einstein condensate [6] were explored. Furthermore, the importance of these mathematical tools in studying superconductors and their phase transitions were highlighted in Refs. [7,8].

From the cosmological point of view, existence of the topological defects could be traced back into early universe and also phase transitions in the early universe [9]. The existence of topological defects is originated from breaking down the symmetry in phase transitions that has taken place in the early universe. Speaking more precisely, regions of the universe which are separated by more than the distance $d = ct$ (in which c is speed of light and t is time), could not know anything about each other. During the phase transition, different regions choose different minima in the set of possible states to fall in. The topological defects are placed at the boundaries of these regions which have chosen different minima. Therefore, one can state that topological defects are the results of disagreement between different choices of different regions. In the context of cosmology, there are different types of the topological defects. Depending on the dimensionality and structural properties, these defects could be categorized into; (i) Domain walls which are due to a broken discrete symmetry and divide universe into blocks. (ii) Cosmic strings which are due axial or cylindrical symmetry breaking and are related to grand unified particle physics models/electroweak scale. (iii) Monopoles which are super massive and carry magnetic charge and are formed when a spherical

* Corresponding author.
E-mail addresses: hendi@shirazu.ac.ir (S.H. Hendi), behzad.eslampanah@gmail.com (B. Eslam Panah), shahram.panahiyan@uni-jena.de (S. Panahiyan), m.momennia@shirazu.ac.ir (M. Momennia).

symmetry is broken. (iv) Textures which are formed due to the breaking of several symmetries. These topological defects carry information regarding the early universe. In addition, it was proposed that they could have specific roles in the large-scale structures [9], anisotropy in the Cosmic Microwave Background (CMB) [10] and dark matter [11]. Besides, these topological defects could be used as cosmological lenses [12]. In other words, the trajectory of the photon on these topological defects are affected depending on deficit angle. This highlights the importance of analyzing deficit angle in the properties of topological defects.

The cosmic strings in the presence of Maxwell field have been investigated [13]. Furthermore, the superconducting property of these topological defects has been explored in Einstein [14], dilaton [15] and Brans–Dicke [16] theories. In addition, the QCD applications of the magnetic strings [17] and their roles in quantum theories [18] have been investigated before. The stability of the cosmic strings through quantum fluctuations has been analyzed in Ref. [19]. The limits on the cosmic string tension have been studied by extracting signals of cosmic strings from CMB temperature anisotropy maps [20]. The spectrum of gravitational wave background produced by cosmic strings is obtained in Ref. [21]. For further investigations regarding cosmic strings, we refer the reader to an incomplete list of references [22].

Domain walls and their evolution in de Sitter universe have been studied in [23]. In addition, the gravitational waves produced from decaying domain walls are investigated in Ref. [24]. The localization of the fields on the dynamical domain wall was investigated and it was shown that the chiral spinor can be localized on the domain walls [25]. For further studies regarding this class of topological defects, we refer the reader to Ref. [26].

On the other hand, considering most of physical systems in nature, one finds that they exhibit nonlinear behavior, and therefore, the nonlinear field theories are of importance in physical researches. There are many motivations for studying the nonlinear electrodynamics (NED) such as; (i) These theories are the generalizations of Maxwell field and reduce to linear Maxwell theory in the special cases (weak nonlinearity). (ii) These nonlinear theories can describe the radiation propagation inside specific materials [27]. (iii) Some special NED models can describe the self-interaction of virtual electron-positron pairs [28]. (iv) Theories of NED can remove the problem of point-like charge self-energy. (v) From the standpoint of quantum gravity and its coupling with these nonlinear theories, we can obtain more information and deep insight regarding the nature of gravity [29]. (vi) Compatibility with AdS/CFT correspondence and string theory are other properties of NED theories. (vii) NED theory improves the basic concept of gravitational redshift and its dependency of any background magnetic field as compared to the well-established method introduced by general relativity. (viii) From the perspective of cosmology, it was shown that NED theories can remove both of the big bang and black hole singularities [30]. (ix) From astrophysical point of view, it was found that the effects of NED become indeed quite important in super-strong magnetized compact objects, such as pulsars and particular neutron stars [31].

There are different models of NED, such as Born–Infeld form [32], logarithmic form [33], exponential form [34], arcsin-electrodynamics form [35] and etc. One of the interesting branches of the nonlinear electrodynamics is power Maxwell invariant (PMI) theory. The Lagrangian of PMI theory is an arbitrary power of the Maxwell Lagrangian [36] which could reduce to the Maxwell field by choosing the unit power. In addition, the PMI theory has an interesting consequence which distinguishes this NED theory from other theories; this theory enjoys conformal invariancy when the power of Maxwell invariant is a quarter of space-time dimensions (power = dimensions/4). In other words, in this case, the energy-momentum tensor will be a traceless tensor which leads to conformal invariancy and also an inverse square law of the electric field for the point-like charge in arbitrary dimensions [36].

Recent observations of gravitational waves from a binary black hole merger in LIGO and Virgo collaboration provided a deep insight to general relativity (GR) and existence of massive gravitons [37]. However, gravitons are massless particles with spin 2 in GR which have two degrees of freedom. Since the quantum theory of massless gravitons is non-renormalizable [38], in order to remove this problem, one may modify GR to massive gravity by adding a mass term to the Einstein-Hilbert action. Therefore, considering this action, the graviton will have a mass of m which in case of $m \to 0$, the effect of massive gravity will be vanished. In other words, massive gravity is a modification of GR that gravitons have mass. Among the motivations of the massive gravity one can mention description of accelerating expansion of universe without considering the cosmological constant [39]. It was shown that the terms of massive gravitons can be equivalent to a cosmological constant [40]. This theory modifies gravity compared with GR which allows the universe to accelerate at the large scale, however at small scale, this theory reduces to GR as well. This theory of gravity may illustrate the dark energy problem [41]. In addition, the existence of massive gravitons provides extra polarization for gravitational waves, and affects the propagation's speed of the gravitational waves [42], hence, the production of gravitational waves during inflation [43]. By adding the interaction terms to GR, massive gravity with flat background was investigated by Fierz and Pauli [44]. However, this theory suffers a van Dam–Veltman–Zakharov (vDVZ) discontinuity [45]. Generalization of massive theory to curved background was done by Boulware–Deser. This generalization leads to the existence of a typical ghost, the so-called Boulware–Deser ghost [46]. Several models of massive theory were proposed by some authors in order to avoid discontinuity and ghost problems [47]. One of the ghost-free massive theories in three dimensions was introduced by Bergshoeff, Hohm and Townsend (new massive gravity (NMG)) [48]. However, NMG has ghost problem in four and higher dimensions. Therefore, in order to resolve ghost problem in diverse dimensions, a new theory of massive gravity was proposed by de Rham, Gabadadze and Tolley (dRGT) in 2011 [49]. The stability of dRGT massive theory was studied and it was shown that such theory enjoys absence of the Boulware–Deser ghost [50]. Black hole and cosmological solutions have been investigated in dRGT massive gravity [51–59]. Also, reentrant phase transitions of higher-dimensional AdS black holes and behavior of quasinormal modes and van der Waals like phase transition of charged AdS black holes in massive gravity have been studied in Refs. [60,61].

It is notable that in massive gravity theory, the mass terms are produced by consideration of a reference metric. Considering the reference metric in massive gravity, one finds that it plays a crucial role in construction of exact solutions [62]. In this regard, Vegh introduced a new reference metric which was motivated by applications of gauge/gravity duality [63]. It is believed that the graviton may behave like a lattice and exhibits a Drude peak in this model of massive theory [63]. Another property of this model is related to ghost-free and stability for arbitrary singular metric [64]. The action of massive gravity in an arbitrary d-dimensions is given by

$$\mathcal{I}_G = -\frac{1}{16\pi} \int_{\mathcal{M}} d^d x \sqrt{-g} \left[\mathcal{R} + m^2 \sum_{i=1}^{4} c_i \mathcal{U}_i(g, f) \right], \tag{1}$$

where $\mathcal{R}$ is the scalar curvature and m^2 is related to the mass of gravitons. In addition, f is a fixed symmetric tensor, c_i's are some constants, and $\mathcal{U}_i$'s are symmetric polynomials of the eigenvalues of matrix $\mathcal{K}^{\mu}_{\nu} = \sqrt{g^{\mu\alpha} f_{\alpha\nu}}$ which are as follows

$$\mathcal{U}_1 = [\mathcal{K}], \quad \mathcal{U}_2 = [\mathcal{K}]^2 - \left[\mathcal{K}^2\right],$$
$$\mathcal{U}_3 = [\mathcal{K}]^3 - 3[\mathcal{K}]\left[\mathcal{K}^2\right] + 2\left[\mathcal{K}^3\right],$$
$$\mathcal{U}_4 = [\mathcal{K}]^4 - 6\left[\mathcal{K}^2\right][\mathcal{K}]^2 + 8\left[\mathcal{K}^3\right][\mathcal{K}] + 3\left[\mathcal{K}^2\right]^2 - 6\left[\mathcal{K}^4\right].$$

Charged black hole solutions with (non)linear field and the existence of van der Waals like behavior in extended phase space and also geometrical thermodynamics by considering dRGT massive gravity have been studied [65–67]. Moreover, the hydrostatic equilibrium equation of neutron stars by using this theory of massive gravity was obtained and it was shown that the maximum mass of neutron stars can be about $3.8M_{\odot}$ (where $M_{\odot}$ is mass of the sun) [68]. Also, holographic conductivity in this gravity with PMI field has been investigated in Ref. [69]. Besides, the generalization of this theory to include higher derivative gravity [70] and gravity's rainbow [71] has been done in literatures. In addition, three dimensional (BTZ) charged black hole solutions with (non)linear field have been studied in Ref. [72].

By adding an electromagnetic Lagrangian ($\mathcal{L}(\mathcal{F})$) and the cosmological constant (Λ) to the action (1) with $d=3$, we have

$$\mathcal{I}_G = -\frac{1}{16\pi}\int_{\mathcal{M}} d^3x\sqrt{-g}\left[\mathcal{R} - 2\Lambda + \mathcal{L}(\mathcal{F}) + m^2\sum_{i=1}^{4} c_i\mathcal{U}_i(g,f)\right]. \tag{2}$$

Varying the action (2) with respect to the gravitational and gauge fields, one can obtain the following field equations

$$R_{\mu\nu} - \frac{1}{2}g_{\mu\nu}(\mathcal{R} - 2\Lambda) + m^2\chi_{\mu\nu} = T_{\mu\nu}, \tag{3}$$
$$\partial_\mu\left(\sqrt{-g}\mathcal{L}_{\mathcal{F}}F^{\mu\nu}\right) = 0, \tag{4}$$

in which $\mathcal{L}_{\mathcal{F}} = d\mathcal{L}(\mathcal{F})/d\mathcal{F}$ where $\mathcal{F} = F_{\mu\nu}F^{\mu\nu}$ is the Maxwell invariant, $F_{\mu\nu} = \partial_\mu A_\nu - \partial_\nu A_\mu$ is the Faraday tensor and A_μ is the gauge potential. In addition, $\chi_{\mu\nu}$ is the massive term with the following form

$$\begin{aligned}\chi_{\mu\nu} &= -\frac{c_1}{2}\left(\mathcal{U}_1 g_{\mu\nu} - \mathcal{K}_{\mu\nu}\right) - \frac{c_2}{2}\left(\mathcal{U}_2 g_{\mu\nu} - 2\mathcal{U}_1\mathcal{K}_{\mu\nu} + 2\mathcal{K}^2_{\mu\nu}\right) \\ &- \frac{c_3}{2}(\mathcal{U}_3 g_{\mu\nu} - 3\mathcal{U}_2\mathcal{K}_{\mu\nu} + 6\mathcal{U}_1\mathcal{K}^2_{\mu\nu} - 6\mathcal{K}^3_{\mu\nu}) \\ &- \frac{c_4}{2}(\mathcal{U}_4 g_{\mu\nu} - 4\mathcal{U}_3\mathcal{K}_{\mu\nu} + 12\mathcal{U}_2\mathcal{K}^2_{\mu\nu} - 24\mathcal{U}_1\mathcal{K}^3_{\mu\nu} + 24\mathcal{K}^4_{\mu\nu}),\end{aligned} \tag{5}$$

and the energy-momentum tensor of Eq. (3) is

$$T_{\mu\nu} = \frac{1}{2}g_{\mu\nu}\mathcal{L}(\mathcal{F}) - 2\mathcal{L}_{\mathcal{F}}F_{\mu\lambda}F^{\lambda}_{\nu}. \tag{6}$$

Here, we want to obtain the magnetic solutions of Eqs. (3) and (4) by considering the Maxwell electromagnetic field ($\mathcal{L}(\mathcal{F}) = -\mathcal{F}$).

Magnetic branes (or horizonless solution) are interesting objects which have been investigated by many authors [73–80]. Our main motivation here is to understand the effects of two generalizations on the magnetic horizonless solutions with interpretation of topological defects. These two generalizations include massive gravity and PMI electromagnetic field. Considering the applications of topological defects in dark matter, CMB, gravitational waves, large scale structure and etc., it is necessary to investigate the effects of the massive gravitons on the structure and formation of topological defects. Here, we intend to show how generalization to massive gravity would modify geometrical structure of the magnetic solutions. To do so, we apply the massive gravity generalization and investigate geometrical properties such as deficit angle. Considering the electromagnetically charged aspect of the objects of interest in this paper (magnetic solutions), we will take two cases of linear and nonlinear electromagnetic fields into account. Here, we would investigate the effects of Maxwell and PMI electromagnetic fields on the deficit angle, hence, geometrical structure of the topological defects known as horizonless magnetic solutions. The combinations of massive gravity and PMI theory is another subject of interest which would be addressed. It is notable that such magnetic source was interpreted as a kind of magnetic monopole reminiscent of a Nielson–Oleson vortex solution [81], while Dias and Lemos interpreted it as a composition of two symmetric and superposed electric charges [13]. In other words, one of the mentioned electric charges is at rest and the other is rotating, and therefore, there is no electric field since the total electric charge is zero, but angular electric current produces a magnetic field.

Now, we use the new metric of three dimensional spacetime with $(-++)$ signature which was introduced in Ref. [80]

$$ds^2 = -\frac{\rho^2}{l^2}dt^2 + \frac{d\rho^2}{g(\rho)} + l^2 g(\rho)d\varphi^2, \tag{7}$$

where $g(\rho)$ is an arbitrary function of radial coordinate ρ which should be determined. The scale length factor l is related to the cosmological constant Λ, and the angular coordinate φ is dimensionless as usual and ranges in $[0, 2\pi]$. The motivation of considering the metric gauge [$g_{tt} \propto -\rho^2$ and $\left(g_{\rho\rho}\right)^{-1} \propto g_{\varphi\varphi}$] instead of the usual Schwarzschild like gauge [$\left(g_{\rho\rho}\right)^{-1} \propto g_{tt}$ and $g_{\varphi\varphi} \propto \rho^2$] comes from the fact that we are looking for magnetic solutions without curvature singularity. It is easy to show that using a suitable transformation, the metric (7) can be mapped to 3-dimensional Schwarzschild like spacetime locally, but not globally [80].

In order to obtain exact solutions, we should make a choice for the reference metric. We consider the following ansatz metric

$$f_{\mu\nu} = diag(\frac{-c^2}{l^2}, 0, 0), \tag{8}$$

where in the above equation c is a positive constant. Using the metric ansatz (8), $\mathcal{U}_i$'s are [65,72]

$$\mathcal{U}_1 = \frac{c}{\rho}, \quad \mathcal{U}_2 = \mathcal{U}_3 = \mathcal{U}_4 = 0, \tag{9}$$

which indicate that the only contribution of massive gravity comes from $\mathcal{U}_1$ in three dimensions. Before proceeding we give a reason for such choice of the reference metric (8). For three dimensional black holes, the spacetime metric with $(-,+,+)$ signature has the following explicit form

$$ds^2 = -g(\rho)dt^2 + \frac{d\rho^2}{g(\rho)} + \rho^2 d\varphi^2. \tag{10}$$

In order to obtain exact black hole solutions, we consider the ansatz metric as $f_{\mu\nu} = diag(0, 0, c^2)$ (see Refs. [63], [65] and [66], for more details). Here, the metric function ($g(\rho)$) is factors of radial and spatial coordinates in magnetic spacetime metric (Eq. (7)). In order to have exact solutions in an axially symmetric spacetime with the form (7), it is necessary to consider the reference metric as $f_{\mu\nu} = diag(\frac{-c^2}{l^2}, 0, 0)$. This form of reference metric is expectable. Comparing black hole metric, Eq. (10), with magnetic spacetime, Eq. (7), we find that Eq. (7) can be reproduced from Eq. (10) by the following local transformations:

$$t \longrightarrow il\varphi \quad \& \quad \varphi \longrightarrow it/l. \tag{11}$$

Since we changed the role of t and φ coordinates, the nonzero component of the reference metric should be changed accordingly.

Since we are going to study the linearly magnetic solutions, we choose the Lagrangian of Maxwell field $\mathcal{L}(\mathcal{F}) = -\mathcal{F}$ for Eqs. (2), (4), and (6). It is well-known that the electric field is associated with the time component of the vector potential A_t, while the magnetic field is associated with the angular component A_φ. Due to our interest to investigate the magnetic solutions, we assume the vector potential as

$$A_\mu = h(\rho)\delta_\mu^\varphi. \tag{12}$$

Using the Maxwell equation (4) with $\mathcal{L}(\mathcal{F}) = -\mathcal{F}$, and the metric (7), one finds the following differential equation

$$F_{\varphi\rho} + \rho F'_{\varphi\rho} = 0, \tag{13}$$

where $F_{\varphi\rho} = h'(\rho)$ in which the prime denotes differentiation with respect to ρ. Equation (13) has the following solution

$$F_{\varphi\rho} = \frac{q}{\rho}, \tag{14}$$

where q is an integration constant. To find the metric function $g(\rho)$, one may insert Eq. (14) in the field equation (3) by considering the metric (7). After some calculations, one can obtain the following differential equations

$$\begin{cases} g'(\rho) + 2\Lambda\rho - \frac{2}{\rho}\left(\frac{q}{l}\right)^2 - cc_1 m^2 = 0, & \rho\rho\ (\varphi\varphi)\ component \\ g''(\rho) + 2\Lambda + 2\left(\frac{q}{\rho l}\right)^2 = 0, & tt\ component \end{cases}, \tag{15}$$

where the double prime is the second derivative versus ρ. It is straightforward to show that these equations have the following solution

$$g(\rho) = m_0 - \Lambda\rho^2 + \frac{2q^2}{l^2}\ln\left(\frac{\rho}{l}\right) + cc_1 m^2 \rho, \tag{16}$$

which m_0 is an integration constant which is related to the mass parameter, and l is an arbitrary constant with length dimension which is coming from the fact that the logarithmic arguments should be dimensionless. As one can see, the massive parameter appears in the metric function as a factor for the linear function of ρ. We should note that the obtained metric function (16) satisfies all components of the field equation (3), simultaneously. In addition, the asymptotical behavior of the solution (16) is adS or dS provided $\Lambda < 0$ or $\Lambda > 0$. Also, it is worthwhile to mention that in the absence of massive parameter ($m = 0$), the metric function (16) reduces to the result of Ref. [80] for $s = 1$.

1.1. Energy conditions

Now, we examine the energy conditions to find physical solutions. To do so, we consider the orthonormal contravariant basis vectors, and then we obtain the three dimensional energy momentum tensor as $T^{\mu\nu} = diag(\mu, p_r, p_t)$ in which μ, p_r, and p_t are the energy density, the radial pressure and the tangential pressure, respectively. Having the energy momentum tensor at hand, we are in a position to investigate the energy conditions. We use the following known constraints in three dimensions (Table 1).

In order to simplify the mathematics and physical interpretations, we use the following orthonormal contravariant (hatted) basis vectors for diagonal static metric (7)

$$\mathbf{e}_{\hat{t}} = \frac{l}{\rho}\frac{\partial}{\partial t}, \quad \mathbf{e}_{\hat{\rho}} = \sqrt{g}\frac{\partial}{\partial\rho}, \quad \mathbf{e}_{\hat{\phi}} = \frac{1}{l\sqrt{g}}\frac{\partial}{\partial\phi}. \tag{17}$$

Table 1
Energy conditions criteria.

$p_r + \mu \geq 0$, $p_t + \mu \geq 0$	*for null energy condition (NEC)*
$\mu \geq 0$, $p_r + \mu \geq 0$, $p_t + \mu \geq 0$	*for weak energy condition (WEC)*
$\mu \geq 0$, $-\mu \leq p_r \leq \mu$, $-\mu \leq p_t \leq \mu$	*for dominant energy condition (DEC)*
$p_r + \mu \geq 0$, $p_t + \mu \geq 0$, $\mu + p_r + p_t \geq 0$	*for strong energy condition (SEC)*

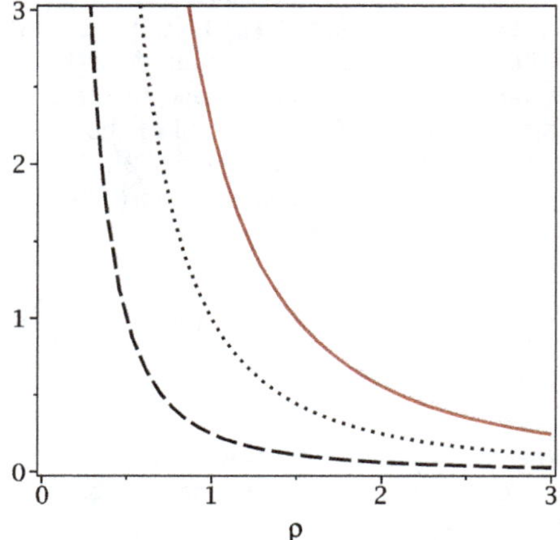

Fig. 1. $T^{\widehat{tt}}$ versus ρ for $l = 1$ and $q = 0.5$ (dashed line), $q = 1.0$ (doted line) and $q = 1.5$ (continuous line).

It is a matter of straightforward calculations to show that the nonzero components of stress-energy tensor are

$$T^{\widehat{tt}} = T^{\widehat{\rho\rho}} = T^{\widehat{\phi\phi}} = \left(\frac{F_{\phi\rho}}{l}\right)^2. \tag{18}$$

All components of stress-energy tensor are the same and positive and it is easy to find that NEC, WEC, DEC and SEC are satisfied, simultaneously.

As one can see, the massive parameter do not contribute to the energy-momentum tensor, so the energy conditions are independent of the massive parameter. In order to investigation the effects of charge on the energy density of the spacetime, we plot the $T^{\widehat{tt}}$ versus ρ. Considering Fig. 1, one can find that the energy density of the spacetime is positive everywhere, and increasing the charge parameter leads to increasing the concentration of energy density.

1.2. Geometric properties

Now, we want to study the properties of spacetime described by Eq. (7) with obtained metric function (16). At first, we calculate $R_{\mu\nu\lambda\kappa}R^{\mu\nu\lambda\kappa}$ for examination of existence of curvature singularity

$$R_{\mu\nu\lambda\kappa}R^{\mu\nu\lambda\kappa} = 12\Lambda^2 - \frac{8\Lambda q^2}{l^2\rho^2} + \frac{2cc_1 m^2(4q^2 + cc_1 m^2 l^2\rho - 4\Lambda l^2\rho^2)}{l^2\rho^3} + \frac{12q^4}{l^4\rho^4}. \tag{19}$$

Considering Eq. (19), the Kretschmann scalar reduces to $12\Lambda^2$ for $\rho \longrightarrow \infty$, which confirms that the asymptotical behavior of this spacetime is (a)dS. It is also obvious that the Kretschmann scalar diverges at $\rho = 0$, and therefore one might think that there

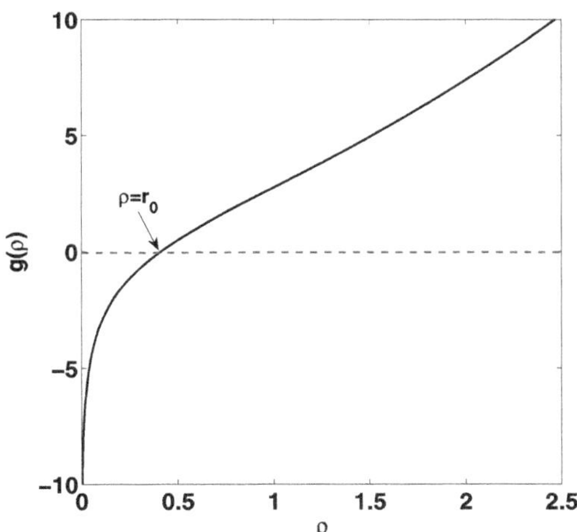

Fig. 2. $g(\rho)$ versus ρ for $l=1$, $q=1$, $\Lambda=-1$, $c=1$, $c_1=1$, $m=0.5$ and $m_0=1.5$.

is a curvature singularity located at $\rho=0$. But as we will see, the spacetime will never achieve $\rho=0$. There are two possible cases for the metric function: the metric function has no real positive root which is interpreted as naked singularity (this case is not of interest here), or metric function has at least one real positive root. If one considers r_0 as the largest root of metric function, it is clear that for $\rho<r_0$ there will be a change in signature of metric (see Fig. 2). In other words, for $\rho<r_0$ the metric function is negative, hence metric signature is $(-,-,-)$, and for $\rho>r_0$ the metric function is positive, therefore metric signature is legal $(-,+,+)$. This change in the metric signature results into a conclusion: it is not possible to extend spacetime to $\rho<r_0$. In order to exclude the forbidden zone ($\rho<r_0$), we introduce a new radial coordinate r as

$$r^2=\rho^2-r_0^2 \Longrightarrow d\rho^2=\frac{r^2}{r^2+r_0^2}dr^2, \tag{20}$$

where for the allowed region, $\rho\geq r_0$, leads to $r\geq 0$ in the new coordinate system. Applying this coordinate transformation, the metric (7) should be written as

$$ds^2=-\frac{r^2+r_0^2}{l^2}dt^2+\frac{r^2}{\left(r^2+r_0^2\right)g(r)}dr^2+l^2g(r)d\varphi^2, \tag{21}$$

in which the coordinate r assumes the values $0\leq r<\infty$, and obtained $g(r)$ (Eq. (16)) is now given by

$$g(r)=m_0-\Lambda\left(r^2+r_0^2\right)+\frac{q^2}{l^2}\ln\left(\frac{r^2+r_0^2}{l^2}\right)+cc_1m^2\sqrt{r^2+r_0^2}. \tag{22}$$

The nonzero component of electromagnetic field in the new coordinate can be given by

$$F_{\varphi r}=\frac{q}{\sqrt{r^2+r_0^2}}, \tag{23}$$

One can show that all curvature invariants do not diverge in the range $0\leq r<\infty$, and $g(r)$ (Eq. (22)) is positive definite for $0\leq r<\infty$. It is evident that for having singular solutions both r and r_0 must be zero whereas this case is never reached due to considering nonzero value for r_0. So, this spacetime has no curvature singularity and horizon. Due to the fact that the limit of the ratio "circumference/radius" is not 2π, the spacetime (21) has a conic geometry and therefore the spacetime has a conical singularity at $r=0$

$$\lim_{r\longrightarrow 0}\frac{1}{r}\sqrt{\frac{g_{\varphi\varphi}}{g_{rr}}}\neq 1. \tag{24}$$

On the other hand, the conical singularity can be removed if one exchanges the coordinate φ with the following period

$$Period_{\varphi}=2\pi\left(\lim_{r\longrightarrow 0}\frac{1}{r}\sqrt{\frac{g_{\varphi\varphi}}{g_{rr}}}\right)^{-1}=2\pi\left(1-4\mu\right), \tag{25}$$

in which the deficit angle is defined as $\delta\varphi=8\pi\mu$, where μ is given by

$$\mu=\frac{1}{4}+\frac{1}{4lr_0\Omega}, \tag{26}$$

where Ω is

$$\Omega=\Lambda-\frac{q^2}{l^2r_0^2}-\frac{cc_1m^2}{2r_0}. \tag{27}$$

In order to have a better insight of the behavior of deficit angle, we calculate the root and divergence points of the deficit angle as

$$r_0|_{\delta\varphi=0}=\left\{\begin{array}{l}\frac{1}{4l\Lambda}\left(cc_1m^2l-2\pm\sqrt{c^2c_1^2m^4l^2-4\left(cc_1m^2l-1-4\Lambda q^2\right)}\right)\\ \frac{1}{4l\Lambda}\left(-cc_1m^2l-2\pm\sqrt{c^2c_1^2m^4l^2-4\left(cc_1m^2l-1-4\Lambda q^2\right)}\right)\end{array}\right., \tag{28}$$

$$r_0|_{\delta\varphi\longrightarrow\infty}=\left\{\begin{array}{l}\frac{1}{4l\Lambda}\left(cc_1m^2l\pm\sqrt{c^2c_1^2m^4l^2+16\Lambda q^2}\right)\\ \frac{1}{4l\Lambda}\left(-cc_1m^2l\pm\sqrt{c^2c_1^2m^4l^2+16\Lambda q^2}\right)\end{array}\right.. \tag{29}$$

Here, we see that the roots are functions of the cosmological constant, massive gravity and electric charge. Existence of the real valued root is restricted to following condition

$$c^2c_1^2m^4l^2-4\left(cc_1m^2l-1-4\Lambda q^2\right)\geq 0. \tag{30}$$

The effects of the massive gravity and electric charge are only observed in numerator of the roots while the effects of the cosmological constant could be observed in both numerator and denominator of the roots. The electric charge is coupled with cosmological constant. While such coupling is not observed for the massive gravity.

As for the divergencies of the deficit angle, one can observe that its existence is also restricted to satisfaction of specific condition in the following form

$$c^2c_1^2m^4l^2+16\Lambda q^2\geq 0 \tag{31}$$

In the absence of the massive gravity, only for dS spacetime divergencies are observable for deficit angle. Generalization to massive gravity provides the possibility of the divergencies for deficit angle in adS spacetime under certain circumstances. This highlights the effects of the massive gravity. Here, similar to the case of roots, a coupling between cosmological constant and electric charge is observed while such coupling could not be seen for massive gravity.

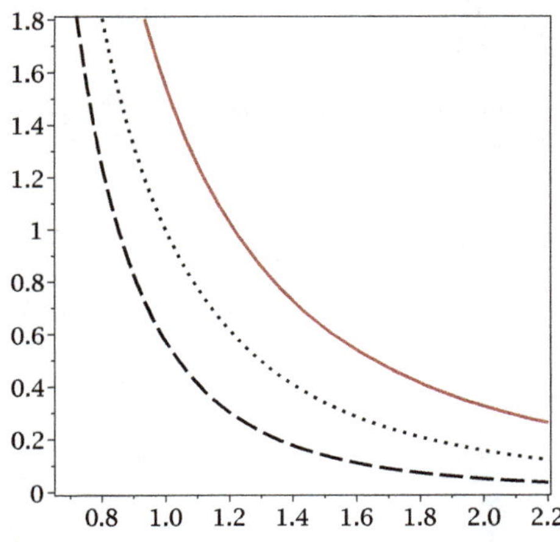

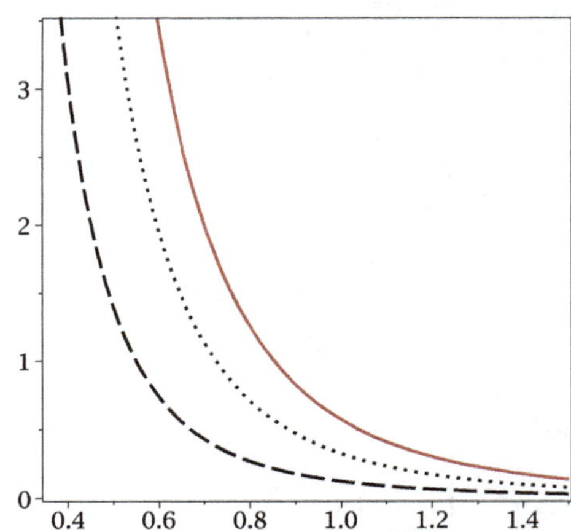

Fig. 3. $T^{\widehat{t}\widehat{t}}$ versus ρ for $l=1$. **Left diagram:** for $q=1.5$, $s=0.7$ (dashed line), $s=0.8$ (doted line) and $s=0.9$ (continuous line). **Right diagram:** for $s=0.7$, $q=0.5$ (dashed line), $q=1.0$ (doted line) and $q=1.5$ (continuous line).

2. Generalization of achievements to the case of nonlinear electrodynamics: PMI theory

In this section, we are going to obtain the solutions in presence of PMI source and investigate the properties. We start with the following PMI Lagrangian

$$\mathcal{L}_{PMI}(\mathcal{F}) = (-\kappa\mathcal{F})^s, \tag{32}$$

where κ and s are coupling and power constants, respectively. Obviously, the PMI Lagrangian (32) reduces to the standard Maxwell Lagrangian ($\mathcal{L}_{Maxwell}(\mathcal{F}) = -\mathcal{F}$) for $s=1$ and $\kappa=1$ which we have investigated before.

Following the method of previous section and considering Eqs. (4), (7) and (32), one can obtain the following differential equation for nonzero component of Faraday tensor

$$F_{\varphi\rho} + (2s-1)\rho F'_{\varphi\rho} = 0, \tag{33}$$

with the following solution

$$F_{\varphi\rho} = q\rho^{1/(1-2s)}, \tag{34}$$

in which q is an integration constant. In order to have a physical asymptotical behavior, we should consider $s > 1/2$. On the other hand, one can easily show that the vector potential A_φ, is

$$A_\varphi = q\rho^{\frac{2(s-1)}{2s-1}}, \tag{35}$$

the electromagnetic gauge potential should be finite at infinity ($\rho \to \infty$), therefore, one should impose following restriction to have this property, so we have

$$\frac{2(s-1)}{2s-1} < 0. \tag{36}$$

The above equation leads to the following restriction on the range of s, as

$$\frac{1}{2} < s < 1. \tag{37}$$

Here, one can insert Eq. (34) in the gravitational field equation (3) by considering the metric (7) to obtain the metric function $g(\rho)$ as

$$g(\rho) = m_0 - \Lambda\rho^2 + cc_1m^2\rho + \frac{(2s-1)^2\rho^2}{2(s-1)}\chi(\rho), \tag{38}$$

where

$$\chi(\rho) = \left(\frac{2q^2}{l^2}\rho^{2/(1-2s)}\right)^s. \tag{39}$$

It is notable that, the obtained metric function in Eq. (38) is related to $s \neq 1$. Also, m_0 is an integration constant which is related to the mass of solutions.

Now, one can calculate the nonzero components of stress-energy tensor by using the introduced basis vectors in Eq. (17) as

$$T^{\widehat{t}\widehat{t}} = \frac{1}{2}\left(\frac{2F_{\phi\rho}^2}{l^2}\right)^s, \tag{40}$$

$$T^{\widehat{\rho}\widehat{\rho}} = T^{\widehat{\phi}\widehat{\phi}} = \left(s - \frac{1}{2}\right)\left(\frac{2F_{\phi\rho}^2}{l^2}\right)^s. \tag{41}$$

According to the above equation, μ ($T^{\widehat{t}\widehat{t}}$) is positive, and so the NEC, WEC, and SEC are satisfied, simultaneously. In addition, in order to satisfy the DEC, the parameter of PMI (s) must be in the range $\frac{1}{2} < s < 1$. As we have mentioned before, the energy conditions do not depend on the massive parameter. Here, we want to investigate the effects of PMI parameter (s) and electrical charge (q) on the energy conditions, so we plot T^{tt} versus ρ in Fig. 3. As one can see, increasing the parameter of PMI theory and electrical charge leads to increasing the concentration of energy density.

One can show that the metric (7) with the metric function (38) has a singularity at $\rho = 0$ by calculating the Kretschmann scalar as

$$\begin{aligned} R_{\mu\nu\lambda\kappa}R^{\mu\nu\lambda\kappa} &= 12\Lambda^2 - \frac{4cc_1m^2}{\rho}\left[2\Lambda - (2s-1)\chi_{q,\rho,s} - \frac{cc_1m^2}{2\rho}\right] \\ &\quad + \left[(8s^2-8s+3)\chi_{q,\rho,s} - 4(4s-3)\Lambda\right]\chi_{q,\rho,s}. \end{aligned} \tag{42}$$

From Eq. (42), it is obvious that the Kretschmann scalar reduces to $12\Lambda^2$ for $\rho \longrightarrow \infty$ and diverges at $\rho = 0$. On the other hand, as we mentioned before, it is not possible to extend spacetime to $\rho < r_0$ because of signature changing. Also, one can apply the coordinate transformation (20) to the metric (7) and find the metric function as

$$\begin{aligned} g(r) &= m_0 - \Lambda\left(r^2 + r_0^2\right) + cc_1m^2\left(r^2 + r_0^2\right)^{1/2} \\ &\quad + \frac{(2s-1)^2\left(r^2 + r_0^2\right)}{2(s-1)}\chi(r), \end{aligned} \tag{43}$$

where

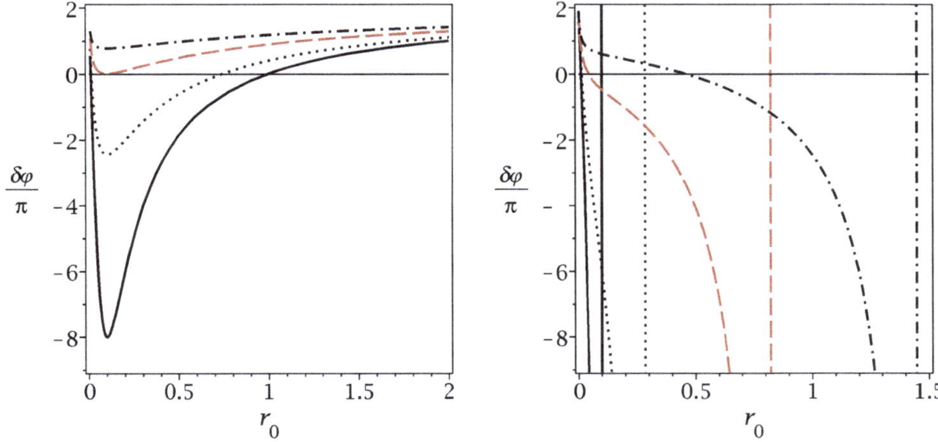

Fig. 4. ***Maxwell solutions:*** $\delta\varphi$ versus r_0 for $q = 0.1$, $c = 1$ and $c_1 = 2$, $m = 0$ (continuous line), $m = 0.5$ (dotted line), $m = 0.9$ (dashed line) and $m = 1.2$ (dashed–dotted line). **Left diagram:** $\Lambda = -1$; **Right diagram:** $\Lambda = 1$.

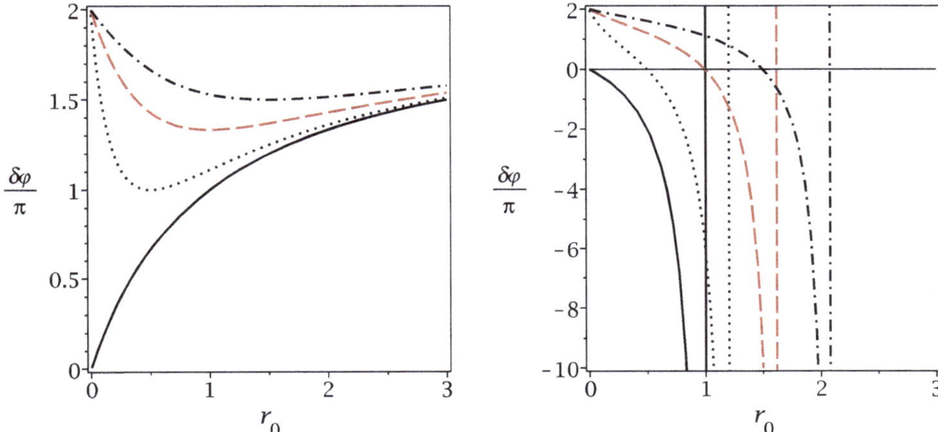

Fig. 5. ***Maxwell solutions:*** $\delta\varphi$ versus r_0 for $m = 1$, $c = 1$ and $c_1 = 2$, $q = 0$ (continuous line), $q = 0.5$ (dotted line), $q = 1$ (dashed line) and $q = 1.5$ (dashed–dotted line). **Left diagram:** $\Lambda = -1$; **Right diagram:** $\Lambda = 1$.

$$\chi(r) = \left(\frac{2q^2}{l^2}\left(r^2 + r_0^2\right)^{1/(1-2s)}\right)^s, \tag{44}$$

and the electromagnetic field in the new coordinate is

$$F_{\varphi r} = q\left(r^2 + r_0^2\right)^{1/2(1-2s)}. \tag{45}$$

Since all curvature invariants do not diverge in the range $0 \leq r < \infty$, one finds that there is no essential singularity. But like the Maxwell case, this spacetime has a conical singularity at $r = 0$ with the deficit angle $\delta\varphi = 8\pi\mu$ where μ is given by Eq. (26) and Ω has the following form

$$\Omega = \Lambda - \frac{cc_1 m^2}{2r_0} - \frac{(2s-1)}{2}\left(\frac{2q^2}{l^2} r_0^{2/(1-2s)}\right)^s. \tag{46}$$

Due to complexity of obtained relation in Eq. (46), it is not possible to calculate the root and divergence points of deficit angle analytically, therefore, we study them in some graphs in next section.

3. Deficit angle diagrams

In order to study the effects of different parameters on the properties of deficit angle for the Maxwell and PMI cases, we have plotted various diagrams (Figs. 4–6 for Maxwell case and Figs. 7–10 for PMI case). The left panels are dedicated to adS spacetime while the right ones are related to dS spacetime. In Ref. [82], it was pointed out that in order to remove ensemble dependency, l should be replaced by following relation

$$\Lambda = \pm\frac{1}{l}, \tag{47}$$

where the positive branch is related to dS spacetime and the opposite is for adS solutions. Hereafter, we employ Eq. (47) to plot deficit angle diagrams. It is notable to highlight a few remarks regarding to values of deficit angle. The deficit angle is restricted by an upper limit provided by geometrical properties of the solutions. Its value could not exceed 2π, and more precisely, deficit angle could have values in range of $-\infty < \delta\varphi \leq 2\pi$.

Depending on the choices of different parameters, the deficit angle of Maxwell-adS and PMI-adS solutions could have a minimum. In adS case, except for neutral solutions, the deficit angle could have; I) Two roots with one region of negativity located between these two roots. II) One extreme root located at the minimum with deficit angle being only positive. III) No root and deficit angle is always positive. The minimum is an increasing function of the massive parameter (left panels of Figs. 4 and 7), electric charge (left panels of Fig. 5 and 8) and c_1 (left panels of Fig. 6 and 9). By considering negative values for c_1, it is possible to have one of the following cases; I) One divergency located between two roots.

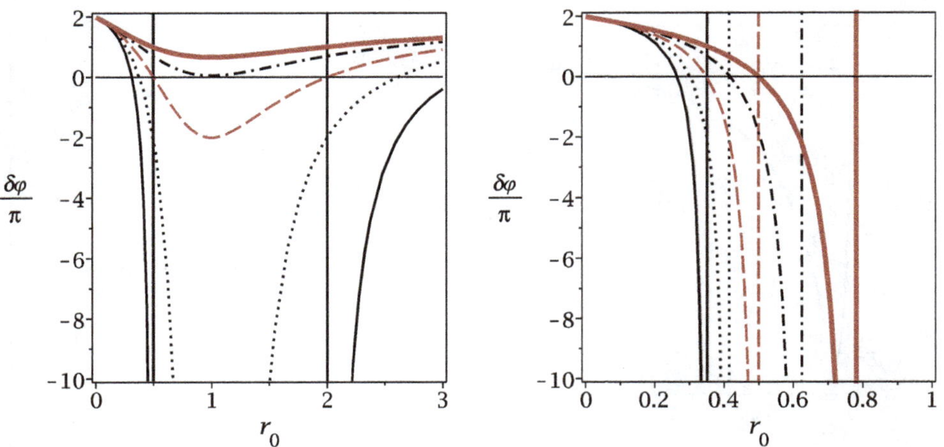

Fig. 6. ***Maxwell solutions:*** $\delta\varphi$ versus r_0 for $q = 0.1$, $c = 1$ and $m = 1$, $c_1 = -5$ (continuous line), $c_1 = -4$ (dotted line), $c_1 = -3$ (dashed line), $c_1 = -1.95$ (dashed-dotted line) and $c_1 = -1$ (bold line). **Left diagram:** $\Lambda = -1$; **Right diagram:** $\Lambda = 1$.

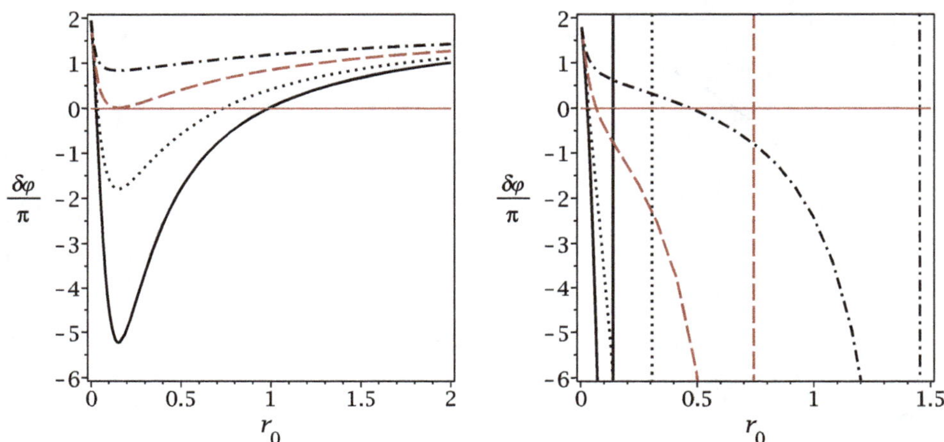

Fig. 7. ***PMI solutions:*** $\delta\varphi$ versus r_0 for $q = 0.1$, $c = 1$, $c_1 = 2$ and $s = 0.9$, $m = 0$ (continuous line), $m = 0.5$ (dotted line), $m = 0.85$ (dashed line) and $m = 1.2$ (dashed-dotted line). **Left diagram:** $\Lambda = -1$; **Right diagram:** $\Lambda = 1$.

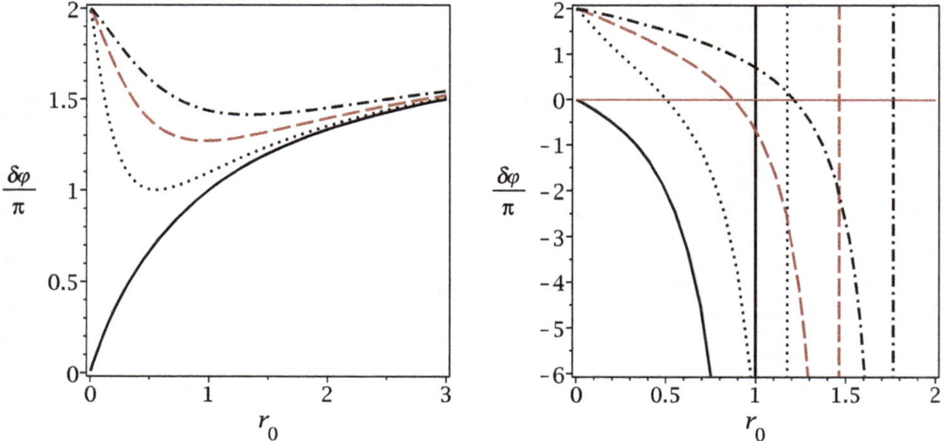

Fig. 8. ***PMI solutions:*** $\delta\varphi$ versus r_0 for $m = 1$, $c = 1$, $c_1 = 2$ and $s = 0.9$, $q = 0$ (continuous line), $q = 0.5$ (dotted line), $q = 1$ (dashed line) and $q = 1.5$ (dashed-dotted line). **Left diagram:** $\Lambda = -1$; **Right diagram:** $\Lambda = 1$.

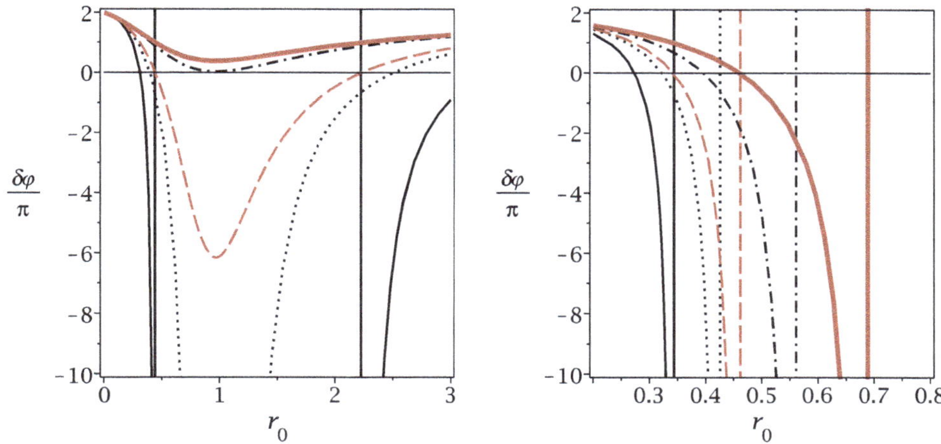

Fig. 9. ***PMI solutions:*** $\delta\varphi$ versus r_0 for $q = 0.1$, $c = 1$, $m = 1$ and $s = 0.9$, $c_1 = -5$ (continuous line), $c_1 = -3.49$ (dotted line), $c_1 = -3$ (dashed line), $c_1 = -1.46$ (dashed-dotted line) and $c_1 = -1$ (bold line). **Left diagram:** $\Lambda = -1$; **Right diagram:** $\Lambda = 1$.

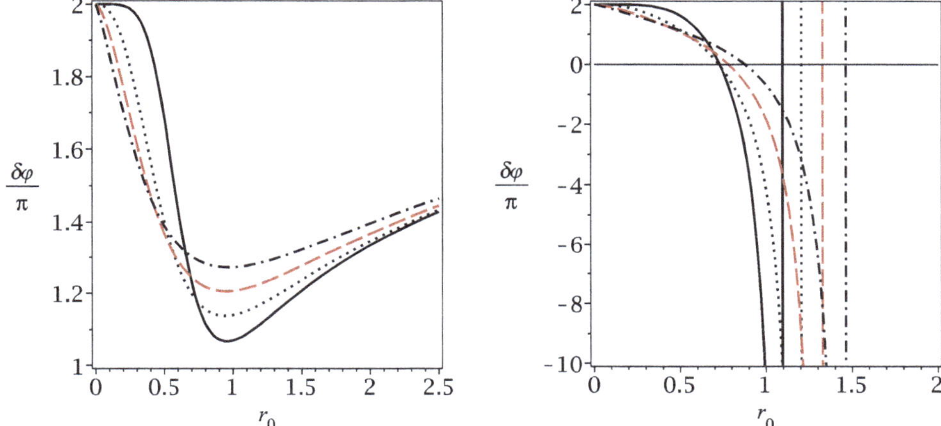

Fig. 10. ***PMI solutions:*** $\delta\varphi$ versus r_0 for $q = 0.1$, $c = 1$, $c_1 = 2$ and $m = 1$, $s = 0.6$ (continuous line), $s = 0.7$ (dotted line), $s = 0.8$ (dashed line) and $s = 0.9$ (dashed-dotted line). **Left diagram:** $\Lambda = -1$; **Right diagram:** $\Lambda = 1$.

II) Two divergencies which are located between two roots. Between the divergencies, the deficit angle is positive but its value is out of the permitted values. In these two cases, the positive deficit angle could only be observed before smaller root and after larger root. Interestingly, in the absence of electric charge, the deficit angle in only an increasing function of r_0 (left panels of Figs. 5 and 8).

For the Maxwell-dS and PMI-dS spacetimes, interestingly, only one root and divergency are observed. The root and divergency are increasing functions of the massive gravity (right panels of Fig. 4 and 7), electric charge (right panels of Fig. 5 and 8) and c_1 (right panels of Fig. 6 and 9). The divergency is located after root. The deficit angle is only positive before root. After divergency, the deficit angle is positive but its values are not in permitted region. The only exception is for the absence of electric charge (right panels of Fig. 5 and 8). In this case, no root is observed and deficit angle is negative valued.

In the case of PMI theory, another free parameter (nonlinearity parameter) exists. Evidently, the minimum in adS case is an increasing function of this parameter (left panel of Fig. 10). For dS spacetime, the root and divergency are increasing functions of this parameter.

Depending on values of deficit angle, the geometrical structure of the magnetic solutions will be determined. Our solutions contain a conical singularity. This conical singularity is built by considering a 2-dimensional plane replaced with cutting an arbitrary slice and sewing together the edges. The singular point is located at the apex of cone. Now, considering this concept, one can see that positive values of the deficit angle represent missing segment of the 2-dimensional plane (Fig. 11). On the contrary, the negative values of the deficit angle represent the additional part that we can add to the mentioned plane (Fig. 12). Therefore, the positivity/negativity of the deficit angle plays a crucial role in the topological structure of the solutions. Here, we see that depending on choices of different parameters, it is possible to obtain negative and positive values of the deficit angle. The roots of deficit angle

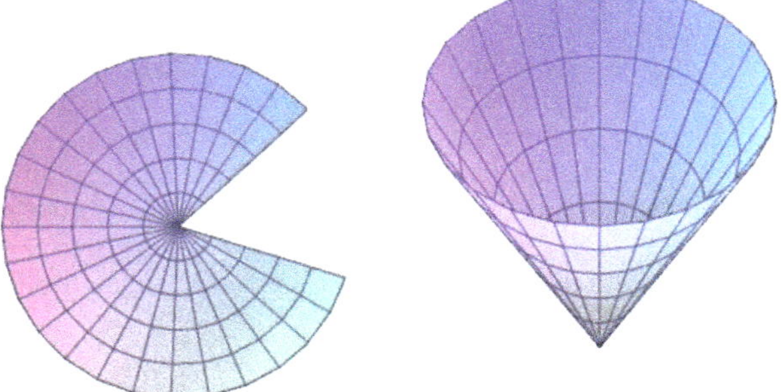

Fig. 11. Deficit angle: by sewing the two edges together (left panel), a cone is formed (right panel).

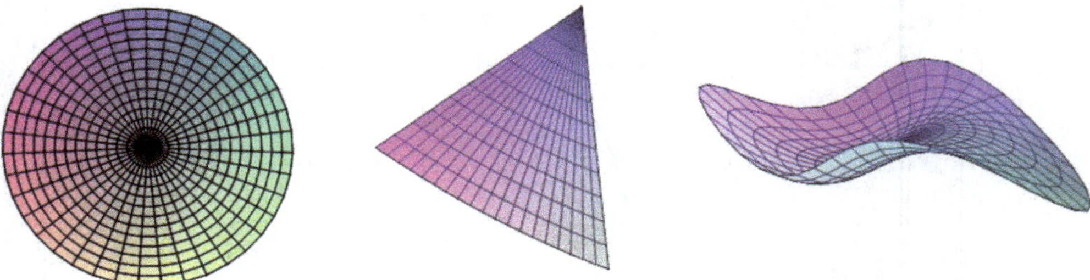

Fig. 12. Surplus angle: by adding additional angle (middle panel) to a circle (left panel), we obtain a new figure (right panel).

could be interpreted as transition points in which the total shape of the object is modified. On the other hand, the existence of divergencies for deficit angle marks the possibility of the absence of magnetic solutions which was observed for both the dS and adS spacetimes. Previously, through several studies, it was shown that existence of deficit/surplus angle enables one to regard the cosmological constant problem [83]. The main motivation of this paper was understanding the effects of massive gravity and PMI theory on the magnetic solutions. The variation in deficit angle shows that the total structure of the solutions depends on contributions of these two generalizations. Specially, we observed that generalization to massive gravity provided the possibility of existence of divergence points for adS spacetime. It is worthwhile to mention that for adS case, between two divergencies, the values of deficit angle are within prohibited range. This indicates that there is no acceptable deficit angle between the divergencies in adS case.

4. Conclusions

In this paper, we have considered magnetic solutions which contain a conical singularity without any event horizon and curvature singularity. The set up for the gravity and energy momentum tensor were consideration of two generalizations: massive gravity and PMI nonlinear electromagnetic field.

The geometrical properties of the solutions were obtained and deficit angle for the two cases of Maxwell-massive and PMI-massive were extracted. It was shown that the general structure of the solutions depends on choices of different parameters through positivity and negativity of the deficit angle. Existence of root and divergency were reported and it was shown that these properties of the solutions depend on the choices of different parameters, such as massive gravity and nonlinearity parameter. In addition, it was shown that depending on the nature of background (being dS or adS), deficit angle, hence geometrical structure of the solutions would be different. The difference was highlighted analytically and numerically through several diagrams.

The existence of both root and divergency for deficit angle was reported which indicates that under certain conditions, suitable choices of different parameters, topological defects known as magnetic solutions would enjoy geometrical phase transition. The dependency of geometrical phase transition on nonlinearity parameter and massive gravity highlighted the importance and roles of massive gravity and also nonlinear electromagnetic field generalizations. Especially, the existence of divergency for adS spacetime in the presence of massive gravity could be pointed out.

The existence of deficit and surplus angles results into two completely different astrophysical objects which essentially requires different methods for detection (see Figs. 11 and 12). In fact, when we are talking about deficit angle, it means that the geometrical structure of the solutions enjoys a positive tension in their structures. On the contrary, existence of the surplus angle corresponds to presence of the negative tension [62]. In this paper, we showed that depending on choices of different parameter, the possibility of both are provided for our magnetic solutions. In fact, in some cases, the existence of discontinuity, hence phase transition between deficit angle and surplus angle was reported for our solutions. Considering the important applications of the deficit/surplus angle in the context of cosmology and cosmological constant problem, one can employ the results of present paper to understand the roles of massive gravity and nonlinear electromagnetic field on these applications and their corresponding results. We leave these matters for future works.

Acknowledgements

The authors wish to thank Shiraz University Research Council. This work has been supported financially by Research Institute for Astronomy and Astrophysics of Maragha.

References

[1] J.M. Kosterlitz, D.J. Thouless, J. Phys. C, Solid State Phys. 5 (1972) L124.
[2] A. Mesaros, et al., Science 333 (2011) 426.
[3] N.D. Mermin, Rev. Mod. Phys. 51 (1979) 591.
[4] H. Braun, Adv. Phys. 61 (2012) 1;
D.M. Stamper-Kurn, M. Ueda, Rev. Mod. Phys. 85 (2013) 1191.
[5] G. Aad, et al., Phys. Rev. D 90 (2014) 052004.
[6] J. Sabbatini, W.H. Zurek, M.J. Davis, Phys. Rev. Lett. 107 (2011) 230402.
[7] J. Kierfeld, T. Nattermann, T. Hwa, Phys. Rev. B 55 (1997) 626.
[8] E. Babaev, Nucl. Phys. B 686 (2004) 3.
[9] T.W.B. Kibble, J. Phys. A 09 (1976) 1387;
A.H. Guth, Phys. Rev. D 23 (1981) 347;
A. Vilenkin, E.P.S. Shellard, Topological Defects and Cosmology, Cambridge Univ. Press, Cambridge, 1994.
[10] J.P. Uzan, N. Deruelle, A. Riazuelo, arXiv:astro-ph/9810313;
A. Riazuelo, N. Deruelle, P. Peter, Phys. Rev. D 61 (2000) 123504.
[11] M. Hindmarsh, R. Kirk, S.M. West, J. Cosmol. Astropart. Phys. 03 (2014) 037;
Y.V. Stadnik, V.V. Flambaum, Phys. Rev. Lett. 113 (2014) 151301.
[12] M.V. Sazhin, et al., Mon. Not. R. Astron. Soc. 376 (2007) 1731.
[13] O.J.C. Dias, J.P.S. Lemos, Class. Quantum Gravity 19 (2002) 2265.
[14] E. Witten, Nucl. Phys. B 249 (1985) 557;
I. Moss, S. Poletti, Phys. Lett. B 199 (1987) 34.
[15] C.N. Ferreira, M.E.X. Guimaraes, J.A. Helayel-Neto, Nucl. Phys. B 581 (2000) 165;
M.H. Dehghani, Phys. Rev. D 71 (2005) 064010.
[16] A.A. Sen, Phys. Rev. D 60 (1999) 067501.
[17] L. Del Debbio, A. Di Giacomo, Y.A. Simonov, Phys. Lett. B 332 (1994) 111.
[18] Y.A. Sitenko, Nucl. Phys. B 372 (1992) 622.
[19] H. Weigel, Russ. Phys. J. 59 (2017) 1768.
[20] L. Hergt, A. Amara, R. Brandenberger, T. Kacprzak, A. Refregier, J. Cosmol. Astropart. Phys. 06 (2017) 004.
[21] Y. Matsui, K. Horiguchi, D. Nitta, S. Kuroyanagi, J. Cosmol. Astropart. Phys. 11 (2016) 005.
[22] R.H. Brandenberger, A.C. Davis, A.M. Matheson, M. Trodden, Phys. Lett. B 293 (1992) 287;
M. Giovannini, Lect. Notes Phys. 737 (2008) 863.
[23] A.D. Dolgov, S.I. Godunov, A.S. Rudenko, J. Cosmol. Astropart. Phys. 10 (2016) 026.
[24] T. Krajewski, Z. Lalak, M. Lewicki, P. Olszewski, J. Cosmol. Astropart. Phys. 12 (2016) 036.
[25] Y. Toyozato, M. Higuchi, S. Nojiri, Phys. Lett. B 754 (2016) 139.
[26] K. Skenderis, P.K. Townsend, A.V. Proeyen, J. High Energy Phys. 08 (2001) 036;
J.C.R.E. Oliveira, C.J.A.P. Martins, P.P. Avelino, Phys. Rev. D 71 (2005) 083509;

D.P. George, M. Trodden, R.R. Volkas, J. High Energy Phys. 02 (2009) 035;
M.C.B. Abdalla, P.F. Carlesso, J.M. hoff da Silva, Eur. Phys. J. C 74 (2014) 2709.
[27] V.A. De Lorenci, M.A. Souza, Phys. Lett. B 512 (2001) 417;
M. Novello, et al., Class. Quantum Gravity 20 (2003) 859.
[28] W. Heisenberg, H. Euler, Consequences of Dirac's Theory, Z. Phys. 98 (1936) 714, arXiv:physics/0605038 (Translation by: W. Korolevski, H. Kleinert, Consequences of Dirac's Theory of the Positron);
J. Schwinger, Phys. Rev. 82 (1951) 664.
[29] N. Seiberg, E. Witten, J. High Energy Phys. 09 (1999) 032.
[30] E. Ayon-Beato, A. Garcia, Phys. Lett. B 464 (1999) 25;
V.A. De Lorenci, R. Klippert, M. Novello, J.M. Salim, Phys. Rev. D 65 (2002) 063501;
C. Corda, H.J. Mosquera Cuesta, Astropart. Phys. 34 (2011) 587.
[31] Z. Bialynicka-Birula, I. Bialynicka-Birula, Phys. Rev. D 2 (1970) 2341;
H.J. Mosquera Cuesta, J.M. Salim, Mon. Not. R. Astron. Soc. 354 (2004) L55.
[32] M. Born, L. Infeld, Proc. R. Soc. Lond. A 144 (1934) 425.
[33] H.H. Soleng, Phys. Rev. D 52 (1995) 6178.
[34] S.H. Hendi, J. High Energy Phys. 03 (2012) 065.
[35] S.I. Kruglov, Commun. Theor. Phys. 66 (2016) 59.
[36] M. Hassaine, C. Martinez, Phys. Rev. D 75 (2007) 027502;
H. Maeda, M. Hassaine, C. Martinez, Phys. Rev. D 79 (2009) 044012;
S.H. Hendi, B. Eslam Panah, Phys. Lett. B 684 (2010) 77.
[37] B.P. Abbott, et al., Phys. Rev. Lett. 116 (2016) 061102.
[38] S. Deser, R. Jackiw, G. 't Hooft, Ann. Phys. 152 (1984) 220.
[39] C. Deffayet, G. Dvali, G. Gabadadze, Phys. Rev. D 65 (2002) 044023.
[40] A.E. Gumrukcuoglu, C. Lin, S. Mukohyama, J. Cosmol. Astropart. Phys. 11 (2011) 030;
P. Gratia, W. Hu, M. Wyman, Phys. Rev. D 86 (2012) 061504.
[41] G. Dvali, G. Gabadadze, M. Shifman, Phys. Rev. D 67 (2003) 044020.
[42] C.M. Will, Living Rev. Relativ. 17 (2014) 4.
[43] M. Mohseni, Phys. Rev. D 84 (2011) 064026;
A.E. Gumrukcuoglu, S. Kuroyanagi, C. Lin, S. Mukohyama, N. Tanahashi, Class. Quantum Gravity 29 (2012) 235026.
[44] M. Fierz, W. Pauli, Proc. R. Soc. Lond. A 173 (1939) 211.
[45] H. van Dam, M.J.G. Veltman, Nucl. Phys. B 22 (1970) 397;
V.I. Zakharov, JETP Lett. 12 (1970) 312.
[46] D.G. Boulware, S. Deser, Phys. Rev. D 6 (1972) 3368.
[47] T. Kugo, N. Ohta, Prog. Theor. Exp. Phys. 2014 (2014) 043B04.
[48] E.A. Bergshoeff, O. Hohm, P.K. Townsend, Phys. Rev. Lett. 102 (2009) 201301.
[49] C. de Rham, G. Gabadadze, A.J. Tolley, Phys. Rev. Lett. 106 (2011) 231101.
[50] S.F. Hassan, R.A. Rosen, Phys. Rev. Lett. 108 (2012) 041101.
[51] M. Fasiello, A.J. Tolley, J. Cosmol. Astropart. Phys. 12 (2013) 002.
[52] E. Babichev, A. Fabbri, Class. Quantum Gravity 30 (2013) 152001.
[53] K. Bamba, Md. Wali Hossain, R. Myrzakulov, S. Nojiri, M. Sami, Phys. Rev. D 89 (2014) 083518.
[54] Y.F. Cai, E.N. Saridakis, Phys. Rev. D 90 (2014) 063528.
[55] G. Goon, A.E. Gumrukcuoglu, K. Hinterbichler, S. Mukohyama, M. Trodden, J. Cosmol. Astropart. Phys. 08 (2014) 008.
[56] A.R. Solomon, J. Enander, Y. Akrami, T.S. Koivisto, F. Konnig, E. Mortsell, J. Cosmol. Astropart. Phys. 04 (2015) 027.
[57] S. Pan, S. Chakraborty, Ann. Phys. 360 (2015) 180.
[58] P. Li, X.zh. Li, P. Xi, Phys. Rev. D 93 (2016) 064040.
[59] L.M. Cao, Y. Peng, Y.L. Zhang, Phys. Rev. D 93 (2016) 124015.
[60] D.C. Zou, R. Yue, M. Zhang, Eur. Phys. J. C 77 (2017) 256.
[61] D.C. Zou, Y. Liu, R.H. Yue, Eur. Phys. J. C 77 (2017) 365.
[62] C. de Rham, Living Rev. Relativ. 17 (2014) 7.
[63] D. Vegh, arXiv:1301.0537.
[64] H. Zhang, X.Z. Li, Phys. Rev. D 93 (2016) 124039.
[65] R.G. Cai, Y.P. Hu, Q.Y. Pan, Y.L. Zhang, Phys. Rev. D 91 (2015) 024032.
[66] S.H. Hendi, B. Eslam Panah, S. Panahiyan, J. High Energy Phys. 11 (2015) 157.
[67] S.H. Hendi, S. Panahiyan, B. Eslam Panah, M. Momennia, Ann. Phys. (Berlin) 528 (2016) 819.
[68] S.H. Hendi, G.H. Bordbar, B. Eslam Panah, S. Panahiyan, J. Cosmol. Astropart. Phys. 07 (2017) 004.
[69] A. Dehyadegari, M. Kord Zangeneh, A. Sheykhi, Phys. Lett. B 773 (2017) 344.
[70] S.H. Hendi, S. Panahiyan, B. Eslam Panah, J. High Energy Phys. 01 (2016) 129.
[71] S.H. Hendi, B. Eslam Panah, S. Panahiyan, Phys. Lett. B 769 (2017) 191.
[72] S.H. Hendi, B. Eslam Panah, S. Panahiyan, J. High Energy Phys. 05 (2016) 029.
[73] E.W. Hirschmann, D.L. Welch, Phys. Rev. D 53 (1996) 5579.
[74] T. Koikawa, T. Maki, A. Nakamula, Phys. Lett. B 414 (1997) 45.
[75] O.J.C. Dias, J.P.S. Lemos, J. High Energy Phys. 01 (2002) 006.
[76] W.A. Sabra, Phys. Lett. B 545 (2002) 175.
[77] M. Cataldo, J. Crisostomo, S. del Campo, P. Salgado, Phys. Lett. B 584 (2004) 123.
[78] Th. Grammenos, Mod. Phys. Lett. A 20 (2005) 1741.
[79] S.H. Hendi, S. Panahiyan, B. Eslam Panah, Eur. Phys. J. C 75 (2015) 296.
[80] S.H. Hendi, B. Eslam Panah, M. Momennia, S. Panahiyan, Eur. Phys. J. C 75 (2015) 457.
[81] E.W. Hirschmann, D.L. Welch, Phys. Rev. D 53 (1996) 5579.
[82] S.H. Hendi, S. Panahiyan, R. Mamasani, Gen. Relativ. Gravit. 47 (2015) 91.
[83] I. Navarro, Class. Quantum Gravity 20 (2003) 3603;
V.M. Gorkavenko, A.V. Viznyuk, Phys. Lett. B 604 (2004) 103;
A. Collinucci, P. Smyth, A. Van Proeyen, J. High Energy Phys. 02 (2007) 060;
G. de Berredo-Peixoto, M.O. Katanaev, J. Math. Phys. 50 (2009) 042501;
A. Ozakin, A. Yavari, J. Math. Phys. 51 (2010) 032902.

28

Tracing the evolution of nuclear forces under the similarity renormalization group

Calvin W. Johnson

Department of Physics, San Diego State University, 5500 Campanile Drive, San Diego, CA 92182-1233, United States

ARTICLE INFO

Editor: W. Haxton

ABSTRACT

I examine the evolution of nuclear forces under the similarity renormalization group (SRG) using traces of the many-body configuration-space Hamiltonian. While SRG is often said to "soften" the nuclear interaction, I provide numerical examples which paint a complementary point of view: the primary effect of SRG, using the kinetic energy as the generator of the evolution, is to shift downward the diagonal matrix elements in the model space, while the off-diagonal elements undergo significantly smaller changes. By employing traces, I argue that this is a very natural outcome as one diagonalizes a matrix, and helps one to understand the success of SRG.

1. Introduction

Nuclear structure theory has undergone a renaissance, driven by advances in high performance computing as well as by rigorous and systematic methodologies for both *ab initio* nuclear forces, such as chiral effective field theory [1–3] and for application of those forces to many-body calculations. Included in the latter are the no-core shell model [4,5] and the similarity renormalization group [6–10], which together have been very successful in calculating properties of light nuclei starting primarily from two-nucleon data. It is often said that the similarity renormalization group (SRG) "softens" the nuclear interaction, improving convergence with model space size. In this short paper I demonstrate that, at a gross level, the dominant effect of SRG on no-core shell model (NCSM) calculations is to shift low-lying energies down, with much smaller effects on wave functions. Using traces over the many-body model space, I argue that, in retrospect, the weakening of off-diagonal matrix elements and a much larger downward shift in diagonal elements go hand-in-hand.

The no-core shell model is a configuration–interaction method, whereby one solves the nonrelativistic nuclear Schrödinger equation $\hat{H}|\Psi_i\rangle = E_i|\Psi_i\rangle$ as a matrix eigenvalue problem, typically in a basis of Slater determinants, that is, antisymmetrized products of single-particle states in the lab frame. An important goal of modern nuclear structure theory is to carry out many-body calculations, in the NCSM or other methodologies, using nucleon–nucleon forces fitted with high precision to experimental phase shifts [11–13]. These forces are generated in relative coordinates and then transformed to the lab frame via Talmi–Moshinsky–Brody brackets [14–16]. Now come two crucial concepts I rely upon. The first is that finding eigenpairs involves a unitary transformation to a diagonal matrix. (Because one only wants low-lying eigenpairs and not all of them, one uses the Lanczos algorithm [17], but the basic idea remains.) To diagonalize the "full" matrix is impractical, hence we must diagonalize in a smaller, truncated model space. Yet *ab initio* nuclear forces have large matrix elements connecting states of low and high relative momentum, which historically and phenomenologically was interpreted as a hard repulsive core. In the shell model configuration basis this "hard core" becomes a strong coupling between the truncated model space and the excluded space, driving the inclusion of many configurations to converge results as a function of model space size, typically described by $N_{\max}$ (the number of excitations in an noninteracting harmonic oscillator space).

In order to improve solutions in the model space, one turns to effective interaction theories. Most of these are unitary or quasi-unitary transformations, and one of the most widely applied is the similarity renormalization group, where one evolves a Hamiltonian by the differential equation

$$\frac{d}{ds}\hat{H}(s) = \left[\hat{H}(s), \left[\hat{H}(s), \hat{G}\right]\right]. \tag{1}$$

Here $\hat{G}$ is the generator of the evolution and is often picked to be the kinetic energy $\hat{T}$. If fully carried out, (1) is a unitary transformation of the Hamiltonian. (It also induces, however, many-body forces [9,18], and as one typically carries out (1) in just the two-

E-mail address: cjohnson@mail.sdsu.edu.

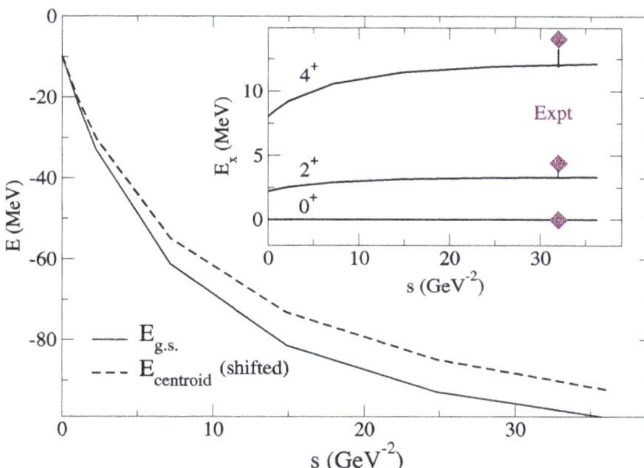

Fig. 1. (Color online.) Evolution under the similarity renormalization group of ground state energy of ^{12}C (black solid line) and excitation energies of the first 2^+, 4^+ states (inset), where the (maroon) diamonds show the experimental values for comparison. The calculations were carried out in an $N_{max} = 8$ many-body space in a harmonic oscillator basis frequency of $\hbar\Omega = 20$ MeV, using the Entem–Machleidt chiral effective interaction evaluated to N3LO. Also shown (red dashed line) is the evolution of the energy of the centroid, as defined in Eq. (3); the centroid energy is shifted down by 221.88 MeV to show it tracking the ground state energy.

or three-body systems, higher rank forces are dropped and one has a loss of unitarity. A closely related methodology is the in-medium SRG [19,20], which by normal ordering approximately accounts for higher rank forces, although I do not consider it further.) In other words, when a matrix is diagonal, each eigenstate is decoupled from all the rest, while SRG and related techniques seek to improve the solution in the model space by approximately decoupling the model space from the excluded space. In what follows I will gives examples of the effect of SRG on solutions in the model space, and then interpret those effects in terms of traces on the many-body Hamiltonian matrix.

As an example of the NCSM using SRG-evolved forces, I consider ^{12}C in a Slater determinant (lab frame) basis built from single-particle harmonic oscillator states with $\hbar\Omega = 20$ MeV, allowing up to $N_{max} = 8$ excitations; such a space allows one to exactly separate out spurious center-of-mass motion using the Palumbo-Glockle–Lawson method [21,22]. The interaction was derived in chiral effective field theory in next-to-next-to-next-to-leading order [13] or N3LO; it was then evolved in relative momentum space via SRG (with the kinetic energy as the generator $\hat{G}$ in Eq. (1)) to various values of s, and then transformed to the lab frame via Talmi-Moshinsky–Brody brackets [14–16]. (The representation of the interaction in relative coordinates has a finite cut-off; in harmonic oscillator states it is typically truncated at 100–300 $\hbar\Omega$ [5], which corresponds to $N_{max} \sim 100$–300. In the lab frame, the non-spurious many-body states are coupled to the ground state of center-of-mass motion. Thus the non-spurious Hamiltonian in a many-body basis in the lab frame is also finite, though $N_{max} \sim$ 100–300 is much larger than any configuration-interaction code can fully handle, and so the Hamiltonian in single-particle coordinates, or the lab frame, is always truncated.) Often one reinterprets the evolution in terms of a wavenumber cutoff, $\lambda = \frac{\sqrt{M_N}}{\hbar} s^{-1/4}$, which has units of fm^{-1}; here M_N is the nucleon mass. Although λ suggests the "resolution" of the renormalized interaction, I retain s to emphasize the evolutionary nature of SRG, so that $s = 0$ is the original or unevolved interaction; this corresponds to $\lambda = \infty$. The shell model calculations are carried out using the BIGSTICK code [23].

Fig. 1 shows how the ground state energy plunges as one goes from the bare force, $s = 0$, to one evolved to $s = 36.3$ GeV^{-2} which is equivalent to $\lambda = 2.0$ fm^{-1}. While the ground state energy changes by nearly 90 MeV, the inset shows the excitation energies change hardly at all. Table 1 shows $\Delta E_{g.s.} = E_{g.s.}(s = 0) - E_{g.s.}(s = 36.3\ \text{GeV}^{-2})$, the change in the ground state energy under SRG, for several different p-shell nuclides, including different N_{max} truncations.

Table 1
Changes in ground state energies and centroids for selected p-shell nuclides, including for different N_{max} truncations, under SRG evolution from $s = 0$ ($\lambda = \infty$) to $s = 36.3$ GeV^{-2} ($\lambda = 2.0$ fm^{-1}). Also shown is the overlap, $|\langle \Psi_{g.s.}(s = 36.3\ \text{GeV}^{-2}) | \Psi_{g.s.}(s = 0)\rangle|^2$ of the ground state wave function vectors in configuration space.

Nuclide	N_{max}	$\Delta E_{g.s.}$ (MeV)	$\Delta E_{centroid}$ (MeV)	overlap
^{6}Li	8	25.97	16.81	0.948
^{6}Li	10	21.54	14.37	0.956
^{6}Li	12	17.40	12.46	0.960
^{7}Li	8	33.58	24.13	0.940
^{7}Li	10	28.182	21.03	0.950
^{8}Be	8	46.33	33.36	0.927
^{8}Be	10	39.48	29.52	0.937
^{9}Be	8	52.96	43.45	0.915
^{9}Be	10	45.33	38.93	0.927
^{10}B	8	62.88	55.40	0.896
^{12}C	8	89.55	82.65	0.860

One can take this further; in the BIGSTICK code, as in all shell-model diagonalization codes, wave functions are represented as vectors, whose components are amplitudes in the basis of Slater determinants. Formally, under the unitary transformation induced by SRG, the interpretation of that basis evolves along with the Hamiltonian. But Table 1, which also gives the overlap $|\langle \Psi_{g.s.}(s = 36.3\ \text{GeV}^{-2}) | \Psi_{g.s.}(s = 0)\rangle|^2$, shows that the vectors change only a small amount. Of course, even small changes in the wave function can lead to large changes in specific matrix elements, particularly if there are large cancellations. But at a gross level, *the largest effect of SRG evolution is to shift down the low-lying energies*, with smaller changes to the wave functions in occupation space and to excitation energies.

While this is gratifying, and perhaps what we would most want from any effective theory, is there someway we can understand this?

2. Spectral distribution theory

To analyze what is happening under SRG, I turn to spectral distribution theory (SDT), sometimes also called statistical spectroscopy [18,24–26]. The key idea of SDT is the use of the average over an N-dimensional many-body space $\mathcal{S} = \{|i\rangle\}$:

$$\langle \hat{R} \rangle^{(\mathcal{S})} \equiv \frac{1}{N} \sum_{i \in \mathcal{S}} \langle i | \hat{R} | i \rangle. \tag{2}$$

Note that this is *not* an expectation value; following practitioners [25–27], a superscript $(\mathcal{S})$ emphasizes this difference.

Traces are a powerful tool. By taking the trace of Eq. (1), and using the cyclic property of traces, one finds almost trivially that $\frac{d}{ds} \text{tr}\, \hat{H}(s) = 0$, as well as $\frac{d}{ds} \text{tr}\, \hat{H}^2(s) = 0$ and higher moments, proving that SRG, if carried out exactly, is a unitary transformation.

Remember, however, that in the version of SRG discussed here the Hamiltonian is evolved in relative coordinates and when transformed to single-particle, lab-frame coordinates is very large and must be truncated. Thus, most calculations are carried out in a truncated model space $\mathcal{S}$, where the trace is not preserved. (The situation for in-medium SRG is different and not discussed here.)

Nonetheless, one can consider the *centroid*, the average of the many-body Hamiltonian in $\mathcal{S}$:

$$E_{\text{centroid}} = \langle \hat{H} \rangle^{(\mathcal{S})}. \tag{3}$$

Table 1 shows the change $\Delta E_{\text{centroid}} = E_{\text{centroid}}(s = 0) - E_{\text{centroid}}(s = 36.3 \text{ GeV}^{-2})$ for a selection of p-shell nuclides. Note that a significant fraction of the drop of the ground state energy comes from the shift in the centroid. This is shown in detail for ^{12}C in Fig. 1, where the evolution of the centroid with s tracks the evolution of the ground state energy; in that figure I've shifted the centroid down by 221.9 MeV so the tracking is more clear. In other words, a significant effect of SRG evolution is not only to shift downwards the low-lying eigenstates, but to shift *all* the states in the model space downwards.

One of the key tools of carrying out traces in SDT is to rewrite in terms of number operators. With that in mind, the $N_{\max} = 8$ data in Table 1 has $\Delta E_{\text{centroid}} \approx 0.6A(A-1)$ MeV. There is a small but nontrivial additional dependence on T_z. Furthermore, when scaled by $A(A-1)$, the evolution in s of the centroids for fixed $N_{\max}$, shown in Fig. 1 for ^{12}C (dashed line), all fall on the same curve. SDT expects this: there is a operator $\hat{N}(\hat{N}-1)$ and its coefficient evolves with s. Although this 'universal' curve does not have an obvious analytic form, one might calculate directly the evolution of the centroid in terms of number operators, including higher order terms. Note that while the bulk of the change in the ground state energy comes from the shift in the centroid, the remainder is non-trivial and is *not* 'universal.'

The centroid is just the average of the diagonal elements in the model space. One can track the evolution in more detail by using a finer tool, *configuration centroids*. A *configuration* is the occupancy of different orbits, for example: $(0s_{1/2})^2(0p_{3/2})^2$, $(0s_{1/2})^2(0p_{3/2})^1$, $(0p_{1/2})^1$, etc. Then one can define a subspace, call the *configuration partition* but sometimes just the configuration, which is the set of all states described by the same orbital occupancies. It turns out to be easy to compute the trace of the Hamiltonian within any configuration partition so defined [26]. If we label configurations by α with a projection operator $\hat{P}_\alpha = \sum_{i \in \alpha} |i\rangle\langle i|$, then the configuration dimension is just $N_\alpha = \text{tr}\, \hat{P}_\alpha$ and the configuration centroid is $\bar{E}_\alpha = N_\alpha^{-1} \text{tr}\, \hat{P}_\alpha \hat{H}$. By subdividing the trace into configuration partitions, we can use configuration centroids to follow the evolution of the diagonal matrix elements.

Fig. 2 plots the distribution of configurations centroids for ^{12}C, both unevolved ($s = 0$, top panel) and highly evolved (bottom panel) Hamiltonians, with the centroids combined into bins of width 2 MeV. The y-axis is the dimension of the binned configuration subspaces, on a log plot. The total centroid for each plot are aligned. Here N is the number of excitations in the harmonic oscillator space. To be clear, here N is fixed, not an upper limit; while CI calculations labeled by $N_{\max}$ include states with $N = N_{\max}, N_{\max}-2, N_{\max}-4, \ldots$ ($\Delta N = 2$ preserves parity), each curve in Fig. 2 is for all configurations of a single fixed N. Despite an overall shift of about 82 MeV in the centroids, the distribution of configuration centroids is remarkably unchanged. The main visible difference is the evolved configuration centroids are "stretched out."

What about off-diagonal elements? Here I exploit another idea from SDT, that by using traces one can establish an inner product on operators [24]:

$$\begin{aligned} \left(\hat{H}_1, \hat{H}_2\right) &\equiv \\ \left\langle \left(\hat{H}_1 - \langle \hat{H}_1 \rangle^{(\mathcal{S})}\right)\left(\hat{H}_2 - \langle \hat{H}_2 \rangle^{(\mathcal{S})}\right)\right\rangle & \\ = \langle \hat{H}_1 \hat{H}_2 \rangle^{(\mathcal{S})} - \langle \hat{H}_1 \rangle^{(\mathcal{S})} \, \langle \hat{H}_2 \rangle^{(\mathcal{S})}. & \end{aligned} \tag{4}$$

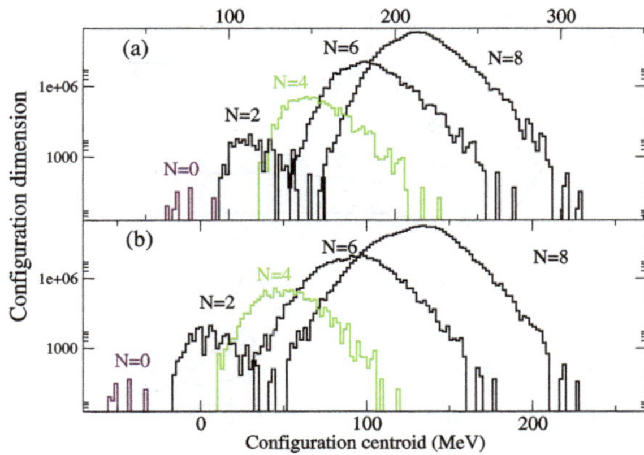

Fig. 2. Evolution of configuration centroids for ^{12}C under the similarity renormalization group, from $s = 0$, or $\lambda = \infty$ (top panel) to $s = 36.3$ GeV^{-2}, or $\lambda = 2.0$ fm^{-1} (bottom panel). The abscissa is the configuration centroid, while the ordinate is the dimension of each configuration subspace. The two plots are shifted so their respective total centroids (3) are aligned. Here N is the *fixed* number of excitations in harmonic oscillator many-body space, so that $N = 4$ includes only 4 $\hbar\Omega$ excitation but no 0$\hbar\Omega$, no 2$\hbar\Omega$, etc.

With an inner product so defined, one has a metric on the space of Hamiltonians and can compare how close or distant two Hamiltonians are; the magnitude of a Hamiltonian, $\|\hat{H}\| = \sqrt{(\hat{H}, \hat{H})}$, is its width, and the "angle" between two Hamiltonians by $\cos\theta = (\hat{H}_1, \hat{H}_2)\|\hat{H}_1\|^{-1}\|\hat{H}_2\|^{-1}$. Note that this definition of the metric is independent of the centroids, which is sensible as centroids affect only absolute energies and not excitation energies nor wave functions.

Using Eq. (1) and the cyclic property of traces,

$$\frac{d}{ds} \text{tr}\, \hat{H}(s)\hat{G} = \text{tr}\, [\hat{H}(s), \hat{G}]^\dagger [\hat{H}(s), \hat{G}]. \tag{5}$$

But the righthand side is the trace of an operator of the form $\hat{A}^\dagger \hat{A}$, which manifestly has real and nonnegative eigenvalues and a real, nonnegative trace. Thus, under generic SRG, $\text{tr}\, \hat{H}(s)\hat{G}$ can only increase; as the magnitudes of $\hat{H}(s)$ and, trivially, $\hat{G}$ are invariant, this means the "angle" defined by the inner product (4) between them can only get smaller and they become more and more parallel. Furthermore, this evolution stops when $[\hat{H}(s), \hat{G}] = 0$, that is, the evolved Hamiltonian commutes with the generator. Thus SRG drives a Hamiltonian "towards" its generator. $\hat{H}(s)$ cannot become proportional to $\hat{G}$, as the eigenvalues, invariant under a unitary transformation, are different, but if one lets $s \to \infty$ they will have the same eigenvectors.

3. Interpretation using traces in the model space

Now let's discuss these empirical results–the large, coherent shifts in low-lying energies, and the relatively modest changes in the wave function vectors–through the lens of spectral distribution theory, using traces as a fundamental tool to investigate what happens under the unitary transformation induced by SRG. To do so, one must pay attention to is the difference between taking traces in the full space and in the much smaller model space.

Let $\hat{H}$ be the original Hamiltonian in the full space, and let $\hat{H}' = \hat{U}^\dagger \hat{H} \hat{U}$ be the transformed Hamiltonian. Since the full space is too large to work in, let the much smaller model space be $\mathcal{S}$ with a projection operator $\hat{P}_\mathcal{S}$.

The trace of any matrix (and of any power of that matrix) is preserved under unitary transformations: $\text{tr}\, \hat{H} = \text{tr}\, \hat{U}^\dagger \hat{H} \hat{U} = \text{tr}\, \hat{H}'$, and similarly $\text{tr}\, \hat{H}^2 = \text{tr}\, (\hat{H}')^2$. Let's divide up the contributions to the trace of $\hat{H}^2$ into diagonal and off-diagonal pieces: $\text{tr}\, \hat{H}^2 =$

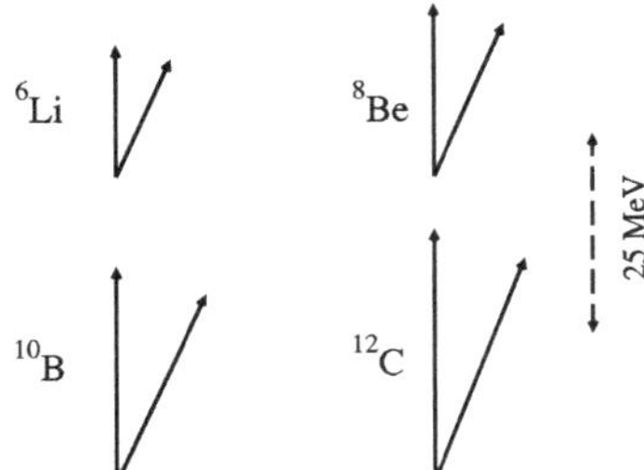

Fig. 3. (Color online.) A representation of the evolution of interactions under SRG, as measured by the spectral distribution theory inner product, Eq. (4), for selected nuclides. The black, vertical arrows represent the unevolved interaction, at $s=0$ or $\lambda=\infty$, while the blue, tilted arrows are the interactions evolved to $s=36.3$ GeV^{-2} or $\lambda=2.0$ fm^{-1}. Here the evolved interaction is defined as $\hat{V}(s)=\hat{H}(s)-\hat{T}$, where $\hat{T}$ is the kinetic energy. The red dashed line gives the scale. Although not shown, results for $A=2,3$ are very similar, although with smaller absolute magnitudes.

$\sum_i H_{ii}^2 + \sum_{i\neq j} H_{ij}^2$. If after a unitary transformation the matrix $\hat{H}'$ is diagonal, then $\mathrm{tr}\,(\hat{H}')^2$ has no off-diagonal contributions. Thus $\mathrm{tr}\,\hat{H}^2 = \mathrm{tr}\,\hat{H}'^2 = \sum_i (H')_{ii}^2 \geq \sum_i H_{ii}^2$. That is, on average, the magnitude of the diagonal elements of the transformed matrix $\hat{H}'$ must be larger than those of the original matrix $\hat{H}$.

Furthermore, the change in the diagonal matrix elements will generally be much larger than that of the off-diagonal matrix elements. How so? Even for sparse matrices, there will be many more off-diagonal matrix elements than diagonal. Suppose in an N-dimensional space there are $\sim N\times M$ non-zero off-diagonal matrix elements, so that $M/N\sim$ the sparsity. Further suppose the off-diagonal matrix elements all have roughly the same magnitude: call it γ. On average for every diagonal matrix element there are $\sim M$ non-zero matrix elements, and $\mathrm{tr}\,\hat{H}^2 \sim \sum_i H_{ii}^2 + NM\gamma^2$. Then on average $(H'_{ii})^2 \sim (H_{ii})^2 + M\gamma^2$ so that the root-mean-square change in the diagonal matrix elements, $\delta H_{ii} \equiv \sqrt{(H'_{ii})^2 - (H_{ii})^2} \sim \sqrt{M}\gamma$. Even for tiny sparsities, M will be a large number. Hence the average changes of diagonal matrix elements will be substantially larger than the changes to the off-diagonal matrix elements.

This argument suggests that despite the large shift in the centroid in the model space, the changes to the off-diagonal matrix elements are much smaller. I argue this is true even when one is not fully diagonalizing but only applying a unitary transformation which "softens" the interaction. As further evidence, I use the metric introduced in Eq. (4). Fig. 3 graphically compares two Hamiltonian as the two vectors, with the magnitudes of each operator and the angle between them defined by (4), for four p-shell nuclides. I considered isospin zero many-body states; the results depend only weakly upon isospin. To focus on the actual evolution of the operators, I consider only the "interaction," defined as $\hat{V}(s)=\hat{H}(s)-\hat{T}$, although such an interaction has components evolved from the original kinetic energy. The inner products were calculated using a publically available code [27]; because this code does not allow for $N_{\max}$ truncations, I instead truncated on the number of harmonic oscillator shells. I show the results for 7 major harmonic oscillator shells, that is, for maximum principal quantum number $N=6$, or up through the $3s$–$2d$–$1g$–$0i$ shell. The results, however, do not change very much as one goes from 5 to 7 major harmonic oscillator shells, and one can understand this as the higher excitations being nearly completely dominated by kinetic energy [28].

Fig. 3 shows the change in the interactions is modest. (Furthermore, although not included in Fig. 3, the angle between evolved and unevolved interactions in this space for $A=2,3$ is nearly identical to the cases shown.) This is in concordance with the previous empirical observation that under SRG evolution the change of the wave functions, as vectors in an abstract space of Slater determinants, is modest. Let me emphasize, however, that changes to the off-diagonal matrix elements are not negligible. If I approximate the SRG transformation by taking the unevolved Hamiltonian and simply swapping in the evolved configuration centroids, the shift in the ground state energy is overestimated by several MeV.

Eq. (5) proves that SRG evolves a Hamiltonian towards the generator. That proof, however, only applies in the full space. In the truncated model space, I find little change in the angle between the interaction and the kinetic energy, and in fact the angle increases, though by less than a degree.

Because the "full" Hamiltonian in the lab frame is generally far too large to be tractable, it must be truncated, which leads to loss of unitarity. But truncation is not the only cause of loss of unitarity. Three-body and higher-rank forces, induced by SRG [9,18] also contribute to unitarity of the transformation. Previous work [29] as well as the success of the in-medium SRG [19,20] has suggested the most important contributions of these many-body forces are those written as density- or state-dependent operators, especially those which contribute to the monopole terms [29], that is, the centroid and the configuration centroids. This is completely consistent with the results presented here.

4. Conclusions

Using traces or averages of operators in a many-body space, also called spectral distribution theory, is a powerful tool for understanding the evolution of nuclear forces under the similarity renormalization group. The primary effect of SRG is to simply to shift the model space centroid downwards. In contrast configuration centroids, or averages within occupancy-defined subspaces, are changed only a little relative to the total centroid. The inner product in SDT also allows one to look at off-diagonal matrix elements, which change only a modest amount under SRG. This picture–large, coherent shifts in the diagonal matrix elements and relatively much smaller changes to off-diagonal matrix elements–arises naturally when diagonalizing a Hamiltonian and is complementary to the conception of SRG "softening" the nuclear interaction. Finally, as the centroid and configuration centroids are written in terms of number operators, and the shift in the centroid is proportional to $A(A-1)$, I suggest one might compute separately the evolution of the centroid, not only as a function of $A(A-1)$ but also of higher powers of the number operator. Induced higher-particle-rank forces are important, and pieces proportional to powers of number operators account for the bulk of the effect [29].

Acknowledgements

I thank P. Navrătil for his code to generate the Entem-Machleidt N3LO force and apply SRG to it, and K.D. Launey for helpful discussions. This material is based upon work supported by the U.S. Department of Energy, Office of Science, Office of Nuclear Physics, under Award Number DE-FG02-03ER41272. This research used resources of the Argonne Leadership Computing Facility, which is a DOE Office of Science User Facility supported under Contract DE-AC02-06CH11357, and of the National Energy Research Scientific Computing Center, a DOE Office of Science User Facility supported by the Office of Science of the U.S. Department of Energy under Contract No. DE-AC02-05CH11231.

References

[1] S. Weinberg, Nucl. Phys. B 363 (1991) 3.
[2] U. Van Kolck, Nucl. Phys. A 645 (1999) 273.
[3] R. Machleidt, D.R. Entem, Phys. Rep. 503 (2011) 1.
[4] P. Navrátil, J. Vary, B. Barrett, Phys. Rev. C 62 (2000) 054311.

[5] B.R. Barrett, P. Navrátil, J.P. Vary, Prog. Part. Nucl. Phys. 69 (2013) 131.
[6] S.D. Głazek, K.G. Wilson, Phys. Rev. D 48 (1993) 5863.
[7] F. Wegner, Ann. Phys. 506 (1994) 77.
[8] S.K. Bogner, R.J. Furnstahl, R.J. Perry, Phys. Rev. C 75 (2007) 061001.
[9] E.D. Jurgenson, P. Navrátil, R.J. Furnstahl, Phys. Rev. Lett. 103 (2009) 082501.
[10] S. Bogner, R. Furnstahl, A. Schwenk, Prog. Part. Nucl. Phys. 65 (2010) 94.
[11] V. Stoks, R. Klomp, C. Terheggen, J. De Swart, Phys. Rev. C 49 (1994) 2950.
[12] R.B. Wiringa, V. Stoks, R. Schiavilla, Phys. Rev. C 51 (1995) 38.
[13] D. Entem, R. Machleidt, Phys. Rev. C 68 (2003) 041001.
[14] I. Talmi, R. Thieberger, K. Ford, E. Konopinski, M. Moshinsky, R. Lawson, M. Goeppert-Mayer, V. Balashov, V. Eltekov, A. Arima, et al., Helv. Phys. Acta 25 (1952) 185.
[15] M. Moshinsky, Nucl. Phys. 13 (1959) 104.
[16] T.A. Brody, M. Moshinsky, Tables of Transformation Brackets for Nuclear Shell-Model Calculations, Gordon and Breach, 1967.
[17] R.R. Whitehead, A. Watt, B.J. Cole, I. Morrison, Adv. Nucl. Phys. 9 (1977) 123–176.
[18] K.D. Launey, T.c.v. Dytrych, J.P. Draayer, Phys. Rev. C 85 (2012) 044003.
[19] K. Tsukiyama, S. Bogner, A. Schwenk, Phys. Rev. Lett. 106 (2011) 222502.
[20] H. Hergert, S. Bogner, T. Morris, A. Schwenk, K. Tsukiyama, Phys. Rep. 621 (2016) 165.
[21] F. Palumbo, Nucl. Phys. A 99 (1967) 100.
[22] D. Gloeckner, R. Lawson, Phys. Lett. B 53 (1974) 313.
[23] C.W. Johnson, W.E. Ormand, P.G. Krastev, Comput. Phys. Commun. 184 (2013) 2761.
[24] J.B. French, Phys. Lett. B 26 (1967) 75.
[25] J.B. French, Nucl. Phys. A 396 (1983) 87.
[26] S.S.M. Wong, Nuclear Statistical Spectroscopy, Oxford University Press, 1986.
[27] K.D. Launey, S. Sarbadhicary, T. Dytrych, J.P. Draayer, Comput. Phys. Commun. 185 (2014) 254.
[28] A.M. Shirokov, A.I. Mazur, I.A. Mazur, J.P. Vary, Phys. Rev. C 94 (2016) 064320.
[29] G. Hagen, T. Papenbrock, D.J. Dean, A. Schwenk, A. Nogga, M. Włoch, P. Piecuch, Phys. Rev. C 76 (2007) 034302.

Two-point functions of SU(2)-subsector and length-two operators in dCFT

Erik Widén [a,b,*]

[a] Nordita, KTH Royal Institute of Technology and Stockholm University, Roslagstullsbacken 23, SE-106 91 Stockholm, Sweden
[b] Department of Physics and Astronomy, Uppsala University, SE-751 08 Uppsala, Sweden

ARTICLE INFO

Editor: M. Cvetič

ABSTRACT

We consider a particular set of two-point functions in the setting of $\mathcal{N}=4$ SYM with a defect, dual to the fuzzy-funnel solution for the probe D5-D3-brane system. The two-point functions in focus involve a single trace operator in the SU(2)-subsector of arbitrary length and a length-two operator built out of any scalars. By interpreting the contractions as a spin-chain operator, simple expressions were found for the leading contribution to the two-point functions, mapping them to earlier known formulas for the one-point functions in this setting.

1. Introduction

Integrable structures in $\mathcal{N}=4$ SYM have been explored extensively since they were first noted in [1] and have provided a useful tool for both deeper field theoretic understanding and numerous tests of the AdS/CFT correspondence. For a pedagogical overview of the first decade, see [2]. Among other directions, the work has lead on to look for, and to employ, surviving integrability in similar theories, departing in different ways from $\mathcal{N}=4$ SYM. One particular branch of this focus is the study of various CFTs with defects (dCFTs).

The setting for these notes is $\mathcal{N}=4$ SYM with a codimension-one defect residing at the coordinate value $z=0$. The theory is the field theory dual of the probe D5-D3-brane system in $AdS_5\times S^5$, in which the probe-D5-brane has a three-dimensional intersection (the defect) with a stack of N D3-branes. We will study the dual of the so called fuzzy-funnel solution [3–6], in which a background gauge field has k units of flux through an S^2-part of the D5-brane geometry, meaning that k D3-branes dissolve into the D5-brane. These parameters appear on the field theory side as the rank N of the gauge group which is broken down to $N-k$ by the defect.

The dCFT action is built out of the regular $\mathcal{N}=4$ SYM field content plus additional fields constrained to the three dimensional defect. These additional fields interact both within themselves and with the bulk[1] fields. However, only the six scalars from $\mathcal{N}=4$ SYM will play a role within these notes.

The defect breaks the 4D conformal symmetry down to those transformations that leave the boundary intact (*i.e.* that map $z=0$ onto itself). Its presence thus changes many of the general statements about CFTs, such as allowing for non-vanishing one-point functions and two-point functions between operators of different conformal dimensions. These new features were first studied in [7, 8] and within the described setting, they have been the topic of a series of recent works. Tree-level one-point functions in the SU(2)- and SU(3)-subsectors where considered in [9–11] while bulk propagators and loop corrections to the one-point functions where worked out in [12–14]. Two-point functions were very recently addressed in [15] and earlier in [16].[2]

The underlaying idea of all this business is to interpret single-trace operators as states in a spin-chain and employ the Bethe ansatz from within this context. The one-point functions were in this spirit found to be expressible in a compact determinant formula, making use of a special spin-chain state, called the Matrix Product State (MPS), and Gaudin norm for Bethe states. The end result for the tree-level one-point functions of operators

$$\mathcal{O}_L \sim \mathrm{Tr}\Big(\overbrace{Z\dots ZXZ\dots ZXZ\dots}^{L\text{ complex scalars out of which }M\text{ are }X}\Big)$$

in the SU(2)-subsector was

* Correspondence to: Nordita, KTH Royal Institute of Technology and Stockholm University, Roslagstullsbacken 23, SE-106 91 Stockholm, Sweden.
E-mail address: erik.widen@nordita.org.

[1] Meaning the region $z>0$.

[2] Wilson loops in these settings with a defect have also attracted attention, see *e.g.* [17–19].

$$\langle \mathcal{O}_L\rangle_{\text{tree}} = \frac{2^{L-1}}{z^L} C_2(\mathbf{u}) \times \sum_{j=\frac{1-k}{2}}^{\frac{k-1}{2}} j^L \prod_{i=1}^{\frac{M}{2}} \frac{u_i^2\left(u_i^2+\frac{k^2}{4}\right)}{\left[u_i^2+(j-\frac{1}{2})^2\right]\left[u_i^2+(j+\frac{1}{2})^2\right]},$$

under the condition that both the length L and the number of excitations M are even and that the set of M Bethe rapidities has the special form $\mathbf{u} = \{u_1, -u_1, u_2, -u_2, \dots\}$. The parameter k can be any positive integer and

$$C_2(\mathbf{u}) = 2\left[\left(\frac{2\pi^2}{\lambda}\right)^L \frac{1}{L}\prod_j \frac{u_j^2+\frac{1}{4}}{u_j^2}\frac{\det G^+}{\det G^-}\right]^{\frac{1}{2}},$$

where $G^\pm$ are $\frac{M}{2}\times\frac{M}{2}$ matrices with matrix elements

$$G^\pm_{jk} = \left(\frac{L}{u_j^2+\frac{1}{4}} - \sum_n K^+_{jn}\right)\delta_{jk} + K^\pm_{jk},$$

within which, in turn,

$$K^\pm_{jk} = \frac{2}{1+(u_j-u_k)^2} \pm \frac{2}{1+(u_j+u_k)^2}.$$

The expression for C_2 was obtained from the spin-chain overlap

$$C_2 = \left(\frac{8\pi^2}{\lambda}\right)^{L/2}\frac{1}{\sqrt{L}}\frac{\langle \text{MPS}|\Psi\rangle}{\sqrt{\langle\Psi|\Psi\rangle}}$$

which is the form we will mostly refer to here. $|\Psi\rangle$ is the spin-chain Bethe state corresponding to the operator $\mathcal{O}_L$; the MPS will be defined below in equation (2).

1.1. The goal of the present notes

These notes consider the leading contribution, in the 't Hooft coupling λ, to the specific two-point function $\langle \mathcal{O}_L\,\mathcal{O}_2\rangle_{1\text{ contr.}}$, where

- both $\mathcal{O}_L$ and $\mathcal{O}_2$ are single-trace scalar operators of length L and 2, respectively, and
- $\mathcal{O}_L$ is restricted to the SU(2)-subsector while $\mathcal{O}_2$ can be built out of any pair of scalars.

We do this by interpreting the contraction as a spin-chain operator Q acting on the Bethe state corresponding to $\mathcal{O}_L$, whence re-expressing the two-point function in terms of the previously known one-point functions.

2. The particular two-point functions

We define the complex scalar fields as

$$Z = \phi_1 + i\phi_4, \quad X = \phi_2 + i\phi_5, \quad W = \phi_3 + i\phi_6,$$
$$\overline{Z} = \phi_1 - i\phi_4, \quad \overline{X} = \phi_2 - i\phi_5, \quad \overline{W} = \phi_3 - i\phi_6,$$

which in the dual fuzzy-funnel solution each has the non-zero classical expectation value

$$\phi_I^{\text{cl}} = \frac{1}{z} t_I \oplus \mathbb{0}_{(N-k)}, \quad I = 1,2,3; \qquad \phi_{\tilde{J}}^{\text{cl}} = 0, \quad \tilde{J} = 4,5,6,$$

where $\{t_1, t_2, t_3\}$ forms a $k\times k$ unitary representation of SU(2) and the $\mathbb{0}_{(N-k)}$ pads the rest of the matrix to the full dimensions $N\times N$.

For definiteness, we choose $Z \sim |\uparrow\rangle$ and $X \sim |\downarrow\rangle$ as the SU(2)-subsector.

We now set out to calculate

$$\langle \mathcal{O}_L \mathcal{O}_{Y_1Y_2}\rangle_{1\text{ contr.}} = \sum_{l=1}^{L} \Psi^{i_1\dots i_L}\,\text{Tr}\left(X^{\text{cl}}_{i_1}\cdots X_{i_l}\cdots X^{\text{cl}}_{i_L}\right)\text{Tr}\left(Y_1 Y_2^{\text{cl}}\right) + (Y_1 \leftrightarrow Y_2), \qquad i_\ell = \uparrow,\downarrow \tag{1}$$

where $X_\uparrow = Z$, $X_\downarrow = X$, $Y_{1,2}$ can be any complex scalar and the coefficients $\Psi^{i_1\dots i_L}$ of $\mathcal{O}_L$ are chosen such that they map to a Bethe state $|\Psi\rangle$ in the spin-chain picture.

We will express it by help of the MPS, which is the following state in the spin-chain Hilbert space:

$$\langle \text{MPS}| = \text{Tr}\left[\left(\langle\uparrow|t_1 + \langle\downarrow|t_2\right)^{\otimes L}\right], \tag{2}$$

where the trace is over the resulting product of t's.

2.1. Scalar propagators

The defect mixes the scalar propagator in both color and flavor indices, explained in detail in [13]. However, since the contracted fields are multiplied by classical fields from both sides we will only need the upper $(k\times k)$-block. The propagator diagonalization involves a decomposition of these components in terms of fuzzy spherical harmonics $\hat{Y}_\ell^m$[3]:

$$[\phi]^{s_1}{}_{s_2} = \sum_{\ell=1}^{k-1}\sum_{m=-\ell}^{\ell}\phi_{\ell,m}[\hat{Y}_\ell^m]^{s_1}{}_{s_2}, \qquad s_{1,2} = 1,\dots,k.$$

Translating back to the s-indices, the relevant propagators for $I, J = 1,2,3$ read

$$\left\langle[\phi_I(x)]^{s_1}{}_{s_2}[\phi_J(y)]^{r_1}{}_{r_2}\right\rangle = \delta_{I,J}\sum_{\ell,m}[\hat{Y}_\ell^m]^{s_1}{}_{s_2}[(\hat{Y}_\ell^m)^\dagger]^{r_1}{}_{r_2}K_1^\ell(x,y) - i\epsilon_{IJK}\sum_{\ell,m,m'}[\hat{Y}_\ell^m]^{s_1}{}_{s_2}[(\hat{Y}_\ell^{m'})^\dagger]^{r_1}{}_{r_2}[t_K^{(2\ell+1)}]_{\ell-m+1,\ell-m'+1}K_2^\ell(x,y)$$

where $t_K^{(2\ell+1)}$ is in the $(2\ell+1)$-dimensional representation. The remaining scalars $\tilde{I}, \tilde{J} = 4,5,6$ have the diagonal propagator

$$\left\langle[\phi_{\tilde{I}}]^{s_1}{}_{s_2}[\phi_{\tilde{J}}]^{r_1}{}_{r_2}\right\rangle = \delta_{\tilde{I}\tilde{J}}\sum_{m=-\ell}^{\ell}[\hat{Y}_\ell^m]^{s_1}{}_{s_2}[(\hat{Y}_\ell^m)^\dagger]^{r_1}{}_{r_2}K^{\text{m}^2=\ell(\ell+1)}(x,y).$$

The spacetime dependent factors are

$$K_1^\ell(x,y) = \frac{\ell+1}{2\ell+1}K^{\text{m}^2=\ell(\ell-1)}(x,y) + \frac{\ell}{2\ell+1}K^{\text{m}^2=(\ell+1)(\ell+2)}(x,y),$$
$$K_2^\ell(x,y) = \frac{1}{2\ell+1}\left(K^{\text{m}^2=\ell(\ell-1)}(x,y) - K^{\text{m}^2=(\ell+1)(\ell+2)}(x,y)\right).$$

K^{m^2} is related to the scalar propagator in AdS and reads

$$K^{\text{m}^2}(x,y) = \frac{g_{\text{YM}}^2}{2}(x_3y_3)^{1/2}\int\frac{d^3\vec{k}}{(2\pi)^3}e^{i\vec{k}\cdot(\vec{x}-\vec{y})}I_\nu\left(|\vec{k}|x_3^<\right)K_\nu\left(|\vec{k}|x_3^>\right),$$

in which I and K are modified Bessel functions with $x_3^<$ ($x_3^>$) the smaller (larger) of x_3 and y_3, and lastly where $\nu = \sqrt{\text{m}^2+\frac{1}{4}}$.

We will from now on suppress all spacetime dependence.

[3] See appendices in [13,20]. We use the normalization of [13].

3. The contraction as a spin-chain operator

With the expressions of the propagators, we can now view the contraction in equation (1) as a $(k \times k)$-matrix

$$[\mathbf{T}_{X_{i_l} Y_1 Y_2}]^{s_1}{}_{s_2} = \langle [X_{i_l}]^{s_1}{}_{s_2} [Y_1]^{r_1}{}_{r_2} \rangle [Y_2^{\mathrm{cl}}]^{r_2}{}_{r_1}$$

replacing the field at site l in the first trace while absorbing the second trace completely.

It turns out that this matrix always is proportional to either t_1, t_2 or t_3. To see this, first use that the fuzzy spherical harmonics are tensor operators, such that

$$\sum_m \hat{Y}_\ell^m [t_K^{(2\ell+1)}]_{\ell-m+1,\ell-m'+1} = [t_K^{(k)}, \hat{Y}_\ell^{m'}] = m' \hat{Y}_\ell^{m'}.$$

Then use the orthogonality of the fuzzy spherical harmonics[4] in the trace by decomposing the t in Y_2^{cl} as

$$t_j = d_j \left(\hat{Y}_1^{-1} + (-1)^j \hat{Y}_1^1 \right), \qquad j = 1, 2,$$

$$t_3 = \sqrt{2} d_1 \hat{Y}_1^0, \qquad d_j = \mathrm{i}^{3+j} \frac{(-1)^{k+1}}{2} \sqrt{k(k^2-1)/6}.$$

Together, these factors in $\mathbf{T}$ then conspire to always give t's for any considered scalar combination. What is left can thus be interpreted as a one-point function of a slightly modified $\mathcal{O}_L$. As such, we can write the two-point function (1) as an operator insertion

$$\langle \mathrm{MPS} | Q_{Y_1 Y_2} | \Psi \rangle$$

in the spin-chain picture, acting on the Bethe state corresponding to $\mathcal{O}_L$.

3.1. The spin-chain operator $Q_{Y_1 Y_2}$

$\mathbf{T}$'s dependence on the involved scalars can be compactly written when expanded in terms of the real scalars:

$$\mathbf{T}_{IJK} = \delta^3_{IJ} K_1^{\ell=1} t_K + (\delta^3_{IK} t_J - \delta^3_{JK} t_I) K_2^{\ell=1} + \delta^6_{IJ} K^{\mathrm{m}^2=2} t_K,$$

$I, J, K = 1, \ldots, 6$ and where the δ^3 (δ^6) is only non-zero for indices 1, 2 and 3 (4, 5, and 6). Taking into account both the sums in the two-point function (1), we can then write the contractions in the spin-chain picture as

$$Q_{Y_1 Y_2} |\Psi\rangle = \sum_{l=1}^{L} \mathbb{1} \otimes \cdots \otimes Q^{(l)}_{Y_1 Y_2} \otimes \cdots \otimes \mathbb{1} |\Psi\rangle,$$

i.e. a linear combination of the spin-chain operators $\{\mathbb{1}^{\otimes L}, S^+, S^-, S^3\}$.[5]

The result arranges itself in the two cases $Y_1^{\mathrm{cl}} = Y_2^{\mathrm{cl}}$ and $Y_1^{\mathrm{cl}} \neq Y_2^{\mathrm{cl}}$, for which[6]

$$Q^{(l)}_{=} = \begin{pmatrix} c^\uparrow & 0 \\ 0 & c^\downarrow \end{pmatrix}, \qquad Q^{(l)}_{\neq} = \begin{pmatrix} 0 & c^+ \\ c^- & 0 \end{pmatrix},$$

and the various coefficients c implicitly depend on Y_1, Y_2. They are listed in Appendix A.

- *Case* $Y_1^{\mathrm{cl}} = Y_2^{\mathrm{cl}}$. The action of $Q_=$ is trivial on any Bethe state. Still denoting the total number of spin-down excitations as M, we immediately get

$$Q_= |\Psi\rangle = \left(c^\uparrow (L-M) + c^\downarrow M\right) |\Psi\rangle.$$

Combining this with the one-point function formula implies

$$\langle \mathcal{O}_L \mathcal{O}_= \rangle_{1\text{ contr.}} = \left(c^\uparrow (L-M) + c^\downarrow M\right) \langle \mathcal{O}_L \rangle_{\text{tree}}.$$

As an example, the Konishi operator has the two-point function $2K^{\mathrm{m}^2=6} L \langle \mathcal{O}_L \rangle_{\text{tree}}$ with any SU(2)-subsector operator.

- *Case* $Y_1^{\mathrm{cl}} \neq Y_2^{\mathrm{cl}}$. In this case we have the spin-flipping operator

$$Q_{\neq} = c^+ S^+ + c^- S^-.$$

Its action simplifies significantly when acting on a Bethe state. First of all, Bethe states with non-zero momenta are highest weight states implying that $S^+ |\Psi\rangle = 0$. Secondly, we have that

$$S^- |\Psi_M\rangle = \lim_{p_{M+1} \to 0} |\Psi_{M+1}\rangle,$$

meaning that acting on a Bethe state with the lowering operator creates a new Bethe state with one more excitation but with the corresponding momentum $p_{M+1} = 0$. All other momenta are the same. These states are called (Bethe) descendants.

It was shown in [9] that only states with L and M both even can have a non-zero overlap with the MPS. Furthermore, by studying the action of Q_3, the third conserved charge in the integrable hierarchy, it was proven that only unpaired[7] states yield finite overlaps. This is true since $Q_3|\mathrm{MPS}\rangle = 0$ and because Q_3 is non-zero on states that are not invariant under parity.

That $Q_{\neq}$ alters the number of excitations now makes it possible to have non-zero overlaps with states with odd M. However, since

$$[Q_3, S^-] = 0$$

the requirement of an unpaired state is still imposed. Hence, the only possible way for the overlap

$$\langle \mathrm{MPS} | Q_{\neq} | \Psi_{\tilde{M}} \rangle$$

to be non-vanishing is that $\tilde{M}$ is *odd* and that the Bethe state is a descendant.

The general expression for such a state is

$$|\Psi_{\tilde{M}=M+n}\rangle = (S^-)^n |\Psi_M\rangle, \quad n \text{ odd}.$$

The two-point function (1) then follows from the commutation relation of the spin-operators, the action of $(S^-)^n$ on the MPS and the norm of the descendants [15,21]:

$$\langle \mathrm{MPS} | (S^-)^n | \Psi_M \rangle = \frac{n! (\frac{L}{2} - M)!}{(\frac{n}{2})! \left(\frac{L-2M-n}{2}\right)!} \langle \mathrm{MPS} | \Psi_M \rangle,$$

$$\langle \Psi_{M+n} | \Psi_{M+n} \rangle = \frac{n! (L-2M)!}{(L-2M-n)!} \langle \Psi_M | \Psi_M \rangle.$$

We find

$$\left\langle \mathcal{O}_{L,M+n} \mathcal{O}_{\neq} \right\rangle_{1\text{ contr.}} = \left(c^+ n (L - 2M - n + 1) C^+_{L,M,n} + c^- C^-_{L,M,n} \right) \left\langle \mathcal{O}_{L,M} \right\rangle_{\text{tree}}$$

with

$$C^{\pm}_{L,M,n} = \frac{(n \mp 1)! (\frac{L}{2} - M)!}{\left(\frac{n \mp 1}{2}\right)! \left(\frac{L-2M-n \pm 1}{2}\right)!} \sqrt{\frac{(L-2M-n)!}{n!(L-2M)!}}. \tag{3}$$

[4] $\mathrm{Tr}\, \hat{Y}_\ell^m (\hat{Y}_{\ell'}^{m'})^\dagger = \delta_{\ell\ell'} \delta_{mm'}$.

[5] This does not explicitly cover the case of $\mathbf{T} \propto t_3$. However, that case eventually yields zero and will be addressed below.

[6] We will denote both the dCFT operator and its spin-chain correspondent with subscripts $_=$ and $_{\neq}$ for these two cases.

[7] "Unpaired" refers to states which are invariant under parity transformation, implying momenta of the form $\{p_1, -p_1, \cdots\}$.

3.2. Remark on $\mathbf{T} \propto t_3$

When *one* of Y_1 or Y_2 is either W or $\overline{W}$, $\mathbf{T}$ is proportional to t_3 and the corresponding $Q_{t_3}^{(l)}$ is no longer a proper spin-chain operator. Insisting on a spin-chain interpretation would describe it as a flip of site $l+1$ followed by a removal of the site l, thus shrinking the length L by one. $Q_{t_3}^{(l)}$ always appears preceded by a projection $\Pi_{\uparrow(\downarrow)}$ on either spin-up or spin-down, depending on the Y which does not involve $W(\overline{W})$. It is straight-forward to show by explicit calculation that

$$\langle \mathrm{MPS}_{L-1}| \sum_{l=1}^{L} \mathbb{1} \otimes \cdots \otimes Q_{t_3}^{(l)} \Pi_{\uparrow(\downarrow)}^{(l)} \otimes \cdots \otimes \mathbb{1} | \updownarrow_L \rangle = 0$$

for any basis vector $| \updownarrow_L \rangle$ of length L.

4. Conclusion

We have studied the $\mathcal{N}=4$ SYM theory with a defect, dual to the probe D5-D3-brane system. Within this theory, the two-point function between a length L operator $\mathcal{O}_L$ in the SU(2)-subsector and any operator $\mathcal{O}_{Y_1Y_2}$ of two scalars can, in the leading order, be written as a spin-chain operator insertion in the scalar product between a matrix product state $\langle \mathrm{MPS}|$ and the Bethe state $|\Psi\rangle$ corresponding to the operator $\mathcal{O}_L$,

$$\langle \mathcal{O}_L \mathcal{O}_{Y_1Y_2} \rangle_{1\text{ contr.}} \propto \langle \mathrm{MPS}| Q_{Y_1Y_2} |\Psi\rangle.$$

The operation of Q depends on the two fields Y_1, Y_2 but is simple for any choice of scalar fields:

- For $Y_1^{\mathrm{cl}} = Y_2^{\mathrm{cl}}$ we get

  $$\langle \mathcal{O}_L \mathcal{O}_{Y_1Y_2} \rangle = \left(c^{\uparrow} L + c^{\downarrow}(L-M)\right) \langle \mathcal{O}_L \rangle_{\mathrm{tree}}$$

  where both L and the number of excitations M need to be *even* and the Bethe state needs to be unpaired.
- For $Y_1^{\mathrm{cl}} \neq Y_2^{\mathrm{cl}}$, the two-point function is zero for any $\mathcal{O}_L$ mapping to a highest weight Bethe state. For operators $\mathcal{O}_{L,M+n}$ mapping to (Bethe) descendants, however, the two-point function is non-vanishing, under the condition that n is odd and that the corresponding Bethe state descends from an unpaired state $|\Psi_{L,M}\rangle$. The result is

  $$\langle \mathcal{O}_{L,M+n} \mathcal{O}_{Y_1Y_2} \rangle_{1\text{ contr.}} = \left(c^{+} n(L-2M-n+1)\, \mathcal{C}^{+}_{L,M,n} + c^{-}\, \mathcal{C}^{-}_{L,M,n}\right) \langle \mathcal{O}_{L,M} \rangle_{\mathrm{tree}},$$

  where the combinatorial factors $\mathcal{C}^{\pm}_{L,M,n}$ can be found in equation (3).

The coefficients c with various indices depend on Y_1, Y_2 and are all spacetime-dependent since they contain expressions of the propagator. See Appendix A below for the full list of coefficients.

These results hold for any k.

Acknowledgements

We would like to thank I. Buhl-Mortenssen, M. de Leeuw, F. Levkovich-Maslyuk, O. Ohlsson Sax, M. Wilhelm and K. Zarembo for guidance and fruitful discussions.

This work was supported by the ERC advanced grant No. 341222.

Appendix A. List of coefficients

Here follows the list of coefficients for the considered two-point functions, written in the form $Q_{Y_1Y_2} : \begin{pmatrix} c^{\uparrow} & c^{+} \\ c^{-} & c^{\downarrow} \end{pmatrix}$.

$$Q_{ZZ} : \begin{pmatrix} \frac{2}{3}(2K^{m^2=0} - 3K^{m^2=2} + K^{m^2=6}) & 0 \\ 0 & -\frac{2}{3}(K^{m^2=0} - K^{m^2=6}) \end{pmatrix}$$

$$Q_{Z\overline{Z}} : \begin{pmatrix} \frac{2}{3}(2K^{m^2=0} + K^{m^2=6}) & 0 \\ 0 & -\frac{2}{3}(K^{m^2=0} - K^{m^2=6}) \end{pmatrix}$$

$$Q_{\overline{ZZ}} : \begin{pmatrix} \frac{2}{3}(2K^{m^2=0} + 3K^{m^2=2} + K^{m^2=6}) & 0 \\ 0 & -\frac{2}{3}(K^{m^2=0} - K^{m^2=6}) \end{pmatrix}$$

$$Q_{ZX} : \begin{pmatrix} 0 & K^{m^2=0} - K^{m^2=2} \\ K^{m^2=0} - K^{m^2=2} & 0 \end{pmatrix}$$

$$Q_{Z\overline{X}} : \begin{pmatrix} 0 & K^{m^2=0} + K^{m^2=2} \\ K^{m^2=0} - K^{m^2=2} & 0 \end{pmatrix}$$

$$Q_{\overline{Z}X} : \begin{pmatrix} 0 & K^{m^2=0} - K^{m^2=2} \\ K^{m^2=0} + K^{m^2=2} & 0 \end{pmatrix}$$

$$Q_{\overline{ZX}} : \begin{pmatrix} 0 & K^{m^2=0} + K^{m^2=2} \\ K^{m^2=0} + K^{m^2=2} & 0 \end{pmatrix}$$

$$Q_{XX} : \begin{pmatrix} -\frac{2}{3}(K^{m^2=0} - K^{m^2=6}) & 0 \\ 0 & \frac{2}{3}(2K^{m^2=0} - 3K^{m^2=2} + K^{m^2=6}) \end{pmatrix}$$

$$Q_{X\overline{X}} : \begin{pmatrix} -\frac{2}{3}(K^{m^2=0} - K^{m^2=6}) & 0 \\ 0 & \frac{2}{3}(2K^{m^2=0} + K^{m^2=6}) \end{pmatrix}$$

$$Q_{\overline{XX}} : \begin{pmatrix} -\frac{2}{3}(K^{m^2=0} - K^{m^2=6}) & 0 \\ 0 & \frac{2}{3}(2K^{m^2=0} + 3K^{m^2=2} + K^{m^2=6}) \end{pmatrix}$$

$$\begin{aligned} Q_{WW} &= Q_{W\overline{W}} \\ &= Q_{\overline{WW}} : \begin{pmatrix} -\frac{2}{3}(K^{m^2=0} - K^{m^2=6}) & 0 \\ 0 & -\frac{2}{3}(K^{m^2=0} - K^{m^2=6}) \end{pmatrix} \end{aligned}$$

References

[1] J.A. Minahan, K. Zarembo, The Bethe ansatz for N=4 superYang-Mills, J. High Energy Phys. 03 (2003) 013, http://dx.doi.org/10.1088/1126-6708/2003/03/013, arXiv:hep-th/0212208.
[2] N. Beisert, et al., Review of AdS/CFT integrability: an overview, Lett. Math. Phys. 99 (2012) 3–32, http://dx.doi.org/10.1007/s11005-011-0529-2, arXiv:1012.3982.
[3] A. Karch, L. Randall, Open and closed string interpretation of SUSY CFT's on branes with boundaries, J. High Energy Phys. 06 (2001) 063, http://dx.doi.org/10.1088/1126-6708/2001/06/063, arXiv:hep-th/0105132.
[4] W. Nahm, A simple formalism for the bps monopole, Phys. Lett. B 90 (4) (1980) 413–414, http://dx.doi.org/10.1016/0370-2693(80)90961-2, http://www.sciencedirect.com/science/article/pii/0370269380909612.
[5] D.-E. Diaconescu, D-branes, monopoles and Nahm equations, Nucl. Phys. B 503 (1997) 220–238, http://dx.doi.org/10.1016/S0550-3213(97)00438-0, arXiv:hep-th/9608163.
[6] N.R. Constable, R.C. Myers, O. Tafjord, The noncommutative bion core, Phys. Rev. D 61 (2000) 106009, http://dx.doi.org/10.1103/PhysRevD.61.106009, arXiv:hep-th/9911136.
[7] K. Nagasaki, S. Yamaguchi, Expectation values of chiral primary operators in holographic interface CFT, Phys. Rev. D 86 (2012) 086004, http://dx.doi.org/10.1103/PhysRevD.86.086004, arXiv:1205.1674.

[8] C. Kristjansen, G.W. Semenoff, D. Young, Chiral primary one-point functions in the D3-D7 defect conformal field theory, J. High Energy Phys. 01 (2013) 117, http://dx.doi.org/10.1007/JHEP01(2013)117, arXiv:1210.7015.

[9] M. de Leeuw, C. Kristjansen, K. Zarembo, One-point functions in defect CFT and integrability, J. High Energy Phys. 08 (2015) 098, http://dx.doi.org/10.1007/JHEP08(2015)098, arXiv:1506.06958.

[10] I. Buhl-Mortensen, M. de Leeuw, C. Kristjansen, K. Zarembo, One-point functions in AdS/dCFT from matrix product states, J. High Energy Phys. 02 (2016) 052, http://dx.doi.org/10.1007/JHEP02(2016)052, arXiv:1512.02532.

[11] M. de Leeuw, C. Kristjansen, S. Mori, AdS/dCFT one-point functions of the SU(3) sector, Phys. Lett. B 763 (2016) 197–202, http://dx.doi.org/10.1016/j.physletb.2016.10.044, arXiv:1607.03123.

[12] I. Buhl-Mortensen, M. de Leeuw, A.C. Ipsen, C. Kristjansen, M. Wilhelm, One-loop one-point functions in gauge-gravity dualities with defects, Phys. Rev. Lett. 117 (23) (2016) 231603, http://dx.doi.org/10.1103/PhysRevLett.117.231603, arXiv:1606.01886.

[13] I. Buhl-Mortensen, M. de Leeuw, A.C. Ipsen, C. Kristjansen, M. Wilhelm, A quantum check of AdS/dCFT, J. High Energy Phys. 01 (2017) 098, http://dx.doi.org/10.1007/JHEP01(2017)098, arXiv:1611.04603.

[14] I. Buhl-Mortensen, M. de Leeuw, A.C. Ipsen, C. Kristjansen, M. Wilhelm, Asymptotic one-point functions in AdS/dCFT, arXiv:1704.07386, 2017.

[15] M. de Leeuw, A.C. Ipsen, C. Kristjansen, K.E. Vardinghus, M. Wilhelm, Two-point functions in AdS/dCFT and the boundary conformal bootstrap equations, arXiv:1705.03898, 2017.

[16] P. Liendo, C. Meneghelli, Bootstrap equations for $\mathcal{N} = 4$ SYM with defects, J. High Energy Phys. 01 (2017) 122, http://dx.doi.org/10.1007/JHEP01(2017)122, arXiv:1608.05126.

[17] K. Nagasaki, H. Tanida, S. Yamaguchi, Holographic interface-particle potential, J. High Energy Phys. 01 (2012) 139, http://dx.doi.org/10.1007/JHEP01(2012)139, arXiv:1109.1927.

[18] J. Aguilera-Damia, D.H. Correa, V.I. Giraldo-Rivera, Circular Wilson loops in defect conformal field theory, J. High Energy Phys. 03 (2017) 023, http://dx.doi.org/10.1007/JHEP03(2017)023, arXiv:1612.07991.

[19] M. de Leeuw, A.C. Ipsen, C. Kristjansen, M. Wilhelm, One-loop Wilson loops and the particle-interface potential in AdS/dCFT, Phys. Lett. B 768 (2017) 192–197, http://dx.doi.org/10.1016/j.physletb.2017.02.047, arXiv:1608.04754.

[20] S. Kawamoto, T. Kuroki, Existence of new nonlocal field theory on noncommutative space and spiral flow in renormalization group analysis of matrix models, J. High Energy Phys. 06 (2015) 062, http://dx.doi.org/10.1007/JHEP06(2015)062, arXiv:1503.08411.

[21] J. Escobedo, N. Gromov, A. Sever, P. Vieira, Tailoring three-point functions and integrability, J. High Energy Phys. 09 (2011) 028, http://dx.doi.org/10.1007/JHEP09(2011)028, arXiv:1012.2475.

Ultra-spinning exotic compact objects supporting static massless scalar field configurations

Shahar Hod [a,b,*]

[a] The Ruppin Academic Center, Emeq Hefer 40250, Israel
[b] The Hadassah Academic College, Jerusalem 91010, Israel

ARTICLE INFO

Editor: M. Cvetič

ABSTRACT

Horizonless spacetimes describing highly compact exotic objects with reflecting (instead of absorbing) surfaces have recently attracted much attention from physicists and mathematicians as possible quantum-gravity alternatives to canonical classical black-hole spacetimes. Interestingly, it has recently been proved that spinning compact objects with angular momenta in the *sub*-critical regime $\bar{a} \equiv J/M^2 \leq 1$ are characterized by an *infinite* countable set of surface radii, $\{r_c(\bar{a};n)\}_{n=1}^{n=\infty}$, that can support asymptotically flat static configurations made of massless scalar fields. In the present paper we study *analytically* the physical properties of *ultra*-spinning exotic compact objects with dimensionless angular momenta in the complementary regime $\bar{a} > 1$. It is proved that ultra-spinning reflecting compact objects with dimensionless angular momenta in the *super*-critical regime $\sqrt{1-[m/(l+2)]^2} \leq |\bar{a}|^{-1} < 1$ are characterized by a *finite* discrete family of surface radii, $\{r_c(\bar{a};n)\}_{n=1}^{n=N_r}$, distributed symmetrically around $r = M$, that can support spatially regular static configurations of massless scalar fields (here the integers $\{l, m\}$ are the harmonic indices of the supported static scalar field modes). Interestingly, the largest supporting surface radius $r_c^{\max}(\bar{a}) \equiv \max_n\{r_c(\bar{a};n)\}$ marks the onset of superradiant instabilities in the composed ultra-spinning-exotic-compact-object-massless-scalar-field system.

1. Introduction

Curved black-hole spacetimes with absorbing event horizons are one of the most exciting predictions of the classical Einstein field equations. The physical and mathematical properties of classical black-hole spacetimes have been extensively explored during the last five decades [1,2], and it is widely believed that the recent detection of gravitational waves [3,4] provides compelling evidence for the existence of spinning astrophysical black holes of the Kerr family. Intriguingly, however, the physical properties of highly compact *horizonless* objects have recently been explored by many physicists (see [5–22] and references therein) in an attempt to determine whether these exotic curved spacetimes can serve as valid alternatives, possibly within the framework of a unified quantum theory of gravity, to canonical black-hole spacetimes.

In a very interesting work, Maggio, Pani, and Ferrari [17] have recently explored the complex resonance spectrum of massless scalar fields linearly coupled to horizonless spinning exotic compact objects. The numerical results presented in [17] have explicitly demonstrated the important physical fact that, for given values $\{l, m\}$ of the scalar field harmonic indices, there is a critical compactness parameter characterizing the central reflecting objects, above which the massless scalar fields grow exponentially in time. This characteristic behavior of the fields in the horizonless spinning curved spacetimes indicates that the corresponding exotic objects may become unstable when coupled to bosonic (integer-spin) fields [23]. In particular, this superradiant instability [24–28] is attributed to the fact that the characteristic absorbing boundary conditions of classical black-hole spacetimes have been replaced in [17] by reflecting boundary conditions at the compact surfaces of the horizonless exotic objects.

The physical properties of *marginally*-stable spinning exotic compact objects were studied analytically in [19]. In particular, it was explicitly proved in [19] that reflecting compact objects with *sub*-critical angular momenta in the regime $0 < \bar{a} \equiv J/M^2 \leq 1$ [29,30] are characterized by an *infinite* countable set of surface radii, $\{r_c(\bar{a};n)\}_{n=1}^{n=\infty}$, which can support spatially regular static (marginally-stable) configurations made of massless scalar fields. The ability of spinning compact objects to support static scalar field configurations is physically interesting from the point of view

* Correspondence to: The Ruppin Academic Center, Emeq Hefer 40250, Israel.
E-mail address: shaharhod@gmail.com.

of the no-hair theorems discussed in [31–33]. In particular, it was proved in [31,32] that spherically-symmetric (non-spinning) horizonless reflecting objects, like black holes with absorbing horizons [34–36], cannot support spatially regular nonlinear massless scalar field configurations [37–39].

Interestingly, the parameter space of the composed spinning-exotic-compact-object-massless-scalar-field system is divided by the outermost supporting radius, $r_c^{max}(\bar{a}) \equiv \max_n\{r_c(\bar{a};n)\}$, to stable and unstable configurations. In particular, horizonless reflecting objects whose surface radii lie in the regime $r_c > r_c^{max}(\bar{a})$ are stable to scalar perturbation modes [17,19], whereas the ergoregion of compact enough spinning objects in the physical regime $r_c < r_c^{max}(\bar{a})$ can trigger superradiant instabilities in the surrounding bosonic clouds [17,19].

The main goal of the present paper is to explore the physical properties of exotic *ultra*-spinning ($\bar{a} > 1$) horizonless compact objects [40–43]. Interestingly, we shall explicitly prove below that spinning compact objects in the *super*-critical $\bar{a} > 1$ regime are characterized by a *finite* discrete family of surface radii, $\{r_c(\bar{a};n)\}_{n=1}^{n=N_r}$ [44], that can support the static (marginally-stable) scalar field configurations. This unique property of the *ultra*-spinning ($\bar{a} > 1$) reflecting compact objects should be contrasted with the previously proved fact [19] that sub-critical ($\bar{a} < 1$) spinning objects are characterized by an *infinite* countable family of surface radii, $\{r_c(\bar{a};n)\}_{n=1}^{n=\infty}$, that can support spatially regular static scalar field configurations.

Using analytical techniques, we shall determine in this paper the characteristic critical (largest) surface radius, $r_c^{max}(\bar{a}) \equiv \max_n\{r_c(\bar{a};n)\}$, of the ultra-spinning reflecting objects that, for given value of the super-critical rotation parameter $\bar{a}$, marks the boundary between stable and superradiantly unstable spinning configurations. In particular, below we shall derive a remarkably compact analytical formula for the discrete (and finite) family of supporting surface radii which characterizes exotic near-critical spinning horizonless compact objects in the physically interesting regime $0 < \bar{a} - 1 \ll 1$.

2. Description of the system

We consider a spatially regular configuration made of a massless scalar field Ψ which is linearly coupled to an ultra-spinning reflecting compact object of radius r_c, mass M, and dimensionless angular momentum in the super-critical regime

$$\bar{a} \equiv \frac{J}{M^2} > 1 \, . \tag{1}$$

Following the interesting physical model of the exotic compact objects discussed by Maggio, Pani, and Ferrari [17] (see also [18–20]), we shall assume that the external spacetime geometry of the spinning compact object is described by the Kerr line element [1,2,29, 45–50]

$$ds^2 = -\frac{\Delta}{\rho^2}(dt - a\sin^2\theta d\phi)^2 + \frac{\rho^2}{\Delta}dr^2 + \rho^2 d\theta^2 + \frac{\sin^2\theta}{\rho^2}\big[adt - (r^2+a^2)d\phi\big]^2 \quad \text{for} \quad r > r_c \, , \tag{2}$$

where the metric functions are given by $\Delta \equiv r^2 - 2Mr + a^2$ and $\rho^2 \equiv r^2 + a^2\cos^2\theta$ with $a \equiv M\bar{a}$.

The spatial and temporal behavior of the massless scalar field configurations in the curved spacetime (2) of the spinning reflecting object is governed by the compact Klein–Gordon wave equation [51,52]

$$\nabla^\nu\nabla_\nu\Psi = 0 \, . \tag{3}$$

Using the spatial-temporal expression [51–53]

$$\Psi(t,r,\theta,\phi) = \sum_{l,m} e^{im\phi} S_{lm}(\theta; a\omega) R_{lm}(r; M, a, \omega) e^{-i\omega t} \tag{4}$$

for the linearized massless scalar field, one finds the ordinary differential equation [51,52]

$$\Delta\frac{d}{dr}\Big(\Delta\frac{dR_{lm}}{dr}\Big) + \Big\{[\omega(r^2+a^2) - ma]^2 + \Delta(2ma\omega - K_{lm})\Big\}R_{lm} = 0 \tag{5}$$

for the radial part $R_{lm}(r; M, a, \omega)$ of the massless scalar eigenfunction. The frequency-dependent eigenvalues $K_{lm}(a\omega)$ of the familiar spheroidal harmonic functions $S_{lm}(\theta; a\omega)$ [51,52,54–58] are given by the small frequency $a\omega \ll 1$ expression

$$K_{lm} - a^2\omega^2 = l(l+1) + \sum_{k=1}^{\infty} c_k (a\omega)^{2k} \, , \tag{6}$$

where the explicit functional expression of the coefficients $\{c_k = c_k(l,m)\}$ is given in [56].

Following the interesting physical models discussed in [17–20] for horizonless curved spacetimes, we shall assume that the scalar fields vanish on the compact reflecting surfaces of the central exotic compact objects [59]:

$$R(r = r_c) = 0 \, . \tag{7}$$

In addition, we consider asymptotically flat linearized scalar field configurations which are characterized by asymptotically decaying radial eigenfunctions:

$$R(r \to \infty) \to 0 \, . \tag{8}$$

3. The resonance condition of the composed ultra-spinning-exotic-compact-object-massless-scalar-field configurations

In the present section we shall derive, for a given set of the dimensionless physical parameters $\{r_c/M, \bar{a}, l, m\}$, the characteristic resonance condition for the existence of ultra-spinning reflecting exotic horizonless objects that support spatially regular *static* (marginally-stable) linearized scalar field configurations.

Substituting into the radial equation (5) the characteristic relation

$$\omega = 0 \tag{9}$$

for the static scalar field configurations, one obtains the ordinary differential equation [19,60]

$$x(1-x)\frac{d^2F}{dx^2} + \{(1-\gamma) - [1 + 2(l+1) - \gamma]x\}\frac{dF}{dx} - [(l+1)^2 - \gamma(l+1)]F = 0 \, , \tag{10}$$

where

$$R(x) = x^{-\gamma/2}(1-x)^{l+1}F(x) \, , \tag{11}$$

$$x \equiv \frac{r - M(1 + i\sqrt{\bar{a}^2 - 1})}{r - M(1 - i\sqrt{\bar{a}^2 - 1})} \, , \tag{12}$$

and

$$\gamma \equiv \frac{m}{\sqrt{1 - \bar{a}^{-2}}} \, . \tag{13}$$

The physically acceptable solution of the characteristic radial scalar equation (10) which respects the asymptotic boundary condition (8) is given by [19,56,61,62]

$$R(x) = A \cdot x^{-\gamma/2}(1-x)^{l+1}{}_2F_1(l+1-\gamma, l+1; 2l+2; 1-x)\ , \tag{14}$$

where A is a normalization constant and ${}_2F_1(a,b;c;z)$ is the hypergeometric function.

Substituting the radial solution (14) into the characteristic inner boundary condition (7) at the surface of the compact reflecting object, one obtains the remarkably compact resonance condition

$${}_2F_1(l+1-\gamma, l+1; 2l+2; 1-x_c) = 0 \tag{15}$$

for the composed ultra-spinning-exotic-compact-object-massless-scalar-field configurations.

As we shall show below, the resonance equation (15) determines the *discrete* set of surface radii $\{r_c = r_c(\bar{a}, l, m; n)\}$ which characterize the unique family of ultra-spinning exotic compact objects that can support the static spatially regular massless scalar field configurations.

4. Generic properties of the composed ultra-spinning-exotic-compact-object-massless-scalar-field configurations

In the present section we shall discuss two important features of the discrete resonance spectrum $\{r_c(\bar{a}, l, m; n)\}$ of surface radii that characterize the composed ultra-spinning-exotic-compact-object-massless-scalar-field configurations: (1) the distribution of the supporting radii, and (2) the (finite) number of supporting radii.

4.1. The resonance spectrum of surface radii is distributed symmetrically around $r = M$

Interestingly, we shall now prove that the discrete set of supporting radii $\{r_c(\bar{a}, l, m; n)\}$, which stems from the characteristic resonance equation (15), is distributed *symmetrically* around $r = M$. To this end, it is convenient to define the dimensionless symmetrical radial coordinate

$$z \equiv \frac{r-M}{M}\ , \tag{16}$$

in terms of which the resonance equation (15) can be written in the form [63]

$${}_2F_1\Big(l+1-\frac{m}{\sqrt{1-\bar{a}^{-2}}}, l+1; 2l+2; \frac{2i\sqrt{\bar{a}^2-1}}{z_c+i\sqrt{\bar{a}^2-1}}\Big) = 0\ . \tag{17}$$

Using the characteristic identity (see Eq. 15.3.15 of [56])

$${}_2F_1(a,b;2b;z) = (1-z)^{-a/2}{}_2F_1(\frac{1}{2}a, b-\frac{1}{2}a; b+\frac{1}{2}; \frac{z^2}{4z-4}) \tag{18}$$

of the hypergeometric function, one can express the resonance condition (17) in the symmetrical form

$${}_2F_1\Big[\frac{1}{2}\Big(l+1-\frac{m}{\sqrt{1-\bar{a}^{-2}}}\Big), \frac{1}{2}\Big(l+1+\frac{m}{\sqrt{1-\bar{a}^{-2}}}\Big); l+\frac{3}{2}; \frac{\bar{a}^2-1}{\bar{a}^2-1+z_c^2}\Big] = 0\ . \tag{19}$$

The resonance equation (19) is obviously invariant under the reflection symmetry $z_c \to -z_c$. We have therefore proved that if the dimensionless surface radius z_c is a solution of the characteristic resonance equation (19), then $-z_c$ is also a valid resonance.

In addition, it is interesting to stress the fact that, for the static ($\omega = 0$) scalar field modes, the radial scalar equation (5) is invariant under the reflection symmetries $a \to -a$ and $m \to -m$ [64]. One therefore deduces that if the dimensionless surface radius z_c characterizes an ultra-spinning exotic compact object with $ma > 0$ that can support a spatially regular static (marginally-stable) scalar field configuration with harmonic indices $\{l, m\}$, then the same supporting radius also characterizes an ultra-spinning exotic compact object with $ma < 0$ that can support the same static scalar field configuration.

Taking cognizance of the three reflection symmetries, $z_c \to -z_c$, $a \to -a$, and $m \to -m$, which characterize the composed ultra-spinning-exotic-compact-object-massless-scalar-field system, we shall henceforth assume, without loss of generality, that

$$a > 0 \ \ ; \ \ m > 0 \ \ ; \ \ z_c \geq 0\ . \tag{20}$$

4.2. The number of discrete supporting radii is finite

As emphasized above, it has recently been proved [19] that exotic compact objects in the *sub*-critical regime $\bar{a} < 1$ are characterized by an *infinite* set of surface radii, $\{r_c(\bar{a}; n)\}_{n=1}^{n=\infty}$, that can support static (marginally-stable) massless scalar field configurations.

On the other hand, we shall now show that *super*-critical ($\bar{a} > 1$) compact reflecting objects are characterized by a *finite* set of surface radii that can support the static massless scalar field configurations. In particular, one finds that, for positive integer values of the dimensionless physical parameter $N(\bar{a}, l, m) \equiv \gamma - (l+1)$, the resonance equation (15), which determines the characteristic spectrum of supporting radii of the *ultra*-spinning exotic compact objects, is a polynomial equation of degree N. Thus, in this case there is a *finite* number N of complex solutions $\{x_c(\bar{a}, l, m; n)\}_{n=1}^{n=N}$ to the resonance condition (15) which in turn, using the relation (12), yield a finite discrete spectrum $\{r_c(\bar{a}, l, m; n)\}_{n=1}^{n=N}$ of supporting surface radii.

In addition, solving numerically the resonance equation (15) we find that, for positive non-integer values of the physical parameter N, the number of discrete surface radii that can support the static (marginally-stable) scalar field configurations is given by (see Tables 1 and 2) $\lfloor N \rfloor$ for even values of $\lfloor N \rfloor$ and by $\lfloor N \rfloor + 1$ for odd values of $\lfloor N \rfloor$ [65].

To summarize, the (*finite*) number $N_r(\bar{a}, l, m)$ of discrete supporting radii that characterize the composed ultra-spinning-exotic-compact-object-massless-scalar-field configurations is given by the simple relations

$$N_r = \begin{cases} \gamma-(l+1) & \text{if} \ \ \gamma-(l+1) \text{ is a positive integer}\ ; \\ \lfloor\gamma-(l+1)\rfloor & \text{if} \ \ \lfloor\gamma-(l+1)\rfloor \\ & \quad \text{is a positive even integer}\ ; \\ \lfloor\gamma-(l+1)\rfloor+1 & \text{if} \ \ \lfloor\gamma-(l+1)\rfloor \\ & \quad \text{is a positive odd integer}\ . \end{cases} \tag{21}$$

[It is important to emphasize that cases 2 and 3 in (21) refer to *non*-integer values of the dimensionless composed parameter $\gamma - (l+1)$.]

5. The regime of existence of the composed ultra-spinning-exotic-compact-object-massless-scalar-field configurations

In the present section we shall derive an upper bound on the characteristic surface radii $\{r_c(\bar{a}, l, m; n)\}_{n=1}^{n=N_r}$ which characterize the ultra-spinning exotic compact objects that can support the static (marginally-stable) configurations of the massless scalar fields.

Substituting the scalar function

$$\Phi(r) \equiv \Delta^{1/2} \cdot R(r) \tag{22}$$

into the characteristic radial equation (5), one obtains the ordinary differential equation

$$\Delta^2 \frac{d^2\Phi}{dr^2} + \Big[(ma)^2 - l(l+1) \cdot \Delta - (a^2 - M^2)\Big]\Phi = 0 \tag{23}$$

for the static ($\omega = 0$) scalar configurations.

Using the characteristic boundary conditions (7) and (8) of the spatially regular linearized scalar field configurations, which are supported in the asymptotically flat curved spacetime (2) of the exotic ultra-spinning reflecting compact object, one deduces that the radial scalar eigenfunction $\Phi(r)$ must have (at least) one extremum point, $r = r_{\text{peak}}$, in the interval

$$r_{\text{peak}} \in (r_c, \infty) \ . \tag{24}$$

In particular, the simple functional relations

$$\{\Phi \neq 0 \ \ ; \ \ \frac{d\Phi}{dr} = 0 \ \ ; \ \ \Phi \cdot \frac{d^2\Phi}{dr^2} < 0\} \quad \text{for} \quad r = r_{\text{peak}} \tag{25}$$

characterize the spatial behavior of the radial scalar eigenfunction at this extremum point.

Taking cognizance of Eqs. (23) and (25), one finds the simple relation

$$(ma)^2 - l(l+1) \cdot \Delta(r_{\text{peak}}) - (a^2 - M^2) > 0 \ . \tag{26}$$

The characteristic inequality (26) implies that r_{peak} is bounded by the relations

$$r_- < r_{\text{peak}} < r_+ \ , \tag{27}$$

where

$$r_{\pm} = M \pm \sqrt{M^2 - \frac{a^2[1 + l(l+1) - m^2] - M^2}{l(l+1)}} \ . \tag{28}$$

Using Eqs. (16), (24), (27), and (28), one deduces that the composed ultra-spinning-exotic-compact-object-massless-scalar-field configurations are characterized by the simple dimensionless upper bound

$$|z_c| < \sqrt{1 - \frac{\bar{a}^2[1 + l(l+1) - m^2] - 1}{l(l+1)}} \ . \tag{29}$$

In particular, from the requirement $\bar{a}^2[1+l(l+1)-m^2]-1 \leq l(l+1)$ [see the r.h.s. of (29)], one finds that the static (marginally-stable) massless scalar field configurations in the curved spacetimes of the ultra-spinning ($\bar{a} > 1$) exotic compact objects are characterized by the compact inequalities

$$\sqrt{1 - \frac{m^2}{1 + l(l+1)}} < |\bar{a}|^{-1} < 1 \ . \tag{30}$$

Interestingly, a stronger upper bound on the dimensionless angular momentum parameter $\bar{a}$, which characterizes the unique family of ultra-spinning exotic compact objects that can support the spatially regular static (marginally-stable) massless scalar field configurations, can be obtained from the observations that [see Eq. (15)] [66]

$$ {}_2F_1[l + 1 - \gamma, l + 1; 2l + 2; 1 - x(r)] \neq 0$$
$$\text{for} \ \ \{r \in \mathbb{R} \ \ \text{and} \ \ -1 < l + 1 - \gamma < 2l + 3\} \tag{31}$$

and

$${}_2F_1(-1, l + 1; 2l + 2; 2) = {}_2F_1(2l + 3, l + 1; 2l + 2; 2) = 0 \ . \tag{32}$$

From Eqs. (13), (31), and (32), one deduces that the composed ultra-spinning-exotic-compact-object-massless-scalar-field configurations exist in the dimensionless physical regime

$$\sqrt{1 - \Big(\frac{m}{l+2}\Big)^2} \leq |\bar{a}|^{-1} < 1 \ , \tag{33}$$

where the equality sign in (33) corresponds to exotic ultra-spinning objects with $r_c = M$ [or, equivalently, $1 - x_c = 2$ and $z_c = 0$, see Eqs. (12) and (16)].

6. The resonance spectrum of the composed ultra-spinning-exotic-compact-object-massless-scalar-field configurations

As mentioned above, the *infinite* countable spectrum of supporting surface radii $\{r_c(\bar{a}, l, m; n)\}_{n=1}^{n=\infty}$ which characterizes the composed spinning-exotic-compact-object-massless-scalar-field configurations in the *sub*-critical regime $\bar{a} < 1$ has been determined in [19]. In the present section we shall explicitly show that *ultra*-spinning exotic compact objects in the complementary regime $\bar{a} > 1$ of super-critical angular momenta are characterized by a *finite* [see Eq. (21)] discrete set $\{r_c(\bar{a}, l, m; n)\}_{n=1}^{n=N_r}$ of surface radii that can support the asymptotically flat static scalar field configurations.

The compact resonance equation (15) can be solved numerically, for a given set $\{\bar{a}, l, m\}$ of the dimensionless physical parameters that characterize the composed compact-object-scalar-field system, to yield the discrete resonant spectrum $\{r_c(\bar{a}, l, m; n)\}_{n=1}^{n=N_r}$ of supporting radii. In Table 1 we present, for various super-critical values of the dimensionless angular momentum parameter $\bar{a}$, the smallest and largest dimensionless surface radii $\{z_c^{\text{min}}(\bar{a}, l, m), z_c^{\text{max}}(\bar{a}, l, m)\}$ of the ultra-spinning exotic compact objects that can support the static spatially regular configurations of the massless scalar fields [67]. We also present in Table 1 the (*finite*) number $N_r(\bar{a}, l, m)$ [see Eq. (21)] of these unique supporting surface radii [68].

The data presented in Table 1 demonstrate the fact that, for given integer values $\{l, m\}$ of the angular harmonic indices of the static (marginally-stable) massless scalar fields, the dimensionless supporting radius $z_c^{\text{max}}(\bar{a})$ of the ultra-spinning exotic compact objects is a monotonically decreasing function of the dimensionless physical parameter $\bar{a}$. As a consistency check, it is worth noting that the numerically computed values $\{z_c^{\text{max}}(\bar{a})\}$ of the characteristic surface radii of the ultra-spinning reflecting compact objects, as displayed in Table 1, conform to the analytically derived upper bound (29).

We would like to emphasize again that, for a given set of the physical parameters $\{\bar{a}, l, m\}$, the critical supporting radius $r_c^{\text{max}}(\bar{a})$ marks the boundary between stable and superradiantly unstable composed ultra-spinning-exotic-compact-object-massless-scalar-field configurations. In particular, the numerical results presented in the interesting work of Maggio, Pani, and Ferrari [17] indicate that ultra-spinning reflecting compact objects which are

Table 1
Marginally-stable ultra-spinning ($\bar{a} > 1$) reflecting compact objects. We present, for various super-critical values of the dimensionless physical parameter $\bar{a}$, the smallest and largest dimensionless radii, $\{z_c^{\min}(\bar{a}, l, m), z_c^{\max}(\bar{a}, l, m)\}$ [see Eq. (16)], of the ultra-spinning exotic compact objects that can support the static (marginally-stable) massless scalar field configurations [67]. Also presented is the *finite* number [see Eq. (21)] of these unique supporting surface radii. The data presented is for the case $l = m = 1$. The critical supporting radii $\{z_c^{\max}(\bar{a})\}$, which characterize the marginally-stable ultra-spinning reflecting compact objects, are found to be a monotonically decreasing function of the dimensionless angular momentum parameter $\bar{a}$. As a consistency check we note that the supporting radii of the ultra-spinning exotic compact objects conform to the analytically derived upper bound (29).

$\sqrt{1-\bar{a}^{-2}}$	$\bar{a}$	# of resonances	$z_c^{\min}(\bar{a})$	$z_c^{\max}(\bar{a})$
1/3	1.0607	1	0	0
0.3	1.0483	2	0.05488	0.05488
0.25	1.0328	2	0.11547	0.11547
0.2	1.0206	3	0	0.15811
0.15	1.0114	4	0.06455	0.18788
0.1	1.0050	8	0.01608	0.20760

Table 2
Marginally-stable ultra-spinning ($\bar{a} > 1$) reflecting compact objects. We present, for various equatorial ($l = m$) modes of the supported scalar fields, the smallest and largest dimensionless surface radii $\{z_c^{\min}(\bar{a}, l, m), z_c^{\max}(\bar{a}, l, m)\}$ [see Eq. (16)] of the ultra-spinning exotic compact objects that can support the spatially regular static (marginally-stable) massless scalar field configurations [67]. We also present the *finite* number of these unique supporting radii. The data presented is for the case $\sqrt{1-\bar{a}^{-2}} = 1/4$. The critical surface radii $\{z_c^{\max}(l)\}$, which characterize the marginally-stable ultra-spinning exotic compact objects, are found to be a monotonically increasing function of the dimensionless harmonic index l of the supported static scalar field configurations.

l	# of resonances	$z_c^{\min}(l)$	$z_c^{\max}(l)$
1	2	0.11547	0.11547
2	5	0	0.28705
3	8	0.03553	0.38489
4	11	0	0.45263
5	14	0.02113	0.50342
6	17	0	0.54338

characterized by the inequality $r_c < r_c^{\max}(\bar{a})$ are superradiantly unstable to massless scalar perturbation modes, whereas ultra-spinning exotic compact objects which are characterized by the relation $r_c > r_c^{\max}(\bar{a})$ are stable.

In Table 2 we present, for various equatorial ($l = m$) modes of the supported static scalar fields, the smallest and largest dimensionless surface radii $\{z_c^{\min}(\bar{a}, l, m), z_c^{\max}(\bar{a}, l, m)\}$ of the supporting marginally-stable ultra-spinning exotic compact objects [67]. Also displayed is the (*finite*) number [see Eq. (21)] of these unique supporting surface radii [68]. The data presented in Table 2 reveal the fact that, for a given value of the dimensionless physical parameter $\bar{a}$, the critical (largest) supporting radius $z_c^{\max}(l)$ of the reflecting exotic compact objects is a monotonically increasing function of the harmonic index l which characterizes the static massless scalar field mode. It is worth noting that the numerically computed surface radii $z_c^{\max}(l)$ of the ultra-spinning marginally-stable exotic compact objects, as presented in Table 2, conform to the analytically derived upper bound (29).

7. The resonance spectrum of near-critical ultra-spinning exotic compact objects

7.1. An analytical treatment

Interestingly, as we shall explicitly show in the present section, the compact resonance equation (15), which determines the discrete family $\{x_c(\bar{a}, l, m; n)\}$ of dimensionless surface radii that characterize the marginally-stable ultra-spinning exotic compact objects, is amenable to an *analytical* treatment in the physically interesting regime

$$0 < \bar{a} - 1 \ll 1 \tag{34}$$

of *near-critical* horizonless spinning configurations.

In particular, in the near-critical regime

$$\frac{m}{\sqrt{1-\bar{a}^{-2}}} \gg l \tag{35}$$

one may use the large-$|b|$ asymptotic expansion [69]

$$\begin{aligned} {}_2F_1(a, b; c; z) = \; & \frac{\Gamma(c)}{\Gamma(c-a)}(-bz)^{-a}[1 + O(|bz|^{-1})] \\ & + \frac{\Gamma(c)}{\Gamma(a)}(bz)^{a-c}(1-z)^{c-a-b}[1 + O(|bz|^{-1})] \end{aligned} \tag{36}$$

of the hypergeometric function in order to express the resonance condition (15) in the remarkably compact form [70]

$$x^{m/\sqrt{1-\bar{a}^{-2}}} = (-1)^{-l} \quad \text{for} \quad \sqrt{1-\bar{a}^{-2}} \ll \frac{m}{l} \, . \tag{37}$$

From the asymptotic relation (37) one finds the set of complex solutions [71]

$$x_c(n) = e^{-i\pi(l+2n)\sqrt{1-\bar{a}^{-2}}/m} \quad ; \quad n \in \mathbb{Z} \tag{38}$$

which, taking cognizance of Eqs. (12) and (16), yields the discrete real resonance spectrum [63,70,72]

$$z_c(n) = \sqrt{\bar{a}^2 - 1} \cdot \cot\Big[\frac{\pi(l+2n)\sqrt{1-\bar{a}^{-2}}}{2m}\Big] \quad \text{for} \quad \Big\{\sqrt{1-\bar{a}^{-2}} \ll \frac{m}{l} \quad \text{and} \quad \pi(l+2n) \gg 1\Big\} \tag{39}$$

for the dimensionless surface radii which characterize the near-critical ($\bar{a} \gtrsim 1$) exotic compact objects that can support the static massless scalar field configurations. Interestingly, the analytically derived resonance formula (39) can be further simplified in the $\pi(l+2n)\sqrt{1-\bar{a}^{-2}}/2m \ll 1$ regime, in which case one finds the remarkably compact expression [72,73]

$$z_c(n) = \frac{2m\bar{a}}{\pi(l+2n)} \quad \text{for} \quad 1 \ll \pi(l+2n) \ll \frac{2m}{\sqrt{1-\bar{a}^{-2}}} \tag{40}$$

for the characteristic radii of the ultra-spinning exotic compact objects that can support the static (marginally-stable) configurations of the massless scalar fields.

7.2. Numerical confirmation

It is of physical interest to verify the accuracy of the approximated (analytically derived) resonance spectrum (39) for the surface radii of the near-critical ($0 < \bar{a} - 1 \ll 1$) ultra-spinning exotic compact objects that can support the spatially regular static (marginally-stable) configurations of the massless scalar fields. In Table 3 we present the dimensionless discrete surface radii $z_c^{\text{analytical}}(n)$ of the supporting near-critical ultra-spinning exotic reflecting objects as obtained from the analytically derived resonance spectrum (39). We also present in Table 3 the corresponding surface radii $z_c^{\text{numerical}}(n)$ of the ultra-spinning exotic compact objects as computed numerically from the exact characteristic resonance equation (15).

The data presented in Table 3 nicely demonstrate the important fact that there is a good agreement between the approximated surface radii $\{z_c^{\text{analytical}}(n)\}$ of the ultra-spinning exotic compact objects that can support the static massless scalar field configurations [as calculated from the compact analytically derived resonance formula (39)] and the corresponding exact surface radii $\{z_c^{\text{numerical}}(n)\}$ of the reflecting compact objects [as determined numerically directly from the resonance equation (15)].

Table 3
Near-critical ultra-spinning ($\bar{a} \gtrsim 1$) exotic compact objects. Displayed are the analytically calculated discrete surface radii $\{z_c^{\text{analytical}}(n)\}$ which characterize the ultra-spinning exotic compact objects that can support the static (marginally-stable) spatially regular configurations of the massless scalar fields. Also displayed are the corresponding supporting radii $\{z_c^{\text{numerical}}(n)\}$ of the near-critical exotic compact objects as obtained numerically directly from the characteristic resonance equation (15). The data presented is for static massless scalar field modes with $l = m = 1$ linearly coupled to near-critical ultra-spinning exotic compact objects with $\sqrt{1-\bar{a}^{-2}} = 10^{-2}$. The displayed data reveal a remarkably good agreement between the exact characteristic surface radii $\{z_c^{\text{numerical}}(n)\}$ of the ultra-spinning exotic compact objects [as determined numerically from the resonance condition (15)] and the corresponding approximated radii $\{z_c^{\text{analytical}}(n)\}$ of the near-critical ultra-spinning compact objects [as calculated analytically from the compact resonance formula (39)].

Formula	$z_c(n=1)$	$z_c(n=2)$	$z_c(n=3)$	$z_c(n=4)$	$z_c(n=5)$
Analytical [Eq. (39)]	0.21206	0.12707	0.09058	0.07027	0.05730
Numerical [Eq. (15)]	0.22240	0.12919	0.09135	0.07062	0.05818

8. The resonance spectrum of ultra-spinning exotic compact objects with $r_c = M$

Interestingly, as we shall now prove, the characteristic resonance equation (15) [or, equivalently, the symmetrical form (19) of the resonance condition] can also be solved *analytically* for the dimensionless angular momentum parameter $\bar{a}$ in the physically interesting case of horizonless ultra-spinning exotic compact objects of mass M whose compact reflecting surfaces coincide with the corresponding horizon radius $r_c = M$ of extremal Kerr black holes with the same mass parameter.

Substituting $r_c = M$ [which corresponds to $z_c = 0$, see Eq. (16)] into the resonance condition (19), and using the characteristic identity (see Eq. 15.1.20 of [56])

$$ {}_2F_1(a, b; c; 1) = \frac{\Gamma(c)\Gamma(c-a-b)}{\Gamma(c-a)\Gamma(c-b)} \tag{41}$$

of the hypergeometric function, one finds the compact resonance equation

$$ \frac{\Gamma(l+\frac{3}{2})\Gamma(\frac{1}{2})}{\Gamma\big(\frac{1}{2}l+1+\frac{m}{2\sqrt{1-\bar{a}^{-2}}}\big)\Gamma\big(\frac{1}{2}l+1-\frac{m}{2\sqrt{1-\bar{a}^{-2}}}\big)} = 0 \quad \text{for} \quad r_c = M\,. \tag{42}$$

Using the well known pole structure of the Gamma functions [namely, $1/\Gamma(-n) = 0$ for $n = 0, 1, 2, \ldots$ [74]], one obtains from (42) the remarkably simple discrete resonance spectrum

$$ \sqrt{1-\bar{a}^{-2}} = \frac{m}{l+2+2n} \quad ; \quad n = 0, 1, 2, \ldots \tag{43}$$

for the ultra-spinning ($\bar{a} > 1$) exotic compact objects whose reflecting surfaces coincide with the corresponding horizon radius $r_c = M$ of extremal ($\bar{a} = 1$) Kerr black holes [75]. Interestingly, one finds that the ultra-spinning exotic compact objects described by the resonance formula (43) with $n = 0$ saturate the previously derived bound (33).

9. Summary and discussion

The physical and mathematical properties of horizonless highly compact exotic reflecting objects have recently been studied by some physicists (see [5–22] and references therein). The main motivation behind these diverse studies has been to examine the intriguing possibility that these exotic horizonless objects may serve as quantum-gravity alternatives to classical black-hole spacetimes.

Interestingly, Maggio, Pani, and Ferrari [17] have recently provided compelling evidence that sub-critical ($\bar{a} \equiv J/M^2 < 1$) horizonless spinning spacetimes, in which the characteristic absorbing boundary conditions of classical black-hole spacetimes have been replaced by reflective boundary conditions at the surfaces of the exotic compact objects, may become superradiantly unstable [24–28] when linearly coupled to massless scalar (bosonic) field modes [23]. In particular, it has been explicitly proved in [19] that, in the sub-critical regime $\bar{a} < 1$ of the spinning reflecting objects and for given harmonic indices $\{l, m\}$ of the massless scalar field, there exists an *infinite* countable set of surface radii, $\{r_c(\bar{a}, l, m; n)\}_{n=1}^{n=\infty}$, which can support spatially regular static (marginally-stable) massless scalar field configurations.

In the present paper we have explored the physical and mathematical properties of marginally-stable composed ultra-spinning-exotic-compact-object-massless-scalar-field configurations which are characterized by super-critical ($\bar{a} > 1$) dimensionless rotation parameters. The following are the main results derived in this paper and their physical implications:

(1) It has been explicitly proved that, for given dimensionless physical parameters $\{\bar{a}, l, m\}$, the unique discrete family $\{r_c(\bar{a}, l, m; n)\}$ of surface radii that characterize the ultra-spinning ($\bar{a} > 1$) exotic compact objects that can support the static (marginally-stable) massless scalar field configurations is determined by the resonance condition [see Eqs. (1), (16), and (19)]

$$ {}_2F_1\Big[\frac{1}{2}\Big(l+1-\frac{ma}{\sqrt{a^2-M^2}}\Big), \frac{1}{2}\Big(l+1+\frac{ma}{\sqrt{a^2-M^2}}\Big); l+\frac{3}{2}; \frac{a^2-M^2}{a^2-M^2+(r_c-M)^2}\Big] = 0\,. \tag{44}$$

(2) We have shown that the composed ultra-spinning-exotic-compact-object-massless-scalar-field configurations, as determined by the resonance condition (44), are restricted to the physical regime [see Eq. (33)]

$$ \sqrt{1-\Big(\frac{m}{l+2}\Big)^2} \le \frac{M}{a} < 1 \tag{45}$$

of the dimensionless super-critical rotation parameter a/M. In addition, it has been proved that, for a given set $\{\bar{a}, l, m\}$ of the dimensionless physical parameters that characterize the composed compact-object-scalar-field system, the simple relation [see Eqs. (16) and (29)]

$$ \Big|\frac{r_c-M}{M}\Big| < \sqrt{1-\frac{\bar{a}^2[1+l(l+1)-m^2]-1}{l(l+1)}} \tag{46}$$

provides an upper bound on the surface radii of the supporting ultra-spinning exotic compact objects.

(3) It has been pointed out that the analytically derived resonance condition in its symmetrical form (44) reveals the fact that, for ultra-spinning exotic compact objects, the discrete resonant spectrum of supporting surface radii is invariant under the reflection symmetries

$$ r_c - M \to -(r_c - M) \quad ; \quad a \to -a \quad ; \quad m \to -m\,. \tag{47}$$

The symmetry transformations (47) imply, in particular, that if z_c [see Eq. (16)] is a dimensionless supporting radius of a composed exotic-object-scalar-field system with dimensionless physical parameters $\{\bar{a}, l, m\}$, then: (1) $-z_c$ is also a valid supporting radius of the same composed physical system, and (2) z_c is also a valid supporting radius of a composed exotic-object-scalar-field system with dimensionless parameters $\{\pm\bar{a}, l, \pm m\}$.

(4) It has been shown that, for ultra-spinning exotic compact objects in the dimensionless physical regime (45), the *finite* number $N_{\rm r}(\bar{a},l,m)$ of surface radii that can support the spatially regular static (marginally-stable) scalar field configurations is given by [see Eqs. (13) and (21)] [65,75,76]

$$N_{\rm r}=\begin{cases} N & \text{if} \quad N \text{ is a positive integer }; \\ \lfloor N\rfloor & \text{if} \quad \lfloor N\rfloor \text{ is a positive even integer }; \\ \lfloor N\rfloor+1 & \text{if} \quad \lfloor N\rfloor \text{ is a positive odd integer }, \end{cases} \tag{48}$$

where

$$N(\bar{a},l,m)\equiv\frac{ma}{\sqrt{a^2-M^2}}-(l+1)\ . \tag{49}$$

Interestingly, the fact that ultra-spinning ($\bar{a}>1$) exotic compact objects are characterized by a *finite* discrete family $\{r_{\rm c}(\bar{a},l,m;n)\}_{n=1}^{n=N_{\rm r}}$ of surface radii that can support the static massless scalar field configurations should be contrasted with the complementary case of sub-critical spinning objects in the $\bar{a}<1$ regime which, as previously proved in [17,19], are characterized by an *infinite* countable family $\{r_{\rm c}(\bar{a},l,m;n)\}_{n=1}^{n=\infty}$ of surface radii that can support the spatially regular static scalar fields [77].

(5) The ability of *spinning* objects to support spatially regular static scalar field configurations is physically intriguing from the point of view of the no-hair theorems that have recently been discussed in [31–33] for horizonless regular spacetimes. In particular, it has been proved in [31,32] that spherically-symmetric (*non*-spinning) horizonless reflecting stars cannot support nonlinear configurations made of massless scalar fields.

(6) It is important to stress the fact that, as shown in [17,19], the outermost (largest) surface radius $r^{\rm max}_{\rm c}(\bar{a})\equiv\max_n\{r_{\rm c}(\bar{a};n)\}$ that can support the static scalar field configurations is of central physical importance since it marks the boundary between stable $[r_{\rm c}>r^{\rm max}_{\rm c}(\bar{a})]$ and unstable $[r_{\rm c}<r^{\rm max}_{\rm c}(\bar{a})]$ composed ultra-spinning-exotic-compact-object-massless-scalar-field configurations.

(7) Solving numerically the analytically derived resonance equation (15), we have demonstrated that the characteristic supporting radius $r^{\rm max}_{\rm c}(\bar{a},l,m)$ is a monotonically decreasing function of the dimensionless rotation parameter $\bar{a}$ of the ultra-spinning exotic compact objects (see Table 1). Likewise, it has been demonstrated that the critical (outermost) supporting surface radius $r^{\rm max}_{\rm c}(\bar{a},l,m)$ is a monotonically increasing function of the harmonic parameter l which characterizes the massless scalar field modes (see Table 2).

(8) We have explicitly shown that the characteristic resonance equation (15) for the discrete family of supporting surface radii is amenable to an analytical treatment in the physically interesting regime $0<\bar{a}-1\ll 1$ of *near-critical* horizonless spinning objects. In particular, the remarkably compact resonance formula [see Eqs. (16) and (40)]

$$r_{\rm c}(n)=M+\frac{2ma}{\pi(l+2n)}\quad ;\quad n\in\mathbb{Z} \tag{50}$$

has been derived analytically for near-critical ($\bar{a}\gtrsim 1$) composed ultra-spinning-exotic-compact-object-massless-scalar-field configurations in the $1\ll\pi(l+2n)\ll 2m/\sqrt{1-\bar{a}^{-2}}$ regime.

(9) We have verified that the predictions of the analytically derived resonance formula (50), which determines the unique family of surface radii of the near-critical ultra-spinning ($0<\bar{a}-1\ll 1$) compact reflecting objects that can support the static (marginally-stable) massless scalar field configurations, agree remarkably well (see Table 3) with the corresponding exact values of the supporting surface radii as determined numerically from the characteristic resonance condition (15).

(10) Finally, it has been proved that the resonance equation (15) can be solved *analytically* in the physically interesting case of ultra-spinning ($\bar{a}>1$) exotic compact objects whose reflecting surfaces coincide with the corresponding horizon radius $r_{\rm c}=M$ of extremal ($\bar{a}=1$) Kerr black holes with the same mass parameter. In particular, we have derived the remarkably compact discrete resonance spectrum [see Eq. (43)]

$$\frac{a}{M}=\frac{1}{\sqrt{1-\big(\frac{m}{l+2+2n}\big)^2}}\quad ;\quad n=0,1,2,... \tag{51}$$

for the ultra-spinning compact configurations with $r_{\rm c}=M$ [78]. Interestingly, one finds from the analytically derived resonance formula (51) that, in the $l+2n\gg m$ regime, the dimensionless angular momenta $\{\bar{a}\}_{n=0}^{n=\infty}$ of the exotic ultra-spinning *reflecting* objects with physical parameters $\{M,r_{\rm c}=M\}$ can be made arbitrarily close [79] to the corresponding dimensionless angular momentum $\bar{a}_{EK}=1$ of an *absorbing* extremal Kerr black hole with the *same* mass and radius.

Acknowledgements

This research is supported by the Carmel Science Foundation. I thank Yael Oren, Arbel M. Ongo, Ayelet B. Lata, and Alona B. Tea for stimulating discussions.

References

[1] C.W. Misner, K.S. Thorne, J.A. Wheeler, Gravitation, W.H. Freeman, San Francisco, 1973.
[2] S. Chandrasekhar, The Mathematical Theory of Black Holes, Oxford University Press, New York, 1983.
[3] L.I.G.O. Virgo, B.P. Abbott, et al., Scientific Collaboration, Phys. Rev. Lett. 116 (2016) 061102.
[4] L.I.G.O. Virgo, B.P. Abbott, et al., Scientific Collaboration, Phys. Rev. Lett. 116 (2016) 241103.
[5] P.O. Mazur, E. Mottola, arXiv:gr-qc/0109035.
[6] S.D. Mathur, Fortschr. Phys. 53 (2005) 793.
[7] C.B.M.H. Chirenti, L. Rezzolla, Class. Quantum Gravity 24 (2007) 4191.
[8] K. Skenderis, M. Taylor, Phys. Rep. 467 (2008) 117.
[9] P. Pani, E. Berti, V. Cardoso, Y. Chen, R. Norte, Phys. Rev. D 80 (2009) 124047.
[10] A. Almheiri, D. Marolf, J. Polchinski, J. Sully, J. High Energy Phys. 02 (2013) 062.
[11] V. Cardoso, L.C.B. Crispino, C.F.B. Macedo, H. Okawa, P. Pani, Phys. Rev. D 90 (2014) 044069.
[12] M. Saravani, N. Afshordi, R.B. Mann, Int. J. Mod. Phys. D 23 (2015) 1443007.
[13] C. Chirenti, L. Rezzolla, Phys. Rev. D 94 (2016) 084016.
[14] J. Abedi, H. Dykaar, N. Afshordi, arXiv:1612.00266.
[15] B. Holdom, J. Ren, arXiv:1612.04889.
[16] C. Barcelo, R. Carballo-Rubio, L.J. Garay, arXiv:1701.09156.
[17] E. Maggio, P. Pani, V. Ferrari, arXiv:1703.03696.
[18] S. Hod, Phys. Lett. B 770 (2017) 186.
[19] S. Hod, J. High Energy Phys. 06 (2017) 132, arXiv:1704.05856.
[20] V. Cardoso, P. Pani, arXiv:1707.03021.
[21] R.A. Konoplya, A. Zhidenko, J. Cosmol. Astropart. Phys. 1612 (2016) 043.
[22] P.V.P. Cunha, E. Berti, C.A.R. Herdeiro, arXiv:1708.04211.
[23] It is important to note that it has recently been pointed out [22] that exotic compact objects which respect the null energy condition may be characterized by nonlinear instabilities to massless perturbation fields which are related to the existence of stable light rings in the corresponding horizonless spacetimes. Thus, in order to be considered as physically acceptable quantum-gravity alternatives to the canonical classical black-hole spacetimes, ultra-compact exotic compact objects must violate the null energy condition or alternatively have long instability time scales.
[24] Ya.B. Zel'dovich, Pis'ma Zh. Eksp. Teor. Fiz. 14 (1971) 270, JETP Lett. 14 (1971) 180;
Ya.B. Zel'dovich, Zh. Eksp. Teor. Fiz. 62 (1972) 2076, Sov. Phys. JETP 35 (1972) 1085.
[25] W.H. Press, S.A. Teukolsky, Nature 238 (1972) 211;
W.H. Press, S.A. Teukolsky, Astrophys. J. 185 (1973) 649.
[26] A.V. Vilenkin, Phys. Lett. B 78 (1978) 301.
[27] J.L. Friedman, Commun. Math. Phys. 63 (1978) 243.
[28] R. Brito, V. Cardoso, P. Pani, Lect. Notes Phys. 906 (2015) 1.
[29] We shall use natural units in which $G=c=\hbar=1$.
[30] Here M and J are respectively the mass and angular momentum of the spinning exotic compact object.
[31] S. Hod, Phys. Rev. D 94 (2016) 104073, arXiv:1612.04823.
[32] S. Hod, Phys. Rev. D 96 (2017) 024019.

[33] S. Bhattacharjee, S. Sarkar, Phys. Rev. D 95 (2017) 084027.
[34] J.E. Chase, Commun. Math. Phys. 19 (1970) 276.
[35] J.D. Bekenstein, Phys. Rev. Lett. 28 (1972) 452.
[36] C. Teitelboim, Lett. Nuovo Cimento 3 (1972) 326.
[37] It is worth noting that, while spinning exotic compact objects with reflecting boundary conditions can support static (marginally-stable) configurations of massless scalar fields [19], these spatially regular static fields cannot be supported by spinning Kerr black holes with regular event horizons. Interestingly, however, it has recently been revealed that stationary configurations made of massive scalar fields can be supported in the exterior regions of spinning black-hole spacetimes [38,39].
[38] S. Hod, Phys. Rev. D 86 (2012) 104026, arXiv:1211.3202;
S. Hod, Eur. Phys. J. C 73 (2013) 2378, arXiv:1311.5298;
S. Hod, Phys. Rev. D 90 (2014) 024051, arXiv:1406.1179;
S. Hod, Phys. Lett. B 739 (2014) 196, arXiv:1411.2609;
S. Hod, Class. Quantum Gravity 32 (2015) 134002, arXiv:1607.00003;
S. Hod, Phys. Lett. B 751 (2015) 177;
S. Hod, Class. Quantum Gravity 33 (2016) 114001;
S. Hod, Phys. Lett. B 758 (2016) 181, arXiv:1606.02306;
S. Hod, O. Hod, Phys. Rev. D 81 (2010) 061502, Rapid communication, arXiv:0910.0734;
S. Hod, Phys. Lett. B 708 (2012) 320, arXiv:1205.1872;
S. Hod, J. High Energy Phys. 01 (2017) 030, arXiv:1612.00014.
[39] C.A.R. Herdeiro, E. Radu, Phys. Rev. Lett. 112 (2014) 221101;
C.L. Benone, L.C.B. Crispino, C. Herdeiro, E. Radu, Phys. Rev. D 90 (2014) 104024;
C.A.R. Herdeiro, E. Radu, Phys. Rev. D 89 (2014) 124018;
C.A.R. Herdeiro, E. Radu, Int. J. Mod. Phys. D 23 (2014) 1442014;
Y. Brihaye, C. Herdeiro, E. Radu, Phys. Lett. B 739 (2014) 1;
J.C. Degollado, C.A.R. Herdeiro, Phys. Rev. D 90 (2014) 065019;
C. Herdeiro, E. Radu, H. Rúnarsson, Phys. Lett. B 739 (2014) 302;
C. Herdeiro, E. Radu, Class. Quantum Gravity 32 (2015) 144001;
C.A.R. Herdeiro, E. Radu, Int. J. Mod. Phys. D 24 (2015) 1542014;
C.A.R. Herdeiro, E. Radu, Int. J. Mod. Phys. D 24 (2015) 1544022;
P.V.P. Cunha, C.A.R. Herdeiro, E. Radu, H.F. Rúnarsson, Phys. Rev. Lett. 115 (2015) 211102;
B. Kleihaus, J. Kunz, S. Yazadjiev, Phys. Lett. B 744 (2015) 406;
C.A.R. Herdeiro, E. Radu, H.F. Rúnarsson, Phys. Rev. D 92 (2015) 084059;
C. Herdeiro, J. Kunz, E. Radu, B. Subagyo, Phys. Lett. B 748 (2015) 30;
C.A.R. Herdeiro, E. Radu, H.F. Rúnarsson, Int. J. Mod. Phys. D 25 (2016) 1641014;
Y. Brihaye, C. Herdeiro, E. Radu, Phys. Lett. B 760 (2016) 279;
Y. Ni, M. Zhou, A.C. Avendano, C. Bambi, C.A.R. Herdeiro, E. Radu, J. Cosmol. Astropart. Phys. 1607 (2016) 049;
M. Wang, arXiv:1606.00811.
[40] It is worth emphasizing that *ultra*-spinning (with $\bar{a} > 1$) hairy black-hole configurations are known to exist as stationary solutions of the nonlinearly coupled Einstein-scalar [38,39] and Einstein–Proca [41] field equations. Moreover, using fully nonlinear numerical (dynamical) simulations [42,43], it has recently been demonstrated explicitly that these ultra-spinning ($\bar{a} > 1$) hairy configurations can be formed dynamically from the complete classical gravitational collapse of spatially regular matter fields. To the best of our knowledge, the analogous dynamical formation of horizonless *ultra*-spinning exotic compact objects has not been demonstrated numerically thus far. (It is worth noting, however, that spinning weakly self-gravitating mundane objects are usually characterized by the relation $\bar{a} > 1$. In particular, planet Earth itself is characterized by the dimensionless relation $\bar{a} \gg 1$.) As previously suggested in [5–22], it may be the case that some quantum mechanism could halt the gravitational collapse of the ultra-spinning matter configurations, thus yielding horizonless quantum exotic compact objects instead of the ultra-spinning classical black holes discussed in [42,43].
[41] C.A.R. Herdeiro, E. Radu, H.F. Rúnarsson, Class. Quantum Gravity 33 (2016) 154001.
[42] W.E. East, F. Pretorius, Phys. Rev. Lett. 119 (2017) 041101.
[43] C.A.R. Herdeiro, E. Radu, arXiv:1706.06597.
[44] See Eq. (21) below for the precise value of the integer parameter N_r which, for super-critical ($\bar{a} > 1$) spinning compact objects, characterizes the *finite* family of supporting surface radii.
[45] Here (t, r, θ, ϕ) are the Boyer–Lindquist spacetime coordinates.
[46] Following [17,19,20], we shall assume that the contributions of the compact reflective surface to the energy and angular momentum of the curved spacetime are negligible.
[47] As nicely emphasized in [17,20] (see also [48–50]), the Birkhoff theorem does not apply in non-spherically symmetric spacetimes, and thus the exterior metric of the spinning compact object is not unique. Following [17,19,20], we shall assume that the external spacetime of the horizonless exotic object is characterized by the familiar metric components (2) of the spinning Kerr spacetime.
[48] P. Pani, Phys. Rev. D 92 (2015) 124030.
[49] N. Uchikata, S. Yoshida, P. Pani, Phys. Rev. D 94 (2016) 064015.
[50] K. Yagi, N. Yunes, Phys. Rev. D 91 (2015) 123008.
[51] S.A. Teukolsky, Phys. Rev. Lett. 29 (1972) 1114;
S.A. Teukolsky, Astrophys. J. 185 (1973) 635.
[52] T. Hartman, W. Song, A. Strominger, J. High Energy Phys. 1003 (2010) 118.
[53] Here ω is the proper frequency of the massless scalar field mode. The dimensionless parameters $\{l, m\}$ are the harmonic (spheroidal and azimuthal) indices of the linearized field with the mathematical property $l \geq |m|$.
[54] A. Ronveaux, Heun's Differential Equations, Oxford University Press, Oxford, UK, 1995;
C. Flammer, Spheroidal Wave Functions, Stanford University Press, Stanford, 1957.
[55] P.P. Fiziev, e-print arXiv:0902.1277;
R.S. Borissov, P.P. Fiziev, e-print arXiv:0903.3617;
P.P. Fiziev, Phys. Rev. D 80 (2009) 124001;
P.P. Fiziev, Class. Quantum Gravity 27 (2010) 135001.
[56] M. Abramowitz, I.A. Stegun, Handbook of Mathematical Functions, Dover Publications, New York, 1970.
[57] S. Hod, Phys. Rev. Lett. 100 (2008) 121101, arXiv:0805.3873.
[58] S. Hod, Phys. Lett. B 717 (2012) 462, arXiv:1304.0529;
S. Hod, Phys. Rev. D 87 (2013) 064017, arXiv:1304.4683;
S. Hod, Phys. Lett. B 746 (2015) 365, arXiv:1506.04148.
[59] For brevity, we shall henceforth omit the dimensionless harmonic indices $\{l, m\}$.
[60] Here we have used the simple relation $K_{lm}(a\omega = 0) = l(l+1)$ for the static (marginally-stable) scalar field modes [see Eq. (6)].
[61] P.M. Morse, H. Feshbach, Methods of Theoretical Physics, McGraw–Hill, New York, 1953.
[62] Using Eq. 15.1.1 of [56], one finds the asymptotic functional behavior $R(r \to \infty) \sim r^{-(l+1)} \to 0$ in the $r \to \infty$ $(x \to 1)$ limit.
[63] Here we have used the relation $x = (z - i\sqrt{\bar{a}^2 - 1})/(z + i\sqrt{\bar{a}^2 - 1})$ [see Eqs. (12) and (16)].
[64] It is worth pointing out that these $a \to -a$ and $m \to -m$ reflection symmetries also manifest themselves in the analytically derived resonance equation (19) [note that ${}_2F_1(a, b; c; z) \equiv {}_2F_1(b, a; c; z)$].
[65] Here $\lfloor x \rfloor$ denotes the floor function (the largest integer less than or equal to x).
[66] We are not aware of any general proof in the mathematical or physical literature for this interesting property of the hypergeometric functions. However, we have verified numerically the validity of the relation (31) for several values of the arguments of the hypergeometric function ${}_2F_1[l + 1 - \gamma, l + 1; 2l + 2; 1 - x_c(r)]$.
[67] It is worth emphasizing again that, as discussed above (see section 4.1), the resonance equation (19) is invariant under the reflection symmetry $z_c \to -z_c$. That is, if z_c is a solution of the characteristic resonance equation (19), then $-z_c$ is also a valid resonance. Thus, without loss of generality, we consider non-negative surface radii with $z_c \geq 0$. In particular, the physical parameter $z_c^{\text{min}}(\bar{a}, l, m)$ refers to the smallest *non*-negative dimensionless surface radius that can support the static scalar field configurations.
[68] It is worth pointing out that the data presented in Tables 1 and 2 reveal the interesting fact that composed ultra-spinning-exotic-compact-object-massless-scalar-field configurations with an *odd* number of resonances are characterized by the simple relation $z_c^{\text{min}} = 0$. This interesting physical property will be discussed in section 8 below. In particular, from the resonance formula (43) one deduces that a *new* resonant mode (which is characterized by the simple dimensionless relation $z_c = 0$) is added to the discrete set of resonances each time the non-negative integer parameter $[\gamma - (l + 2)]/2$ [see Eq. (13)] increases by one. When the value of $[\gamma - (l + 2)]/2$ is further increased, this new supporting radius splits into *two* distinct supporting radii which are distributed symmetrically around $r = M$ [see the discussion in section 4.1]. Thus, composed ultra-spinning-exotic-compact-object-massless-scalar-field configurations with the property $z_c^{\text{min}} = 0$ are characterized by an odd number of resonances (that is, for these systems, there is one special resonant mode with $z_c = 0$ and an additional even number of resonant supporting radii which are distributed symmetrically around $z = 0$).
[69] See http://link.springer.com/content/pdf/bbm%3A978-3-540-40914-4%2F1.pdf;
F.W.J. Olver, Unsolved problems in the asymptotic estimation of special functions, in: R. Askey (Ed.), Theory and Application of Special Functions, Academic Press, New York, 1975, pp. 99–142.
[70] It is worth emphasizing that the asymptotic expansion (36) of the hypergeometric function ${}_2F_1(a, b; c; z)$ is valid in the regime $|b| \gg \max\{|a|, |c|, |z|\}$ [69]. This strong inequality is satisfied in the near-critical regime $\sqrt{1 - \bar{a}^{-2}} \ll m/l$.
[71] Here we have used the relation $1 = e^{-2i\pi n}$, where the integer n is the resonance parameter of the near-critical composed ultra-spinning-exotic-compact-object-massless-scalar-field configurations.
[72] It is important to stress the fact that the asymptotic expansion (36) of the hypergeometric function ${}_2F_1(a, b; c; y)$ also requires the strong inequality $|by| \gg 1$ [69]. If $\pi(l + 2n)\sqrt{1 - \bar{a}^{-2}}/m \ll 1$ then we find from Eqs. (13), (15), and (38) the simple relation $|by| = O[\pi(l + 2n)]$ (note that $y \equiv 1 - x_c$ in our case), which implies that the asymptotic approximation (36) is valid in the regime $\pi(l + 2n) \gg 1$.

[73] Here we have used the small-x relation $\cot(x) = 1/x + O(x)$.

[74] See Eq. 6.1.7 of [56].

[75] An independent way to deduce the (*finite*) number $N_r(\bar{a}, l, m)$ of discrete supporting radii which characterize the composed ultra-spinning-exotic-compact-object-massless-scalar-field configurations is to use the analytically derived resonance spectrum (43) for the $r_c = M$ case. In particular, from Eq. (43) one finds that a *new* resonant supporting radius [which is characterized by the simple dimensionless relation $z_c = 0$, see Eq. (16)] is added to the discrete family of supporting surface radii each time the composed non-negative integer parameter $[\gamma - (l+2)]/2$ [see Eq. (13)] increases by one. When we further increase the value of $[\gamma - (l+2)]/2$, this new resonant supporting radius splits into *two* distinct supporting radii which, as explicitly proved in section 4.1, are distributed symmetrically around $z = 0$. Thus, the number $N_r(\bar{a}, l, m)$ of discrete supporting radii can be expressed in the compact form $N_r = 2\lfloor[\gamma - (l+2)]/2\rfloor + 1 + \delta$, where $\delta = 0$ for non-negative integer values of the dimensionless physical parameter $[\gamma - (l+2)]/2$ and $\delta = 1$ otherwise. It is worth emphasizing that this analytical formula for the finite number N_r of discrete resonant modes is confirmed by the numerical data presented in Tables 1 and 2.

[76] It is worth emphasizing again that cases 2 and 3 in (48) refer to *non*-integer values of the dimensionless parameter N.

[77] The fact that spinning exotic compact objects in the sub-critical $\bar{a} < 1$ regime are characterized by an infinite set of surface radii, $\{r_c(\bar{a}, l, m; n)\}_{n=1}^{n=\infty}$, that can support the spatially regular static scalar fields can be attributed to the existence of an infinitely blue-shifted surface (a classical horizon which, in our model [17–20], is covered by the external reflecting surface of the corresponding ultra-compact object) at $r = r_+ \equiv M[1 + (1 - \bar{a}^2)^{1/2}]$. In this case, the supported fields are blue-shifted in the near-horizon region in the sense that their radial profile oscillates infinitely many times in the $r/r_+ \to 1^+$ limit (with decreasing wavelengths as the $r/r_+ \to 1^+$ limit is approached). Thus, in the $\bar{a} < 1$ regime, the supported fields are characterized by an *infinite* number of nodes that can approach arbitrarily close to the classical horizon at $r = r_+$. On the other hand, the curved spacetimes of ultra-spinning objects in the complementary regime $\bar{a} > 1$ do not contain an infinitely blue-shifted surface (a covered horizon), and thus the external supported fields are no longer characterized by an infinite set of blue-shifted (arbitrarily dense) resonant nodes.

[78] It is worth pointing out that the ultra-spinning exotic configurations described by the resonance formula (51) with $n = 0$ saturate the bound (45).

[79] That is, $\bar{a} \to 1^+$ in the asymptotic $(l + 2n)/m \to \infty$ limit.

Distillation of scalar exchange by coherent hypernucleus production in antiproton–nucleus collisions

A.B. Larionov [a,b,c,*], H. Lenske [a]

[a] Institut für Theoretische Physik, Universität Giessen, D-35392 Giessen, Germany
[b] National Research Center "Kurchatov Institute", 123182 Moscow, Russia
[c] Frankfurt Institute for Advanced Studies (FIAS), D-60438 Frankfurt am Main, Germany

ARTICLE INFO

Editor: J.-P. Blaizot

Keywords:
$\bar{p} + {}^{40}\text{Ar} \to {}^{40}_{\Lambda}\text{Cl} + \bar{\Lambda}$
Meson-exchange model
κ meson

ABSTRACT

The total and angular differential cross sections of the coherent process $\bar{p} + {}^{A}Z \to {}^{A}_{\Lambda}(Z-1) + \bar{\Lambda}$ are evaluated at the beam momenta $1.5 \div 20$ GeV/c within the meson exchange model with bound proton and Λ-hyperon wave functions. It is shown that the shape of the beam momentum dependence of the hypernucleus production cross sections with various discrete Λ states is strongly sensitive to the presence of the scalar κ-meson exchange in the $\bar{p}p \to \bar{\Lambda}\Lambda$ amplitude. This can be used as a clean test of the exchange by scalar πK correlation in coherent $\bar{p}A$ reactions.

1. Introduction

Light scalar mesons represent one of the most puzzling areas in the quark/hadron physics as their structure is different from $q\bar{q}$. In particular, identifying the κ ($K_0^*(800)$) [1] in various hadronic processes is of importance since this meson is a candidate member of the hypothetical SU(3) octet of scalar mesons with masses below 1 GeV. Apart from $\bar{\kappa}$, the octet-partners of κ are the non-strange isoscalar σ ($f_0(500)$) and isovector δ ($a_0(980)$) mesons. So far the corresponding particles are only seen as broad resonance-like structures in the 0^+-meson spectrum. The σ, δ, κ and $\bar{\kappa}$ are likely the $\pi\pi$, $\pi\eta$, πK and $\pi\bar{K}$ resonance states, respectively. In that sense, the κ exchange can be regarded as an economical way to take into account the correlated πK channel. The κ exchange channel is of particular interest for multi-strangeness baryonic matter in heavy ion collisions and in neutron stars. This channel is also an indispensable part of baryon–baryon interaction approaches utilizing the SU(3) flavour [2] and SU(6) spin-flavour [3] group structures.

The purpose of this letter is to study Λ-hypernucleus production in antiproton–nucleus reactions initiated by the $\bar{p}p \to \bar{\Lambda}\Lambda$ process on the bound proton. The produced Λ is captured to one of the shells of the outgoing hypernucleus. The main emphasis is put on the influence of the inclusion of the κ-meson on the reaction cross section.

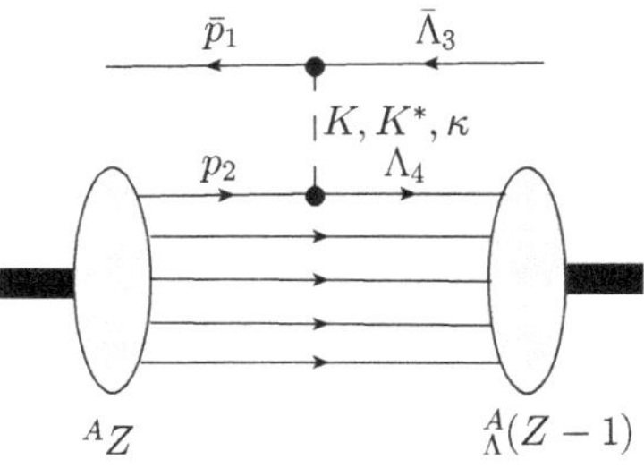

Fig. 1. The Feynman graph of the ${}^{A}Z(\bar{p}, \bar{\Lambda}){}^{A}_{\Lambda}(Z-1)$ process. Dashed line represents the propagator of the exchange meson. The grey ellipsoids correspond to the wave functions of the initial ground state nucleus ${}^{A}Z$ and final hypernucleus ${}^{A}_{\Lambda}(Z-1)$.

The full process is sketched in Fig. 1. At momentum transfers $\lesssim 1$ GeV/c, the amplitude of the $\bar{p}p \to \bar{\Lambda}\Lambda$ reaction can be described by t-channel exchanges of the light mesons with strangeness $|S| = 1$, such as pseudoscalar K, vector K^*, and, probably, hypothetical scalar κ mesons, respectively. If the initial proton and final Λ are bound, the unnatural parity K-exchange is strongly suppressed, since it is pure relativistic effect proceeding through the lower wave function components of either proton or Λ Dirac spinors. Hence the coherent hypernucleus production should filter-out the kaon exchange allowing to exclusively address the exchange of the natural parity K^* and κ. As it will be

* Corresponding author.
E-mail addresses: larionov@fias.uni-frankfurt.de (A.B. Larionov), Horst.Lenske@theo.physik.uni-giessen.de (H. Lenske).

shown, the scalar κ-exchange contribution to the hypernucleus production cross section is very well visible on the background of smoothly growing with beam momentum vector K^*-exchange contribution. Therefore the nuclear binding helps to effectively distillate the κ meson from its mixture with kaon exchange in free-space $\bar{p}p \to \bar{\Lambda}\Lambda$ amplitude.

The paper is structured as follows. In sec. 2 we describe the strange meson exchange model starting from the basic Lagrangians. The amplitude of the hypernucleus production is derived in the impulse approximation (IA) and afterwards corrected for the initial and final state interactions (ISI/FSI) of $\bar{p}$ and $\bar{\Lambda}$ in the nucleus. We introduce two sets of model parameters, both of which describe the momentum dependence of the total $\bar{p}p \to \bar{\Lambda}\Lambda$ cross section. The first set does not include κ-meson exchange and the second set includes it. The wave functions of the bound proton and Λ states are obtained by solving the Dirac equation with relativistic mean fields (RMF). Section 3 contains numerical results. The angular differential and total cross sections of the hypernucleus production with Λ on various shells are calculated. We emphasize that the shape of the momentum dependence of total cross sections is qualitatively different in calculations with and without κ. The summary and outlook are given in sec. 4.

2. The model

We will introduce the K, K^* and κ exchanges by using the following interaction Lagrangians [4–6]:

$$\mathcal{L}_{KN\Lambda} = -ig_{KN\Lambda}\bar{N}\gamma^5\Lambda K + \text{h.c.}\,, \tag{1}$$

$$\mathcal{L}_{K^*N\Lambda} = \bar{N}\left(G_v\gamma^\mu - \frac{G_t\sigma^{\mu\nu}\partial_\nu^{K^*}}{m_N+m_\Lambda}\right)\Lambda K^*_\mu + \text{h.c.}\,, \tag{2}$$

$$\mathcal{L}_{\kappa N\Lambda} = -g_{\kappa N\Lambda}\bar{N}\Lambda\kappa + \text{h.c.}\,. \tag{3}$$

In the case of the bound proton and Λ we include their wave functions in the field operators of the Lagrangians Eqs. (1)–(3) and calculate the S-matrix in the second order perturbation theory using Wick's theorem (cf. Ref. [7]):

$$S = \frac{2\pi\delta(E_1+E_2-E_3-E_4)}{(2E_1V2E_3V)^{1/2}}i(\mathcal{M}_K+\mathcal{M}_{K^*}+\mathcal{M}_\kappa)\,, \tag{4}$$

where E_i, $i=1,2,3,4$ are particle energies (see Fig. 1 for notation) and V is the normalization volume. The K, K^* and κ exchange (noninvariant) partial matrix elements are expressed as

$$\mathcal{M}_K = -g^2_{KN\Lambda}F^2_K(t)\sqrt{\Omega}\,\bar{u}_{-p_1,-\lambda_1}\gamma^5u_{-p_3,-\lambda_3}\frac{1}{t-m_K^2} \times \int d^3r e^{-i\mathbf{qr}}\bar{\psi}_4(\mathbf{r})\gamma^5\psi_2(\mathbf{r})\,, \tag{5}$$

$$\mathcal{M}_{K^*} = -F^2_{K^*}(t)\sqrt{\Omega}\,\bar{u}_{-p_1,-\lambda_1}\Gamma^\mu(-q)u_{-p_3,-\lambda_3}G_{\mu\nu}(q) \times \int d^3r e^{-i\mathbf{qr}}\bar{\psi}_4(\mathbf{r})\Gamma^\nu(q)\psi_2(\mathbf{r})\,, \tag{6}$$

$$\mathcal{M}_\kappa = g^2_{\kappa N\Lambda}F^2_\kappa(t)\sqrt{\Omega}\,\bar{u}_{-p_1,-\lambda_1}u_{-p_3,-\lambda_3}\frac{1}{t-m_\kappa^2} \times \int d^3r e^{-i\mathbf{qr}}\bar{\psi}_4(\mathbf{r})\psi_2(\mathbf{r})\,. \tag{7}$$

Here, $\psi_2(\mathbf{r})$ and $\psi_4(\mathbf{r})$ are the relativistic wave functions of the bound proton and Λ, respectively. They satisfy the normalization conditions:

$$\int d^3r\psi_i^\dagger(\mathbf{r})\psi_i(\mathbf{r}) = 1\,, \quad i=2,4\,. \tag{8}$$

p_i is the four-momentum and $\lambda_i = \pm 1/2$ is the spin projection of an antiproton ($i=1$) and antilambda ($i=3$). $q=p_3-p_1$ is the four-momentum transfer, $t=q^2$. In Eq. (6),

$$G_{\mu\nu}(q) = \frac{-g_{\mu\nu}+q_\mu q_\nu/m^2_{K^*}}{t-m^2_{K^*}} \tag{9}$$

is the K^* meson propagator. We neglected the widths of the K^* and κ mesons in their propagators as the momentum transfers are space-like (e.g. $-t=0.08 \div 1.7$ GeV^2 at $p_{\text{lab}}=2$ GeV/c). The $K^*N\Lambda$ vertex function is defined as

$$\Gamma^\mu(q) = iG_v\gamma^\mu + \frac{G_t}{m_N+m_\Lambda}\sigma^{\mu\nu}q_\nu\,. \tag{10}$$

The vertex form factors are chosen in the monopole form:

$$F_j(t) = \frac{\Lambda_j^2-m_j^2}{\Lambda_j^2-t}\,, \quad j=K, K^*, \kappa\,. \tag{11}$$

Similar to Refs. [8–10] we included in Eqs. (5)–(7) the attenuation factor $\sqrt{\Omega}$ to describe the modification of the elementary $\bar{p}p \to \bar{\Lambda}\Lambda$ amplitude due to ISI in the $\bar{p}p$ channel and FSI in the $\bar{\Lambda}\Lambda$ channel. This effectively takes into account the absorptive $\bar{p}p$ and $\bar{\Lambda}\Lambda$ potentials. For simplicity, we assume the attenuation factor Ω to be energy independent. With $\Omega=1$, Eqs. (5)–(7) correspond to the Born approximation. The Dirac spinors of the antiproton and antilambda plane waves are normalized according to Ref. [7]: $\bar{u}_{-p,-\lambda}u_{-p,-\lambda} = -2m_{N(\Lambda)}$.

The differential cross section in the rest frame of the target nucleus is written as

$$d\sigma = \frac{2\pi\delta^{(4)}(p_1+p_A-p_3-p_B)}{2p_{\text{lab}}} \times \frac{1}{2}\sum_{\lambda_1,m_2,\lambda_3,m_4}|\mathcal{M}_K+\mathcal{M}_{K^*}+\mathcal{M}_\kappa|^2\frac{d^3p_3d^3p_B}{(2\pi)^32E_3}\,, \tag{12}$$

where p_A and p_B are the four-momenta of the initial nucleus (A) and final hypernucleus (B). The δ function in Eq. (12) takes into account the recoil of the hypernucleus. The cross section is summed over the total angular momentum projections m_2 and m_4 of the bound target proton and of the Λ hyperon, respectively, and over the spin projection λ_3 of the outgoing $\bar{\Lambda}$. The averaging is taken over the spin projection λ_1 of the incoming antiproton, as expressed by the factor 1/2.

The choice of coupling constants is based on SU(3) relations [11]:

$$g_{KN\Lambda} = -g_{\pi NN}\frac{3-2\alpha_{PS}}{\sqrt{3}}\,, \tag{13}$$

$$G_{v,t} = -G^\rho_{v,t}\frac{3-2\alpha_{E,M}}{\sqrt{3}}\,, \tag{14}$$

$$g_{\kappa N\Lambda} = -g_{\sigma NN}\frac{3-2\alpha_S}{3-4\alpha_S}\,, \tag{15}$$

where α's are the D-type coupling ratios. The πNN coupling constant is very well known, $g_{\pi NN}=13.4$ [12]. The vector ρNN coupling constant is also fixed, $G^\rho_v=2.66$, however, the tensor ρNN coupling constant is quite uncertain, $G^\rho_t = 10.9 \div 20.6$ [4]. The σNN coupling constant can be estimated either from the Bonn model [13] or from the Walecka-type models (cf. [14]). In both cases one obtains $g_{\sigma NN} \simeq 10$. The α's for the octets of light pseudoscalar and vector mesons are reasonably well determined [4,6]: $\alpha_{PS} \simeq 0.6$, $\alpha_E \simeq 0$, $\alpha_M \simeq 3/4$. However, safe phenomenological information on α_S is lacking.

Table 1
Parameters of the $\bar{p}p \to \bar{\Lambda}\Lambda$ amplitude. The value of $g_{KN\Lambda}$ slightly differs from -13.3 as given by Eq. (13) and is taken from K^+N scattering analysis of Ref. [15]. The cutoff parameters Λ_K, Λ_{K^*} and Λ_κ are in GeV. The attenuation factors are shown in the last column.

Set	$g_{KN\Lambda}$	G_v	G_t	$g_{\kappa N\Lambda}$	Λ_K	Λ_{K^*}	Λ_κ	Ω
1	−13.981	−4.6	−8.5	-	2.0	1.6	-	0.0150
2	−13.981	−4.6	−9.0	−7.5	1.8	2.0	1.8	0.0044

Thus, the coupling constants G_t and $g_{\kappa N\Lambda}$, the cutoff parameters Λ_K, Λ_{K^*} and Λ_κ, and the attenuation factor Ω remain to be determined from comparison with experimental data. We adjusted these parameters to describe the beam momentum dependence of the $\bar{p}p \to \bar{\Lambda}\Lambda$ cross section. The two sets of parameters, (1) without κ meson and (2) with κ meson, are listed in Table 1. In the calculations we used the mass $m_\kappa = 682$ MeV [1].

As we see from Fig. 2, in the calculation with set 1 the peak of the $\bar{p}p \to \bar{\Lambda}\Lambda$ cross section at $p_{\rm lab} \simeq 2$ GeV/c is saturated by the K exchange. In contrast, in the case of set 2 the peak is saturated mostly by the κ exchange. The K^*-exchange contribution grows monotonically with beam momentum and becomes dominant at $p_{\rm lab} > 3 \div 4$ GeV/c. We note that set 2 gives steeper increasing angular differential cross section towards $\Theta_{\rm c.m.} = 0$ at $p_{\rm lab} = 2.060$ GeV/c in a better agreement with experimental data [17] than set 1.

The matrix elements Eqs. (5)–(7) are obtained in the impulse approximation (IA). More realistic calculation should take into account the distortion of the incoming $\bar{p}$ and outgoing $\bar{\Lambda}$ waves, mostly due to strong absorption of the antibaryons in the nucleus. In the eikonal approximation the incoming $\bar{p}$ wave is multiplied by the factor

$$F_{\bar{p}}(\mathbf{r}) = \exp\left(-\frac{1}{2}\sigma_{\bar{p}N}(1 - i\alpha_{\bar{p}N})\int\limits_{-\infty}^{0} d\xi\,\rho(\mathbf{r} + \frac{\mathbf{p}_{\bar{p}}}{p_{\bar{p}}}\xi)\right), \tag{16}$$

and the outgoing $\bar{\Lambda}$ wave is multiplied by

$$F_{\bar{\Lambda}}(\mathbf{r}) = \exp\left(-\frac{1}{2}\sigma_{\bar{\Lambda}N}(1 - i\alpha_{\bar{\Lambda}N})\int\limits_{0}^{+\infty} d\xi\,\rho(\mathbf{r} + \frac{\mathbf{p}_{\bar{\Lambda}}}{p_{\bar{\Lambda}}}\xi)\right), \tag{17}$$

where $\rho(\mathbf{r})$ is the nucleon density, σ_{jN} is the total jN cross section, $\alpha_{jN} = {\rm Re}f_{jN}(0)/{\rm Im}f_{jN}(0)$ is the ratio of the real-to-imaginary part of the forward jN amplitude ($j = \bar{p}, \bar{\Lambda}$). Equations (16), (17) can be obtained by applying the eikonal approximation to solve the Schrödinger equation for the scattering of a particle in the external potential (cf. Ref. [18]) which is then replaced by the optical potential in the low-density approximation. The integrands in the matrix elements Eqs. (5)–(7) are then multiplied by $F_{\bar{p}}(\mathbf{r})F_{\bar{\Lambda}}(\mathbf{r})$ (cf. Refs. [19,20]). In numerical calculations we applied the momentum dependent total $\bar{p}N$ cross section and the ratio $\alpha_{\bar{p}N}$ as described in Ref. [21]. We assumed that $\sigma_{\bar{\Lambda}N} = \sigma_{\bar{p}N}$ at the same beam momenta which is supported by experimental data on the total $\bar{\Lambda}p$ cross section at $p_{\rm lab} = 4 \div 14$ GeV/c [22]. For simplicity we have set $\alpha_{\bar{\Lambda}N} = 0$.

The nucleon and hyperon ($B = N, \Lambda$) single particle bound state wave functions are determined as solutions of a static Dirac equation with scalar and vector potentials (cf. Refs. [23–25]):

$$\left(-i\boldsymbol{\alpha}\cdot\nabla + \beta m_B^*(r) + V_B(r) + V_C(r) - \varepsilon\right)\psi_B(\mathbf{r}) = 0\,, \tag{18}$$

where $m_B^*(r) = m_B + S_B(r)$ is the effective (Dirac) mass. Both the scalar (S_B) and nuclear vector (V_B) potentials are chosen in the form of superpositions of the classical meson fields,

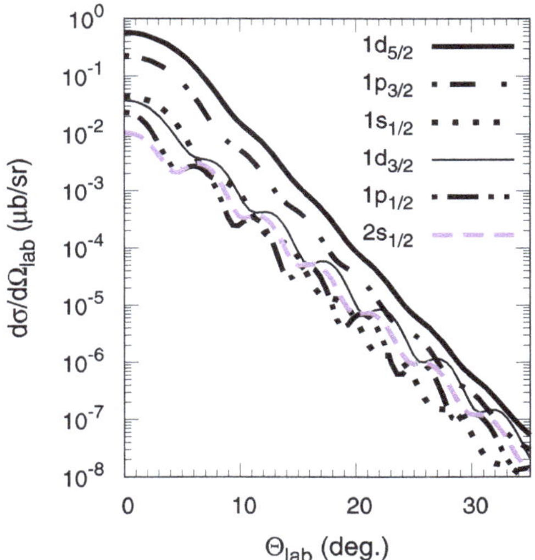

Fig. 3. Angular differential cross section of the reaction ^{40}Ar($\bar{p}$, $\bar{\Lambda}$)$^{40}_{\Lambda}$Cl at $p_{\rm lab} = 2$ GeV/c. Lines show the calculations for Λ in various states, as indicated. Calculations include κ exchange.

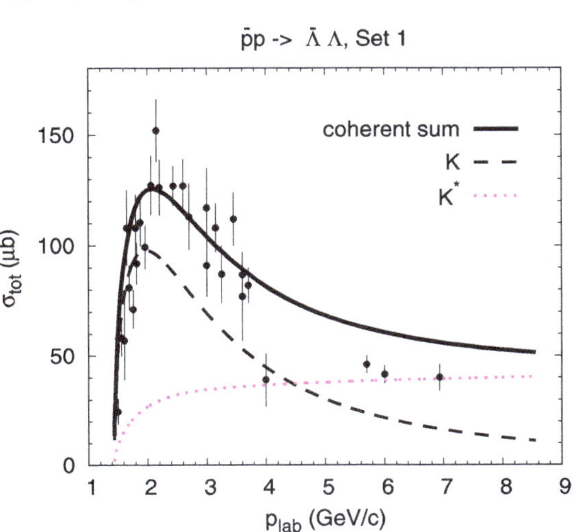

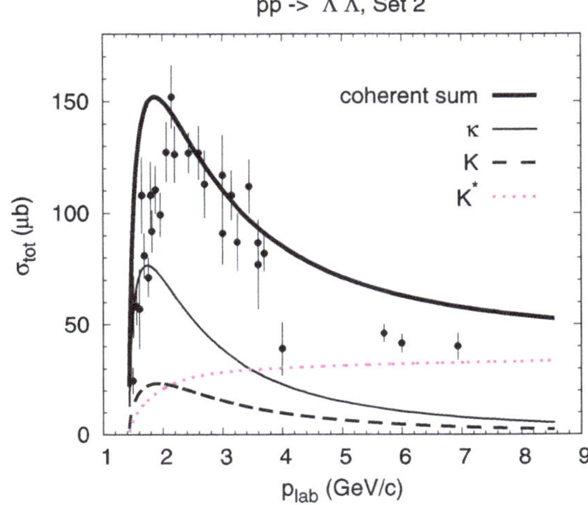

Fig. 2. Angle-integrated cross section of the process $\bar{p}p \to \bar{\Lambda}\Lambda$ as a function of the beam momentum calculated without (Set 1) and with (Set 2) inclusion of the κ meson. Full calculations are shown by thick solid lines. Other lines, as indicated, show the partial contributions of the K, K^* and κ exchanges. Experimental data are from Ref. [16].

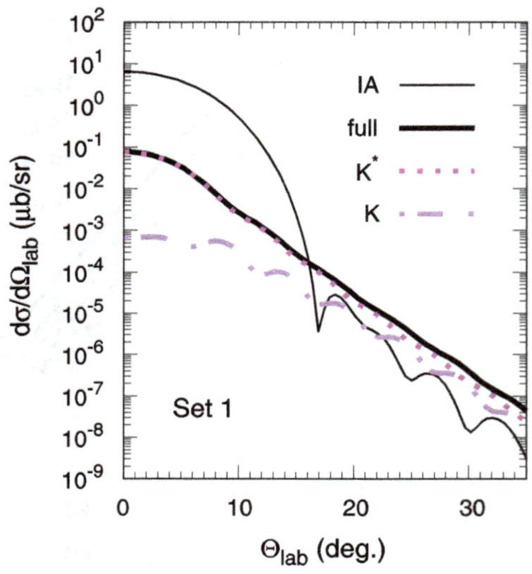

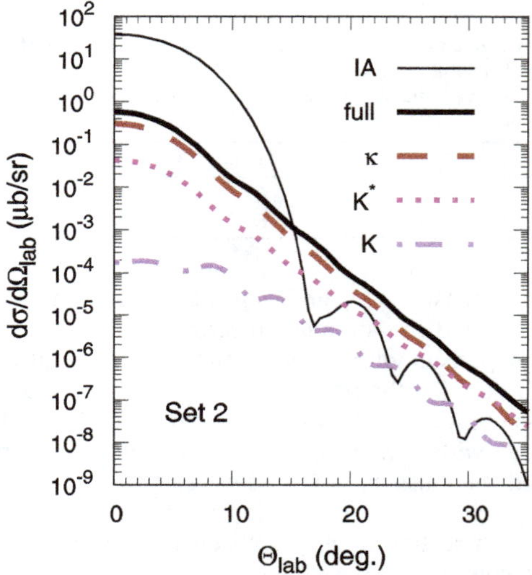

Fig. 4. Angular differential cross section of the reaction ${}^{40}\mathrm{Ar}(\bar{p},\bar{\Lambda}){}^{40}_{\Lambda}\mathrm{Cl}$ at $p_{lab}=2$ GeV/c with $1d_{5/2}$ Λ state. The IA calculation, the full calculation (with absorption), and the separate meson contributions to the full calculation are shown by different lines. Results without (Set 1, left panel) and with (Set 2, right panel) κ meson are displayed as indicated.

$\sigma(I=0, J^P=0^+)$, $\omega(0,1^-)$, $\delta(1,0^+)$ and $\rho(1,1^-)$, weighted by the strong interaction coupling constants appropriate for the given baryon. The meson fields are parameterized by Woods–Saxon form factors. For protons also the static Coulomb potential (V_C) contributes [25]. We will consider the case of a spherical nucleus. Hence the eigenfunctions of the Dirac equation are characterized by radial, orbital and total angular momentum quantum numbers, n, l, j, respectively, and the magnetic quantum number $m \equiv j_z$.

In the future P̄ANDA experiment at FAIR the noble gases will be used as targets. Thus as a representative case we consider the reaction $\bar{p} + {}^{40}\mathrm{Ar} \to {}^{40}_{\Lambda}\mathrm{Cl} + \bar{\Lambda}$. Using the RMF approach the parameters of the Woods–Saxon form factors have been chosen to fit the binding energy, single-particle separation energies, and the root-mean-square radii of the nucleon density distributions in the ^{40}Ar nucleus. Under the assumption that the nuclear potentials do not change after a sudden removal of the valence proton, the Λ-hyperon scalar and vector potentials in the ${}^{40}_{\Lambda}$Cl hypernucleus were obtained by multiplying the scalar and vector nucleon potentials in the ^{40}Ar nucleus by the factors 0.525 and 0.550, respectively. This leads to a good agreement of the Λ energy levels with the empirical systematics and with the previous relativistic mean-field calculations [25]. In order to assure that after the reaction the residual core nucleus carries as little excitation energy as possible, we consider only strangeness creation processes on protons of the $1d_{3/2}$ valence shell in ^{40}Ar.

3. Numerical results

The differential hypernuclear production cross sections with the Λ occupying various shells are displayed in Fig. 3. Irrespective of spin-orbit effects,[1] the cross sections are larger for larger hyperon orbital angular momentum, l_Λ. We have checked that within the IA the cross sections at $\Theta_{lab}=0$ for the $1d_{5/2}$, $1p_{3/2}$ and $1s_{1/2}$ hyperon states are nearly equal. Thus the rise of the cross sections at forward laboratory angles with l_Λ is mostly caused by the nuclear absorption which is diminished with increasing l_Λ due to the shift of the hyperon density distribution to larger radii. The largest cross section is obtained for the ${}^{40}_{\Lambda}$Cl hypernucleus with Λ in the $1d_{5/2}$ state. The differential angular distribution for this case is analyzed in more detail in Fig. 4. From the comparison of the full and IA calculations we observe that the absorption of $\bar{p}$ and $\bar{\Lambda}$ has a quite significant effect: it reduces the cross section drastically, amounting at forward angles to about two orders of magnitude, and smears out the diffractive structures. Similar effects of the absorption are present also for the other Λ states.

A deeper insight into the production mechanism is obtained by decomposing the total reaction amplitude into different meson exchange parts. From the partial meson exchange contributions, shown in Fig. 4, it is remarkable that for Set 1 the kaon contribution is small and the spectrum is dominated by K^*, however, from Fig. 2 one would expect the opposite. This surprising result can be understood by the fact that the momentum transfer to the $\bar{\Lambda}$ is provided by the nucleus as a whole while the produced Λ is almost at rest. The exchange by pseudoscalar meson is suppressed in this case since it proceeds through the lower components of the proton and hyperon Dirac spinors[2] which are suppressed by factors $\sim 1/m_B R$, where R is the nuclear radius. In contrast, in the case of the free space $\bar{p}p \to \bar{\Lambda}\Lambda$ process the Λ is produced with finite momentum. Therefore the upper and lower components of its Dirac spinor are of comparable magnitude which favours the pseudoscalar meson exchange.

The situation is very different in the case of Set 2. Here, κ plays the dominant role both for the free scattering $\bar{p}p \to \bar{\Lambda}\Lambda$ and for the hypernucleus production since the scalar exchange is not suppressed in the recoilless kinematics.

As we see from Fig. 5, the cross section of coherent hypernucleus production in different states is much larger when the κ exchange is included. This is pure quantum coherence effect since the angle-integrated $\bar{p}p \to \bar{\Lambda}\Lambda$ cross sections differ by $\sim 15\%$ only at $p_{lab}=2$ GeV/c (Fig. 2) while the hypernuclear production cross sections differ by almost one order of magnitude for Set 1 and Set 2.

The robust signal of the κ exchange is visible in the beam momentum dependence of the hypernucleus production cross section shown in Fig. 6. In calculations without κ the K^* exchange

[1] There is a strong difference between the cross sections with $j_\Lambda = l_\Lambda \pm 1/2$, so that the cross sections for the larger value of j_Λ are much larger at forward angles. This effect arises from the detailed structure of the Fourier transforms of the transition form factors in the matrix elements Eqs. (5)–(7) and its study is beyond the scope of this work.

[2] Formally this is due to the γ^5 matrix in the integrand of Eq. (5).

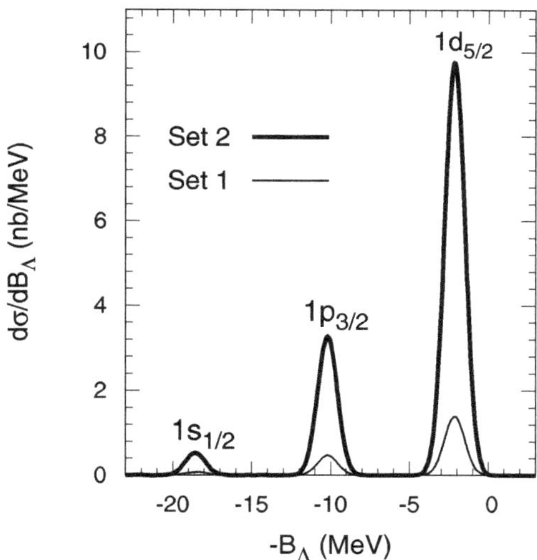

Fig. 5. Spectral distribution of Λ bound states in a ${}^{40}_{\Lambda}$Cl hypernucleus produced in coherent $\bar{p}\,{}^{40}$Ar collisions at $p_{\text{lab}} = 2$ GeV/c. The angle-integrated cross sections for the hypernucleus production in $1s_{1/2}$, $1p_{3/2}$ and $1d_{5/2}$ states have been folded by Gaussians of width $\Gamma_{FWHM} = 1.5$ MeV which is a typical experimental energy resolution. Thick and thin solid lines correspond to the results obtained with (Set 2) and without (Set 1) κ exchange, respectively.

dominates and produces a smoothly growing cross section with increasing beam momentum. The κ meson dominates at moderate beam momenta $\sim 1.5 \div 3$ GeV/c (Fig. 2). This leads to the characteristic shoulder in p_{lab}-dependence of the cross section of the hypernucleus production and even to the appearance of the maximum for the $1d_{5/2}$ Λ state. We have checked that within the IA the maximum in the beam momentum dependence in the calculation including κ becomes even more pronounced. Hence this maximum is a clean manifestation of the κ exchange and not an artifact of particular approximation for the ISI/FSI effects.

4. Summary and outlook

In the present work, the calculations of the coherent hypernucleus production in $\bar{p}A$ collisions have been performed. We have demonstrated that the pseudoscalar K exchange is strongly suppressed for the bound Λ states. Thus the hypernucleus production

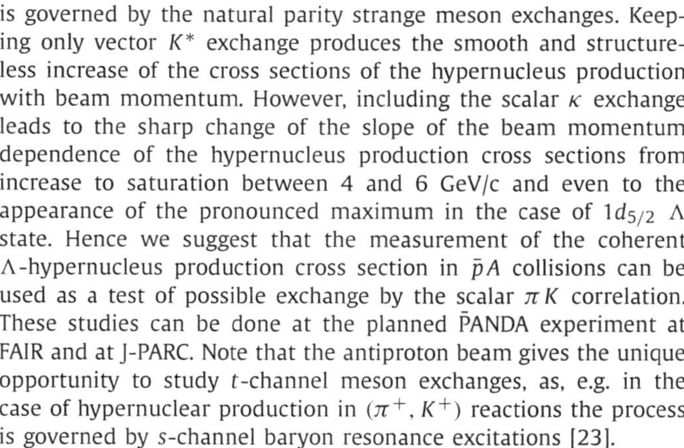

is governed by the natural parity strange meson exchanges. Keeping only vector K^* exchange produces the smooth and structureless increase of the cross sections of the hypernucleus production with beam momentum. However, including the scalar κ exchange leads to the sharp change of the slope of the beam momentum dependence of the hypernucleus production cross sections from increase to saturation between 4 and 6 GeV/c and even to the appearance of the pronounced maximum in the case of $1d_{5/2}$ Λ state. Hence we suggest that the measurement of the coherent Λ-hypernucleus production cross section in $\bar{p}A$ collisions can be used as a test of possible exchange by the scalar πK correlation. These studies can be done at the planned P̄ANDA experiment at FAIR and at J-PARC. Note that the antiproton beam gives the unique opportunity to study t-channel meson exchanges, as, e.g. in the case of hypernuclear production in (π^+, K^+) reactions the process is governed by s-channel baryon resonance excitations [23].

With appropriate extensions the methods used here can be applied to the production of hyperons in low-energy unbound states, thus allowing in principle to explore elastic scattering of Λ and Σ hyperons on nuclei. Experimentally, such reactions could be identified e.g. by observation of the recoiling target fragment of mass number $A - 1$. This may also help to clarify further the still undecided question whether there are bound Σ hypernuclei.

Of large interest is to establish the existence of the $\bar{\Lambda}$-hypernuclei which can be produced in the process $(\bar{p}, \Lambda)$ with the capture of $\bar{\Lambda}$ in the residual nucleus. However, that production process requires a large momentum transfer to the struck proton (backward scattering). In this case, the description based on the reggeized t-channel meson exchange model should be more appropriate for the elementary $\bar{p}p \to \bar{\Lambda}\Lambda$ amplitude.

Charmed hadrons embedded in- or interacting with nuclei represent another related field of studies. The coherent Λ_c^+-hypernuclei production in $(\bar{p}, \bar{\Lambda}_c^-)$ processes has been explored theoretically recently in Ref. [26]. The underlying $\bar{p}p \to \bar{\Lambda}_c^- \Lambda_c^+$ reaction on the bound proton has been described with a t-channel exchange by pseudoscalar D^0 and vector D^{*0} [9]. Overall, the uncertainties in the charm sector are quite large due the lack of experimental data on the cross sections of the elementary production processes and on nuclear FSI of the emitted charmed hadrons. In particular, also the exchange by the scalar $D_0^*(2400)$ meson is possible in this case and can be distilled by using the Λ_c^+-hypernuclei production in a similar mechanism.

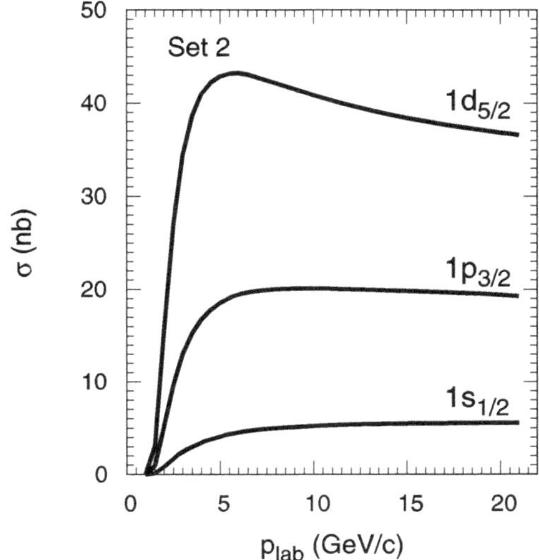

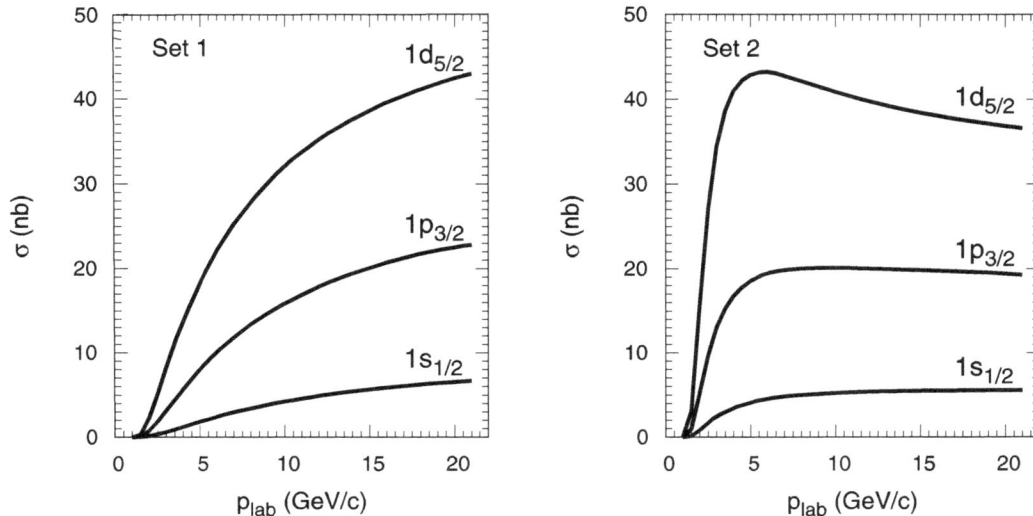

Fig. 6. Beam momentum dependence of the ${}^{40}_{\Lambda}$Cl hypernucleus production cross section in $\bar{p}\,{}^{40}$Ar collisions for Λ in $1s_{1/2}$, $1p_{3/2}$ and $1d_{5/2}$ states. The total cross sections without (Set 1) and with (Set 2) κ exchange are shown in the left and the right panel, respectively.

Acknowledgements

This work was supported by the Deutsche Forschungsgemeinschaft (DFG) under Grant No. Le439/9. Stimulating discussions with M. Bleicher and M. Strikman are gratefully acknowledged.

References

[1] C. Patrignani, et al., Particle Data Group, Chin. Phys. C 40 (2016) 100001.
[2] R.G.E. Timmermans, T.A. Rijken, J.J. de Swart, Phys. Rev. D 45 (1992) 2288.
[3] J. Haidenbauer, U.-G. Meissner, Phys. Rev. C 72 (2005) 044005.
[4] M.K. Cheoun, B.S. Han, I.T. Cheon, B.G. Yu, Phys. Rev. C 54 (1996) 1811.
[5] K. Tsushima, A. Sibirtsev, A.W. Thomas, G.Q. Li, Phys. Rev. C 59 (1999) 369, Erratum: Phys. Rev. C 61 (2000) 029903.
[6] B.S. Han, M.K. Cheoun, K.S. Kim, I.-T. Cheon, Nucl. Phys. A 691 (2001) 713.
[7] V.B. Berestetskii, E.M. Lifshitz, L.P. Pitaevskii, Relativistic Quantum Theory, Pergamon Press, 1971.
[8] N.J. Sopkovich, Nuovo Cimento 26 (1962) 186.
[9] R. Shyam, H. Lenske, Phys. Rev. D 90 (2014) 014017.
[10] R. Shyam, H. Lenske, Phys. Rev. D 93 (2016) 034016.
[11] J.J. de Swart, Rev. Mod. Phys. 35 (1963) 916, Erratum: Rev. Mod. Phys. 37 (1965) 326.
[12] O. Dumbrajs, R. Koch, H. Pilkuhn, G.c. Oades, H. Behrens, J.j. De Swart, P. Kroll, Nucl. Phys. B 216 (1983) 277.
[13] R. Machleidt, K. Holinde, C. Elster, Phys. Rep. 149 (1987) 1.
[14] G.A. Lalazissis, J. König, P. Ring, Phys. Rev. C 55 (1997) 540.
[15] R. Büttgen, K. Holinde, A. Müller-Groeling, J. Speth, P. Wyborny, Nucl. Phys. A 506 (1990) 586.
[16] A. Baldini, et al., Landolt–Börnstein, vol. 12, Springer, Berlin, 1987.
[17] B. Jayet, M. Gailloud, P. Rosselet, V. Vuillemin, S. Vallet, M. Bogdanski, E. Jeannet, C.J. Campbell, J. Dawber, D.N. Edwards, Nuovo Cimento A 45 (1978) 371.
[18] L.D. Landau, E.M. Lifshitz, Quantum Mechanics, Pergamon Press, 1965.
[19] H. Bando, T. Motoba, J. Zofka, Int. J. Mod. Phys. A 5 (1990) 4021.
[20] L.L. Frankfurt, M.I. Strikman, M.B. Zhalov, Phys. Rev. C 50 (1994) 2189.
[21] A.B. Larionov, H. Lenske, Nucl. Phys. A 957 (2017) 450.
[22] F. Eisele, et al., Phys. Lett. B 60 (1976) 297.
[23] S. Bender, R. Shyam, H. Lenske, Nucl. Phys. A 839 (2010) 51.
[24] N.K. Glendenning, D. Von-Eiff, M. Haft, H. Lenske, M.K. Weigel, Phys. Rev. C 48 (1993) 889.
[25] C.M. Keil, F. Hofmann, H. Lenske, Phys. Rev. C 61 (2000) 064309.
[26] R. Shyam, K. Tsushima, Phys. Lett. B 770 (2017) 236.

Casimir scaling and Yang–Mills glueballs

Deog Ki Hong [a,*], Jong-Wan Lee [a], Biagio Lucini [b], Maurizio Piai [b], Davide Vadacchino [b]

[a] Department of Physics, Pusan National University, Busan 46241, Republic of Korea
[b] College of Science, Swansea University, Singleton Park, Swansea SA2 8PP, UK

ARTICLE INFO

Editor: B. Grinstein

Keywords:
Glueballs
Yang–Mills theories
Confinement
Casimir scaling

ABSTRACT

We conjecture that in Yang–Mills theories the ratio between the ground-state glueball mass squared and the string tension is proportional to the ratio of the eigenvalues of quadratic Casimir operators in the adjoint and the fundamental representations. The proportionality constant depends on the dimension of the space-time only, and is henceforth universal. We argue that this universality, which is supported by available lattice results, is a direct consequence of area-law confinement. In order to explain this universal behavior, we provide three analytical arguments, based respectively on a Bethe–Salpeter analysis, on the saturation of the scale anomaly by the lightest scalar glueball and on QCD sum rules, commenting on the underlying assumptions that they entail and on their physical implications.

1. Introduction

Yang–Mills (YM) theories without matter fields are believed to exhibit a confining phase at low energies, in which all bound states (glueballs) are gapped and color-singlet. Confinement in YM theories is supported by lattice studies [1]. However, since glueballs are nonperturbative objects, we do not have yet good understanding of the properties of glueballs such as their mass spectrum or decay widths.

It has been suggested that color confinement can be described in terms of a dual Higgs mechanism or monopole condensation [2–4]. In this picture, monopoles, dual to color charges, condense in the color-confined phase, and 't Hooft operators develop a vacuum expectation value. The dynamical scale κ is set by the condensate, which should be responsible for all other dimensional quantities in the confined phase. Monopole condensation implies a linear potential between a pair of static color charges, or equivalently an area law for the Wilson loop.

In this letter we provide theoretical arguments and numerical evidence for the existence of a new universal law. The law states that the ratio of ground state glueball mass squared and the string tension is universally proportional to the ratio of the eigenvalues of quadratic Casimir operators for all confining gauge theories. The proportionality constant is independent of the gauge group and the strength of coupling as long as the area law arises. It depends only on the dimensionality of the space-time.

* Corresponding author.
E-mail address: dkhong@pusan.ac.kr (D.K. Hong).

2. Glueball mass

Calculating the ground state glueball mass is tantamount to showing that there is a gap in the ground state of pure YM theory, which has never been proved analytically except in three dimensions [5]. Numerical calculations of glueball masses on the lattice show the existence of a gap in YM theories [6–13].

Asymptotically, for confining YM theories, the expectation value of rectangular Wilson loops $\mathcal{C}$ can be written as

$$\langle W(\mathcal{C})\rangle = \left\langle \frac{1}{N} e^{i \oint_{\mathcal{C}} A} \right\rangle = \exp\left[-\sigma LT + \cdots\right], \tag{1}$$

where LT is the area of $\mathcal{C}$, σ is the string tension between a static quark–antiquark pair, and the ellipsis includes subleading corrections such as the Lüscher term. Following the area law confinement, we write the string tension σ as to define κ via the proportionality to the quadratic Casimir operator on the fundamental representation

$$\sigma = \kappa^2 C_2(F), \tag{2}$$

which is consistent with lattice results [8,14–18]. The glueball is a bound state of adjoint gluons. On dimensional grounds, its mass should be proportional to κ. For the ground state glueball we conjecture

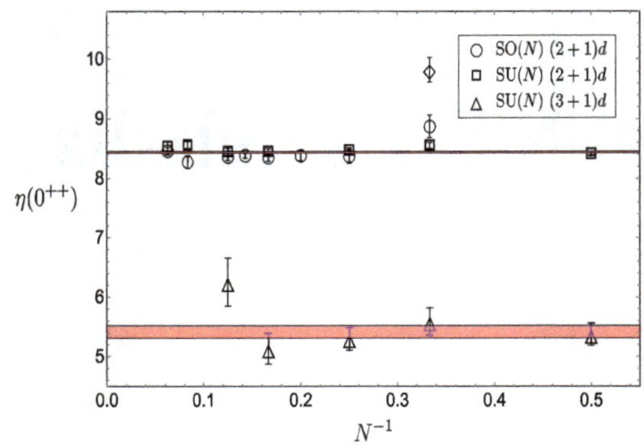

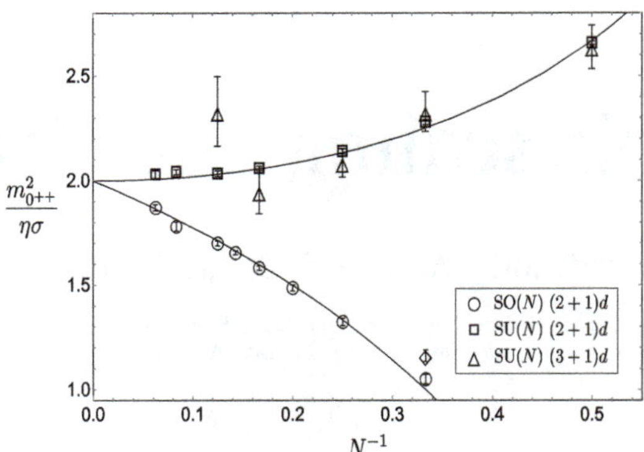

Fig. 1. The universal ratio η (left panel), and glueball masses squared in units of the string tension (right panel), for various YM theories as a function of $1/N$. The solid curves are the Casimir ratio $C_2(A)/C_2(F)$ for SU(N) (upper curve) and SO(N) (lower curve), respectively. The value of η from the tension of the SO(3) fundamental string is marked as $\diamond$.

$$m^2_{0^{++}} = \eta\,\kappa^2\,C_2(A)\,, \tag{3}$$

where η is a universal ratio and $C_2(A)$ the quadratic Casimir for the adjoint representation. The existence of the universal ratio η is consistent with the large-N universality of YM theories, supported by Wilson loop calculations [19] and gauge-gravity dualities [20]. At finite N, the ratio of the eigenvalues of the relevant quadratic Casimir operators is [21]

$$\frac{C_2(A)}{C_2(F)} = \begin{cases} \frac{2N^2}{N^2-1} & \text{for SU}(N) \\ \frac{2(N-2)}{N-1} & \text{for SO}(N) \\ \frac{4(N+1)}{2N+1} & \text{for Sp}(2N)\,, \end{cases} \tag{4}$$

and approaches 2 in the large-N limit.

Glueball masses and string tensions have been calculated by various collaborations for YM theories in $3+1$ and $2+1$ dimensions [6–13]. From the continuum-extrapolated lattice results of glueball mass and string tension, taking the data from the most recent large-N calculations available in the literature [8,11,13] (Fig. 1), we find[1]

$$\eta(0^{++}) \equiv \frac{m^2_{0^{++}}}{\sigma}\cdot\frac{C_2(F)}{C_2(A)} = \begin{cases} 5.41(12), & (d=3+1)\,, \\ 8.440(14)(76), & (d=2+1)\,. \end{cases} \tag{5}$$

For $3+1$ dimensions Eq. (5) is the constant fit of SU(N) results over $2\le N\le 8$, with $\chi^2/\text{d.o.f.}\simeq 1$. For $2+1$ dimensions, lattice results are available for SU(N), as well for SO(N), with $2\le N\le 16$, hence we performed a constant fit for the universal ratio η of both data sets.[2] The resulting statistical error is quoted in the first parentheses in Eq. (5), with somewhat larger value of $\chi^2/\text{d.o.f.}\simeq 1.9$.[3]

Deviations from universality in $2+1$ dimensions between two classes of gauge groups are assessed by calculating η separately. We find $\eta = 8.386(25)$ ($\chi^2/\text{d.o.f.}\simeq 1.3$) for SO($N$) and $\eta = 8.462(16)$ ($\chi^2/\text{d.o.f.}\simeq 1.9$) for SU($N$). Given the expectation that the large-N limit of the two sets should coincide, this difference of 3σ level is probably due to the systematic errors in the lattice data. We account for the discrepancy with a systematic error reported in the second parenthesis in Eq. (5). We also studied two heavier states, the 2^{++} glueball and the first excited scalar glueball, 0^{*++}. The excited states start to see the deviation from the area-law confinement, hence it is not surprising that the 0^{*++} does not show universal behavior. (See Fig. 2.) For the 2^{++}, however, it is inclusive, because the constant fit gives a poor $\chi^2/\text{d.o.f.}\simeq 19$ for the 2^{++} tensor glueballs in $2+1$ dimensions, while it fits much better in $3+1$ dimensions with $\chi^2/\text{d.o.f.}\simeq 1.1$.

3. Glueball mass and Casimir scaling

Motivated by the strong numerical evidence for Casimir scaling, we provide three analytical arguments to explain its origin. None of the arguments is fully conclusive, as they all rely on specific dynamical assumptions that we highlight explicitly, yet the picture that emerges is that Casimir scaling of ground state mass should capture much of the essence of the confinement properties of YM theories.

3.1. Bethe–Salpeter equation

The amplitude for creating two gluons out of vacuum to form a color-singlet bound state of momentum P with a polarization λ can be defined as

$$\Gamma_R^{\mu\nu}(x_1,x_2;P,\lambda) = \langle 0|\,\mathrm{T}A^{\mu\,a}(x_1)A^{\nu\,a}(x_2)\,|R(P,\lambda)\rangle\,, \tag{6}$$

where T denotes the time-ordered product and $\langle 0|$ is the vacuum. Summation over color indices a is understood.

The bound state amplitude satisfies the Bethe–Salpeter (BS) equations, obtained from the gluon four-point scattering amplitude near the pole, which are diagrammatically shown for the amputated BS amplitude in Fig. 3.

From the BS equation, the scalar (amputated) amplitude χ_P obeys, in Euclidean space,

$$\left[\partial^2 - P^2\right]\chi_P(x) = \int \mathrm{d}^4y\, V(x-y)\chi_P(y)\,, \tag{7}$$

with $x = x_1 - x_2$ the displacement of two external gluons.

The area law for confinement is associated with the Regge behavior of the spectrum: $M_n^2 \sim n$, where $n = 1,2,\cdots$ are the radial quantum numbers, reproduced by the approximate BS kernel

$$V(x-y) \approx \frac{1}{2}\omega^2 x^2\,\delta^4(x-y)\,. \tag{8}$$

---

[1] Our conjecture for the universal ratio is also supported by the analytic calculation of the ground-state glueball mass in $2+1$ dimensional SU(N) gauge theories [22], which finds $\eta(0^{++})\simeq 8.41$, and suspected in the constituent gluon model in [23].

[2] The string tension can be defined also for SO(3) by considering distances of the order of the confinement scale. Yet, it is affected by large systematic uncertainties due to its instability [11,13]. To mitigate the systematics, instead of this quantity, we use the string tension obtained from the fundamental of SU(2), assuming Casimir scaling for the string tension. We checked that by using the measured value of the string tension of SO(3), the value of η does not change but yields a poor $\chi^2/\text{d.o.f}\simeq 4.8$.

[3] The χ^2 distribution does not improve significantly, even if the data for the lowest N is excluded.

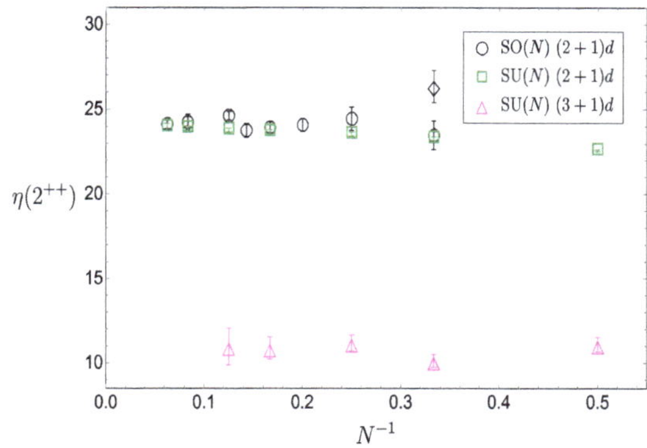

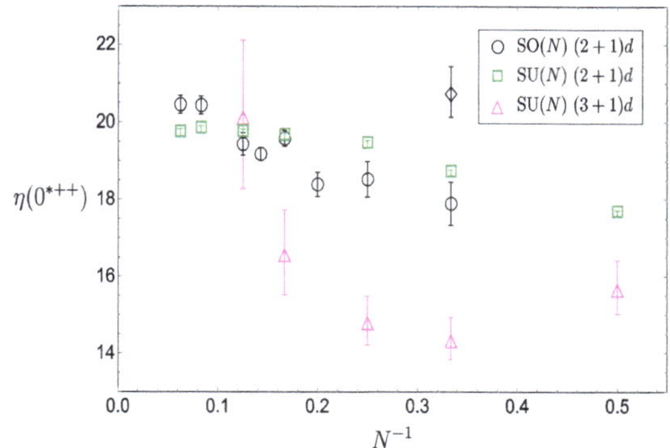

Fig. 2. The ratio η for the lowest-lying 2^{++} and 0^{*++} (first excitation in the scalar channel) as a function of $1/N$. The value of η from the tension of the SO(3) fundamental string is marked as $\diamond$.

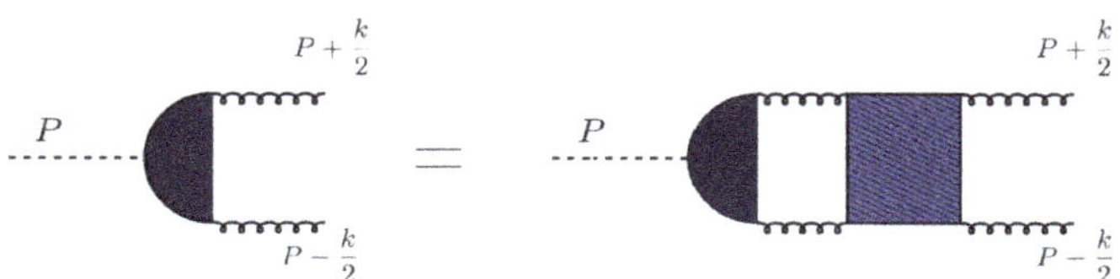

Fig. 3. The BS equation for the glueballs. The half disk denotes the BS amplitude of momentum P and the relative momentum k of two gluons. The box denotes the BS kernel.

The BS kernel is nothing but the four-point function of gluons, properly projected for the spin-0 state. If the string flux picture holds for the glueball states, ω should be the string tension of the Nambu–Goto action for the closed string that describes glueballs, $\omega \sim \sigma(A) = C_2(A)\kappa^2$ (see for example Eq. 2.26 in [24]). The radially excited scalar glueball mass is then (for $n = 1, 2\cdots$)

$$M_n^2 \sim C_2(A)\kappa^2\,(n+1)\,. \tag{9}$$

Since the string tension is $\sigma = \kappa^2 C_2(F)$, for the mass of ground state ($n = 1$) glueball we find

$$\frac{m_{0^{++}}^2}{\sigma} = \eta\,\frac{C_2(A)}{C_2(F)}\,. \tag{10}$$

There are corrections to Casimir scaling, coming from the corrections to the area law in Eq. (1). But such corrections are suppressed, arising at the next-to-next-to-leading order. Namely, the Lüscher term in the expectation value of the Wilson loop in (1) does not modify Casimir scaling, Eq. (10), since the Lüscher term is a universal number [25] correcting the BS kernel by a shift itself proportional to the Casimir; for $|x| \gg \kappa^{-1}$

$$\frac{1}{2}\left(C_2(A)\kappa^2 x - \frac{\alpha}{x}\right)^2 \approx \frac{1}{2}C_2(A)^2\kappa^4 x^2 - \alpha\, C_2(A)\kappa^2\,, \tag{11}$$

where $\alpha = (D-2)\pi/24$ is the universal coefficient of the Lüscher term in D dimensions. The ground state glueball mass then is corrected as

$$m_{0^{++}}^2 \sim C_2(A)\kappa^2\,(2-\alpha)\,, \tag{12}$$

which does not change the universal scaling law.

The corrections become more important at high energies (short distances), in particular for the excited states, for which we expect violations of the Casimir scaling to show. As discussed in following, the characteristic behavior of the 0^{++} ground state may be understood in terms of its special role with respect to the scale symmetry of the system.

3.2. Scale anomaly

Pure Yang–Mills theories in four dimensions are classically scale invariant, but scale symmetry is anomalous, broken by quantum effects. Furthermore, the (anomalous) scale symmetry in YM theory is spontaneously broken as well, since the YM vacuum develops a non-vanishing expectation value for the order parameter for confinement. If the scale anomaly is parametrically small, compared to such vacuum expectation value of the order parameter, then there should be a (pseudo) Nambu–Goldstone boson in YM theory, associated with spontaneous breaking of the scale symmetry. Namely, by Goldstone's theorem, the dilatation current, associated with the scale symmetry $x^\rho \to e^\lambda x^\rho$, creates a state, called dilaton:

$$\langle 0|\, D^\mu(x)\,|D(p)\rangle = -if_D\, p^\mu e^{-ip\cdot x}\,, \tag{13}$$

where the dilatation current $D_\mu = x^\nu\theta_{\mu\nu}$ with the improved energy–momentum tensor $\theta_{\mu\nu}$ [26] and f_D is the dilaton decay constant.

The scale anomaly in pure YM theory is given as

$$\partial^\mu D_\mu = -\frac{\beta(g)}{2g}\,F^a_{\mu\nu}F^{a\,\mu\nu}\,, \tag{14}$$

where $\beta(g)$ is the beta function and $F^a_{\mu\nu}$ the field-strength tensor. Since the divergence of the dilatation current can be written in terms of the trace of the energy–momentum tensor as $\partial_\mu D^\mu = \theta^\mu_\mu$, the anomalous Ward identity (14) relates the two-point function of the trace of the energy–momentum tensor to its one-point function or the scale anomaly:

$$\int_x \langle 0|\,\mathrm{T}\theta^\mu_\mu(x)\theta^\nu_\nu(y)\,|0\rangle = -4\left\langle\theta^\mu_\mu(y)\right\rangle. \tag{15}$$

As all the gluons equally and additively contribute to the vacuum energy, the scale anomaly should be given on dimensional ground as $\langle\theta^\mu_\mu\rangle = -\tilde\beta\, C_2(A)\kappa^4$, after subtracting out the part that is independent of the condensate. If the scale anomaly is parametrically small or $|\tilde\beta| \ll 1$, there should be a light dilaton, defined as in Eq. (13), that saturates the two-point function in (15):

$$\int_x \langle 0|\,\mathrm{T}\theta^\mu_\mu(x)\theta^\nu_\nu(y)\,|0\rangle \approx f_D^2 m_D^2\,. \tag{16}$$

We then have a so-called partially conserved dilatation current (PCDC) relation,

$$f_D^2 m_D^2 = -4\left\langle\partial_\mu D^\mu\right\rangle = -16\,\mathcal{E}_{\rm vac}\,, \tag{17}$$

where $\mathcal{E}_{\rm vac} = -\tilde{\beta}\, C_2(A)\kappa^4/4$ is the vacuum energy density of YM theories in the confined phase. The vacuum energy (density) scales as $C_2(A)$. On the other hand the dilaton decay constant, f_D, measures the strength of the amplitude that creates the dilaton out of vacuum, which should not depend on the number of gluon fields, but only on the characteristic scale κ that defines the scale of spontaneous scale-symmetry breaking. We hence find the ground state glueball mass $m_{0^{++}}^2 \propto \tilde{\beta}\, C_2(A)\kappa^2$, if identified as dilaton, and it becomes parametrically small if $\tilde{\beta} \ll 1$.

The assumption about saturation of the Ward identity by the lightest 0^{++} state, Eq. (16), is equivalent to assuming the existence of a weakly-coupled low-energy effective field theory for the 0^{++} state in terms of the dilaton field, in spite of its mass not being particularly small, compared to other excited states like 0^{*++} glueball states, which implies $\tilde{\beta} \sim 1$. The fact that Casimir scaling holds for the ground state glueball but not for excited states (as hinted also by lattice calculations) is therefore quite intriguing, and very distinctive from analysis based on other approaches that do not differentiate the lightest state.

3.3. Sum rules

The glueball mass can be extracted from the correlators of interpolating operators made of gluons. For scalar glueballs one considers the correlator of the gluonic field strength tensor $\mathcal{O}_S(x) \equiv \alpha_s F^a_{\mu\nu} F^{a\,\mu\nu}$:

$$\Pi_S(x) = \langle 0|\, \mathrm{T}\left[\mathcal{O}_S(x)\mathcal{O}_S(0)\right] |0\rangle = \sum_n c_n\, e^{-m_n|x|} \,, \tag{18}$$

where T is the time-ordering operation and the smallest m_n will be the mass of ground state glueball 0^{++}.

The sum rules, associated with the moments of the correlators, exploit the operator production expansion. For the zero moment, one finds [27]

$$\int \mathrm{d}^4x\, \Pi_S(x) = \frac{32\pi^2}{b} \langle 0|\, \frac{\alpha_s}{\pi} F^a_{\mu\nu} F^{a\,\mu\nu}\, |0\rangle \,, \tag{19}$$

where b is the first coefficient of the beta function and the integral is regularized by subtracting out the perturbative contributions. Assuming single-particle states to be stable and inserting a complete set between the interpolating operators in (18), we have

$$\sum_{n=0}^{\infty} f_n^2\, m_n^2 = \frac{32\pi^2}{b} \langle 0|\, \frac{\alpha_s}{\pi} F^a_{\mu\nu} F^{a\,\mu\nu}\, |0\rangle \,, \tag{20}$$

where the decay constants f_n are normalized by

$$f_n\, m_n^2 \equiv \langle 0|\, \mathcal{O}_S(p)\, |n\rangle_{p^2=0} \,. \tag{21}$$

Because of the summation over gluons in the condensate in (20), we expect the scalar glueball mass squared to be proportional to $C_2(A)$. We note the similarity with the low energy theorem (17), if the sum rule (20) is saturated by the ground state or, equivalently, for the excited states $f_n m_n^2 \ll f_0 m_0^2$, which suggests that the excited states have very little overlap with the operator $\mathcal{O}_S$. The numerical analysis we report in this letter seems to suggest that this is the case, as we do not see evidence of Casimir scaling in the excited states, but only in the ground state.

4. Discussions and conclusion

For Yang–Mills theories, we conjectured that the ground state glueball mass squared, measured in units of string tension, is universally proportional to the ratio of the eigenvalues of the quadratic Casimir operator of the adjoint over that of the fundamental representation. The conjecture relies on the area law for confinement, and the specific coefficient should depend only on the dimensionality of the space-time, but not on the specific group.

We provided three analytical arguments to justify Casimir scaling, based respectively on the Bethe–Salpeter equation, scale anomaly, and sum rules. We tested this law on existing numerical lattice results in pure SU(N) and SO(N) Yang–Mills theories in $3+1$ and $2+1$ dimensions. The data strongly support Casimir scaling for the ground state. The values of the universal constant extracted from lattice data are $\eta(0^{++}) = 5.41(12)$ for $3+1$ dimensions and $\eta(0^{++}) = 8.444(15)(85)$ for $2+1$ dimensions. Numerical results are inconclusive for the 2^{++} state, while showing that universality does not hold for the first excitation in the 0^{++} channel.

If the conjectured universal scaling is confirmed, it would shed light on the mechanism yielding confinement in YM theories. It would be therefore quite interesting to test further numerically our conjecture for other gauge groups such as Sp($2N$) and SO(N) in $3+1$ dimensions, and to extend the arguments discussed here to provide systematic control over sub-leading corrections (if they exist) to exact Casimir scaling.

Acknowledgements

DKH thanks L. Giusti and C. Kim for useful comments and the CERN theory group for the hospitality during his visit. We thank A. Armoni, G. Aarts and F. Buisseret for comments. This research was supported by Basic Science Research Program through the National Research Foundation of Korea (NRF) funded by the Ministry of Education (NRF-2017R1D1A1B06033701) (DKH) and also by Korea Research Fellowship program funded by the Ministry of Science, ICT and Future Planning through the National Research Foundation of Korea (2016H1D3A1909283) and under the framework of international cooperation program (NRF-2016K2A9A1A01952069) (DKH and JWL). The work of BL, MP and DV is supported in part by the STFC grant ST/L000369/1.

References

[1] For a recent review see, for instance: J. Greensite, Prog. Part. Nucl. Phys. 51 (2003) 1, arXiv:hep-lat/0301023.
[2] G. 't Hooft, Nucl. Phys. B 138 (1978) 1.
[3] S. Mandelstam, Phys. Rep. 23 (1976) 245.
[4] N. Seiberg, E. Witten, Nucl. Phys. B 426 (1994) 19, Erratum: Nucl. Phys. B 430 (1994) 485, arXiv:hep-th/9407087.
[5] D. Karabali, C.j. Kim, V.P. Nair, Phys. Lett. B 434 (1998) 103, arXiv:hep-th/9804132;
D. Karabali, V.P. Nair, Nucl. Phys. B 464 (1996) 135, arXiv:hep-th/9510157.
[6] C.J. Morningstar, M.J. Peardon, Phys. Rev. D 60 (1999) 034509, arXiv:hep-lat/9901004.
[7] Y. Chen, et al., Phys. Rev. D 73 (2006) 014516, arXiv:hep-lat/0510074.
[8] B. Lucini, M. Teper, U. Wenger, J. High Energy Phys. 0406 (2004) 012, arXiv:hep-lat/0404008.
[9] B. Lucini, M. Panero, Phys. Rep. 526 (2013) 93, arXiv:1210.4997 [hep-th].
[10] A. Athenodorou, M. Teper, J. High Energy Phys. 1702 (2017) 015, arXiv:1609.03873 [hep-lat].
[11] A. Athenodorou, R. Lau, M. Teper, Phys. Lett. B 749 (2015) 448, arXiv:1504.08126 [hep-lat].
[12] R. Lau, M. Teper, J. High Energy Phys. 1603 (2016) 072, arXiv:1510.07841 [hep-lat].
[13] R. Lau, M. Teper, arXiv:1701.06941 [hep-lat].
[14] B. Lucini, M. Teper, Phys. Rev. D 64 (2001) 105019, arXiv:hep-lat/0107007.
[15] G.S. Bali, Phys. Rev. D 62 (2000) 114503, arXiv:hep-lat/0006022.
[16] S. Deldar, Phys. Rev. D 62 (2000) 034509, arXiv:hep-lat/9911008.
[17] N. Cardoso, M. Cardoso, P. Bicudo, Phys. Lett. B 710 (2012) 343, arXiv:1108.1542 [hep-lat].
[18] B. Bringoltz, M. Teper, Phys. Lett. B 663 (2008) 429, arXiv:0802.1490 [hep-lat].
[19] C. Lovelace, Nucl. Phys. B 201 (1982) 333.
[20] T. Imoto, T. Sakai, S. Sugimoto, Prog. Theor. Phys. 122 (2010) 1433, arXiv:0907.2968 [hep-th].
[21] R. Slansky, Phys. Rep. 79 (1981) 1.

[22] R.G. Leigh, D. Minic, A. Yelnikov, Phys. Rev. D 76 (2007) 065018, arXiv:hep-th/0604060.

[23] F. Buisseret, Eur. Phys. J. C 71 (2011) 1651, arXiv:1101.0907 [hep-ph];
F. Buisseret, V. Mathieu, C. Semay, Eur. Phys. J. C 73 (2013) 2504, arXiv:1301.3247 [hep-ph];
E. Abreu, P. Bicudo, J. Phys. G 34 (2007) 195207, arXiv:hep-ph/0508281;
V. Mathieu, C. Semay, F. Brau, Eur. Phys. J. A 27 (2006) 225, arXiv:hep-ph/0511210;
See also the section 3 in V. Mathieu, N. Kochelev, V. Vento, Int. J. Mod. Phys. E 18 (2009) 1, arXiv:0810.4453 [hep-ph].

[24] H.B. Meyer, arXiv:hep-lat/0508002.

[25] M. Luscher, K. Symanzik, P. Weisz, Nucl. Phys. B 173 (1980) 365.

[26] C.G. Callan Jr., S.R. Coleman, R. Jackiw, Ann. Phys. 59 (1970) 42.

[27] V.A. Novikov, M.A. Shifman, A.I. Vainshtein, V.I. Zakharov, Acta Phys. Pol. B 12 (1981) 399.

A universal constraint on the infrared behavior of the ghost propagator in QCD

Fei Gao [a,b], Can Tang [a,b], Yu-xin Liu [a,b,c,*]

[a] *Department of Physics and State Key Laboratory of Nuclear Physics and Technology, Peking University, Beijing 100871, China*
[b] *Collaborative Innovation Center of Quantum Matter, Beijing 100871, China*
[c] *Center for High Energy Physics, Peking University, Beijing 100871, China*

ARTICLE INFO

Editor: A. Ringwald

Keywords:
Gauge transformation
Ward–Takahashi identity
Ghost dressing function

ABSTRACT

With proposing a unified description of the fields variation at the level of generating functional, we obtain a new identity for the quark–gluon interaction vertex based on gauge symmetry, which is similar to the Slavnov–Taylor Identities (STIs) based on the Becchi–Rouet–Stora–Tyutin transformation. With these identities, we find that in Landau gauge, the dressing function of the ghost propagator approaches to a constant as its momentum goes to zero, which provides a strong constraint on the infrared behaviour of ghost propagator.

1. Introduction

Quantum Chromo-Dynamics (QCD) is a non-Abelian gauge theory, of which the action is built upon the SU(3) gauge symmetry. However, the gauge symmetry of the action is broken after the gauge fixing in practical calculations. Then, in a local covariant operator formalism introduced by Faddeev and Popov [1], the ghost and the anti-ghost fields are needed in the action. Though the SU(N) gauge symmetry has been broken, a new symmetry called the Becchi–Rouet–Stora–Tyutin (BRST) symmetry is found in the complete action [2]. As a consequence, the Ward–Takahashi identities (WTIs) [3,4] based on the gauge symmetry are replaced by the Slavnov–Taylor identities (STIs) [5,6].

The BRST transformation can be considered as the generalization of the SU(N) gauge transformation by generalizing the infinitesimal quantity of the gauge transformation from real number field to the Grassmann number field. For instance, the gauge transformation of the fermion field in the usual SU(N) gauge transformation reads $\delta\psi = -igt^a\theta^a\psi$ with θ^a a real number, the respective BRST transformation is to replace the real number to Grassmann number, i.e. $\delta\psi = -igt^a\lambda c^a\psi$ with c^a the Grassmann number and λ the phase angle. The field spanned by the Grassmann numbers is just the ghost field. The STIs can be obtained with the BRST transformation that keeps the action unchanged.

Generally speaking, the variation of the fields does not need to be constrained by the symmetry of the action, instead, the variation can be employed at the level of generating function. In the generating functional, the fields are variables of the integral, and thus, the variation of the fields can be regarded as the replacement of the integral variables. Such a replacement can be taken to derive the STIs with the BRST symmetry of the action [7]. As long as the functional integral converges, which is a natural demand for the generating functional, such kind technique can be employed. The relation derived from this variation might not be the charge conservation relation as $\partial_\mu J_\mu = 0$, but with the additional source term, for instance, the mass term which breaks the chiral symmetry and leads to the relation: $\partial_\mu J_{\mu 5} = m\Gamma_5$. Here with such kind variation, we derive a new WTI for the quark–gluon interaction vertex with the original gauge transformation. Meanwhile, the conventional STI from the BRST symmetry for the quark–gluon interaction vertex still holds. With these two identities, the ghost propagator can be strongly constrained.

People are interested in the infrared behaviour of the ghost propagator for a long time. Kugo and Ojima have firstly proposed an infrared-enhanced ghost propagator whose dressing function is divergent [8,9]. However, the assumption of the asymptotic fields has been employed in their derivations which might not be appropriate in QCD. Gribov and Zwanziger have also conjectured an infrared enhanced ghost propagator by considering the overcompensation of the configurations close to the Gribov horizon [10,11].

* Corresponding author at: Department of Physics and State Key Laboratory of Nuclear Physics and Technology, Peking University, Beijing 100871, China.
E-mail addresses: hiei@pku.edu.cn (F. Gao), yxliu@pku.edu.cn (Y.-x. Liu).

Nevertheless, a refinement of the Gribov–Zwanziger action has recently been proposed [12,13]. This improvement includes the condensates as the dynamical effects, and obtains that the ghost propagator is infrared-finite. Besides, the numerical results of the recent calculations including those in the Dyson–Schwinger equations of QCD [14–16] and the lattice QCD simulations [17–21] seem to be more supportive to an infrared-finite ghost propagator. Though the numerical computations are convincible, a model-independent analytic proof is still needed.

In this article, we derive the WTI for the quark–gluon interaction vertex which is parallel to the STI. Combining the WTI and the STI for the quark–gluon interaction vertex together, we give a universal conclusion on the infrared behaviour of the ghost propagator, that is, in Landau gauge, the dressing function of the ghost propagator at zero momentum becomes constant. The infrared behaviour of the ghost propagator is then clarified.

2. Derivations and discussions

In the local covariant operator formalism, the action of QCD reads (see, e.g., Ref. [22]):

$$S = \int d^4x \mathcal{L}_{QCD}, \tag{1}$$

with

$$\begin{aligned}\mathcal{L}_{QCD} = &\bar{\psi}(-\not{\partial}+m)\psi - ig\bar{\psi}\gamma_\mu t^a \psi A^a_\mu \\ &+ \frac{1}{2}A^a_\mu\Big(-\partial^2\delta_{\mu\nu} - (\frac{1}{\xi}-1)\partial_\mu\partial_\nu\Big)A^a_\nu \\ &- gf^{abc}(\partial_\mu A^a_\nu)A^b_\mu A^c_\nu \\ &+ \frac{1}{4}g^2 f^{abe} f^{cde} A^a_\mu A^b_\nu A^c_\mu A^d_\nu \\ &+ \bar{c}^a\partial^2 c^a + gf^{abc}\bar{c}^a\partial_\mu(A^c_\mu c^b),\end{aligned} \tag{2}$$

where $\bar{\psi}$, ψ, A^a_μ, $\bar{c}$, c is the anti-quark, quark, gluon, anti-ghost and ghost field, respectively. t^a and f^{abc} are the generators and the structure constants of the SU(3) gauge group, respectively. This action is no longer gauge invariant because the gauge fixing term changes corresponding to the gauge transformation. The action is only invariant under the BRST transformation defined by:

$$\begin{aligned}&\delta\Psi = -igt^a c^a\psi\lambda, \quad \delta A^a_\mu = (\delta^{ab}\partial_\mu + gf^{abc}A^c_\mu)c^b\lambda, \\ &\delta c^a = -\frac{g}{2}f^{abc}c^b c^c\lambda, \quad \delta\bar{c}^a = \frac{1}{\xi}\partial_\mu A^a_\mu\lambda.\end{aligned} \tag{3}$$

From the BRST symmetry, the STI for the quark–gluon interaction vertex can be written as [23,24]:

$$\begin{aligned}G^{-1}(k^2)k_\mu\Gamma^a_\mu(p,q) = &(gt^a - B^a(k,q))S^{-1}(p) \\ &- S^{-1}(q)(gt^a - B^a(k,q)),\end{aligned} \tag{4}$$

where $G(k^2)$ is the dressing function of the ghost propagator, $D^{ab}(k) = -\delta^{ab}G(k^2)/k^2$, and $B^a(k,q)$ the ghost-quark scattering kernel. This STI contains the ghost propagator and the ghost-quark scattering kernel, which is unknown. People usually employ the so called Abelian approximation that eliminates the ghost effect during computations. In Abelian approximation, the STI degenerates into the WTI:

$$k_\mu\Gamma_\mu(p,q) = gS^{-1}(p) - S^{-1}(q)g.$$

With this constraint, people usually take the Ball–Chiu vertex [25] as the longitudinal part of the quark–gluon interaction vertex [26–28].

As we analyzed above, this identity is based on the BRST transformation to guarantee the action invariant. However, the transformation does not need to coincide with this constraint beforehand. To understand this, we resort to the variation of the generating functional. In the path integral scheme, the generating functional is defined [22] as:

$$Z[J] = \int \mathcal{D}\phi_i \exp\{-\int d^4x(\mathcal{L}_{QCD} - J_i\phi_i)\}, \tag{5}$$

with J_i standing for the source of any field ϕ_i in the Lagrangian of QCD. If the generating functional converges, the functional is unchanged under the replacement of the integral variables, that is:

$$\begin{aligned}&\int \mathcal{D}\phi' \exp\Big\{-\int d^4x(\mathcal{L}'_{QCD} - J_i\phi'_i)\Big\} \\ &= \int \mathcal{D}\phi \exp\Big\{-\int d^4x(\mathcal{L}_{QCD} - J_i\phi_i)\Big\}.\end{aligned} \tag{6}$$

The replacement is naive, but will lead to nontrivial results. At first, we can implement it to derive the axial-vector Ward–Takahashi identities (AV-WTIs). Considering the replacement:

$$\begin{aligned}\psi' &= \psi - ig\frac{\tau^i}{2}\gamma_5\theta^i\psi, \\ \bar{\psi}' &= \bar{\psi} - ig\bar{\psi}\frac{\tau^i}{2}\gamma_5\theta^i,\end{aligned} \tag{7}$$

where $\tau^i/2$ is the generator of the group demonstrating the flavour symmetry. Under this transformation, only the terms $\bar{\psi}(-\not{\partial}+m)\psi$ and the external sources change in the generating functional. With the above mentioned principle that the functional is invariant under the replacement of the integral variables, we can get straightforwardly

$$\begin{aligned}&\Big\langle\theta^i\partial_\mu(\bar{\psi}(x)\gamma_\mu\frac{\tau^i}{2}\gamma_5\psi(x))\Big\rangle + 2m\Big\langle\theta^i\bar{\psi}(x)\frac{\tau^i}{2}\gamma_5\psi(x)\Big\rangle \\ &-\Big\langle\bar{\psi}(x)\frac{\tau^i}{2}\gamma_5\theta^i\eta(x) + \bar{\eta}(x)\frac{\tau^i}{2}\gamma_5\theta^i\psi(x)\Big\rangle = 0,\end{aligned} \tag{8}$$

with η, η' being the external sources of $\bar{\psi}$ and ψ, respectively. Deviating it by $\frac{\delta^2}{\delta\eta(x_1)\delta\bar{\eta}(x_2)}$ and converting into the momentum space, Eq. (8) reads

$$\begin{aligned}k_\mu\Gamma^i_{5\mu}(p,q) = &S^{-1}(p)\gamma_5\frac{\tau^i}{2} + \gamma_5\frac{\tau^i}{2}S^{-1}(q) \\ &- 2m\Gamma^i_5(p,q),\end{aligned} \tag{9}$$

where $k = p - q$ is the momentum transfer, S^{-1} is the inverse of the quark propagator, $\Gamma^i_{5\mu}$ and Γ^i_5 stand for the isovector axial-vector vertex, the isovector pseudoscalar vertex, respectively. We can see that it is exactly the AV-WTI with a mass term which is the source term breaking the symmetry. The respective vector Ward–Takahashi identity (V-WTI) can be obtained via the similar transformation

$$\begin{aligned}\psi' &= \psi - ig\frac{\tau^i}{2}\theta^i\psi, \\ \bar{\psi}' &= \bar{\psi} + ig\bar{\psi}\frac{\tau^i}{2}\theta^i.\end{aligned} \tag{10}$$

It is easy to find that only the terms $-\bar{\psi}\not{\partial}\psi$ and the external sources change in the generating functional. With the similar procedure, we get

$$\left\langle\theta^i\partial_\mu(\bar{\psi}(x)\gamma_\mu\frac{\tau^i}{2}\psi(x))\right\rangle + \left\langle\bar{\psi}(x)\frac{\tau^i}{2}\theta^i\eta(x) - \bar{\eta}(x)\frac{\tau^i}{2}\theta^i\psi(x)\right\rangle = 0\,. \tag{11}$$

Deviating it and converting the result into momentum space, we have

$$k_\mu\Gamma^i_\mu(p,q) = S^{-1}(p)\frac{\tau^i}{2} - \frac{\tau^i}{2}S^{-1}(q)\,, \tag{12}$$

where Γ^i_μ is the isovector vector vertex. These two vertices are both color-singlet vertices. The axial-vector vertex contains pion. Therefore, the AV-WTI has been widely used in QCD, succeeded to predict pion as the Goldstone boson and also the particle that reveals the dynamical chiral symmetry breaking [29]. The V-WTI has also been employed as the conditions that must be satisfied when computing the meson's form factor [30]. It can be seen that the variation replacement of the generating functional can be taken to derive the identities no matter the symmetry has been broken or not.

Now considering the color gauge transformation for ψ and $\bar{\psi}$ which reads

$$\begin{aligned}\psi' &= \psi - igt^a\theta^a\psi\,,\\ \bar{\psi}' &= \bar{\psi} + ig\bar{\psi}t^a\theta^a\,,\end{aligned} \tag{13}$$

which could also be regarded as the variables' replacement in the generating functional integral. The Jacobian determinant of this transformation is

$$\begin{aligned}J_{ij}\bar{J}_{mn} &= \det(\delta_{ij} - igt^a_{ij}\theta^a)\det(\delta_{mn} + igt^a_{mn}\theta^a)\\ &= 1 + \mathcal{O}(\theta^2)\,.\end{aligned} \tag{14}$$

Under this transformation, only the terms $\bar{\psi}\partial\!\!\!/\psi$, $g\bar{\psi}\gamma_\mu t^a\psi A^a_\mu$ and the external sources change in the generating functional, we get directly then

$$\begin{aligned}&\langle\partial_\mu(\bar{\psi}(x)\gamma_\mu t^a\psi(x))\rangle - \langle gf^{abc}A^b_\mu\bar{\psi}(x)\gamma_\mu t^c\psi(x)\rangle\\ &\qquad + \langle\bar{\psi}(x)t^a\eta(x) - \bar{\eta}(x)t^a\psi(x)\rangle = 0\,.\end{aligned} \tag{15}$$

Deviated by $\frac{\delta^2}{\delta\eta(x_1)\delta\bar{\eta}(x_2)}$, the expression becomes:

$$\begin{aligned}&-\langle\partial_\mu(\bar{\psi}(x)\gamma_\mu t^a\psi(x))\psi(x_2)\bar{\psi}(x_1)\rangle\\ &\quad + \langle gf^{abc}A^b_\mu\bar{\psi}(x)\gamma_\mu t^c\psi(x)\psi(x_2)\bar{\psi}(x_1)\rangle\\ &= \delta^4(x-x_2)\langle\psi(x)t^a\bar{\psi}(x_1)\rangle\\ &\quad - \langle\psi(x_2)t^a\bar{\psi}(x)\rangle\delta^4(x-x_1)\,.\end{aligned} \tag{16}$$

This identity is exact without including any approximation. The second term in Eq. (16) is unusual, since it owns very similar structure with that in the quark propagator's self-energy. Expanding this term in the momentum space we can obtain

$$\begin{aligned}&\int d^4ke^{-ikx}\langle gf^{abc}A^b_\mu\bar{\psi}(x)\gamma_\mu t^c\psi(x)\psi(p)\bar{\psi}(q)\rangle\\ &= \delta^4(k-p+q)\frac{1}{2}f^{abc}f^{bcd}\\ &\quad\times S(p)\Big\{\Sigma^d(p) - \Sigma^d(q) + \int d^4k'D_{\mu\nu}\gamma_\mu H^d_6(k,p,q)\Gamma_\nu\\ &\quad + \int d^4k'\Sigma^d(p-k')S(p-k')K(p-k',q-k')S(q-k')\\ &\quad - \int d^4k'\Sigma^d(q-k')S(q-k')K(q-k',p-k')\\ &\qquad S(p-k')\Big\}S(q)\,,\end{aligned} \tag{17}$$

where

$$\Sigma^d(p) = g^2\int d^4k'D_{\mu\nu}(k')\gamma_\nu S(p-k')\Gamma^d_\mu(p,p-k')$$

is the self energy of the quark without the sum of color indices, $K(p,q)$ is the four-quark scattering kernel, $H^d_6(k,p,q)$ is in terms of the six-quark Green function. Converting the relation into the momentum space, we have

$$\begin{aligned}k_\mu\tilde{\Gamma}^a_\mu(p,q) &= gt^aS^{-1}(p) - S^{-1}(q)gt^a\\ &\quad - \frac{N_c^2}{(N_c^2-1)}\Big(gt^a\Sigma(p) - \Sigma(q)gt^a\\ &\quad + t^a\overline{K}(p,q) + \overline{K}(q,p)t^a + \overline{H}^a_6(p,q)\Big)\,,\end{aligned} \tag{18}$$

where

$$\overline{K}(p,q) = \int d^4k'\Sigma(p-k')S(q-k')K(p-k',q-k')S(q-k')\,,$$

$$\overline{H}^a_6(p,q) = \int d^4k'D_{\mu\nu}\gamma_\mu H^a_6(k,p,q)\Gamma_\nu\,.$$

The definition of the quark–gluon interaction vertex $\tilde{\Gamma}^a_\mu$ is a little different from above, which is only the vertex that the gluon couples directly with quarks, the difference from the complete vertex is a six-point scattering kernel $B_6(p,q)$ in $\langle\bar{\psi}(x)\gamma_\mu t^a\psi(x)\bar{c}^e(x)\times c^f(0)\psi(x_2)\bar{\psi}(x_1)\rangle$ with four quarks and two ghosts as shown in Ref. [7]. With this correction we get:

$$\begin{aligned}&\frac{1}{1-B_6(p,q)}k_\mu\Gamma^a_\mu(p,q)\\ &= gt^aS^{-1}(p) - S^{-1}(q)gt^a - \frac{N_c^2}{(N_c^2-1)}\Big(gt^a\Sigma(p) - \Sigma(q)gt^a\\ &\quad + gt^a\overline{K}(p,q) + \overline{K}(q,p)gt^a + \overline{H}^a_6(p,q)\Big)\,.\end{aligned} \tag{19}$$

This is another identity for the quark–gluon interaction vertex, which represents the relation between the gauge current and the symmetry breaking source, just like the case of AV-WTI. Noticing that it is exact in QCD, and the STI, Eq. (4), still holds at the same time. These two sets of identities stand for two different transformations, one is the gauge transformation and the other is the BRST transformation. The two transformations are respective to different phase spaces, and then the two sets of identities together describe the complete structure in QCD at the level of generating functional.

Moreover, if combining the WTI for the quark–gluon interaction vertex we derived here with Eq. (4), it gives a strong constraint for the ghost propagator. In Landau gauge, it has been shown [5] perturbatively that $B^a(k,q)\to 0$ for $k\to 0$, but the nonperturbative effect might make it deviate against 0. The analysis in Ref. [24] has shown that the function

$$t^aH(p,q) = gt^a - B^a(p,q)$$

should be infrared finite, and thus it may change the value of $G(k^2=0)$ quantitatively, but not qualitatively. Besides, the six-point kernels B_6 and $\bar{H}^a_6$ are the higher order corrections, which will arise some more dressing effect on the quark–gluon interaction vertex but will not change the behaviour of the ghost propagator quantitatively. In practical computations (e.g., Refs. [27,31]), such higher order corrections are usually neglected. After this, if we take p and q both tend to be zero, we get the value of the dressing function of the ghost propagator at zero momentum in Landau gauge as

$$G(k^2 = 0) = \frac{1/H(0,0)}{1 + \left(N_c^2/(N_c^2 - 1)\right)\Gamma(0)} \tag{20}$$

with

$$\Gamma(p-q) = \frac{tr[\not{k}\left(\Delta\Sigma(0) + \int d^4k'\left(\Delta\Sigma(k)\right)t^a\right)SKS)]}{tr[\not{k}k_\mu\Gamma^a_\mu]},$$

where only the Lorentz indices are taken into account in the trace and $\Delta\Sigma(k) = \Sigma(p-k') - \Sigma(q-k')$. If only considering the leading order of $\not{k}$ in the above formula, we can see that the leading order of $\Sigma(p-k') - \Sigma(q-k')$ is proportional to $(A(k'^2) - 1)\not{k}$ with $A(k'^2)$ being the vector part of the quark propagator. The remanent part $t^a\not{k} + t^a k_\mu \int d^4k'\gamma_\mu SKS$ is just the expansion for the quark–gluon interaction vertex in the Dyson–Schwinger equations and can be equivalently rewritten as $k_\mu\Gamma^a_\mu$. Taking $A(0)$ as the approximation of $A(k'^2)$, $\Gamma(0)$ in the above formula can be simplified as $A(0) - 1$. With typical numerical value obtained in Dyson–Schwinger equation calculations (see, e.g. Refs. [32–35]), $A(0) \approx 1.6$, we can get the value of the ghost dressing function at zero momentum under the SU(3) gauge symmetry as $G(0) = 0.60/H(0,0)$, or with the value obtained in lattice QCD calculation [36], we get $G(0) = 0.78/H(0,0)$. It proves that the ghost propagator is in the so called decoupling solution whose dressing function becomes constant at the deep infrared limit, which is consistent with what the recent lattice QCD calculation [37] hints. Specifically, Ref. [38] claims that $H(0,0)$ might be divergent. It leads then a vanishing value for $G(0)$, which means that the constant of $G(0)$ could be zero. It is known that there may exist two solutions for the ghost and gluon propagators depending on the boundary condition [31], and there can be another solution of which the dressing function is infrared-enhanced if taking the value of $A(0)$ as a small value down to $1/9$. However, this has not yet been obtained in any numerical calculation on the quark propagator with realistic parameters. Therefore, such a relation reveals that the quark–gluon interaction can serve as the boundary condition and gives strong constraints on the properties of the gauge fields.

3. Summary

In our sense, the relations of the Green functions have broader forms which need not to be limited by the symmetry of the original action. For instance, even though the isovector axial-vector symmetry can be broken by the mass term in the action, the AV-WTI with source term can still be obtained. Similarly, even though the gauge symmetry has been broken by the gauge fixing term in QCD's local covariant action, the WTIs based on the gauge symmetry can still be obtained with some additional source terms, which is another set of identities parallel to the STIs based on the BRST symmetry just like the relation between the AV-WTIs and the V-WTIs.

In this paper we give a unified description for the symmetry of the Green functions, i.e. at the level of the generating functional. The key point is that the field variation in the generating functional is regarded as the replacement of the integral variables. Therefore, the replacement can be arbitrary without being constrained by the symmetry. Such replacement under the gauge transformation can be easily implemented in practical calculations. We get, in turn, another WTI with source term for the quark–gluon interaction vertex. By combining the presently obtained WTI with the STI, we find a relation that strongly constrains the infrared behavior of the dressing function of the ghost propagator. The relation proves that in Landau gauge, the dressing function of the ghost propagator becomes constant at zero momentum.

Acknowledgements

We thank Prof. Si-xue Qin, Prof. Lei Chang and Prof. Craig D. Roberts for useful discussion. The work was supported by the National Natural Science Foundation of China under Contract No. 11435001; the National Key Basic Research Program of China under Contracts No. G2013CB834400 and No. 2015CB856900.

References

[1] L.D. Faddeev, V.N. Popov, Phys. Lett. B 25B (1967) 29.
[2] C. Becchi, A. Rouet, R. Stora, Phys. Lett. B 52 (1974) 344; Commun. Math. Phys. 42 (1975) 127.
[3] J.C. Ward, Phys. Rev. 78 (1950) 182.
[4] Y. Takahashi, Nuovo Cimento 6 (1957) 371.
[5] J.C. Taylor, Nucl. Phys. B 33 (1971) 436.
[6] A.A. Slavnov, Theor. Math. Phys. 10 (1972) 99.
[7] Han-xin He, Phys. Rev. D 80 (2009) 016004.
[8] T. Kugo, I. Ojima, Prog. Theor. Phys. Suppl. 66 (1979) 1.
[9] H. Hata, Prog. Theor. Phys. 67 (1982) 1607.
[10] V.N. Gribov, Nucl. Phys. B 139 (1978) 1.
[11] D. Zwanziger, Nucl. Phys. B 323 (1989) 513.
[12] D. Dudal, O. Oliveira, J. Rodriguez-Quintero, Phys. Rev. D 86 (2012) 105005.
[13] D. Zwanziger, Phys. Rev. D 87 (2013) 085039.
[14] A. Aguilar, D. Binosi, J. Papavassiliou, Phys. Rev. D 86 (2012) 014032.
[15] A. Ayala, A. Bashir, D. Binosi, M. Cristoforetti, J. Rodriguez-Quintero, Phys. Rev. D 86 (2012) 074512.
[16] S. Strauss, C.S. Fischer, C. Kellermann, Phys. Rev. Lett. 109 (2012) 252001.
[17] P.O. Bowman, U.M. Heller, D.B. Leinweber, M.B. Parappilly, A.G. Williams, Phys. Rev. D 70 (2004) 034509; P.O. Bowman, U.M. Heller, D.B. Leinweber, M.B. Parappilly, A.G. Williams, J.B. Zhang, Phys. Rev. D 71 (2005) 054507.
[18] I. Bogolubsky, E. Ilgenfritz, M. Muller-Preussker, A. Sternbeck, Phys. Lett. B 676 (2009) 69.
[19] P. Boucaud, M.E. Gomez, J.P. Leroy, A. Le Yaouanc, J. Micheli, O. Pene, J. Rodriguez-Quintero, Phys. Rev. D 82 (2010) 054007.
[20] O. Oliveira, P. Bicudo, J. Phys. G 38 (2011) 045003.
[21] A. Cucchieri, D. Dudal, T. Mendes, N. Vandersickel, Phys. Rev. D 85 (2012) 094513.
[22] R. Alkofer, L. Von Smekal, Phys. Rept. 353 (2001) 281.
[23] E.J. Eichten, F.L. Feinberg, Phys. Rev. D 10 (1974) 3254.
[24] R. Alkofer, C.S. Fischer, F.J. Llanes-Estrada, K. Schwenzer, Ann. Phys. 324 (2009) 106.
[25] J.S. Ball, T.W. Chiu, Phys. Rev. D 22 (1980) 2542; Phys. Rev. D 22 (1980) 2550.
[26] C.D. Roberts, A.G. Williams, Prog. Part. Nucl. Phys. 33 (1994) 477.
[27] S.X. Qin, L. Chang, Y.X. Liu, C.D. Roberts, S.M. Schmidt, Phys. Lett. B 722 (2013) 384.
[28] F. Gao, Y.X. Liu, Phys. Rev. D 94 (2016) 076009.
[29] P. Maris, C.D. Roberts, P.C. Tandy, Phys. Lett. B 420 (1998) 267.
[30] P. Maris, P.C. Tandy, Phys. Rev. C 61 (2000) 045202.
[31] C.S. Fischer, A. Maas, J.M. Pawlowski, Ann. Phys. 324 (2009) 2408.
[32] C.S. Fischer, R. Alkofer, Phys. Rev. D 67 (2003) 094020.
[33] C.S. Fischer, J. Phys. G 32 (2006) R253.
[34] L. Chang, Y.X. Liu, M.S. Bhagwat, C.D. Roberts, S.V. Wright, Phys. Rev. C 75 (2007) 015201.
[35] K.L. Wang, S.X. Qin, Y.X. Liu, L. Chang, C.D. Roberts, S.M. Schmidt, Phys. Rev. D 86 (2012) 114001.
[36] O. Oliveira, A. Kizilersu, P.J. Silva, J.I. Skullerud, A. Sternbeck, A.G. Williams, Acta Phys. Pol. Supp. 9 (2016) 363.
[37] A.G. Duarte, O. Oliveira, P.J. Silva, Phys. Rev. D 94 (2016) 014502.
[38] E. Rojas, J.P.B.C. de Melo, B. El-Bennich, O. Oliveira, T. Frederico, J. High Energy Phys. 10 (2013) 193.

Permissions

All chapters in this book were first published in PLB, by Elsevier; hereby published with permission under the Creative Commons Attribution License or equivalent. Every chapter published in this book has been scrutinized by our experts. Their significance has been extensively debated. The topics covered herein carry significant findings which will fuel the growth of the discipline. They may even be implemented as practical applications or may be referred to as a beginning point for another development.

The contributors of this book come from diverse backgrounds, making this book a truly international effort. This book will bring forth new frontiers with its revolutionizing research information and detailed analysis of the nascent developments around the world.

We would like to thank all the contributing authors for lending their expertise to make the book truly unique. They have played a crucial role in the development of this book. Without their invaluable contributions this book wouldn't have been possible. They have made vital efforts to compile up to date information on the varied aspects of this subject to make this book a valuable addition to the collection of many professionals and students.

This book was conceptualized with the vision of imparting up-to-date information and advanced data in this field. To ensure the same, a matchless editorial board was set up. Every individual on the board went through rigorous rounds of assessment to prove their worth. After which they invested a large part of their time researching and compiling the most relevant data for our readers.

The editorial board has been involved in producing this book since its inception. They have spent rigorous hours researching and exploring the diverse topics which have resulted in the successful publishing of this book. They have passed on their knowledge of decades through this book. To expedite this challenging task, the publisher supported the team at every step. A small team of assistant editors was also appointed to further simplify the editing procedure and attain best results for the readers.

Apart from the editorial board, the designing team has also invested a significant amount of their time in understanding the subject and creating the most relevant covers. They scrutinized every image to scout for the most suitable representation of the subject and create an appropriate cover for the book.

The publishing team has been an ardent support to the editorial, designing and production team. Their endless efforts to recruit the best for this project, has resulted in the accomplishment of this book. They are a veteran in the field of academics and their pool of knowledge is as vast as their experience in printing. Their expertise and guidance has proved useful at every step. Their uncompromising quality standards have made this book an exceptional effort. Their encouragement from time to time has been an inspiration for everyone.

The publisher and the editorial board hope that this book will prove to be a valuable piece of knowledge for researchers, students, practitioners and scholars across the globe.

List of Contributors

L.V. Bork
Institute for Theoretical and Experimental Physics, Moscow, Russia
The Center for Fundamental and Applied Research, All-Russia Research Institute of Automatics, Moscow, Russia

A.I. Onishchenko
Bogoliubov Laboratory of Theoretical Physics, Joint Institute for Nuclear Research, Dubna, Russia
Moscow Institute of Physics and Technology (State University), Dolgoprudny, Russia
Skobeltsyn Institute of Nuclear Physics, Moscow State University, Moscow, Russia

Pieter-Jan De Smet and Chris D.White
Centre for Research in String Theory, School of Physics and Astronomy, Queen Mary University of London, 327 Mile End Road, London E1 4NS, UK

Robert Dickinson and Jeff Forshaw
Consortium for Fundamental Physics, School of Physics and Astronomy, University of Manchester, Manchester M139PL, United Kingdom

Peter Millington
School of Physics and Astronomy, University of Nottingham, Nottingham NG72RD, United Kingdom

Patrick Concha
Instituto de Física, Pontificia Universidad Católica de Valparaíso, Casilla 4059, Valparaiso, Chile

Evelyn Rodríguez
Departamento de Ciencias, Facultad de Artes Liberales, Universidad Adolfo Ibáñez, Av. Padre Hurtado 750, Viña del Mar, Chile

X. Calmet
Department of Physics & Astronomy, University of Sussex, Falmer, Brighton, BN1 9QH, United Kingdom

R. Casadio
Dipartimento di Fisica e Astronomia, Università di Bologna, and INFN, Via Irnerio 46, 40126 Bologna, Italy

A.Yu. Kamenshchik
Dipartimento di Fisica e Astronomia, Università di Bologna, and INFN, Via Irnerio 46, 40126 Bologna, Italy
L.D. Landau Institute for Theoretical Physics of the Russian Academy of Sciences, Kosygin str.2, 119334 Moscow, Russia

O.V. Teryaev
Bogoliubov Laboratory of Theoretical Physics, Joint Institute for Nuclear Research, 141980 Dubna, Russia
Lomonosov Moscow State University, Leninskie Gory1, 119991 Moscow, Russia

M. Ali-Akbari
Department of Physics, Shahid Beheshti University G.C., Evin, Tehran 19839, Iran

F. Charmchi
School of Particles and Accelerators, Institute for Research in Fundamental Sciences (IPM), P.O.Box 19395-5531, Tehran, Iran

Jian-Pin Wu
Institute of Gravitation and Cosmology, Department of Physics, School of Mathematics and Physics, Bohai University, Jinzhou 121013, China

Peng Liu
Department of Physics, Jinan University, Guangzhou 510632, China

Giovanni Amelino-Camelia and Michele Ronco
Dipartimento di Fisica, Università di Roma "La Sapienza", P.le A. Moro 2, 00185 Roma, Italy
INFN, Sez. Roma1, P.le A. Moro 2, 00185 Roma, Italy

Gianluca Calcagni
Instituto de Estructura de la Materia, CSIC, Serrano 121, 28006 Madrid, Spain

P.D. Alvarez
Universidad Técnica Federico Santa María, Santiago, Chile

F. Canfora
Centro de Estudios Científicos (CECS), Casilla 1469, Valdivia, Chile

N. Dimakis
Instituto de Ciencias Fisicas y Matematicas, Universidad Austral de Chile, Valdivia, Chile

A. Paliathanasis
Instituto de Ciencias Fisicas y Matematicas, Universidad Austral de Chile, Valdivia, Chile
Institute of Systems Science, Durban University of Technology, PO Box 1334, Durban 4000, South Africa

Salvador Robles-Pérez
Estación Ecológica de Biocosmología, Pedro de Alvarado 14, 06411 Medellín, Spain
Instituto de Física Fundamental, CSIC, Serrano 121, 28006 Madrid, Spain

Dario Zappalà
INFN, Sezione di Catania, Via Santa Sofia 64, 95123 Catania, Italy

Daniel W.F. Alves and Horatiu Nastase
Instituto de Física Teórica, UNESP-Universidade Estadual Paulista, Rua Dr. Bento T. Ferraz 271, Bl. II, São Paulo 01140-070, SP, Brazil

Carlos Hoyos
Department of Physics, Universidad de Oviedo, Calle Federico García Lorca 18, 33007, Oviedo, Spain

Jacob Sonnenschein
School of Physics and Astronomy, The Raymond and Beverly Sackler Faculty of Exact Sciences, Tel Aviv University, Ramat Aviv 69978, Israel

Imtak Jeon
Harish-Chandra Research Institute, Chhatnag Road, Jhusi, Allahabad 211019, India

Shailesh Lal
LPTHE - UMR 7589, UPMC Paris 06, Sorbonne Universités, Paris 75005, France

Yang-Hui He
Merton College, University of Oxford, UK
Department of Mathematics, City, University of London, UK
School of Physics, NanKai University, China

Ancas Tureanu
Department of Physics, University of Helsinki, P.O. Box 64, FIN-00014 Helsinki, Finland

Kazuo Fujikawa
Department of Physics, University of Helsinki, P.O. Box 64, FIN-00014 Helsinki, Finland
Quantum Hadron Physics Laboratory, RIKEN Nishina Center, Wako 351-0198, Japan

Pujian Mao
Institute of High Energy Physics and Theoretical Physics Center for Science Facilities, Chinese Academy of Sciences, 19B Yuquan Road, Beijing 100049, PRChina

Hao Ouyang
Institute of High Energy Physics and Theoretical Physics Center for Science Facilities, Chinese Academy of Sciences, 19B Yuquan Road, Beijing 100049, PRChina
School of Physical Sciences, University of Chinese Academy of Sciences, 19A Yuquan Road, Beijing 100049, PRChina

A. Gorsky
Institute of Information Transmission Problems of the Russian Academy of Sciences, Moscow, Russia
Moscow Institute of Physics and Technology, Dolgoprudny 141700, Russia

F. Popov
Moscow Institute of Physics and Technology, Dolgoprudny 141700, Russia
Institute of Theoretical and Experimental Physics, Moscow, Russia
Department of Physics, Princeton University, Princeton, NJ 08544, USA

Baocheng Zhang
School of Mathematics and Physics, China University of Geosciences, Wuhan 430074, China

Yoshinori Honma
National Center for Theoretical Sciences, National Tsing-Hua University, Hsinchu 30013, Taiwan

Hajime Otsuka
Department of Physics, Waseda University, Tokyo 169-8555, Japan

Chi-Fang Chen and Andrew Lucas
Department of Physics, Stanford University, Stanford, CA 94305, USA

Peter Lowdon, Kelly Yu-Ju Chiu and Stanley J. Brodsky
SLAC National Accelerator Laboratory, Stanford University, 2575 Sand Hill Rd, CA 94025, USA

Thomas Pappas and Panagiota Kanti
Division of Theoretical Physics, Department of Physics, University of Ioannina, Ioannina GR-45110, Greece

W.A. Sabra
Centre for Advanced Mathematical Sciences and Physics Department, American University of Beirut, Lebanon

Ramit Dey, Stefano Liberati and Daniele Pranzetti
SISSA, Via Bonomea 265, 34136 Trieste, Italy
INFN, Sezione di Trieste, Italy

M. Dehghani
Department of Physics, Ilam University, Ilam, Iran

Alessandro Bravetti
Instituto de Investigaciones en Matemáticas Aplicadas y en Sistemas, Universidad Nacional Autónoma de México, Ciudad Universitaria, Ciudad de México 04510, Mexico A.P. 70-543, 04510 Ciudad de México, Mexico

Christine Gruber and Francisco Nettel
Instituto de Ciencias Nucleares, Universidad Nacional Autónoma de México, A.P. 70-543, 04510 Ciudad de México, Mexico

Cesar S. Lopez-Monsalvo
Conacyt-Universidad Autónoma Metropolitana Azcapotzalco Avenida San Pablo Xalpa 180, Azcapotzalco, Reynosa Tamaulipas, 02200 Ciudad de México, Mexico

M. Momennia
Physics Department and Biruni Observatory, College of Sciences, Shiraz University, Shiraz 71454, Iran

S.H. Hendi
Physics Department and Biruni Observatory, College of Sciences, Shiraz University, Shiraz 71454, Iran
Research Institute for Astrophysics and Astronomy of Maragha (RIAAM), P.O. Box 55134-441, Maragha, Iran

B. Eslam Panah
Physics Department and Biruni Observatory, College of Sciences, Shiraz University, Shiraz 71454, Iran
Research Institute for Astrophysics and Astronomy of Maragha (RIAAM), P.O. Box 55134-441, Maragha, Iran
ICRANet, Piazza della Repubblica 10, I-65122 Pescara, Italy

S. Panahiyan
Physics Department and Biruni Observatory, College of Sciences, Shiraz University, Shiraz 71454, Iran
Helmholtz-Institut Jena, Fröbelstieg 3, Jena D-07743, Germany
Physics Department, Shahid Beheshti University, Tehran 19839, Irans

Calvin W. Johnson
Department of Physics, San Diego State University, 5500 Campanile Drive, San Diego, CA 92182-1233, United States

Erik Widén
Nordita, KTH Royal Institute of Technology and Stockholm University, Roslagstullsbacken 23, SE-106 91 Stockholm, Sweden
Department of Physics and Astronomy, Uppsala University, SE-751 08 Uppsala, Sweden

Shahar Hod
The Ruppin Academic Center, Emeq Hefer 40250, Israel
The Hadassah Academic College, Jerusalem 91010, Israel

A.B. Larionov
Institut für Theoretische Physik, Universität Giessen, D-35392 Giessen, Germany
National Research Center "Kurchatov Institute", 123182 Moscow, Russia
Frankfurt Institute for Advanced Studies (FIAS), D-60438 Frankfurt am Main, Germany

H. Lenske
Institut für Theoretische Physik, Universität Giessen, D-35392 Giessen, Germany

Deog Ki Hong and Jong-Wan Lee
Department of Physics, Pusan National University, Busan 46241, Republic of Korea

Biagio Lucini, Maurizio Piai and Davide Vadacchino
College of Science, Swansea University, Singleton Park, Swansea SA2 8PP, UK

Fei Gao and Can Tang
Department of Physics and State Key Laboratory of Nuclear Physics and Technology, Peking University, Beijing 100871, China
Collaborative Innovation Center of Quantum Matter, Beijing 100871, China

Yu-xin Liu
Department of Physics and State Key Laboratory of Nuclear Physics and Technology, Peking University, Beijing 100871, China
Collaborative Innovation Center of Quantum Matter, Beijing 100871, China
Center for High Energy Physics, Peking University, Beijing 100871, China

Index

www.ingramcontent.com/pod-product-compliance
Lightning Source LLC
Chambersburg PA
CBHW080651200326
41458CB00013B/4813
* 9 7 8 1 6 8 2 8 5 6 4 6 8 *